Nuclear Magnetic Resonance Spectroscopy

Volume III

Author

Professor Pál Sohár, Ph.D., D.Sc.
Head
Spectroscopic Department
EGYT Pharmacochemical Works
Professor
Department of Organic Chemistry
Eötvös Lóránd University of Sciences
Budapest, Hungary

CRC Press, Inc.
Boca Raton, Florida

Library of Congress Cataloging in Publication Data

Sohár, Pál.
Nuclear magnetic resonance spectroscopy.

Bibliography: p.
Includes index.
1. Nuclear magnetic resonance spectroscopy.

I. Title
QC762.S575 1984 543′.0877 82-9524
ISBN 0-8493-5632-6 (v. 1) AACR2
ISBN 0-8493-5633-4 (v. 2)
ISBN 0-8493-5634-2 (v. 3)

Direct all inquiries to CRC Press, Inc., 2000 Corporate Blvd., N.W., Boca Raton, Florida, 33431.

International Standard Book Number 0-8493-5632-6 (v. 1)
International Standard Book Number 0-8493-5633-4 (v. 2)
International Standard Book Number 0-8493-5634-2 (v. 3)

Library of Congress Card Number 82-9524
Printed in the United States

PREFACE

These books are an updated version of a Hungarian handbook the author published in 1976 with the same title. Since it was the first original NMR handbook in the Hungarian language, I endeavored to embrace, even if briefly, all topics of NMR spectroscopy. When the manuscript was finished, at the beginning of the 1970s, FT spectroscopy was only beginning to be studied, and thus the investigation of nuclei other than protons had much less significance than now. Therefore, many parts of the books discuss NMR phenomena primarily from the aspect of proton resonance.

With the very rapid spread of Fourier transform technique and FT instruments, the investigation of other nuclei, first of all carbon, has become feasible. In addition to chemical shifts, coupling constants and intensities are now sources of NMR information and relaxation times have become one of the starting points of structure elucidation. This progress made it necessary to deal more closely with certain theoretical and technical principles as well as with some relationships between spectrum and chemical structure less important from the aspect of proton resonance.

Therefore, some of the problems are touched in more than one chapter. The resulting, somewhat looser structure of the books has, however, the advantage that the special principles and applications pertaining to the same nucleus may be discussed in a more concentrated and unique manner.

The encyclopedic character of the book has claimed a relatively great volume and a necessarily brief, schematic discussion of some details. But perhaps this distinguishes my book from other handbooks and justifies its publication. The books are meant primarily for chemists, undergraduates, and young spectroscopists who wish to get broad but not too detailed information on all important topics of NMR spectroscopy. I hope that scientists working in related fields, e.g., analysts, biochemists, or physicists will also find useful information in the books. At any rate, the books were written keeping in mind the demands of organic chemists who wish to use NMR spectroscopy as a tool in structure elucidation, but who also want to know, at least schematically, the theoretical foundations of the method applied.

The first chapter is a concise but comprehensive discussion, primarily from the aspect of proton resonance, of NMR theory, which assumes only minimum quantum chemical knowledge, and the mathematical apparatus is limited to the level accessible for average chemists. The main purpose of this chapter is to point out how the magnetic nuclei of molecules may be identified with the individual spin systems, and how can one determine the spectrum parameters given the key to structure elucidation, i.e., chemical shifts and coupling constants, from the measured spectrum. As the quantum theory of NMR spectroscopy is a settled, classical branch of science, and as I endeavored to review the most important principles only, the references in this chapter extend only to original publications of fundamental importance, completely neglecting more recent literature pertaining to special details.

Chapter 2 deals with the operation principles and technical problems of traditional (CW) and modern (FT) spectrometers, and also discusses special techniques of measurement. Of the latter, the measurement and use of various double resonance spectra, temperature-dependent measurements important in the investigation of dynamic phenomena, and the measurement of integrated intensities permitting quantitative analysis have the greatest weight. The theoretical foundations of Fourier transform spectroscopy, the experimental determination of relaxation times, the theory of relaxation, and relaxation mechanisms are also discussed here. In connection with relaxation times, a separate section is devoted to the cancer diagnostic application of NMR and tomography.

Chapter 3 reviews the proton resonance parameters of the most important groups of organic compounds. The huge and ever-increasing amount of data collected in this field prohibits any attempt toward completeness, and thus again I restricted myself to classical results, selecting the illustration material from the most characteristic or the first examples. Therefore, more recent literature is represented in the chapter only by some reviews or extremely interesting papers. An exception is the last section of this chapter, which deals with the medium effects, predominantly with the shift reagent technique. This part discusses the exchange phenomena and hydrogen bonds, as well as the NMR investigation of optical purity and radical reactions (CIDNP).

Chapter 4 discusses, on the basis of similar principles, the most characteristic properties of the NMR spectra of carbon nuclei, in much smaller volume those of nitrogen, and very schematically only the properties of ^{17}O, ^{19}F, and ^{31}P nuclei, with hints to the special aspects, pertaining to the given nuclei, of some theoretical problems, principles, and measurement technique.

Chapter 5 is a collection of problems. All problems are concerned with elucidation of chemical structures taken from the practice of the author. Most of the problems were real, occurring in practice. The 66 problems connected with ^{1}H NMR spectra and the 12 problems connected with carbon NMR, together with the roughly 200 spectra provide, hopefully, a good training for the chemist in his own spectrum evaluation practice.

I feel it my duty to express my gratitude to all who contributed to the publication of this book.

My first thanks should go to my students, since the subject of the book was collected first for my lectures. Inspired by their interest and questions, many ideas arose for working out certain topics and selecting the material for illustration.

To my chemist colleagues, whose names are listed in the name index, thanks are due not only for the samples synthesized partly by them and given for recording the spectra of the Problems section, but also for the preparation of model derivatives, or purification of compounds in course of structure elucidation, and also for the assistance provided by chemical information about the compounds studied.

For constructive discussions and suggestions on the completion or modification of certain parts, I am indebted to Professors G. Snatzke (Bochum), and H. Wamhoff (Bonn) and Doctors G. Jalsovszky (Budapest), I. Kövesdi (Budapest), and T. Széll (New York).

For the promotion of the English edition my thanks are due to Professor B. Csakvári and I. Kovács, as well as to Publishing House of the Hungarian Academy of Sciences; I wish to express my thanks to Dr. G. Jalsovszky for the tedious, competent, and unrewarding work of translation.

I am indebted to publishers Academic Press, Elsevier, Heyden and Son Ltd., Pergamon Press and Wiley-Interscience, personally to Professors E. D. Becker and T. C. Farrar (Washington), A. F. Casy (Alberta) and H. Wamhoff (Bonn), Doctors H. Schneiders (Bruker, Karlsruhe), D. Shaw (Oxford), C. S. Springer (New York), F. W. Wehrli (Bruker, USA), T. Wirthlin (Varian, Zug) and to firms Bruker Physik AG, JEOL, Sadtler Research Laboratories Inc., and Varian Associates for placing at my disposal illustration material or, respectively, for permitting my use of this material.

My thanks are due to the managers of my firm, EGYT Pharmacochemical Works, personally to Professor L. Pallos, Scientific Director, for ensuring the conditions for compiling this book.

It gives me great pleasure to acknowledge all my previous and present collaborators their enthusiasm, careful and competent work in all phases of the preparation of this book. I want to mention them by name: late Mrs. Dr. Zs. Méhesfalvy, Mrs. A. Csokán, Mrs. J. Csákvári, Mr. A. Fürjes, Mrs. Á. Kiss-Tamás, Mrs. M. Leszták, Mr. Gy. Mányai, Mrs. É. Mogyorósy, Mrs. Dr. K. Ósapay, Mr. I. Pelczer, Miss V. Windbrechtinger, and Mr. J. Zimonyi.

I release these books in the hope that it will benefit all those colleagues who wish to be acquainted with this singularly many faceted, still rapidly developing, in all fields of chemistry extremely efficient and again and again reviving branch of science: NMR spectroscopy.

Pál Sohár
Budapest, April 1981

THE AUTHOR

Pál Sohár, Ph.D., D.Sc. is the Head of the Spectroscopic Department of EGYT Pharmacochemical Works and Professor of the Eötvös Loránd University of Sciences, Budapest, Hungary. He graduated in 1959 at the Technical University of Budapest and obtained his Ph.D. degree in 1962 in physical chemistry with "Summa cum Laude" qualification.

Professor Sohár received the "Candidate of Sciences" degree in physical chemistry from the Hungarian Academy of Sciences in 1967 on the basis of his Thesis "Investigation of Association Structures by Infrared Spectroscopy" and the D.Sc. degree in 1973 for his research work in the application of IR and NMR spectroscopy in the structure elucidation of organic molecules. He has served as Professor at the Eötvös Loránd University since 1975. He was the Head of the Spectroscopy Department of the Institute of Drug Research. He began his career in this Institute in 1959 and received promotions to scientific assistant in 1962, to senior assistant in 1969, and to scientific counselor from 1974 to 1980. He assumed his present position in 1980.

Professor Sohár is a member of the Committees of Physical and Inorganic Chemistry, Spectroscopy and Theoretical Organic Chemistry of the Hungarian Academy of Sciences and the member of ISMAR (International Society of Magnetic Resonance, Chicago, Illinois). He is the Secretary of the Committee on Molecular and Material Structure of the Hungarian Academy of Sciences.

Professor Sohár is the author of more than 180 scientific papers and has been the author or coauthor of six books, among them the first monographs in Hungarian on infrared spectroscopy and nuclear magnetic resonance. He is the coeditor of the series *Absorption Spectra in the Infrared Region* published by Akadémiai Kiadó, Budapest (Publishing House of the Hungarian Academy of Sciences). He has given more than 150 scientific presentations or invited lectures, and was several times invited lecturer of postgraduate courses at the Technical University of Budapest.

His current major research interests include structure elucidation of organic compounds by nuclear magnetic resonance and infrared spectroscopy.

ACKNOWLEDGMENT

I am indebted to my colleagues listed below for sample preparation and for providing compounds for NMR investigation.

Ambrus, G., Dr., **34**
Ágai, B., Dr., **17**
Bárczay-Beke, M., Dr., **37**
Bernáth, G., Prof., Dr., **35, 45, 75**
Bodnár, J., Dr., **64**
Bordás, B., Dr., **56**
†Bruckner-Wilhelms, A., Dr., **54**
Dudás, T., Dr., **41**
Érczi, I., **31**
Farkas, L., Dr., **5**
†Fehér, Ö., Dr., **72**
Fischer, J., Dr., **6**
Gyimesi, J., Dr., **29**
Hajós, A., Dr., **42, 73**
Hideg, K., Dr., **76**
Horváth, T., Dr., **74**
Kajtár, M., Dr., **4**
Kasztreiner, E., Dr., **28, 55**
Kolonits, P., Dr., **47**
Kosáry, J., Dr., **13**
Kuszmann, J., Dr., **43, 48, 50, 59, 63, 65, 66, 67, 78**
Kuszmann-Borbély, A., Dr., **7, 38**
Lázár, J., Dr., **71**
Lempert, K., Prof., Dr., **10**
Lempert-Sréter, M., Dr., **40, 51**
Löw, M., Dr., **19**
Matolcsy, Gy. Dr., **27**
Máray, P., **44**
Medgyes, G., Dr., **67, 69**
Merész-Márton, M., Dr., **12, 52, 62, 78**
Moravcsik, I., Dr., **34**
Nagy, J., Dr., **22, 44**
Nemes, A., Dr., **44, 60**
Nógrádi, M., Dr., **20**
Nyitrai, J., Dr., **2, 9**
Reiter, J., Dr., **15, 21, 25, 26**
Réffy, J., Dr., **26**
Sipos, Gy., Dr., **36**
Sólyom, S., Dr., **68**
Stájer, G., Dr., **49**
Szabó, J., Dr., **77**
Szántay, Cs., Dr., Prof., **39**
Széll, T., Dr., **14, 57**
Szilágyi, G., Dr., **32, 53**
Szilágyi-Faragó, K., **16, 18,, 24, 70**
Toldy, L., Dr., **1, 11, 33**
Tóth, I., Dr., **23,, 70**
Tóth, Z., Dr., **30**
Turán, A., Dr., **61**
†Vargha, L., Prof., Dr., **64**
Zauer, K., Dr., **8, 9**
Zubovics, Z., **3**

TABLE OF CONTENTS

Volume I

Chapter 1
Theory of Nuclear Magnetic Resonance Spectroscopy

Chapter 2
NMR Spectrometers, Recording Techniques, Measuring Methods

Volume II

Chapter 4
The Resonance Spectra of Nuclei Other Than Hydrogen

TABLE OF CONTENTS

Volume III

Chapter 5

STRUCTURE DETERMINATION PROBLEMS

INTRODUCTION

The following collection of ^{1}H and ^{13}C NMR problems is intended to illustrate the possibilities of application of theoretical principles and empirical rules and experiences discussed in the previous chapters for the solution of structure problems in organic chemistry. The ^{1}H NMR spectra were recorded in the majority of cases on a VARIAN® A-60-D (denoted by ***V***) and in a few cases on a JEOL® C-60 (***J***) and HL-60 (*H*) or BRUKER® Spektrospin 90; the ^{13}C NMR were recorded on a VARIAN® XL-100 FT spectrometer (*S*), usually in the concentration range of 1.0 to 0.1 mol/ℓ at room temperature, using TMS (or in D_2O, DMSO-d_6, and in acid solutions, DSS) as internal standard. The old J-NM-C-60 equipment was unable to provide the specified optimum performance owing to its instability, and thus the spectra recorded with this spectrometer are generally inferior in quality. However, the spectra are by no means model spectra, prepared for reference collections; they have been selected from the material piled up during the everyday practical work of the author and there are among them ones of rather poor quality.

The poor quality is due, in part, to incorrect operation (saturation, too fast recording, spinning side bands, incorrect intensity adjustment, unsatisfactory field homogeneity, strong "drift", poor resolution, etc.) and in part to the specimen (impurities in the sample or the solvent, inhomogeneity of the solution, presence of solid impurities, undissolved or precipitated crystals in the solution, too low concentration, etc.). These "bad" spectra are given to illustrate the effects of the above faults.

The spectra of the collection are reduced in size, and thus it would be impossible to measure the shifts as accurately as in the original spectra and to eliminate the error arising from the drift. Therefore the accurate shifts and frequencies of the lines are tabulated below each spectrum (in parts per million units on the δ-scale or in hertz in italics). For singlets and first-order multiplets, shifts are given (for the multiplets the shift refers to their mid points). The tables also show the multiplicities and the coupling constants: *s* = singlet, *d* = doublet, *t* = triplet, *qa* = quartet, *qi* = quintet, *sx* = sextet, *sp* = septet, *o* = octet, *n* = nonet, *m* = multiplet. Broad signals are denoted by ~. In the case of complex multiplets, the line frequencies, and for broad signals, only the frequency limits or just the frequency of main maxima (MM), are given. For symmetric multiplets the frequency of the midpoint (MP) is tabulated.

The data are given in the order of increasing values or frequency, and the various signals or groups of signals are denoted by the lower case letters of the alphabet. With some exceptions, the tables also contain the rounded integrated intensities. No integrals are given if the integral curve is shown in the spectrum or intentionally, if the problem has to be solved without the integrals. The integrals could not be given when the signals of the sample and the solvent, some impurity, or other component of the specimen overlap.

The solvent and the temperature are specified if the spectrum was not recorded at the usual working temperature. The signals that undergo shifts upon the addition of heavy water or acid are marked. If necessary, the changed spectrum, or a portion of it, is also given.

The following marks are used in the spectra: *, the absorption of the solvent (signals of the light isotope or other impurities water content, etc.);** overlapping signals of the solvent protons and the mobile protons (OH, NH, etc.) of the sample; ○, signal due to the impurity of the sample; ▽, spinning side band; □, signal shifted upon the addition of heavy water, acid, or base.

The problems are numbered in Arabic numerals from **1** to **78** (in the Problems **1** to **66** ^1H and in Problems **67** to **78** ^{13}C spectra are involved), and the spectra within one problem are given a second number. Accordingly, Spectrum **74**/*2* is Spectrum *2* belonging to Problem **74**. The chemical formulas are numbered within the problems in Roman numerals; thus for instance **39/IV** denotes Formula **IV** of Problem **39**. In the text only the Roman numerals are given, and this is preceded by the number of the problem only if this formula is referred to in another problem or in a former section of the book. For the sake of clarity the abbreviations given in the "list of notations"* are also used in the formulas.

PROBLEM 1

On the basis of Spectrum **1**/*1* choose from the possible structural isomers **I**: $CH_3CH_2CHBr_2$; **II**: $CH_2(CH_2Br)_2$; **III**: $CH_3CHBrCH_2Br$; and **IV**: $(CH_3)_2CBr_2$.

ANSWER

The spectra of compounds **I** to **IV** will presumably contain the following signals (in the order of increasing δ). Since Spectrum **1**/*1* contains a triplet and a quintet of 2:1 intensity, the sample is 1,3-dibromopropane (**II**). $\delta CCH_2 = 2.31$ and $\delta CH_2Br = 3.57$ ppm, and $J = 6.5$ Hz. The two and four methylene protons form and A_2X_4 spin system, proving the chemical and magnetic equivalence of the bromomethyl groups. Note that the adjacent bromine atoms cause a paramagnetic shift of 1.26 ppm with respect to the hydrogens of the central methylene group, which is connected only to carbon atoms.

Table 101
THE EXPECTED STRUCTURE OF THE SPECTRA OF COMPOUNDS 1/I to IV

	Expected signals		
Structure	Multiplicity	Relative intensity	Assignment
I	*t*	3	CH_3
	qi[a]	2	CH_2
	t	1	CH
II	*qi*	2	CCH_2
	t	4	CH_2Br
III	*d*	3	CH_3
	d	2	CH_2
	m[b]	1	CH
IV	*s*	—	CH_3

[a] Possibly double quartet.
[b] 4 × 3 = 12 lines, possibly partly coalesced.

* See p. 335.

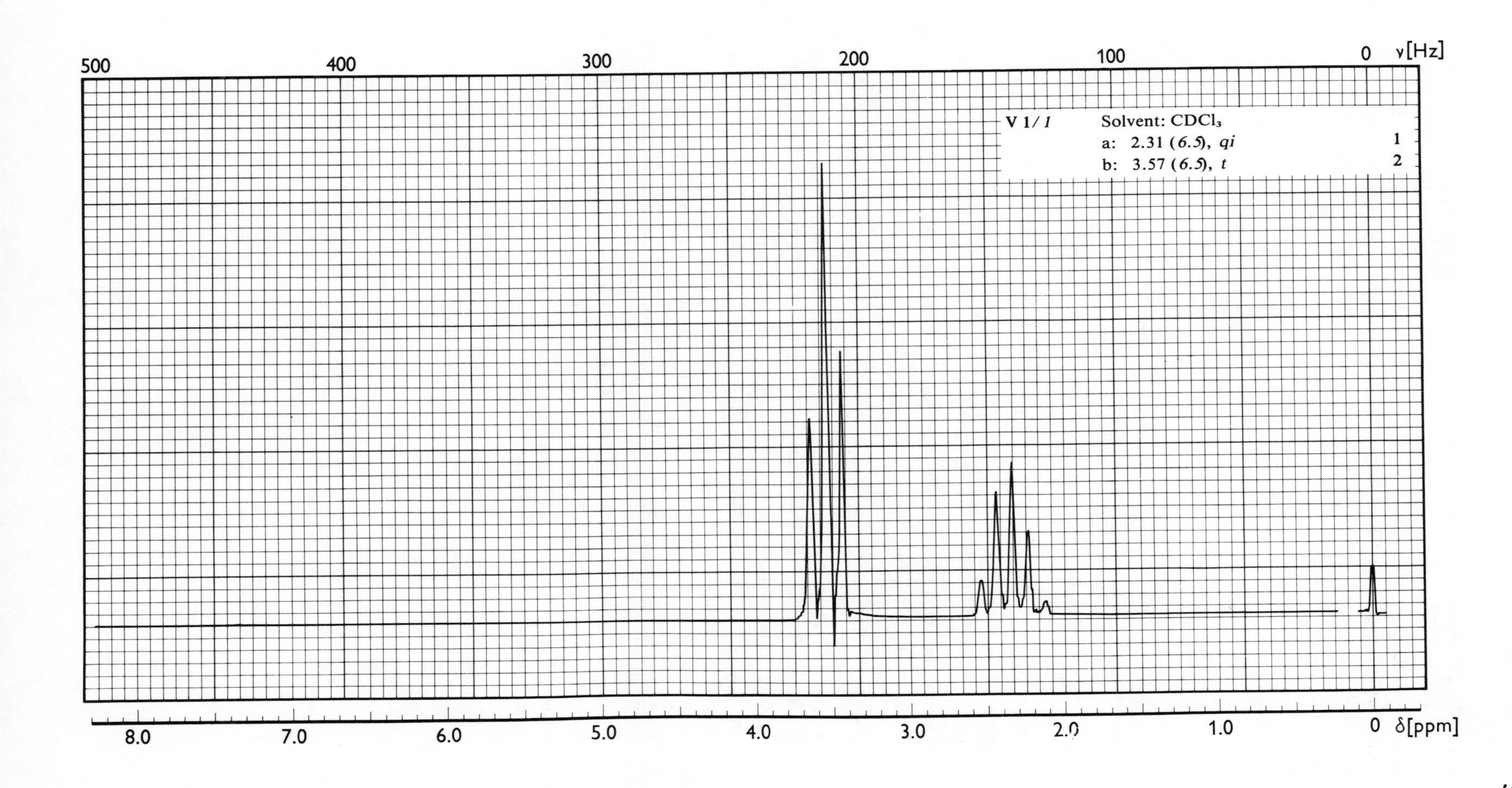

500
400
300
200
100
0
ν[Hz]
V 1/1
Solvent: $CDCl_3$
a: 2.31 (6.5), qi 1
b: 3.57 (6.5), t 2
8.0
7.0
6.0
5.0
4.0
3.0
2.0
1.0
0
δ[ppm]

PROBLEM 2

Determine from the position of the CH_2 signal which, of the Spectra **2**/*1* and **2**/*2*, corresponds to Structures **I** and **II**, respectively. Where it can be expected and why is the signal of NH protons absent? How do the intensities reflect the origin of the signals?

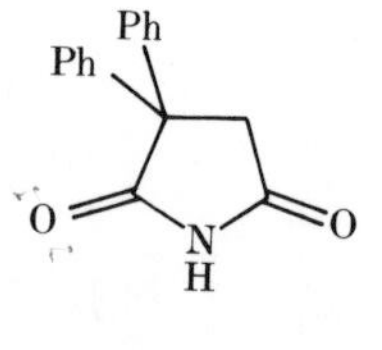

2/I 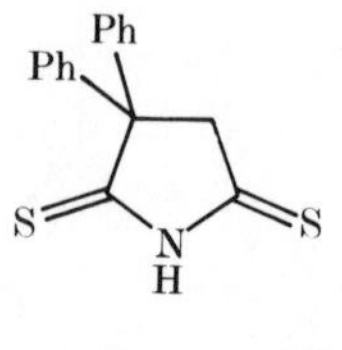 **2/II**

ANSWER

The methylene signal is obviously at higher δ for compound **II**, since the paramagnetic shift due to the anisotropy of thioxo groups is greater than the analogous effect of carbonyl groups.* Consequently, Structure **II** can be assigned to Spectrum **2**/*1*, in which the methylene singlet appears at 3.90 ppm, whereas in Spectrum **2**/*2*, it appears at 3.50 ppm. The phenyl protons have practically the same shift in both cases (7.37 and 7.35 ppm, respectively). Note that the relative heights of ArH and CH_2 signals reflect only qualitatively the real ratio of protons; the ratio of heights is 4:1 instead of 5:1. The reason is that the stronger lines are broader, and thus the true intensities are reflected faithfully only by the integrals.

The expected chemical shift of the imido group is high: $\delta NH > 10$ ppm, and thus the signal is, of course, outside the range shown in the figures. However, this signal cannot be discovered in the offset range, either, since it is very broad and merges into the baseline. This is a frequent phenomenon with the signals of acidic protons.** The maxima at 6.40 and 7.27 ppm are due to the impurities of the solvent; the latter can be assigned to $CHCl_3$, i.e., the light isotope content of the solvent. This can also be seen from Spectrum **2**/*1*, recorded with higher gain owing to the lower concentration of the solution, in which the relative intensities for these signals are higher.

* See Volume II, p. 43 and Problem **9**.

** OH, NH, SH, etc.; see Volume II, p. 106.

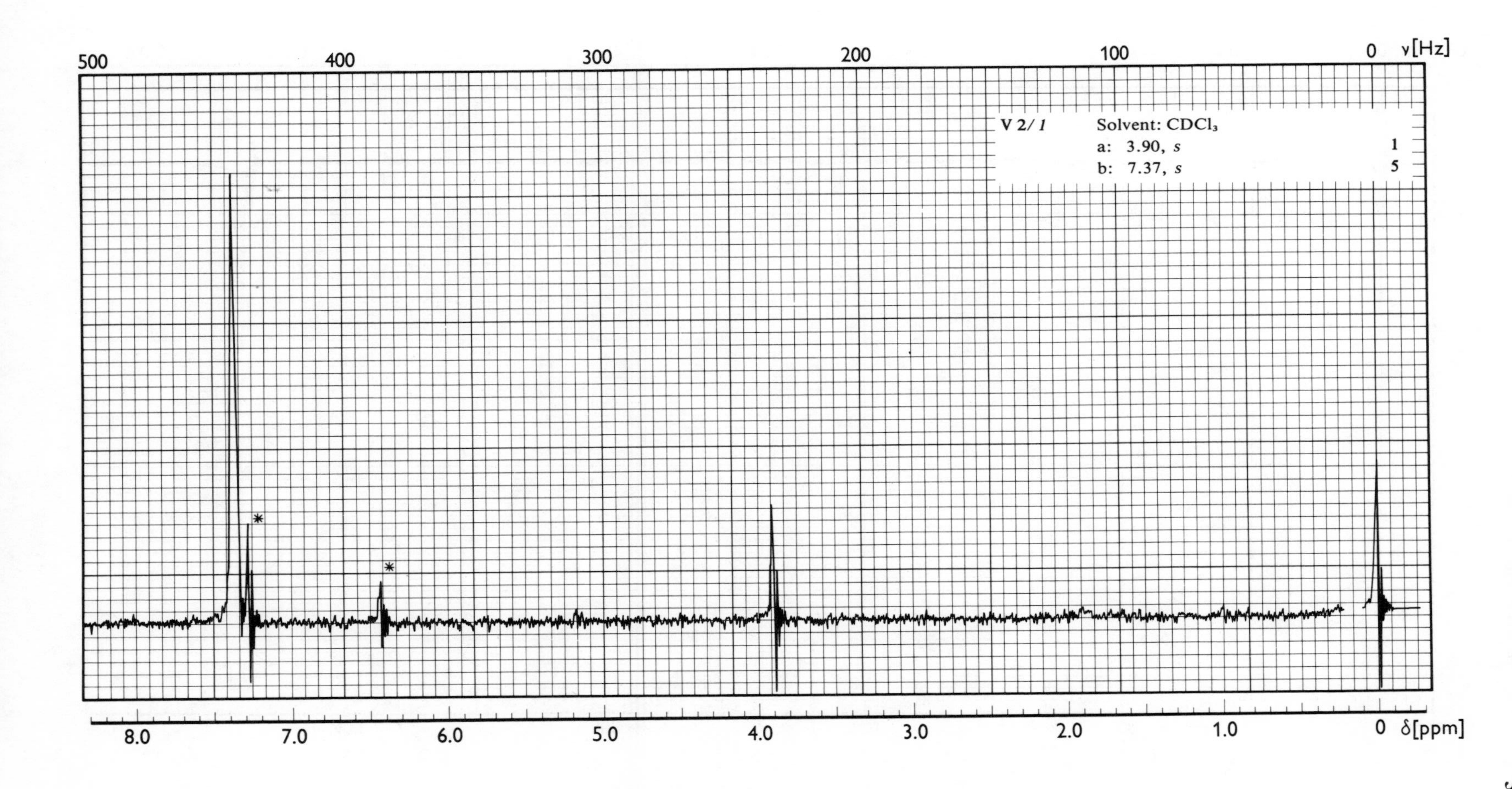
ν[Hz]
500
400
300
200
100
0
V 2/1
Solvent: $CDCl_3$
a: 3.90, *s* 1
b: 7.37, *s* 5
*
*
8.0
7.0
6.0
5.0
4.0
3.0
2.0
1.0
0
δ[ppm]

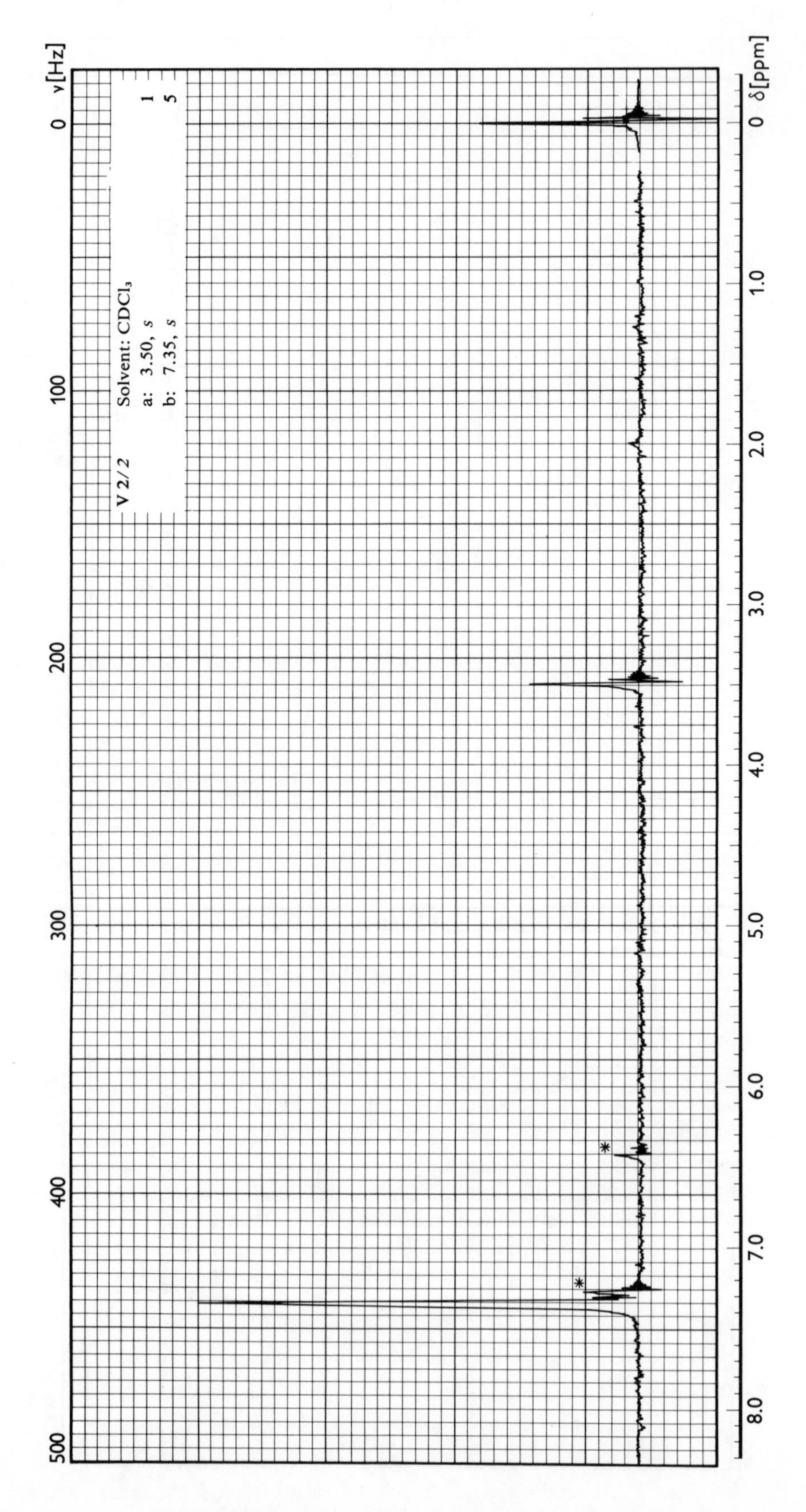
V 2/2
Solvent: CDCl₃
a: 3.50, s 1
b: 7.35, s 5
ν[Hz]
0
100
200
300
400
500
δ[ppm]
0
1.0
2.0
3.0
4.0
5.0
6.0
7.0
8.0
*
*

PROBLEM 3

On the basis of their spectra (**3**/*1* and **3**/*2*) determine the structures of the compounds both having the empirical formula $C_2H_6O_3S$.

ANSWER

As Spectrum **3**/*1* contains only one maximum at 3.64 ppm, the six hydrogens are equivalent. The chemical shift is in the region expected for methoxy groups. This spectrum must therefore correspond to dimethylsulfite, MeO–SO–OMe (**I**).

In Spectrum **3**/*2*, two maxima can be found at 3.00 and 3.92 ppm with equal intensities. Consequently, there are two different methyl groups in the molecule, one of them being attached to the oxygen and the other being attached to the sulfur. Hence, this spectrum is compatible with methanesulfonic acid methyl ester, Me–SO_2–OMe (**II**). The signal with higher chemical shift belongs to the methoxy group (compare Table 22) and the increase of 0.28 ppm with respect to the methoxy signal in **I** is, in part, a consequence of the higher $-I$-effect and the anisotropy of SO_2 group over that of the SO group.

PROBLEM 4

Determine the structure of compound $C_2H_3OF_3$ from Spectra **4**/*1* and **4**/*1a*. The latter spectrum is recorded after the addition of heavy water.

ANSWER

Spectrum **4**/*1* contains a symmetric quartet of 1:3:3:1 intensity at 3.92 ppm and a singlet arising from acidic hydrogens (shifting upon the addition of heavy water)* at 4.45 ppm (4.20 ppm in Spectrum **4**/*1a*). The intensity ratio of the quartet and singlet is 2:1.

The quartet indicates the presence of a CH_2CF_3 group; the singlet must correspond to a hydroxy group. The coupling constant is J(F,H) = 9 Hz.** Accordingly, the unknown is 1,1,1-trifluoroethanol, CF_3CH_2OH.

* See Volume II, p. 97.

** See Volume II, p. 264.

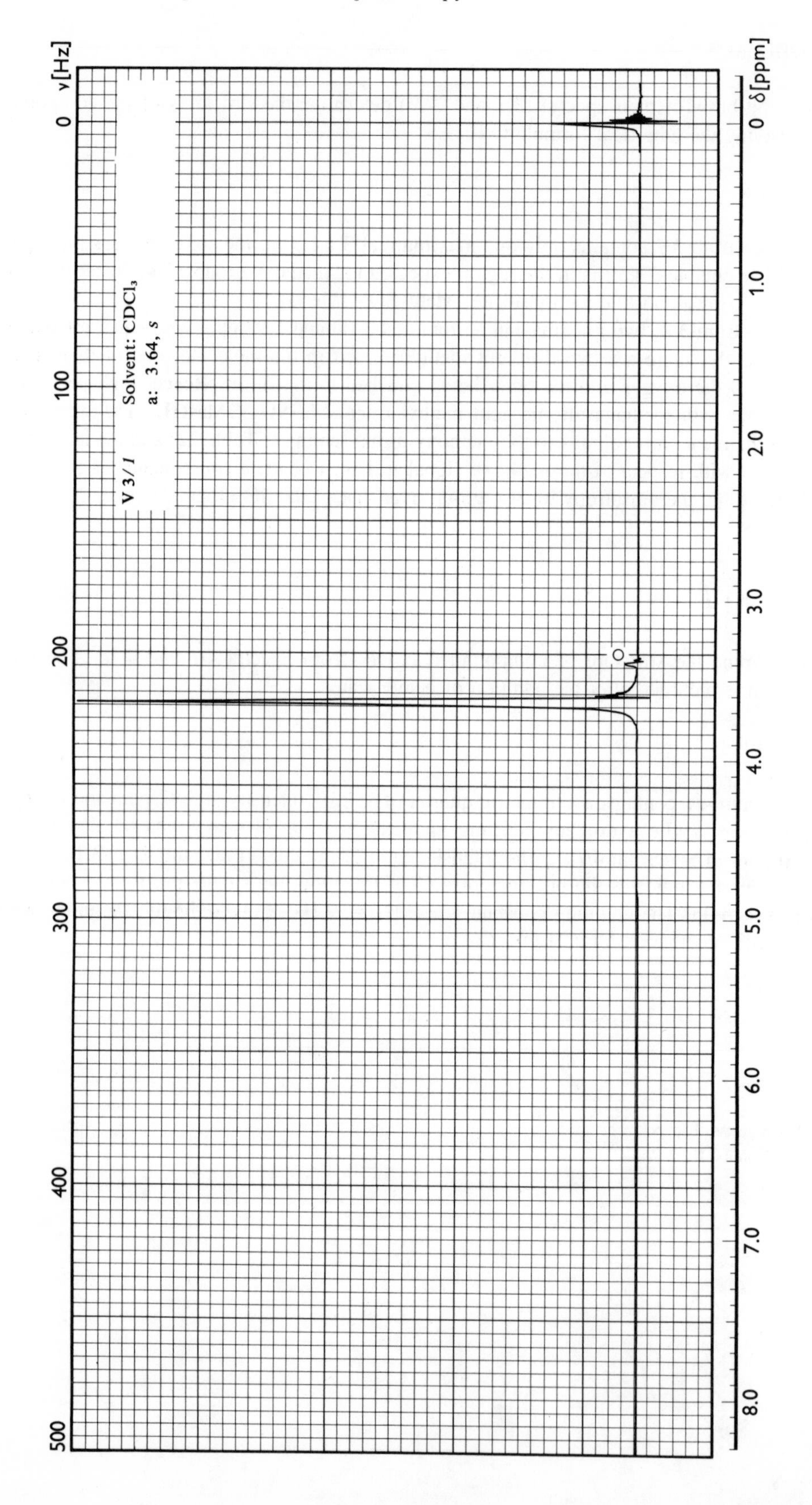

V 3/1
Solvent: $CDCl_3$
a: 3.64, *s*
ν[Hz]
0
100
200
300
400
500
δ[ppm]
0
1.0
2.0
3.0
4.0
5.0
6.0
7.0
8.0

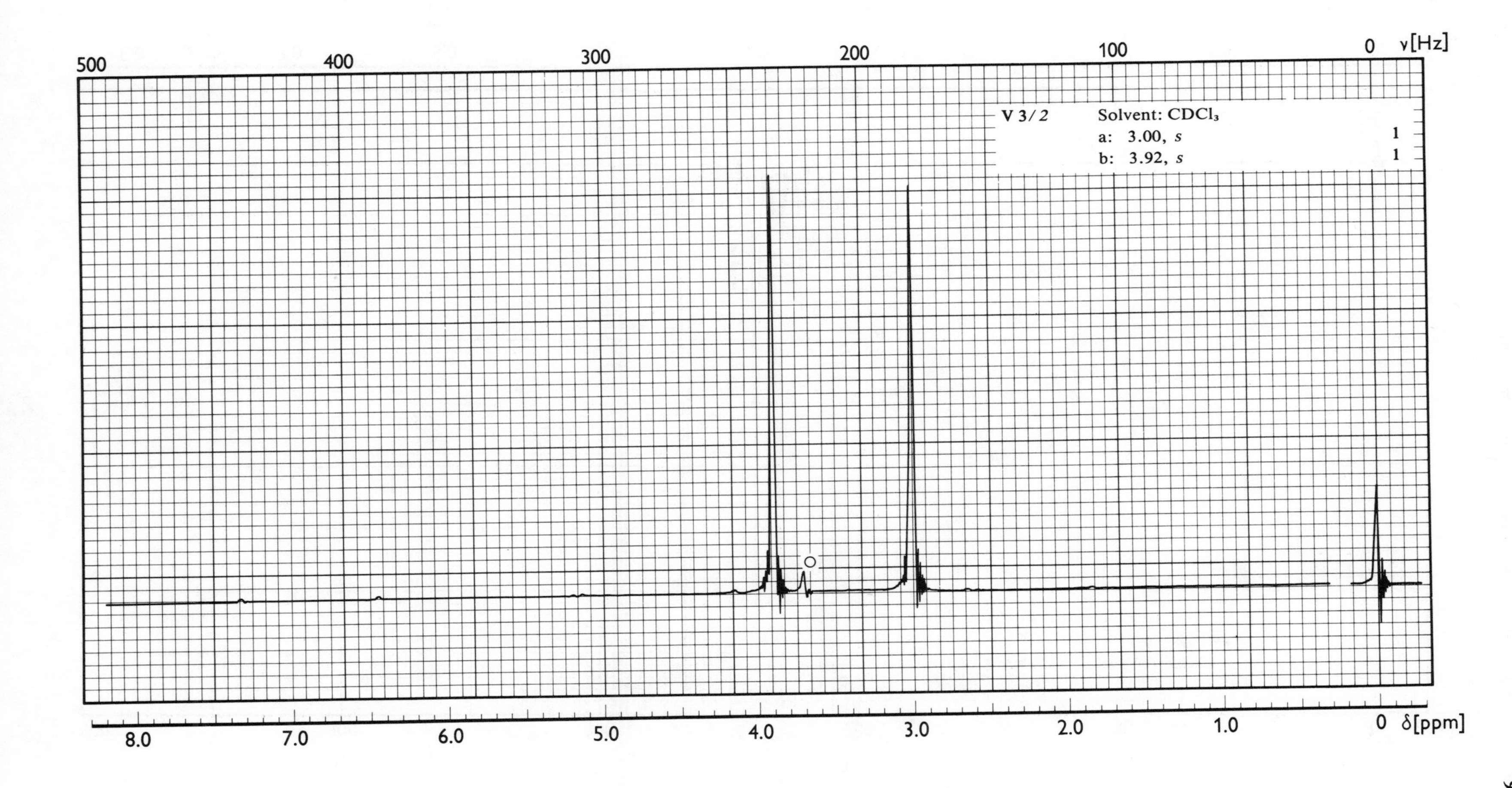
ν[Hz]
500
400
300
200
100
0
V 3/2
Solvent: $CDCl_3$
a: 3.00, *s* 1
b: 3.92, *s* 1
8.0
7.0
6.0
5.0
4.0
3.0
2.0
1.0
0
δ[ppm]

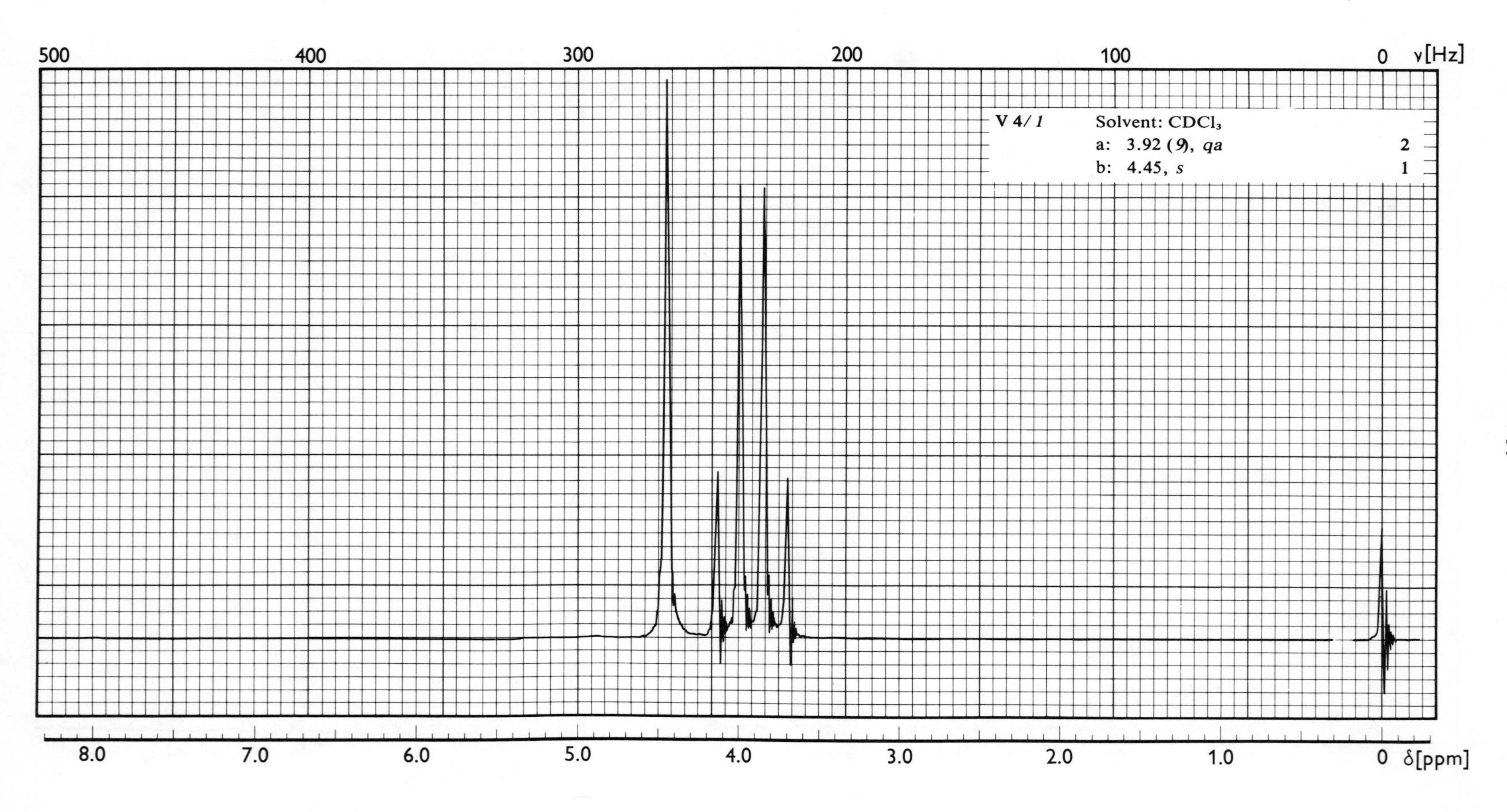
500
400
300
200
100
0 ν[Hz]
V 4/1
Solvent: CDCl₃
a: 3.92 (9), qa 2
b: 4.45, s 1
8.0
7.0
6.0
5.0
4.0
3.0
2.0
1.0
0 δ[ppm]

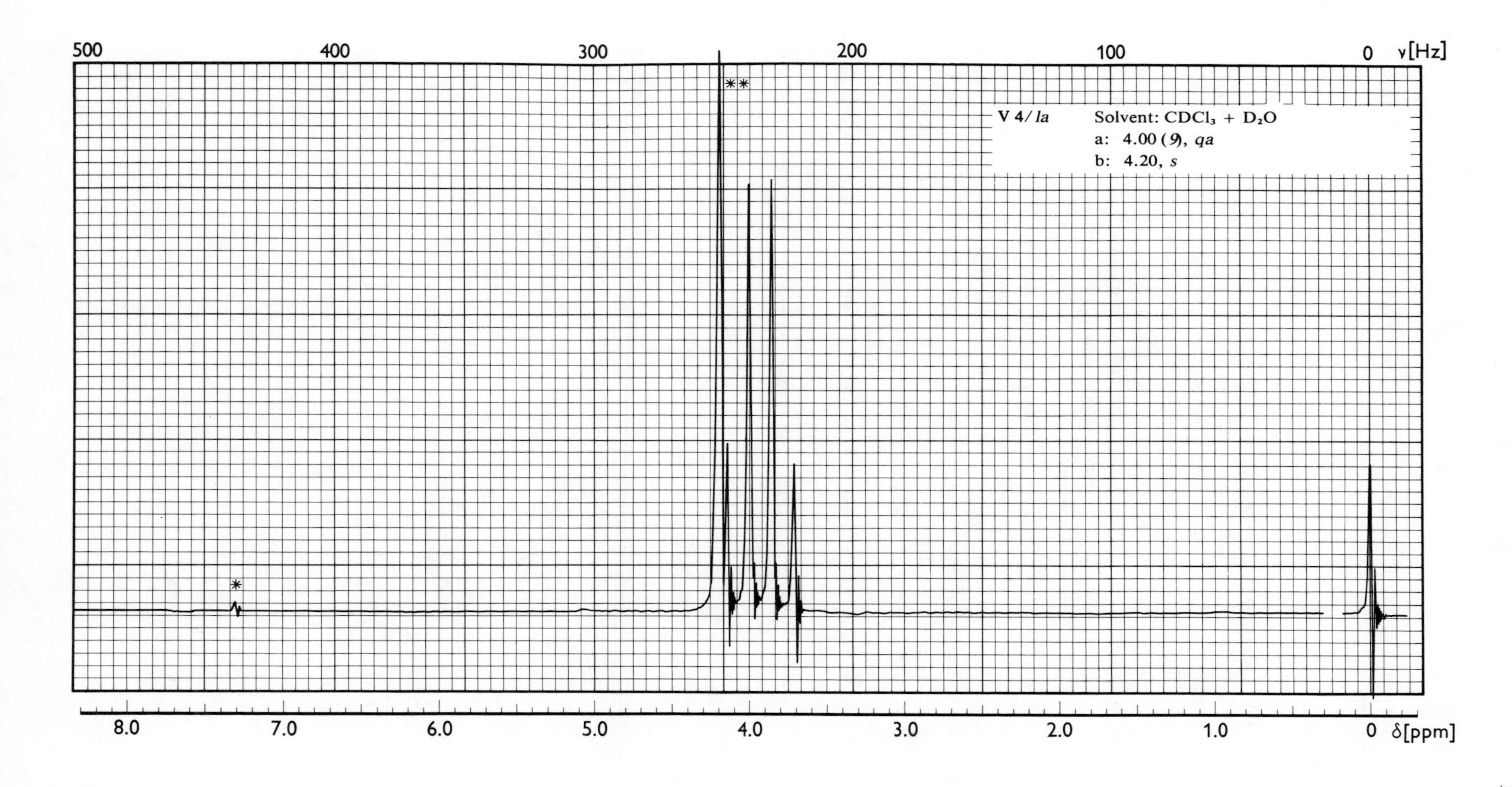

500
400
300
200
100
0 ν[Hz]
V 4/la
Solvent: CDCl3 + D2O
a: 4.00 (9), qa
b: 4.20, s
8.0
7.0
6.0
5.0
4.0
3.0
2.0
1.0
0 δ[ppm]

PROBLEM 5

Select the one of the 11 possible dimethyl dinitrobenzene isomers that corresponds to Spectrum **5**/*1*.

ANSWER

Among the 11 conceivable isomers, there are 6 of intrinsic symmetry in which the ring protons and, with one exception also, the two methyl groups are chemically equivalent. Since Spectrum **5**/*1* consists of two methyl singlets at 2.33 and 2.55 ppm and an *AB* multiplet arising from the ring protons, these isomers are excluded. Among the five asymmetric isomers two-two have ring protons in *ortho* and *meta* position, whereas the fifth isomer has *para* ring hydrogens. For the three cases, according to experience,* J_{AB} must be about 9, 2, and <1 Hz, respectively. The coupling constant J_{AB} is equal to the spacing of the two outer lines of the four-line *AB* spectrum, i.e., $J_{AB} = 464 - 455.5 = 493.5 - 485 = 8.5$ Hz,** suggesting the *ortho* position of the ring protons. Hence the structure is **I** or **II**. The difference between the chemical shifts of the two ring protons is 0.47 ppm on the basis of the lines at 455.5, 464, 485, and 493.5 Hz ($\delta A = 8.14$ and $\delta B = 7.67$ ppm). These values can be calculated as follows:** $\Delta\nu_{AB} = \sqrt{(493.5 - 455.5)(485 - 464)} = \sqrt{38 \cdot 21} = \sqrt{798} \approx 28$ Hz; $\delta AB = 28/60 = 0.47$ ppm. The midpoint of the spectrum is at $(493.5 + 455.5)/2 = (485 + 464)/2 = 474.2$ Hz. Then, if $\delta A > \delta B$, $\nu A = 474.2 + 28/2$ Hz ≈ 488 Hz; $\delta A \approx 8.14$ ppm; and $\nu B = 474.2 - 14 \approx 460$ Hz; $\delta B \approx 7.67$ ppm.

According to Table 39, the expected chemical shift differences are 0.55 for compound **I** and 0.95 ppm for **II**, thus the spectrum corresponds to the former. Note that the more deshielded methyl group of 2.55 ppm shift is adjacent to the electron-attracting nitro group, causing paramagnetic shift.

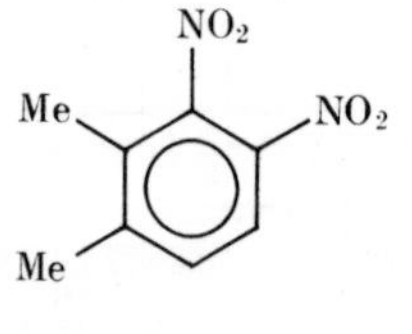

5/I 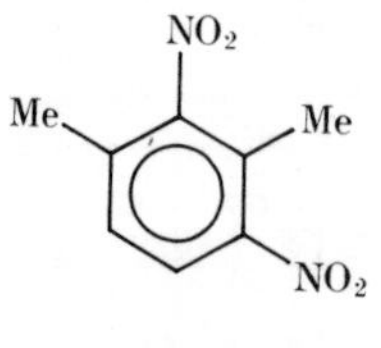 **5/II**

* Compare Volume II, p. 68.

** See Volume I, p. 98.

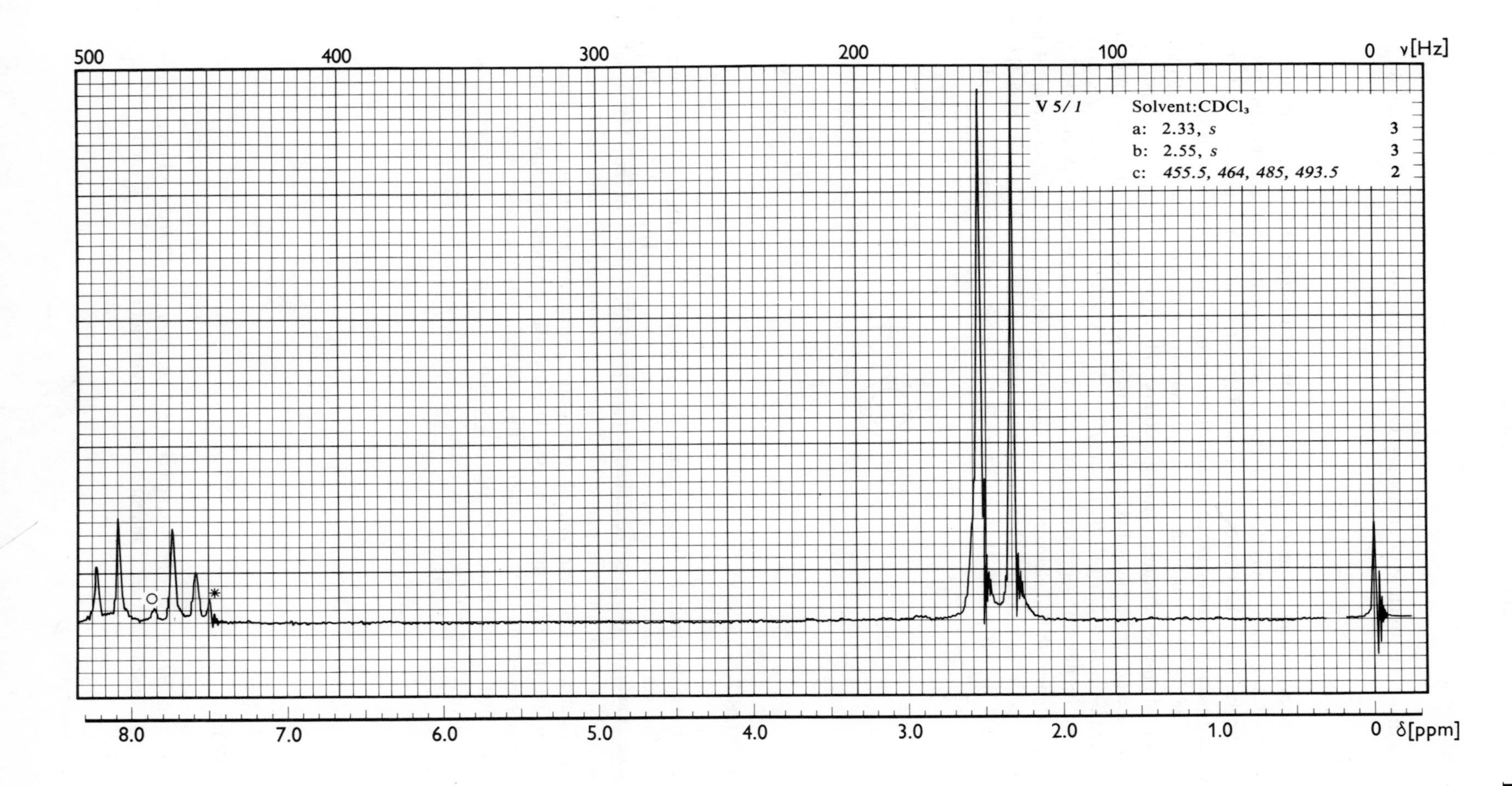

V 5/1
Solvent: $CDCl_3$
a: 2.33, s 3
b: 2.55, s 3
c: 455.5, 464, 485, 493.5 2
500
400
300
200
100
0
ν[Hz]
8.0
7.0
6.0
5.0
4.0
3.0
2.0
1.0
0
δ[ppm]

PROBLEM **6**

Determine the structure of compound $C_{19}H_{13}Cl_3O_3$ on the basis of Spectrum **6**/*1*.

ANSWER

The spectrum contains an $AA'BB'$ multiplet and a singlet at 6.40 ppm of 12:1 intensity ratio, indicating three equivalent 1,4-disubstituted aromatic rings and thereby Structure **I**. Accordingly, as can also be expected (see Table 26), the superimposed $-I$ effects of the three *para* chlorophenoxy groups deshield the methine signals as much as by 6.40 ppm.

When an $AA'BB'$ multiplet has four, extremely strong lines (and this is generally characteristic of *para*-disubstituted aromatic compounds where $J_{AB} > J_{AA'} \approx J_{BB'}$ and $J_{AB} \approx 0$)* J_{AB}, δA, and δB can be calculated like for the AB system, since it is a very good approximation to regard $\delta A(AB) = \delta A(AA'BB')$; $\delta B(AB) = \delta B(AA)'BB')$; and $J_{AB}(AB) \approx J_{AB}(AA)'BB')$. Of course, in this "$AB$-approximation" $J_{AB'}$, $J_{AA'}$, and $J_{BB'}$ cannot be determined.**

The chemical shifts calculated in the "AB-approximation" for the ring protons A and B, adjacent to the oxygen and chlorine, respectively, are $\Delta\nu_{AB} = \sqrt{26{\cdot}8} = 14.5$ Hz; hence $\nu A = 434$ Hz, $\nu B = 420$ Hz, and $\delta A = 7.23$ and $\delta B = 7.00$ ppm. $J_{AB} = 9$ Hz.

$(Cl-C_6H_4-O)_3CH$

6/I

PROBLEM **7**

Determine the structure of compound $C_6H_{16}S_2Si_2$ from Spectrum **7**/*1*. It is known that the two sulfur atoms are attached to silicon atoms, and for thiane (**I**) δH - 3,4,5 = 1.70 ppm.[1305]

Note: Owing to the lines presumably appearing in the vicinity of the TMS signal (occasionally at $\delta < 0$) or overlapping with it, in the case of silicon-containing substances, the internal standard (TMS) can be applied only after preliminary investigations without internal standard (see also Problems **26** and **38**). Usually, the spectrum of a sample containing no TMS is recorded, then TMS is added to the sample in two-three portions, and the spectrum is taken after each addition. The signal which becomes gradually stronger in these spectra can be assigned unambiguously to TMS.

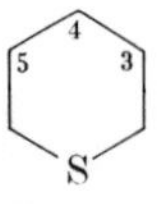

7/I

ANSWER

Spectrum **7**/*1* consists of two singlets of 3:1 intensity at 0.42 and 1.85 ppm. This indicates that all hydrogens are attached to carbon atoms, and 4 and 12 of them,

* See Volume I, p. 133.
** See Volume I, p. 131.

respectively, are equivalent, because the hydrogens attached to silicon atoms would have much higher chemical shifts,* and each of the sulfur atoms could bear only 1 hydrogen. Hence, there are four methyl and two methylene groups in the molecule. They are not adjacent (there is no splitting), and all of them are attached to silicon (according to the low chemical shifts methylmercapto- or $-SCH_2-CH_2S-$ groups may not be present). Therefore, taking also into account the required two Si–S bonds, the presence of two equivalent $-CH_2-Si(Me_2)-S$ groups is certain, thus the structure must be **III** because the chemical shifts (1.85 ppm) of the methylene groups would be lower in Structure **II**, which is otherwise also compatible with the above. The $Si-CH_2-CH_2-Si$ protons could be only more shielded (by the silicons) than the H-3,4,5 atoms of thianes (1.70 ppm).

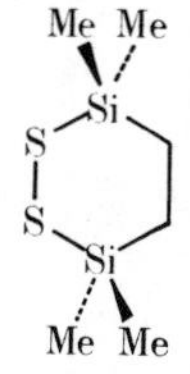

7/II 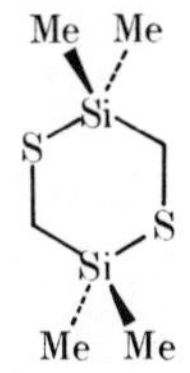 **7/III**

PROBLEM **8**

Determine the structure of compound C_7H_8S on the basis of its spectra in CCl_4 (**8**/*1*) and in TFA (**8**/*1a*).

ANSWER

In Spectrum **8**/*1* a triplet at 1.55 ppm (J = 7.5 Hz), a doublet at 3.57 ppm (J = 7.5 Hz), and a singlet at 7.18 ppm can be found. The intensity ratio is 1:2:5.

The five equivalent protons evidently belong to a phenyl group. The substituent is a methylmercapto-group. The compound is therefore benzylmercaptan, $PhCH_2SH$.

In CCl_4 solution the exchange is so slow that the CH_2SH group yields an AX_2 spectrum with sharp lines. The CH–SH couplings can often be observed (unlike the CH–OH and CH–NH couplings) in CCl_4, and this is characteristic of mercaptans.** Of course, by increasing the temperature or adding D_2O, acid, or base to the solution, the rate of exchange processes can be increased, thereby eliminating the splitting. This is illustrated by Spectrum **8**/*2*, in which the SH signal cannot be seen (since it appears in the low-field region, not shown in the figure, overlapping with the signal of the solvent), and the methylene signal is a singlet at 3.53 ppm.

* See Volume II, the last paragraph of Section 3.7.1.4.

** See Volume II, the first paragraph of Section 3.7.1.5.

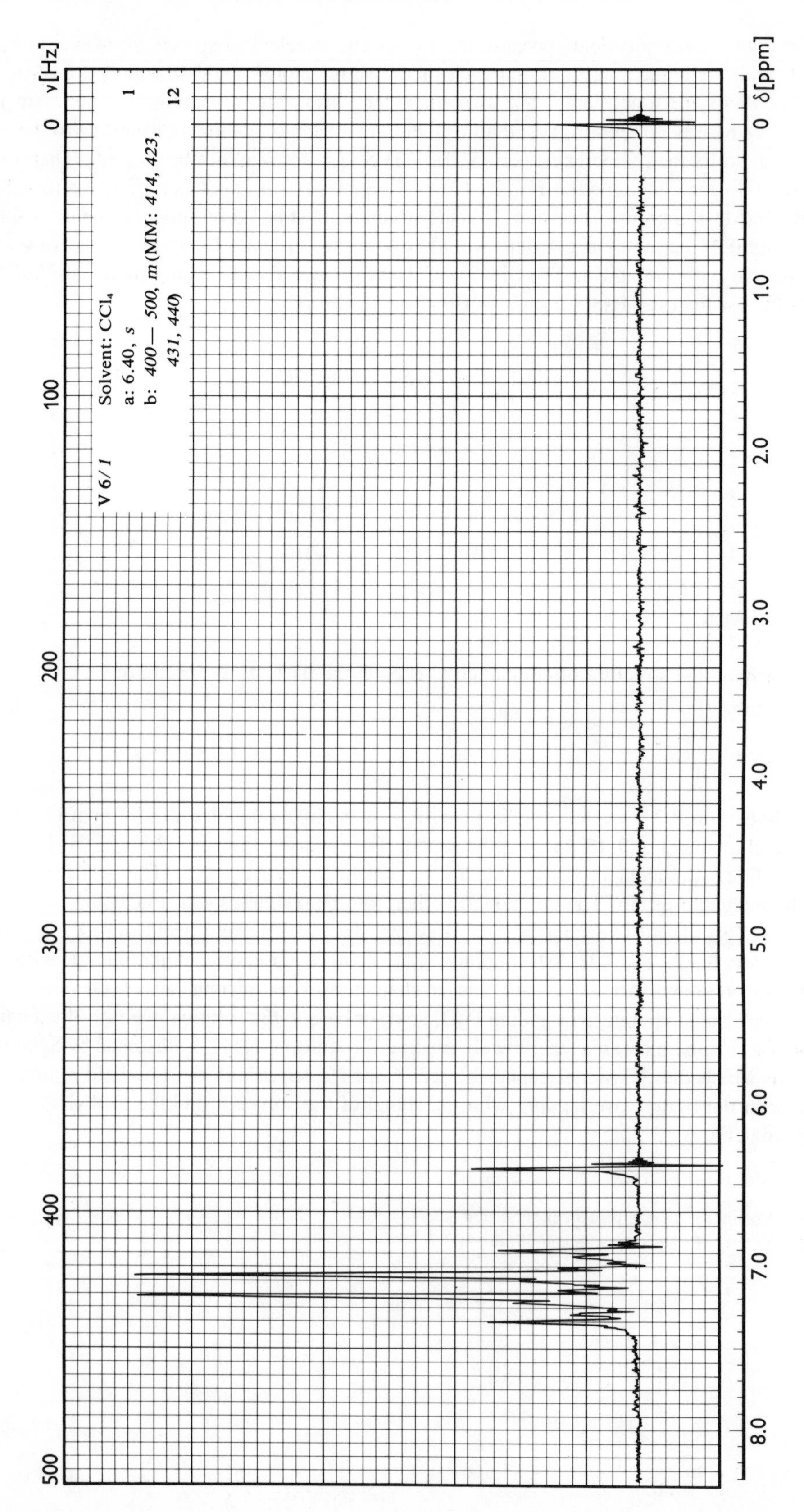

V 6/1
Solvent: CCl_4
a: 6.40, *s* 1
b: *400— 500, m* (MM: *414, 423, 431, 440*) 12
ν[Hz]
0
100
200
300
400
500
δ[ppm]
0
1.0
2.0
3.0
4.0
5.0
6.0
7.0
8.0

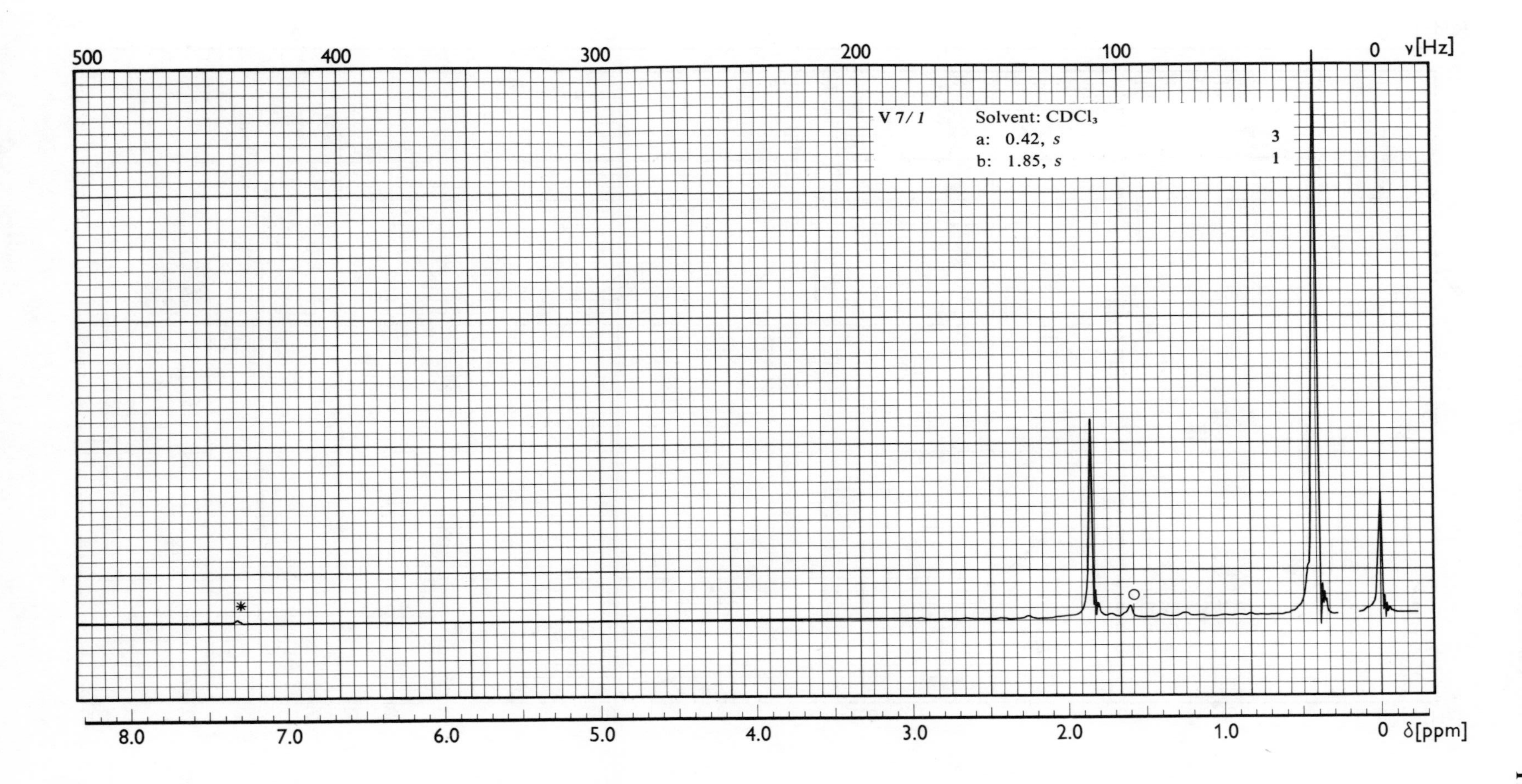
ν[Hz]
500
400
300
200
100
0
V 7/1
Solvent: $CDCl_3$
a: 0.42, *s* 3
b: 1.85, *s* 1
δ[ppm]
8.0
7.0
6.0
5.0
4.0
3.0
2.0
1.0
0

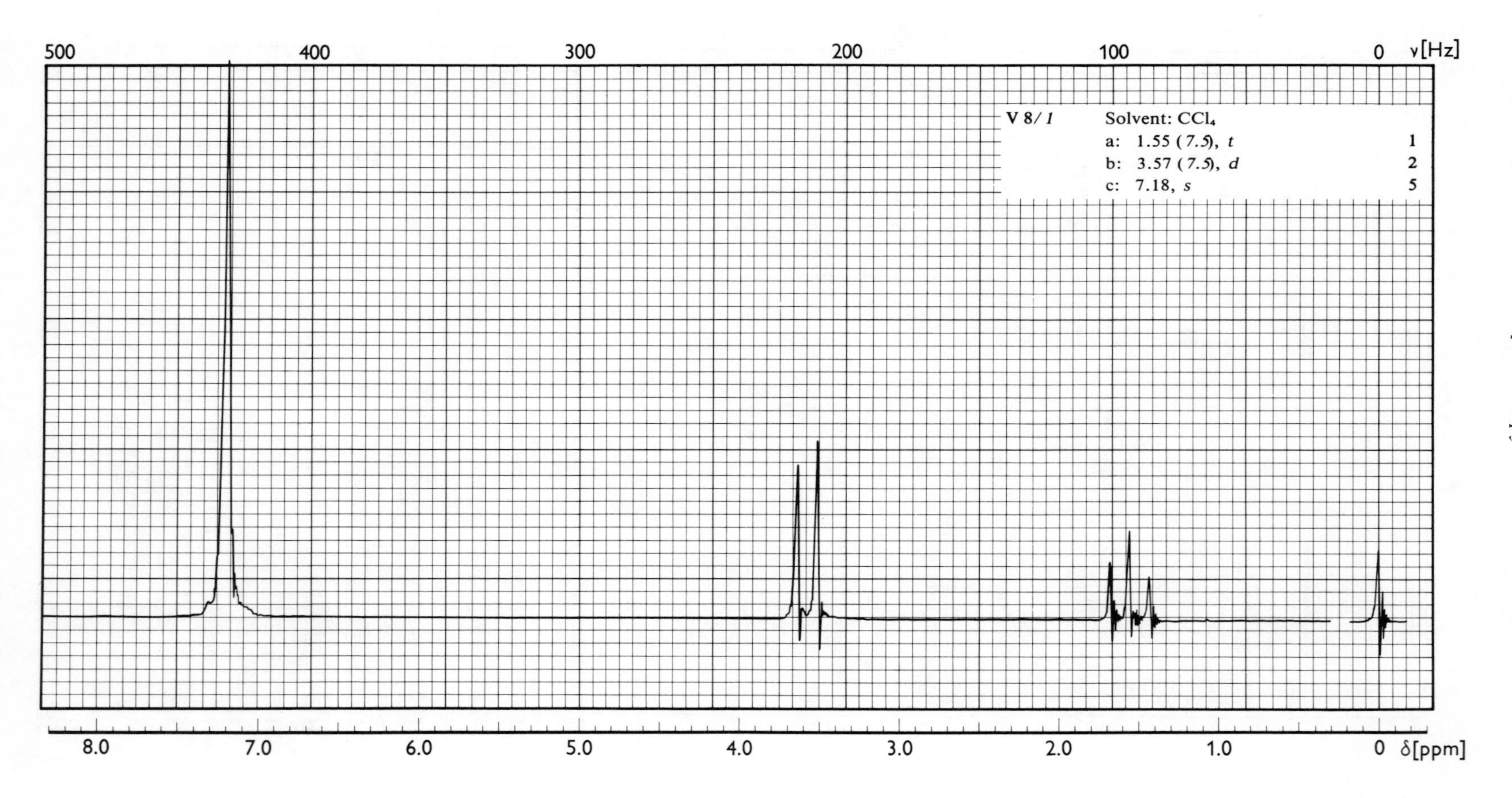
500
400
300
200
100
0 ν[Hz]
V 8/*1*
Solvent: CCl_4
a: 1.55 (7.5), *t* 1
b: 3.57 (7.5), *d* 2
c: 7.18, *s* 5
8.0
7.0
6.0
5.0
4.0
3.0
2.0
1.0
0 δ[ppm]

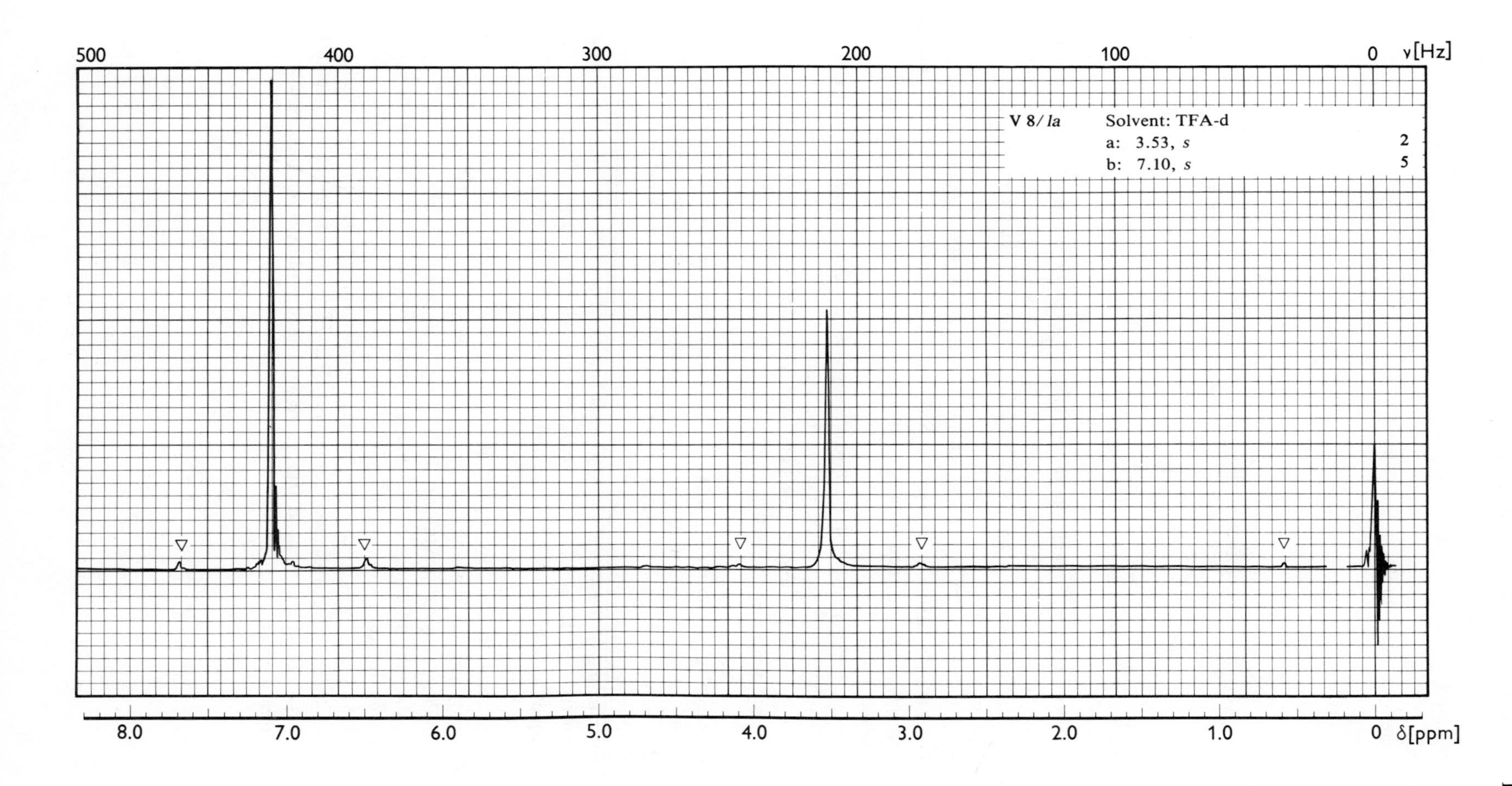
500
400
300
200
100
0
ν[Hz]
V 8/1a
Solvent: TFA-d
a: 3.53, s 2
b: 7.10, s 5
8.0
7.0
6.0
5.0
4.0
3.0
2.0
1.0
0
δ[ppm]

PROBLEM 9[1338]

Assign Spectra **9**/*1* to **9**/*4* to Structures **I** to **IV**.

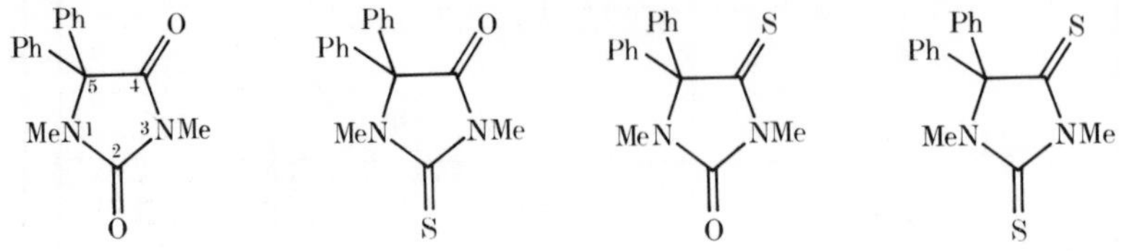

9/I 9/II 9/III 9/IV

ANSWER

The spectra contain, in addition to the signals of aromatic protons at approximately 7.35 ppm, only the singlets of the two methyl groups. The problem must therefore be solved on the basis of their relative chemical shifts, based on the following considerations:

1. The 3-methyl groups are situated between two oxo or thioxo groups, respectively; thus they must be deshielded relative to the 1-methyl substituents. Consequently, the upfield signals can be assigned to the 1-methyl groups, and the downfield ones can be assigned to the 3-methyl groups.
2. The deshielding effect of the thioxo groups on the adjacent coplanar methylhydrogens is stronger than that of the carbonyl groups.*
3. The group in position 2 affects both methyl groups, while group 4 affects only the 3-methyl substituent.

Consequently, Spectrum **9**/*1* must correspond to Structure **IV**, because the two methyl groups are most deshielded in this case (3.10 and 3.70 ppm, respectively). For similar reasons, Structure **I** can be assigned to Spectrum **9**/*3*, giving the smallest δ-values at 2.80 and 3.10 ppm, respectively.

The shielding of the 1-methyl group in Structures **II** and **IV** is similar and smaller than in the spectrum of **III** and **I**, which are similar to each other, again. Therefore Structure **II** corresponds to Spectrum **9**/*4* (with signals of 3.10 and 3.30 ppm) and Structure **III** corresponds to Spectrum **9**/*2*, giving methyl signals at 2.75 and 3.40 ppm. In agreement with this assignment, the 3-methyl signals have similar chemical shifts in these spectra (3.40 and 3.30 ppm), and the 1-methyl shift in Spectrum **9**/*2* of Structure **III** is just as low as in Spectrum **9**/*3*, assigned to Structure **I** (2.75 and 2.80 ppm).

Thus, the shift difference of the methyl signals $\Delta\delta_{1,3}$ is smaller in Spectra **9**/*3* and **9**/*4* than in the other two, and this difference is the greatest in Spectrum **9**/*2*. Since $\delta NCH_3(3) > \delta NCH_3(1)$, $\Delta\delta_{1,3}$ will be greater when the chemical shift of the 3-methyl signal increases, i.e., when there is a 4-thioxo group (Structures **III** and **IV**), and of these, the greater difference can be expected with compound **III**, where the 1-methyl shift is not increased by a neighboring thioxo group.

* Compare Problem **2** and Volume II, p. 43.

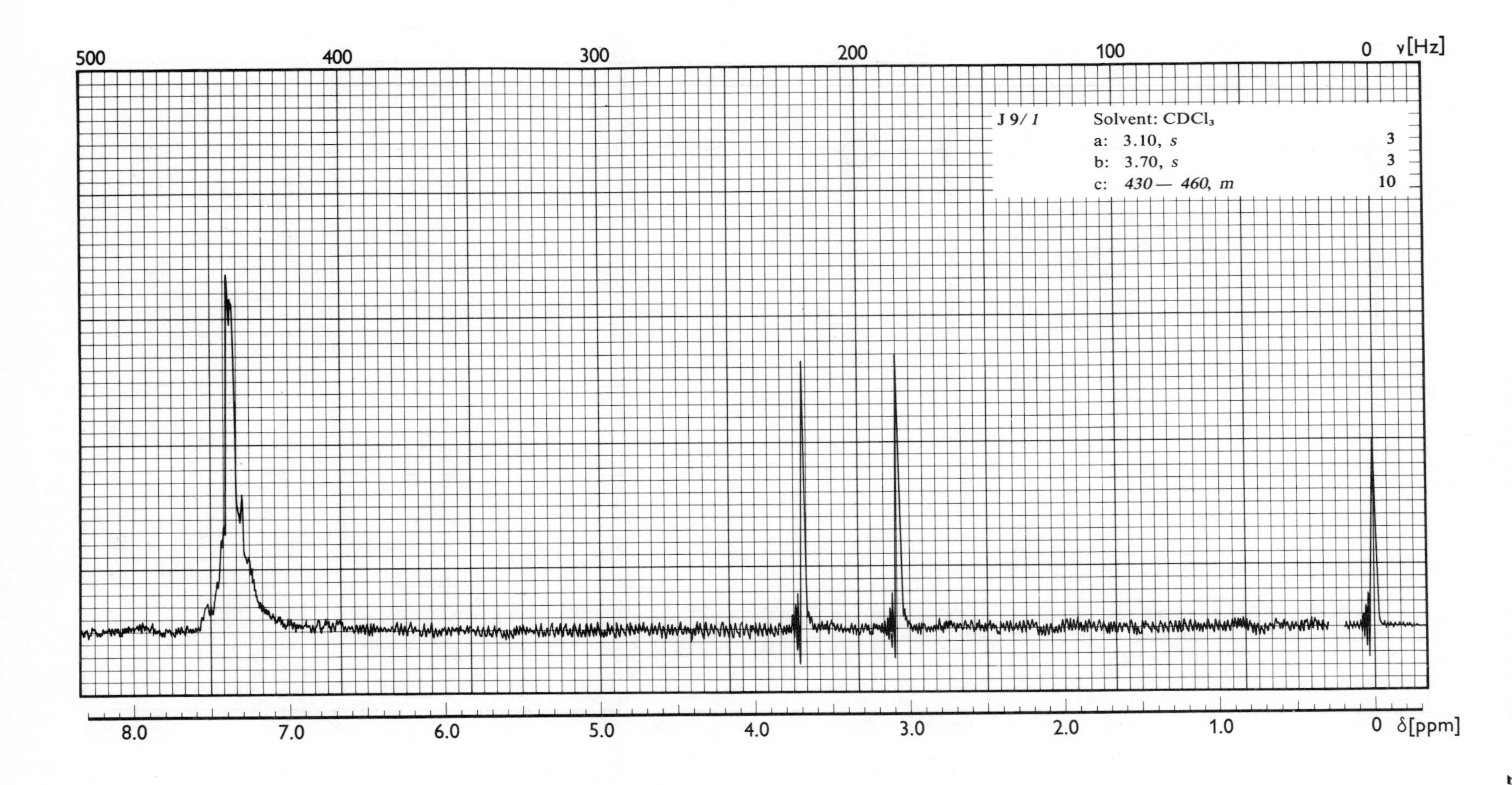
500
400
300
200
100
0 ν[Hz]
J 9/1
Solvent: $CDCl_3$
a: 3.10, s 3
b: 3.70, s 3
c: 430— 460, m 10
8.0
7.0
6.0
5.0
4.0
3.0
2.0
1.0
0 δ[ppm]

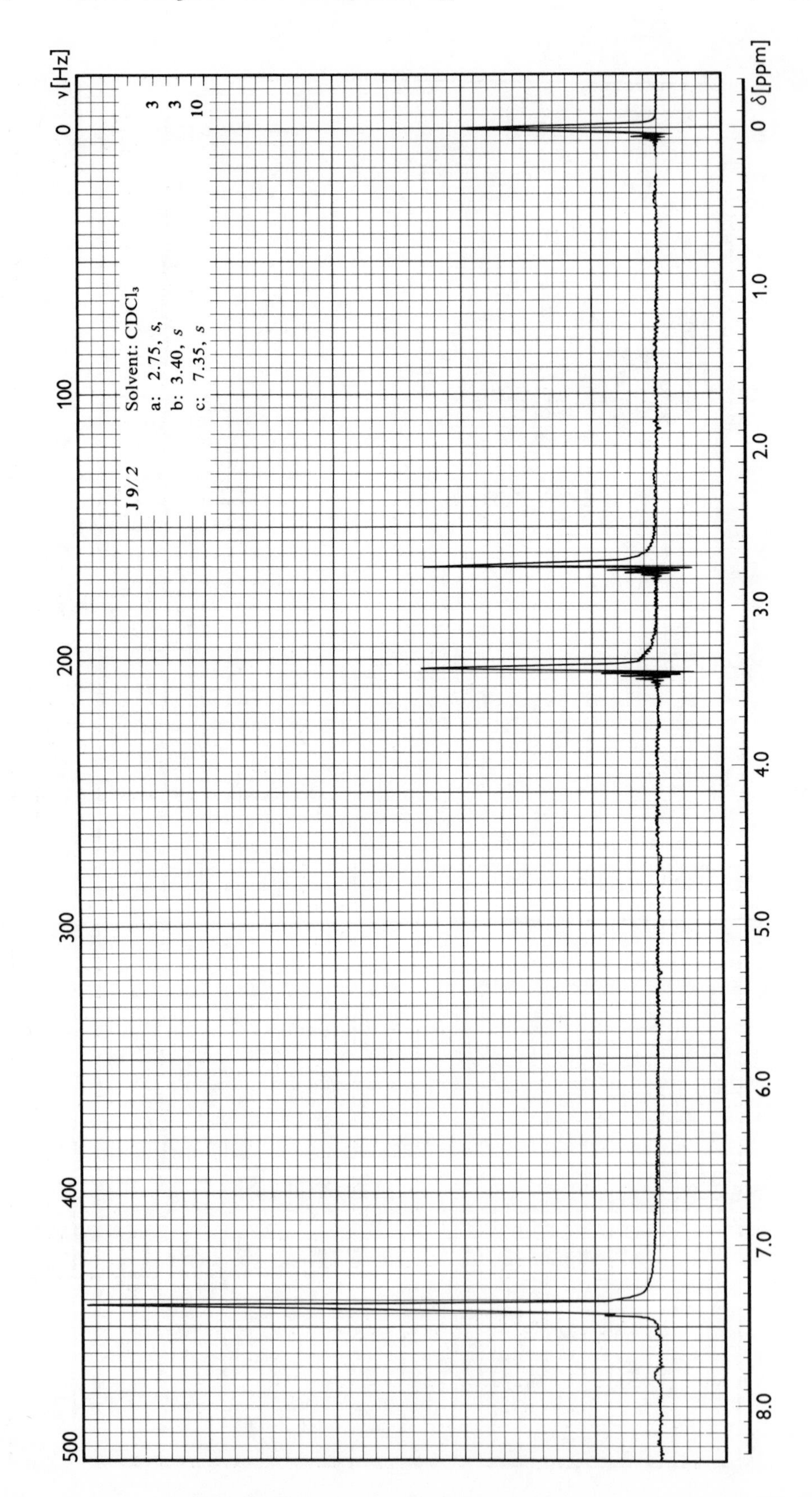
J 9/2
Solvent: CDCl3
a: 2.75, s, 3
b: 3.40, s 3
c: 7.35, s 10
ν[Hz]
0
100
200
300
400
500
δ[ppm]
0
1.0
2.0
3.0
4.0
5.0
6.0
7.0
8.0

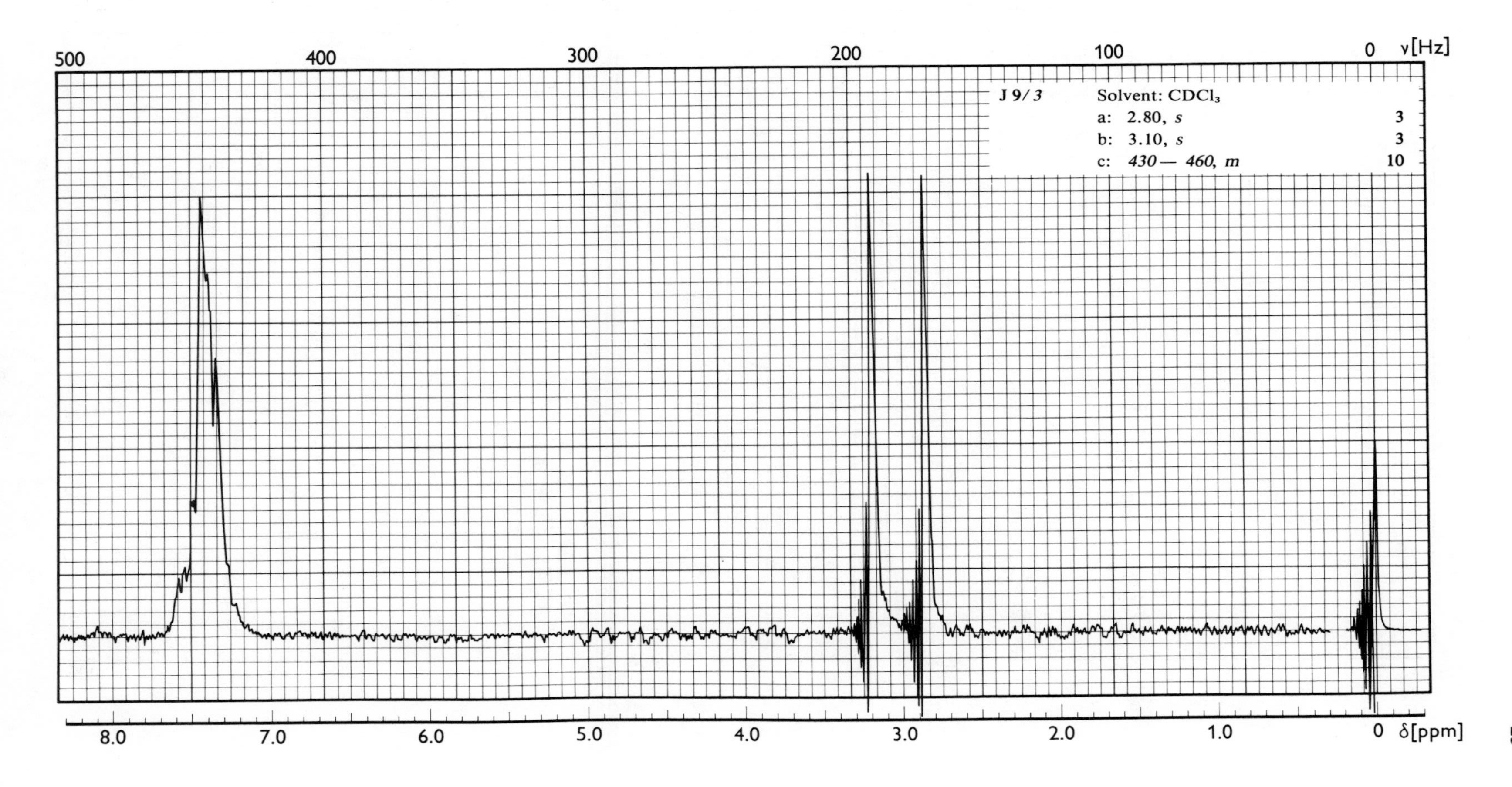

500
400
300
200
100
0
ν[Hz]
J 9/3
Solvent: $CDCl_3$
a: 2.80, *s* 3
b: 3.10, *s* 3
c: *430 — 460, m* 10
8.0
7.0
6.0
5.0
4.0
3.0
2.0
1.0
0
δ[ppm]

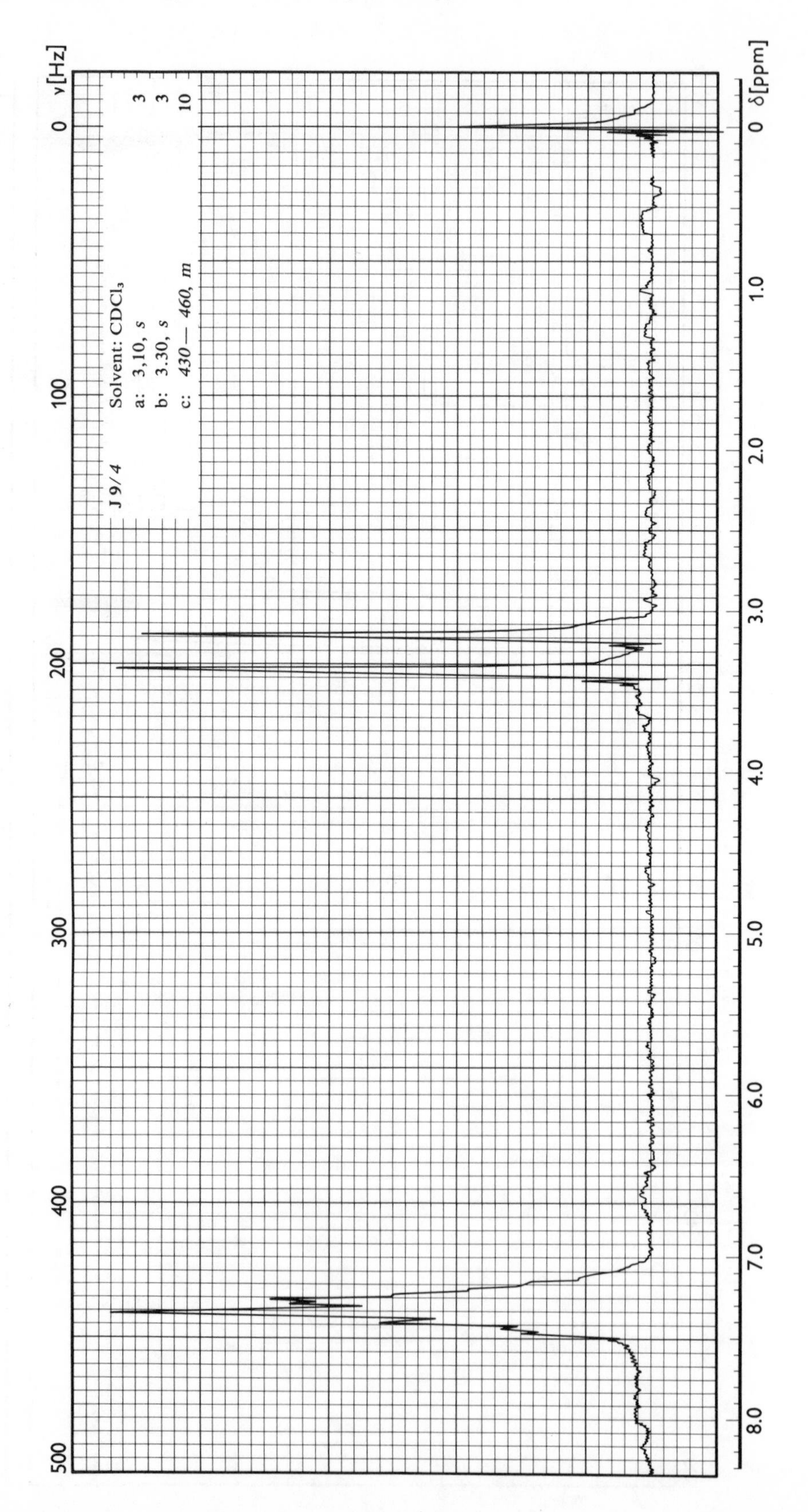
ν[Hz]
0
100
200
300
400
500
J 9/4
Solvent: CDCl₃
a: 3,10, s 3
b: 3.30, s 3
c: 430 — 460, m 10
δ[ppm]
0
1.0
2.0
3.0
4.0
5.0
6.0
7.0
8.0

PROBLEM 10

Decide which of Spectra **10**/*1* and **10**/*2* corresponds to Structures **I** and **II**. Interpret the differences and assign the signals.

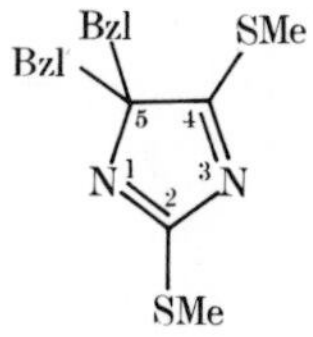

10/I 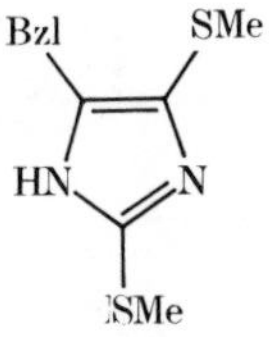 **10/II**

ANSWER

The solution is obvious, e.g., on the basis of the intensity ratio of the SCH_3 and aromatic signals. Accordingly, Spectrum **10**/*1* corresponds to **I**, and Spectrum **10**/*2* corresponds to Structure **II**. The assignment is as follows: **10**/*1*, δCH_3; 2x*s*, 2.40 and 2.45 ppm. The methylene groups give an *AB* quartet (J_{AB} = 13 Hz). δH_A = 3.32 and δH_B = 2.95 ppm. (The frequencies of the lines are 169, 182, 194, and 207 Hz, wherefrom $\Delta\nu AB = \sqrt{38 \cdot 12} \approx 22$ Hz; thus νA = 199 and νB = 177 Hz.) δArH, *s* = 7.08 ppm.

10/*2*, $\delta CH_3(4)$; *s*, 2.31 ppm; $\delta CH_3(2)$, *s*, 2.52 ppm; δCH_2, *s*, 4.05 ppm; δArH, *s*, 7.22 ppm. The NH signal cannot be identified.

The CH_2 signal is an *AB* quartet for Structure **I** because the two pairs of methylene hydrogens are diastereotopic, i.e., nonequivalent due to the prochiral centrum at C-5.*

In the spectrum of compound **II** the chemical shift of the methylene hydrogens increases owing to the vicinity of double bond, and because of the lack of prochirality, the CH_2 signal is a singlet. In the case of **I** the mutual anisotropy of the rings causes a diamagnetic shift of 0.14 ppm in the δArH signal.

* See Volume I, p. 69.

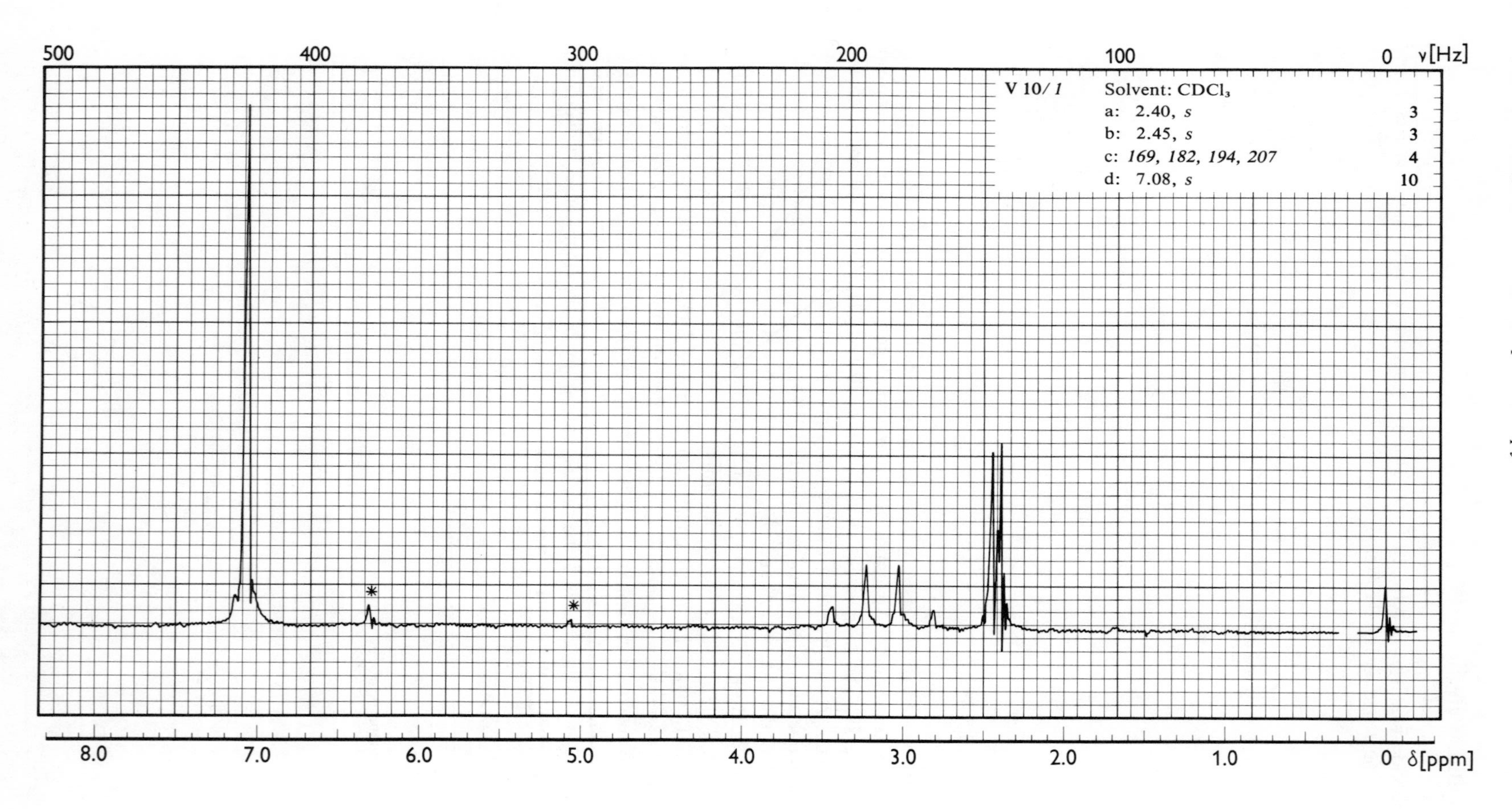
V 10/*1*
Solvent: $CDCl_3$
a: 2.40, *s* 3
b: 2.45, *s* 3
c: *169, 182, 194, 207* 4
d: 7.08, *s* 10
500 400 300 200 100 0 ν[Hz]
8.0 7.0 6.0 5.0 4.0 3.0 2.0 1.0 0 δ[ppm]
*
*

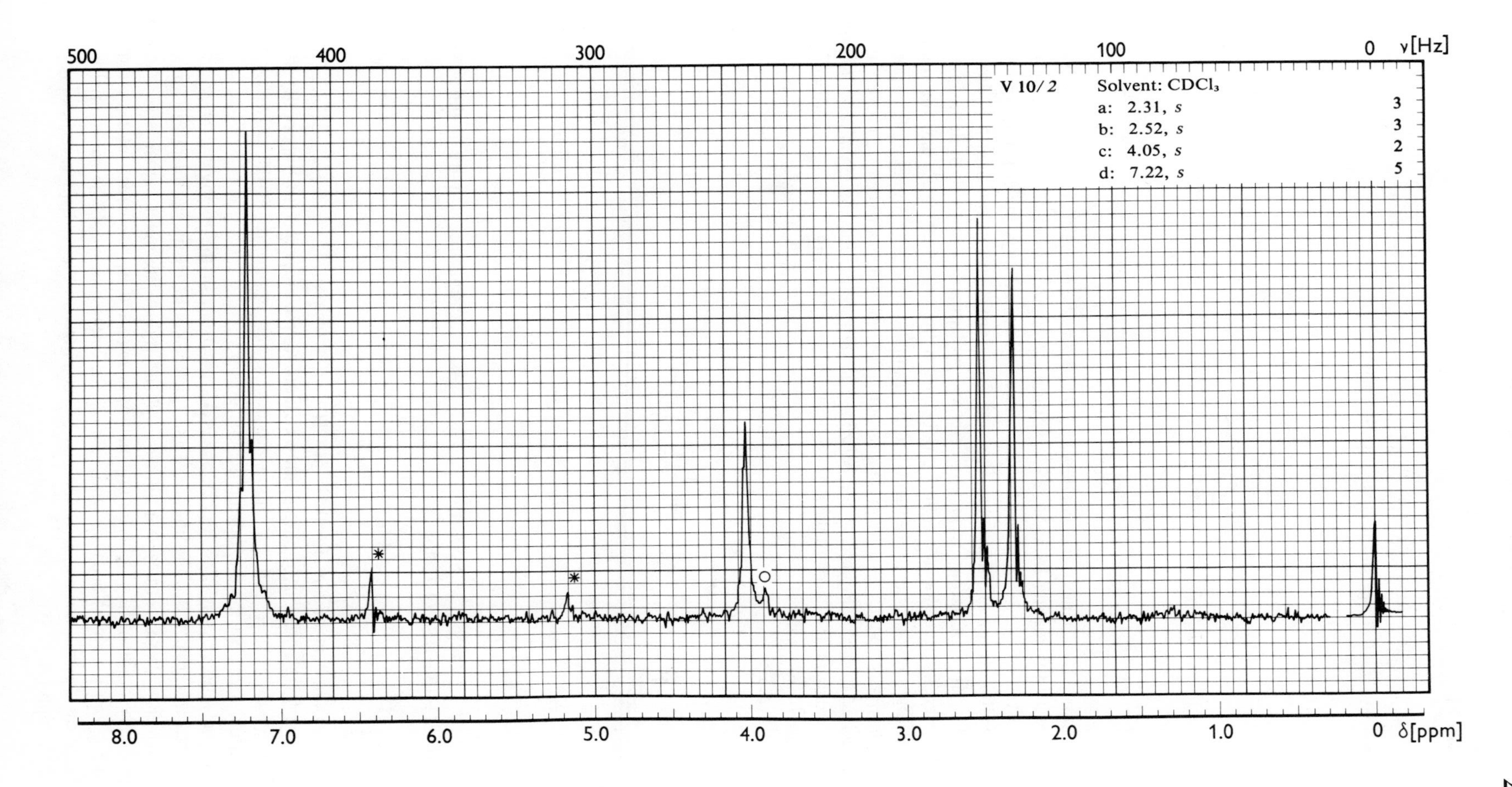
V 10/2
Solvent: CDCl$_3$
a: 2.31, *s* 3
b: 2.52, *s* 3
c: 4.05, *s* 2
d: 7.22, *s* 5
ν[Hz]
500
400
300
200
100
0
δ[ppm]
8.0
7.0
6.0
5.0
4.0
3.0
2.0
1.0
0

PROBLEM 11

Determine on the basis of Spectrum **11**/*1* the substituents (containing only C and H atoms) attached to skeleton **I** at the indicated positions.

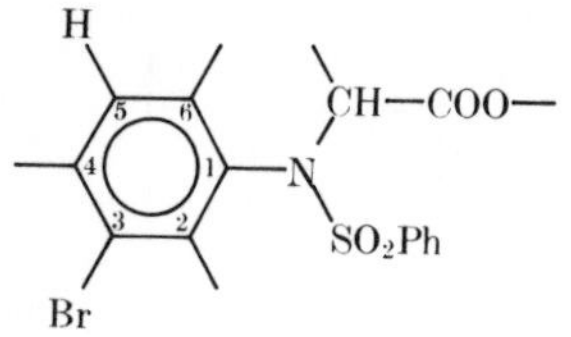

11/I

ANSWER

Spectrum **11**/*1* contains five singlets at 2.03, 2.23, 2.33, 4.29, and 7.07 ppm and a broad signal, evidently arising from an acidic proton, at approximately 9.5 ppm. The intensities correspond to 3, 3, 3, 2, 1, and 1H, respectively. Furthermore, there is a multiplet between 450 and 480 Hz of 5H intensity. The signal of acidic proton requires a hydrogen to be attached to the COO-group to yield a carboxyl substituent. The three upfield singlets, each of 3H intensity, are due to methyl groups. Both the chemical shifts and the multiplicity indicate that they are the substituents of the aromatic ring. The signal at 4.29 ppm of two hydrogens corresponds to a methylene group. The multiplet between 420 and 470 Hz can be assigned to the hydrogens of the benzenesulfonyl group, and the signal at 6.90 ppm can be assigned to the H-5 atom of the pentasubstituted aromatic ring. Consequently, the unknown molecule is Structure **II**.

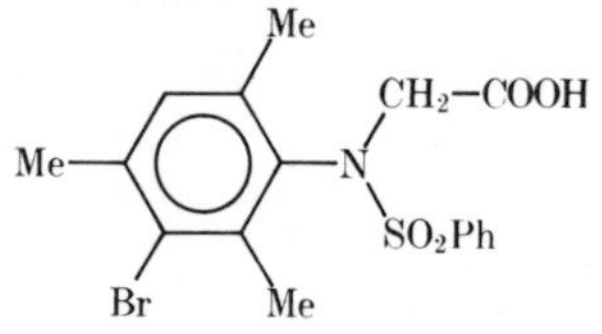

11/II

PROBLEM 12

Decide whether Spectrum **12**/*1* is compatible with the structure $N_2CH–CO–(CH_2)_3–CO–CHN_2$.

ANSWER

The assignment of the signals is δCCH_2, *qi*, 1.95 ppm (J = 6 Hz); δCH_2CO, *t*, 2.40 ppm (J = 6 Hz); δCHN_2, *s*, 5.25 ppm, which proves the proposed structure of the compound. (The signals at 6.38 and 7.27 ppm are due to the impurities of the solvent.) Note the downfield signal of the diazomethylketone group.

As the ratio $J/\Delta\nu$ is high (0.22), the A_2X_4 spectrum of the methylene chain is strongly distorted towards A_2B_4, and therefore the quintet could easily be mistaken for a quartet. Nevertheless, the spectrum is a quintet, but the right-wing outer signal is so weak that it can hardly be identified. The quintet nature is revealed by the equal intensities of lines **2** and **5** (from the right) because with quartets the latter line would be substantially stronger (due to the "roof"-structure, which can be seen also on the A_2 triplet, indicating a higher-order coupling).

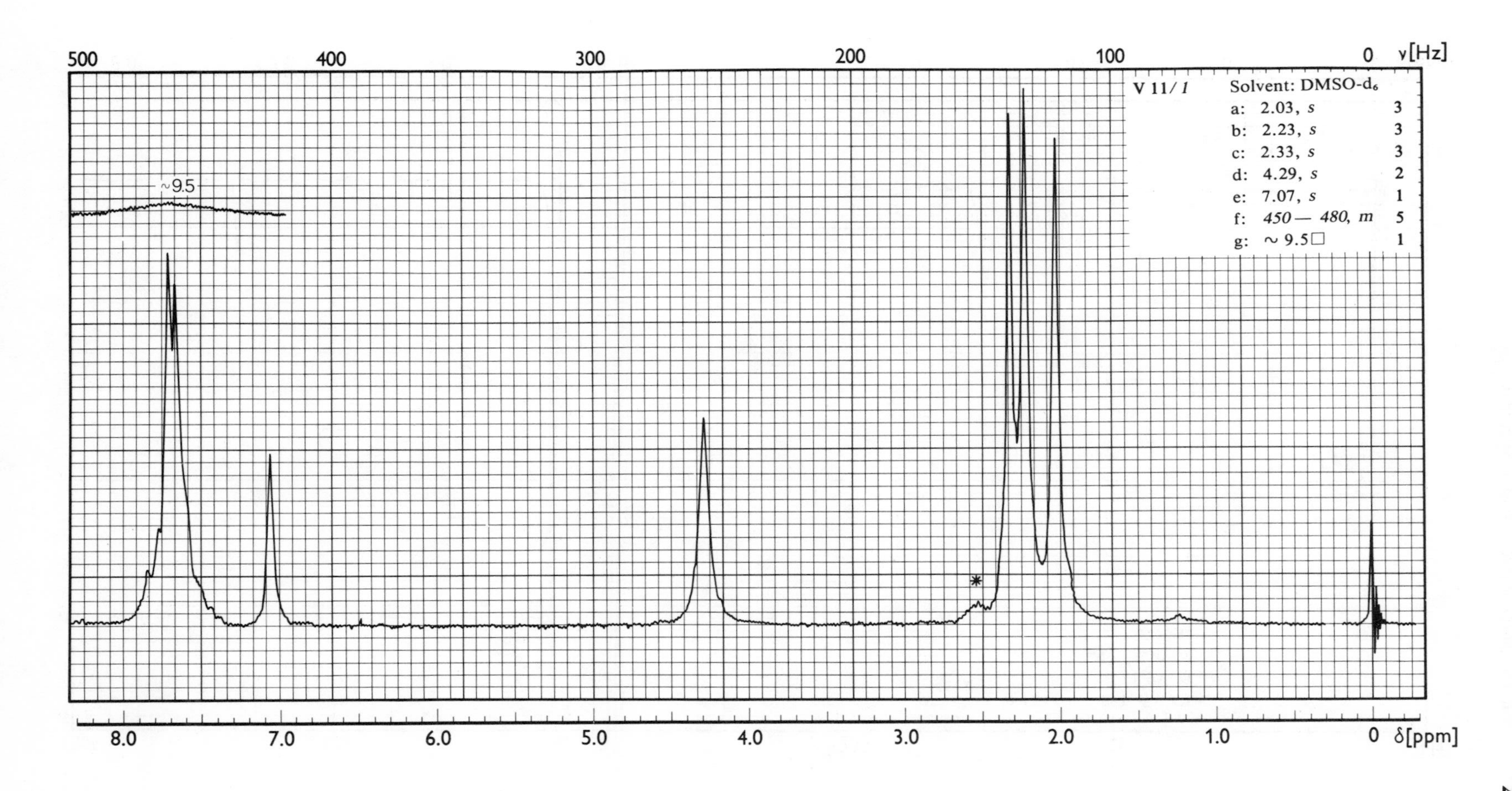
V 11/1
Solvent: DMSO-d₆
a: 2.03, s 3
b: 2.23, s 3
c: 2.33, s 3
d: 4.29, s 2
e: 7.07, s 1
f: 450— 480, m 5
g: ~ 9.5□ 1
ν[Hz]
500
400
300
200
100
0
~9.5
*
8.0
7.0
6.0
5.0
4.0
3.0
2.0
1.0
0
δ[ppm]

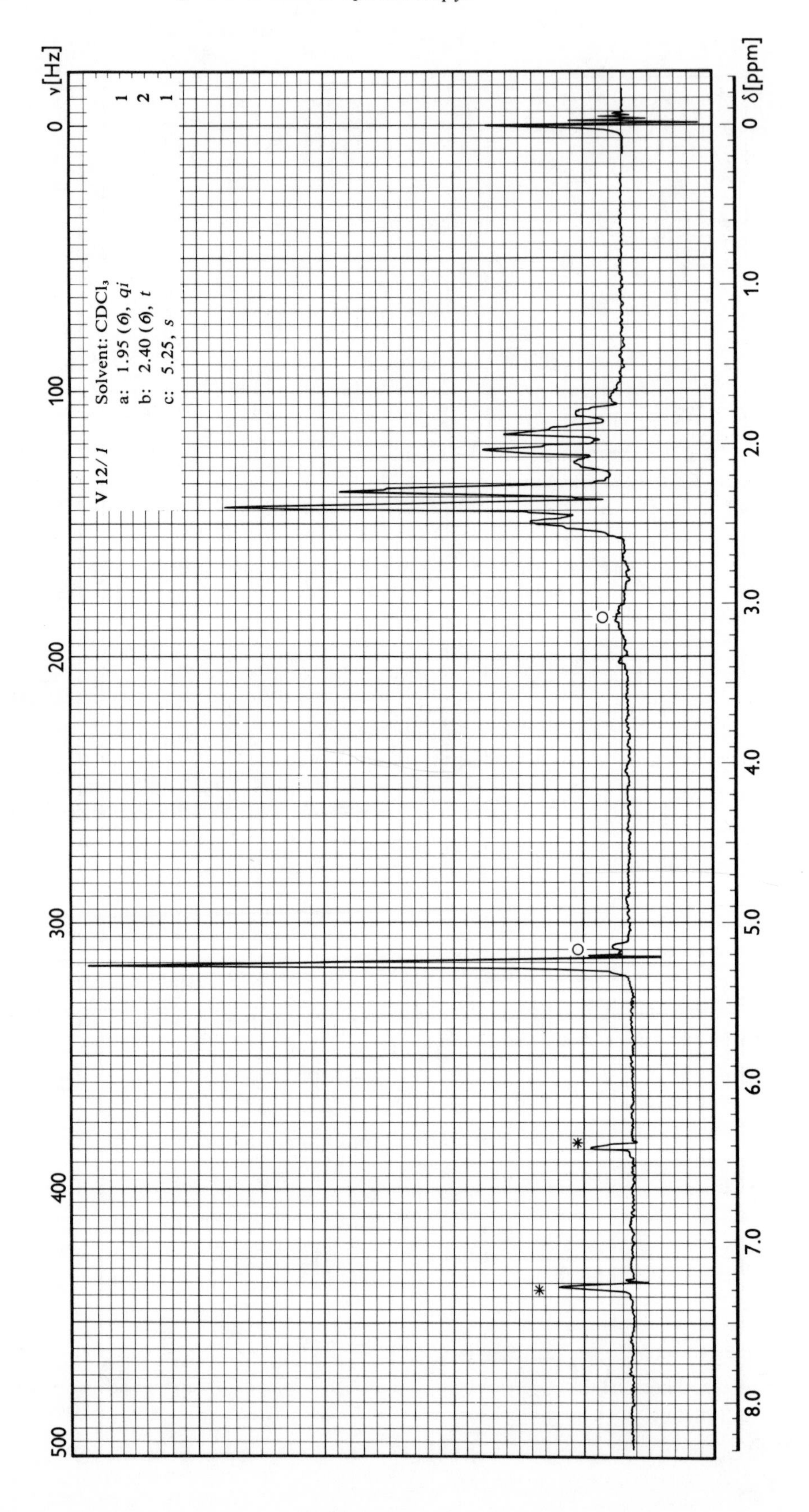
V 12/1
Solvent: CDCl3
a: 1.95 (6), qi 1
b: 2.40 (6), t 2
c: 5.25, s 1
ν[Hz]
0
100
200
300
400
500
δ[ppm]
0
1.0
2.0
3.0
4.0
5.0
6.0
7.0
8.0

PROBLEM 13

Determine substituent R of the phthalimido derivative **I** on the basis of Spectrum **13**/*1* and interpret the structure of the spectrum.

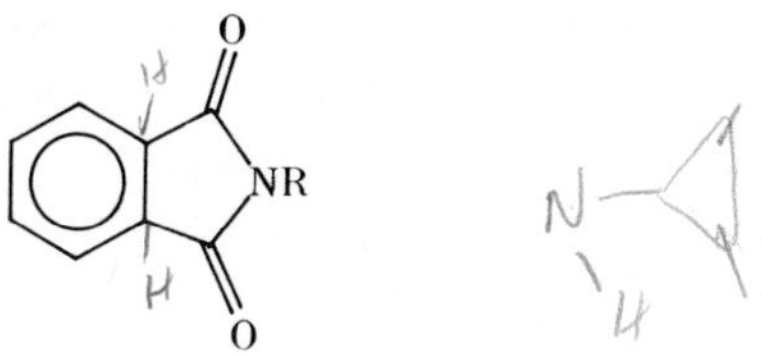

13/I

ANSWER

The signal at 7.67 ppm is the $AA'\ BB'$ multiplet (very close to the limiting A_4 case) of the aromatic hydrogens, therefore its intensity corresponds to four protons. Consequently, group R has five hydrogens, one of which, as follows from the higher chemical shift, is adjacent to the nitrogen atom, and the other four are practically equivalent and strongly shielded, which indicates that R is a cyclopropyl ring.

The doublet and quintet-like structure of the methylene and methine signals, respectively, can be understood by supposing that the coupling constants of the methine hydrogen with the *cis* and *trans* methylene protons are approximately the same, and thus an AX_4 spectrum is obtained in the first approach. It can be seen, however, that this is not true because the quintet is split and the "inner" lines of the multiplets, being closer to one another, are much stronger. The interaction is, actually, of higher order and even more complicated than an AB_4 system despite of the low $J/\Delta\nu$ ratio ($\delta CH_2 \approx 1.03$ ppm, $\delta CH \approx 2.75$ ppm, $J \approx 5$ Hz, and thus $J/\Delta\nu$ $5/100 = 0.05$). Clearly, the reason is that in the cyclic system the methylene protons in the same group are magnetically nonequivalent, and thus the actual spin system is $AA'BB'C$.

PROBLEM 14

The acetylation of compound **I** yields two products. Determine their structure on the basis of Spectra **14**/*1* and **14**/*2*.

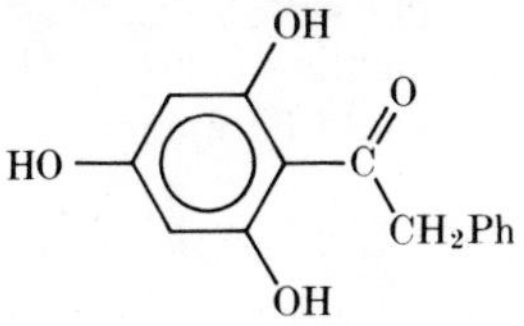

14/I 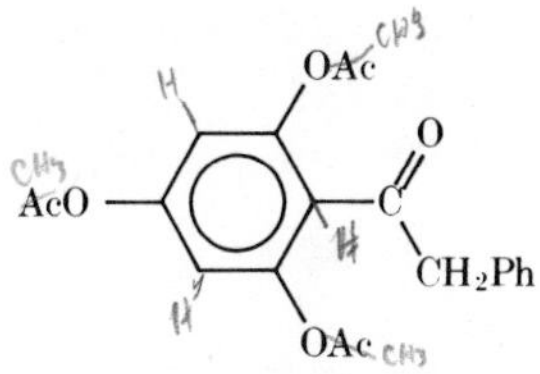 **14/II**

ANSWER

Spectrum **14**/*1* contains five singlets at 2.13, 2.25, 4.07, 6.95, and 7.25 ppm of 6:3:2:2:5 intensity. Thus, three acetyl groups are present in the molecule, of which two are equivalent, suggesting Structure **II**. The assignment is, in the order of increasing shift, $\delta CH_3(2,6)$, $\delta CH_3(4)$, δCH_2, δArH-3,5, δArH (Bzl).

In Spectrum **14**/*2* the methylene signal at approximately 4 ppm is absent. The chemical shifts of the aromatic protons remain practically unchanged: δH-3,5 = 6.95 ppm and δArH(Bzl) ≈ 7.4 ppm (beside the latter at 7.28 ppm the $CHCl_3$ signal of the solvent appears). The relative intensities show that four acetyl groups have been introduced in the molecule. Of these, the singlets of three groups form a common signal at approximately 2.25 ppm, and one group gives a separate line at 2.10 ppm. Moreover, a further signal of 1H intensity can be found at 6.22 ppm. Consequently, this product is the tetraacetoxy derivative **III**, and the maximum at 6.22 ppm can be assigned to the olefinic hydrogen. This chemical shift is in agreement with the expectation.*

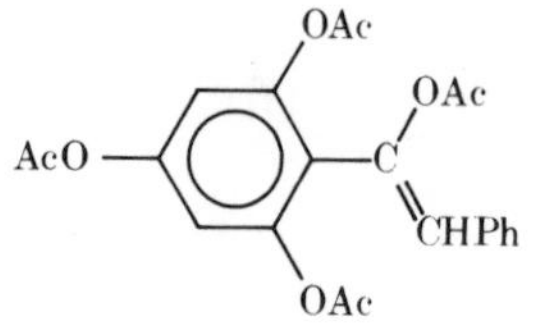

14/III

PROBLEM **15**

Determine the structure of compound $C_{10}H_{12}OCl_2$ from Spectrum **15**/*1*.

ANSWER

In Spectrum **15**/*1*, a triplet, a doublet, a quintet, and a sextet can be found at 0.98, 1.21, 1.68, and 4.45 ppm, respectively, of 3:3:2:1 intensity. All multiplets arise from first-order splitting ($J \approx$ 6 to 7 Hz). Thus, there are two methyl groups adjacent to a methylene and a methine group, respectively, i.e., the molecule contains an ethyl and a CH-CH_3 group. Furthermore, there is an AB_2 multiplet, corresponding to three hydrogens, in the region characteristic of aromatic hydrogens. This signal indicates the presence of a symmetrically 1,2,3- or 1,3,5-trisubstituted aromatic ring. These groups account for all carbons and hydrogens in the empirical formula. The small chemical shift of the methylene group and the sextet splitting of the methine signal prove that they are linked together, i.e., form a secondary butyl group. Therefore, only Structures **I** and **II** are possible, which can be distinguished by the coupling constant, J_{AB}. Deshielding of the methine proton, indicating the vicinity of the oxygen, is also compatible with Structure **I** or **II**.

From the frequencies of the AB lines, we obtain** $\delta A = A_3/60 = 408/60 = 6.80$ ppm, $\delta B = (\nu B_5 + \nu B_7/(2\cdot 60) = (427+433)/(2\cdot 60) = 7.17$ ppm and $J_{AB} = (\nu A_4 + \nu B_7 - \nu B_5 - \nu A_1/3 = (415 + 435 - 427 - 399)/3 = 8$ Hz. The coupling constant of 8 Hz reveals to Structure **II** ($J_{AB} = J^m$ corresponding to compound **I** would be much lower, about 2 to 3 Hz).

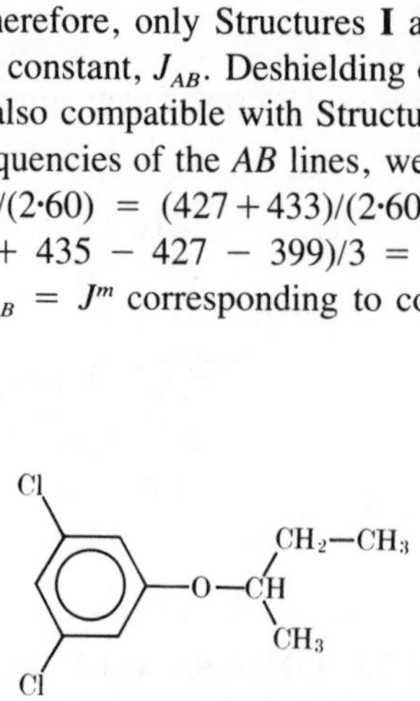

15/I 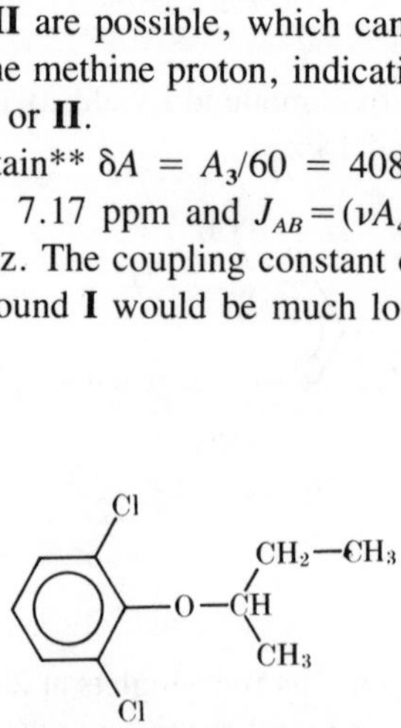 **15/II**

* See Volume II, p. 49, compare also Equation 276 and Table 37.

** See Volume I, p. 106.

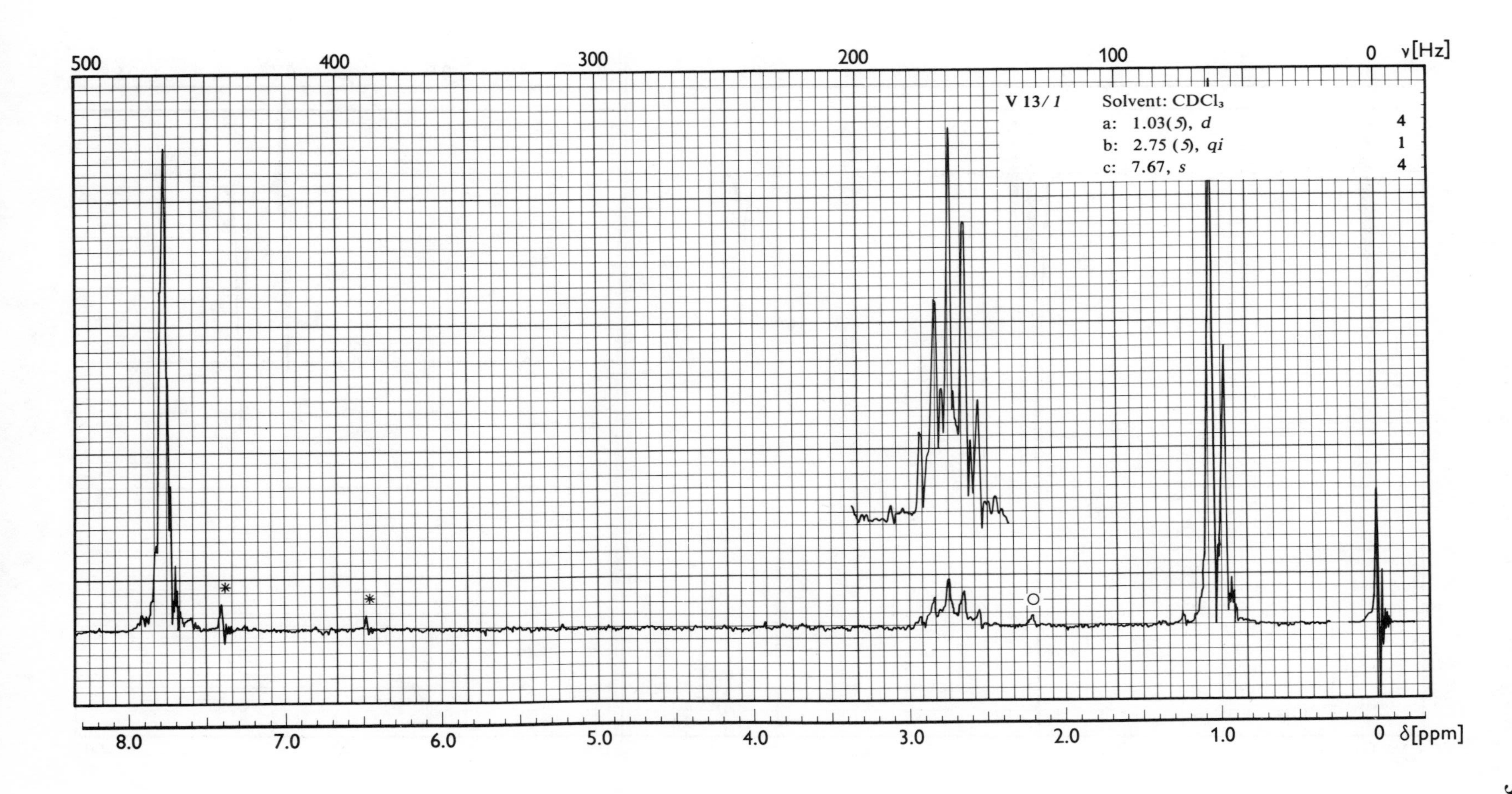

V 13/1
Solvent: CDCl₃
a: 1.03(5), d 4
b: 2.75 (5), qi 1
c: 7.67, s 4
500
400
300
200
100
0
ν[Hz]
8.0
7.0
6.0
5.0
4.0
3.0
2.0
1.0
0
δ[ppm]

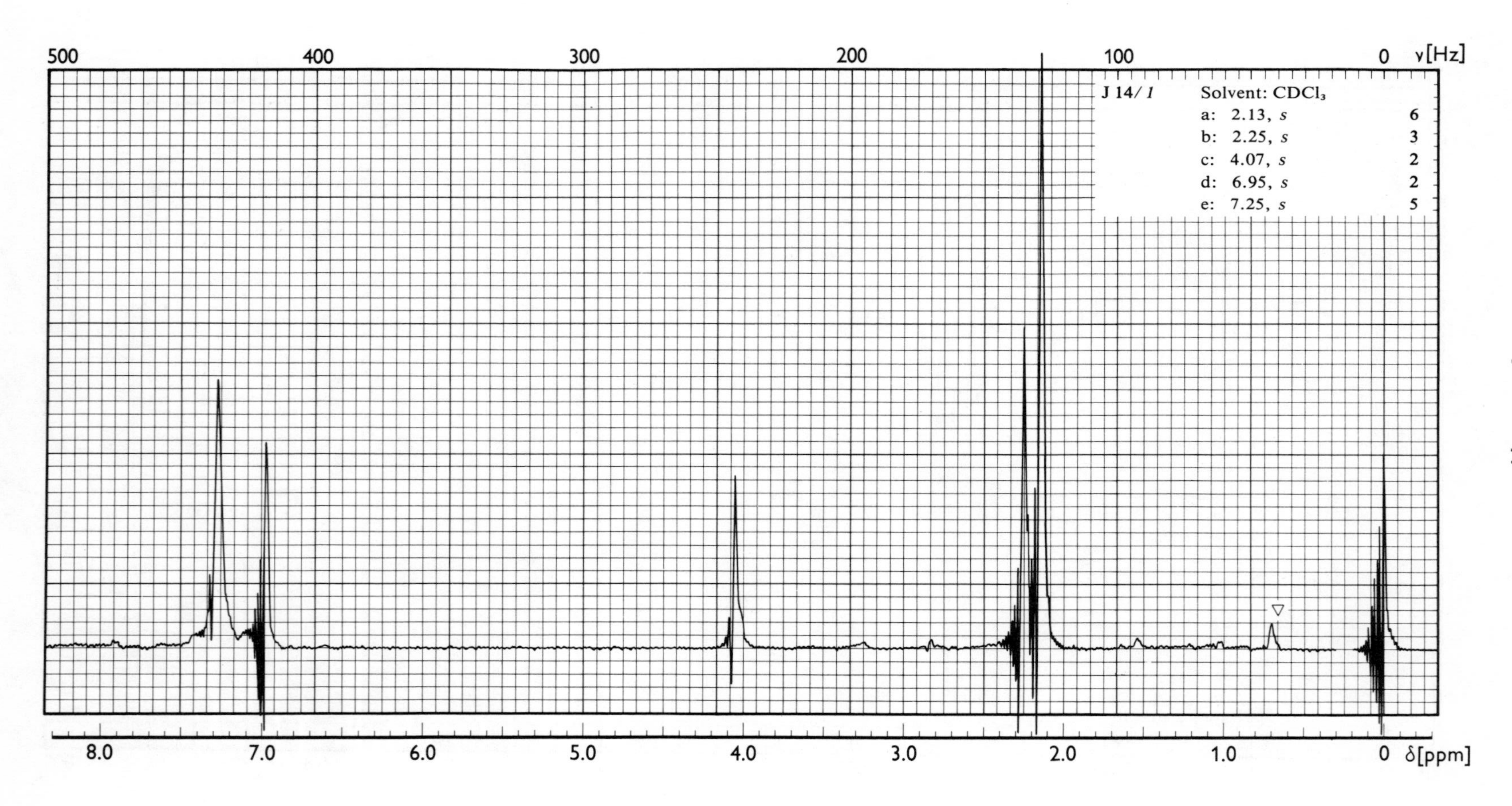
500
400
300
200
100
0 ν[Hz]
J 14/1
Solvent: CDCl3
a: 2.13, s 6
b: 2.25, s 3
c: 4.07, s 2
d: 6.95, s 2
e: 7.25, s 5
8.0
7.0
6.0
5.0
4.0
3.0
2.0
1.0
0 δ[ppm]

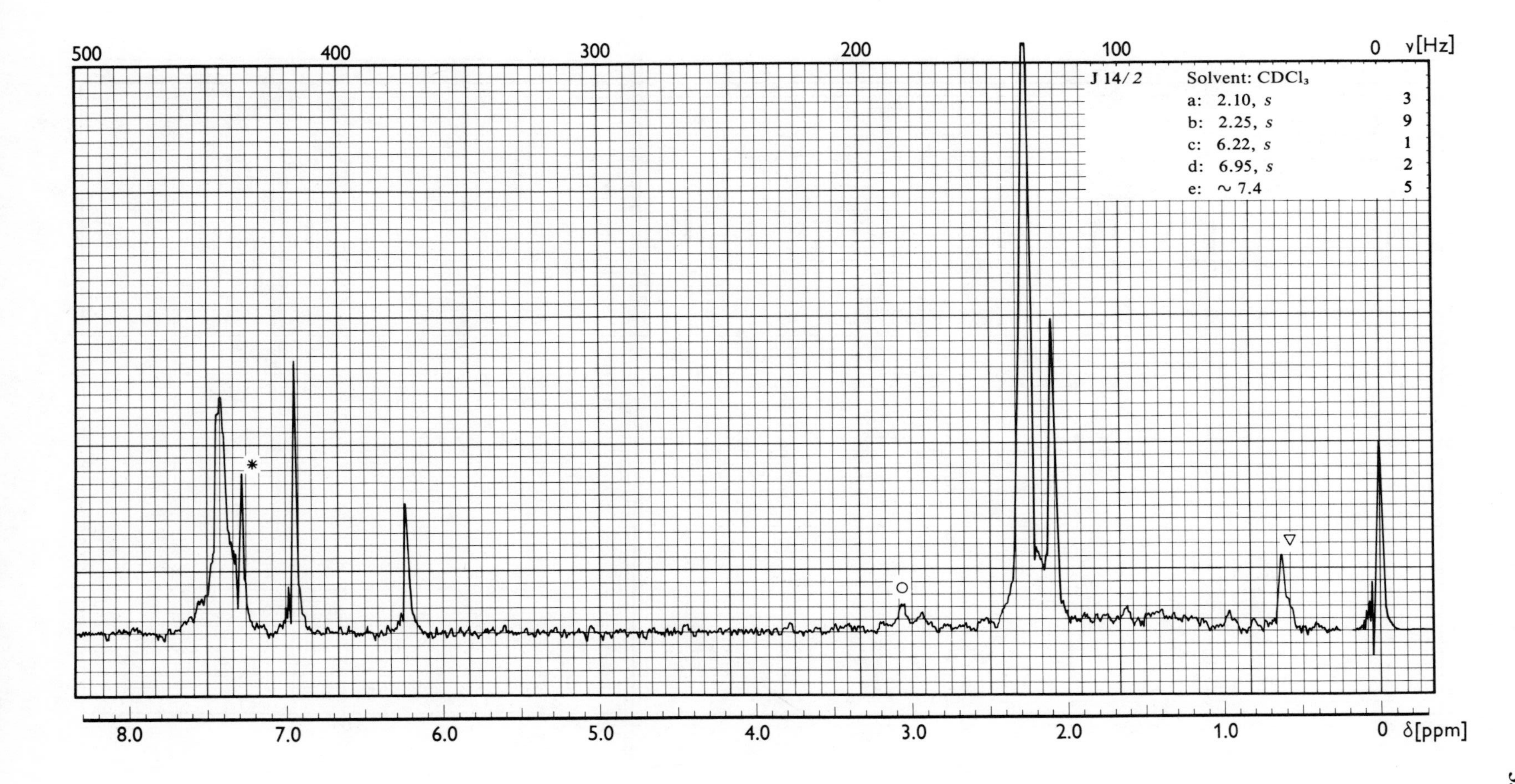

500
400
300
200
100
0 ν[Hz]
J 14/2
Solvent: $CDCl_3$
a: 2.10, *s* 3
b: 2.25, *s* 9
c: 6.22, *s* 1
d: 6.95, *s* 2
e: ~ 7.4 5
8.0
7.0
6.0
5.0
4.0
3.0
2.0
1.0
0 δ[ppm]

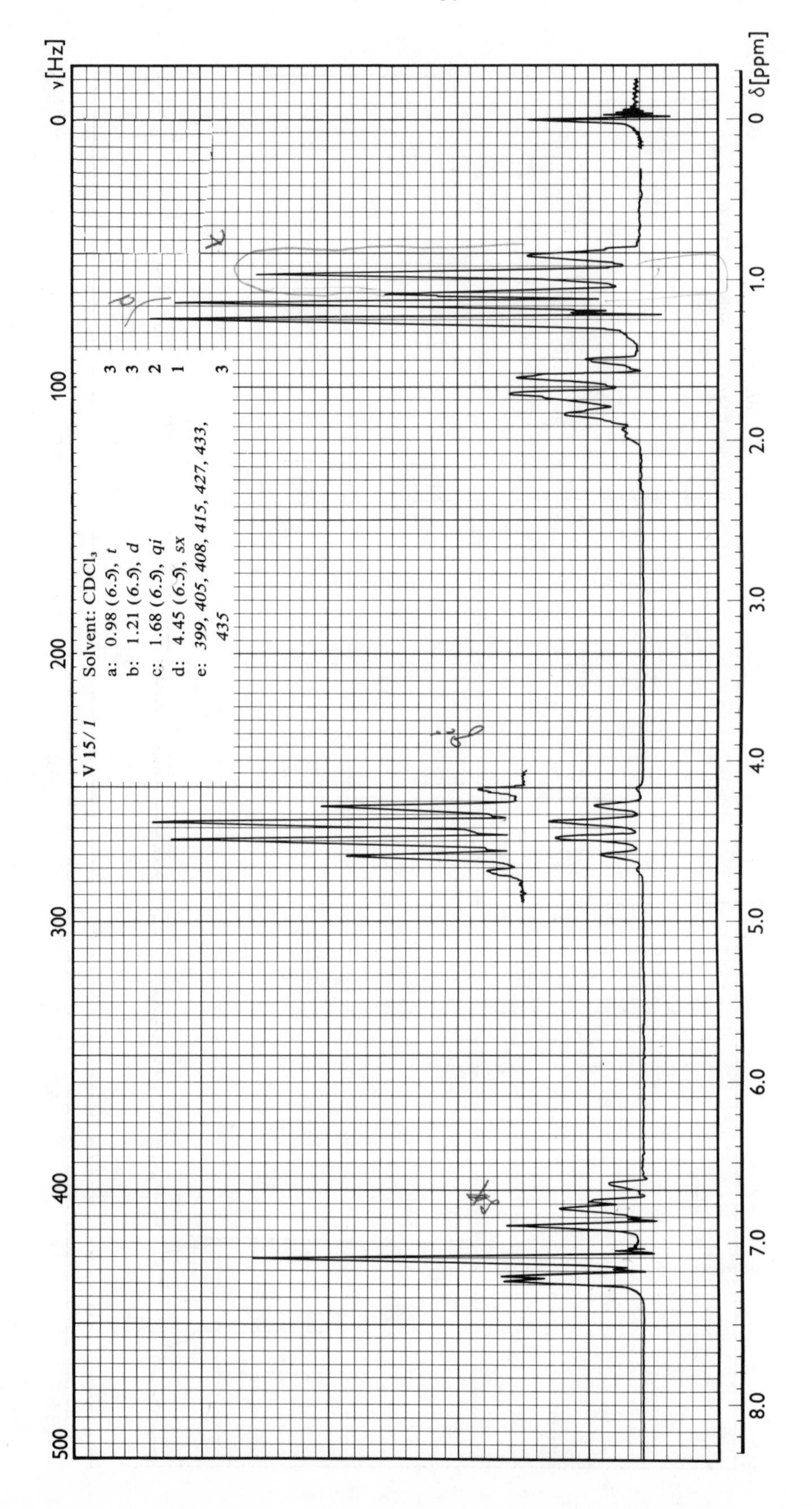
ν[Hz]
0
100
200
300
400
500
δ[ppm]
0
1.0
2.0
3.0
4.0
5.0
6.0
7.0
8.0
V 15/1
Solvent: $CDCl_3$
a: 0.98 (6.5), t 3
b: 1.21 (6.5), d 3
c: 1.68 (6.5), qi 2
d: 4.45 (6.5), sx 1
e: 399, 405, 408, 415, 427, 433, 435 3

PROBLEM 16

The Spectrum **16/*1*** recorded in methanol-d_6 was taken from a guanidinium derivative of type R–NH–C (=NH)NH_2·HCl. Determine substituent R. Explain the changes in the spectrum of acetone-d_6 solution (**16/*1a***) and of the latter solution after acidification with TFA (Spectrum **16/*1b***).

ANSWER

Spectrum **16/*1*** contains two doublets at 1.23 and 2.83 ppm, corresponding to three and two protons, respectively (J = 6.5 Hz), and a multiplet (presumably sextet), corresponding to one proton at 3.90 ppm, with a splitting of again 6.5 Hz. These features suggest the presence of a CH_3–CH–CH_2–type group.

There are two further signals in the spectrum: a broad absorption at approximately 4.75 ppm and a sharp singlet of five hydrogens at 7.28 ppm. The former signal can be assigned to NH protons, and the later can be assigned to a phenyl group. Therefore, one should choose from Structures **I** and **II**.

H_3C–CH(Ph)–CH_2–NH–C(=NH)NH_2 HCl

16/I

Ph–CH_2–CH(CH_3)–NH–C(=NH)NH_2 HCl

16/II

The electron-withdrawing guanidinium ion causes a strong paramagnetic shift in the neighboring group. Consequently, the ionic group is adjacent to the methine group, i.e., the structure is **II**. The position of the δCH_2 signal is also in a good agreement with the expected shifts in benzyl groups.*

In Spectrum **16/*1a*** the methyl doublet appears at 1.20 ppm and the δCH_2 signal is split further. Consequently, in this solution the diastereotopic methylene protons are anisochronic and yield an *AB* multiplet of an *ABX* system, where $\delta A \approx \delta B$ still holds (2.9 ppm). At 3.45 ppm a part of the NH protons absorbs and at approximately 4.20 ppm the methine hydrogen absorbs. The sextet of the latter splits up as well, owing to the coupling with the adjacent NH proton. The corresponding NH doublet can also be identified at 8.20 ppm (J = 9 Hz). In Spectrum **16/*1a*** of acetone-d_6 solution, the accidental isochrony of the aromatic hydrogens in methanol-d_4 is also eliminated and a complex *AA'BB'C* multiplet can be observed.

After acidification (Spectrum **16/*1b***) the NH signals disappear, and the splitting of methine multiplets due to the coupling with the NH proton is also eliminated to yield the original sextet. Furthermore, the accidental isochrony of methylene and aromatic protons occurring in methanol-d_4 is restored. These spectra well illustrate the effects of solvent and pH on the structure of multiplets.

* See Table 25.

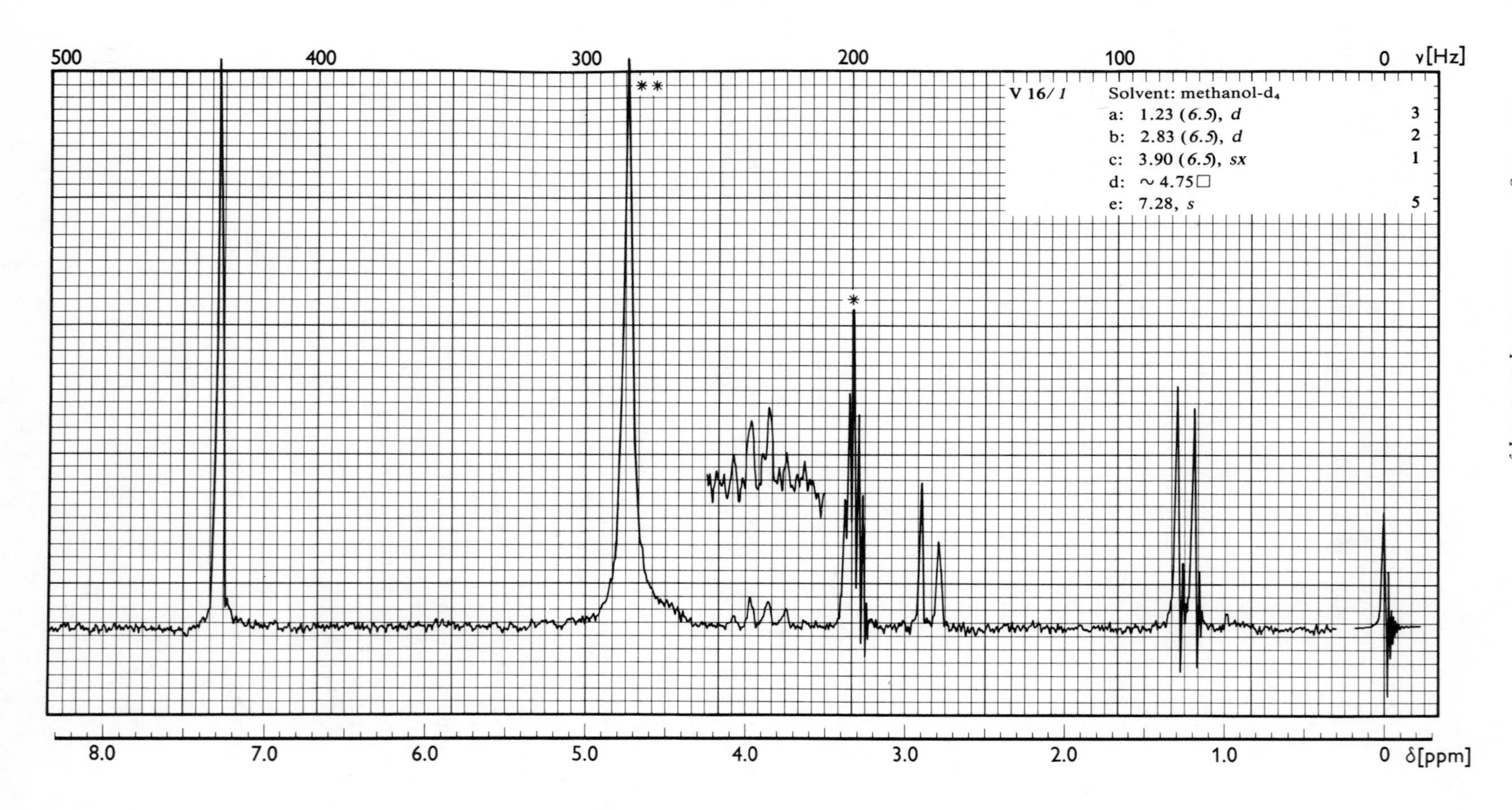
500
400
300
200
100
0 ν[Hz]
V 16/1
Solvent: methanol-d4
a: 1.23 (6.5), d 3
b: 2.83 (6.5), d 2
c: 3.90 (6.5), sx 1
d: ~ 4.75
e: 7.28, s 5
**
*
8.0
7.0
6.0
5.0
4.0
3.0
2.0
1.0
0 δ[ppm]

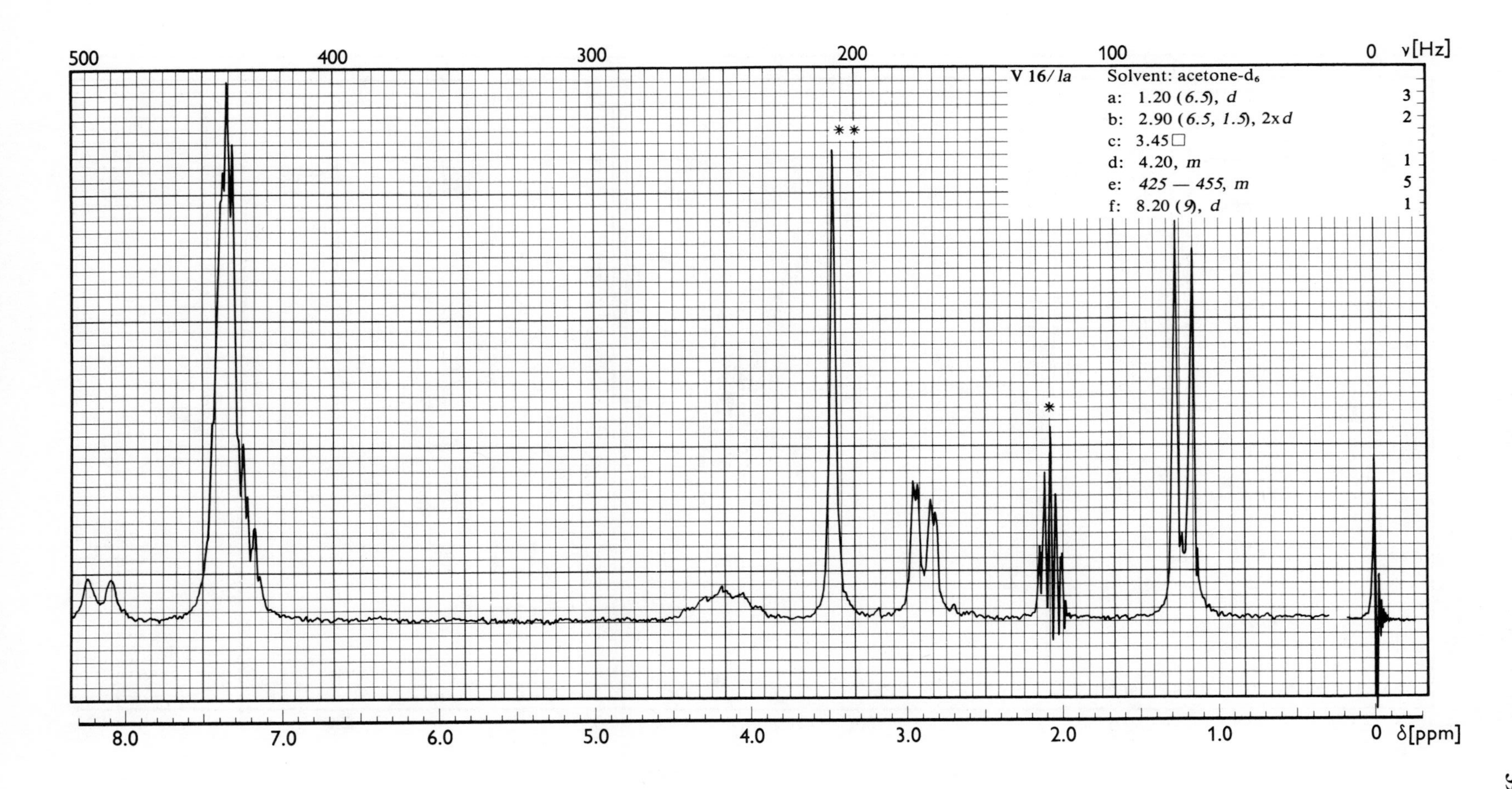

V 16/ *la*
Solvent: acetone-d_6
a: 1.20 (*6.5*), *d* 3
b: 2.90 (*6.5*, *1.5*), 2x*d* 2
c: 3.45□
d: 4.20, *m* 1
e: *425 — 455*, *m* 5
f: 8.20 (*9*), *d* 1
* *
*
500
400
300
200
100
0
ν[Hz]
8.0
7.0
6.0
5.0
4.0
3.0
2.0
1.0
0
δ[ppm]

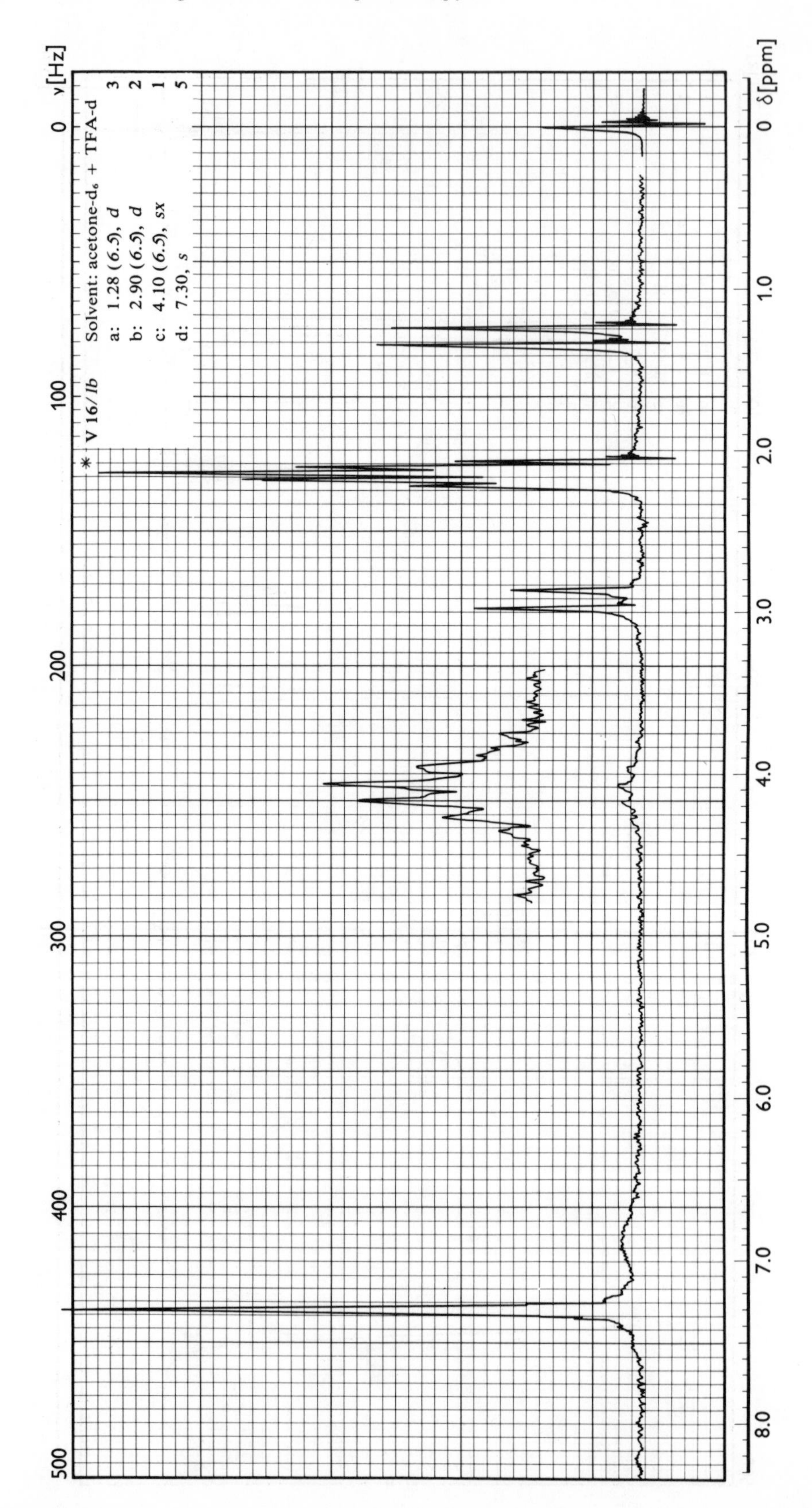

ν[Hz]
0
100
200
300
400
500
δ[ppm]
0
1.0
2.0
3.0
4.0
5.0
6.0
7.0
8.0
*
V 16/Ib
Solvent: acetone-d6 + TFA-d
a: 1.28 (6.5), d 3
b: 2.90 (6.5), d 2
c: 4.10 (6.5), sx 1
d: 7.30, s 5

PROBLEM **17**

By reacting pyridine and azido-formic acid ethyl ester (N_3COOEt), *N*-pyridino-carbethoxyamidate derivatives Py^+–N^-–COOEt can be prepared. Upon the addition of HCl the formation of hydrochloride Py^+–NH–COOEt + Cl^- can be expected. Decide from the spectra of the two compounds (**17**/*1* and **17**/*2*) whether they have the presumed structures and explain the differences between the spectra. For the answer also use the data of pyridine.*

ANSWER

Spectrum **17**/*1* contains the signals of the five protons of the pyridine ring and those of the ethyl group δCH_3, *t*; 1.31 ppm (J = 7 Hz); δCH_2, *qa*: 4.19 ppm (J = 7 Hz); νArH-3-5, *m*: 450 to 485 Hz; νArH-2,6, *m*: 525 to 545 Hz. Since H-3,4,5 of pyridine absorb between 420 and 470 Hz and H-2,6 absorb between 505 and 525 Hz, it is certain that the expected structure is correct. The downfield shift of multiplets is due to the electrophilic substituent and also to that in this compound the lone electron pairs of the exocyclic nitrogen are conjugated rather with the carbonyl than with the heterocyclic group.

The assignment of the spectrum of the hydrochloride (**17**/*2*) can be given as follows: δCH_3, *t*: 1.37 ppm (J = 7 Hz); δCH_2, *qa*: 4.42 ppm (J = 7 Hz); νArH-3,5: 485 to 510 Hz; νArH-4: 520 to 535 Hz; νArH-2,6: 535 to 550 Hz. The protonation is proved by the paramagnetic shifts in the signals. This shift of 0.23 ppm in δCH_2 proves the assumed negative polarization of carbonyl group in the base. It is worth noting that δH-2,6 does not increase significantly, but in the environment of the other protons, primarily of H-4, the electron density decreases. Consequently, similar to pyridine-*N*-oxyde and its salt** in the base, the $-I$ effect of the substituent affects mainly the *ortho* protons, but in the salt it already extends to the whole ring. It is noted that in TFA the spectra of the two compounds are completely identical and very similar to the spectrum of the salt in heavy water solution, proving the proposed structure of the latter.

* Compare Volume II, p. 82.
** Compare Volume II, p. 83.

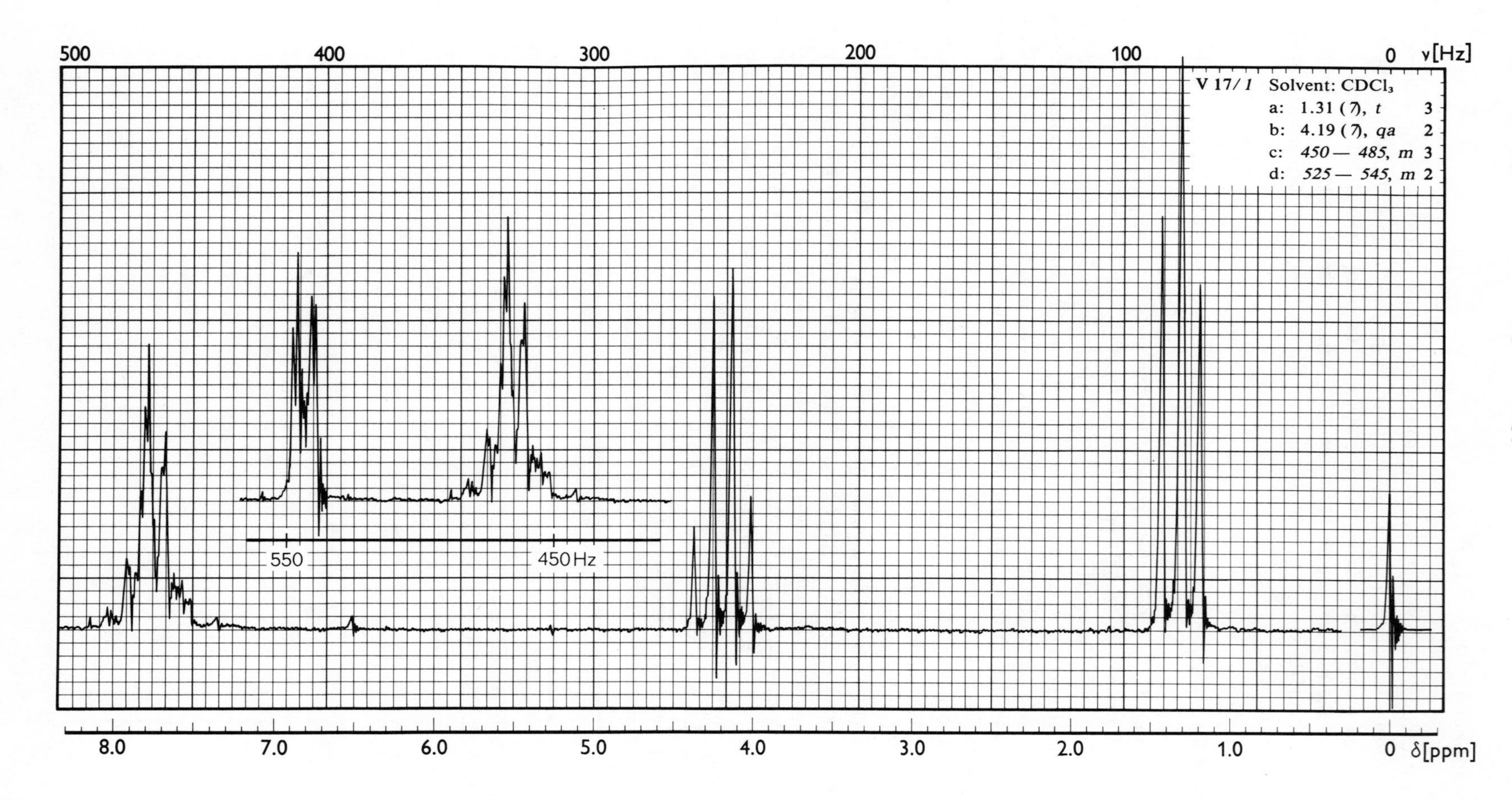
500
400
300
200
100
0 ν[Hz]
V 17/*1* Solvent: $CDCl_3$
a: 1.31 (7), *t* 3
b: 4.19 (7), *qa* 2
c: *450 — 485*, *m* 3
d: *525 — 545*, *m* 2
550
450 Hz
8.0
7.0
6.0
5.0
4.0
3.0
2.0
1.0
0 δ[ppm]

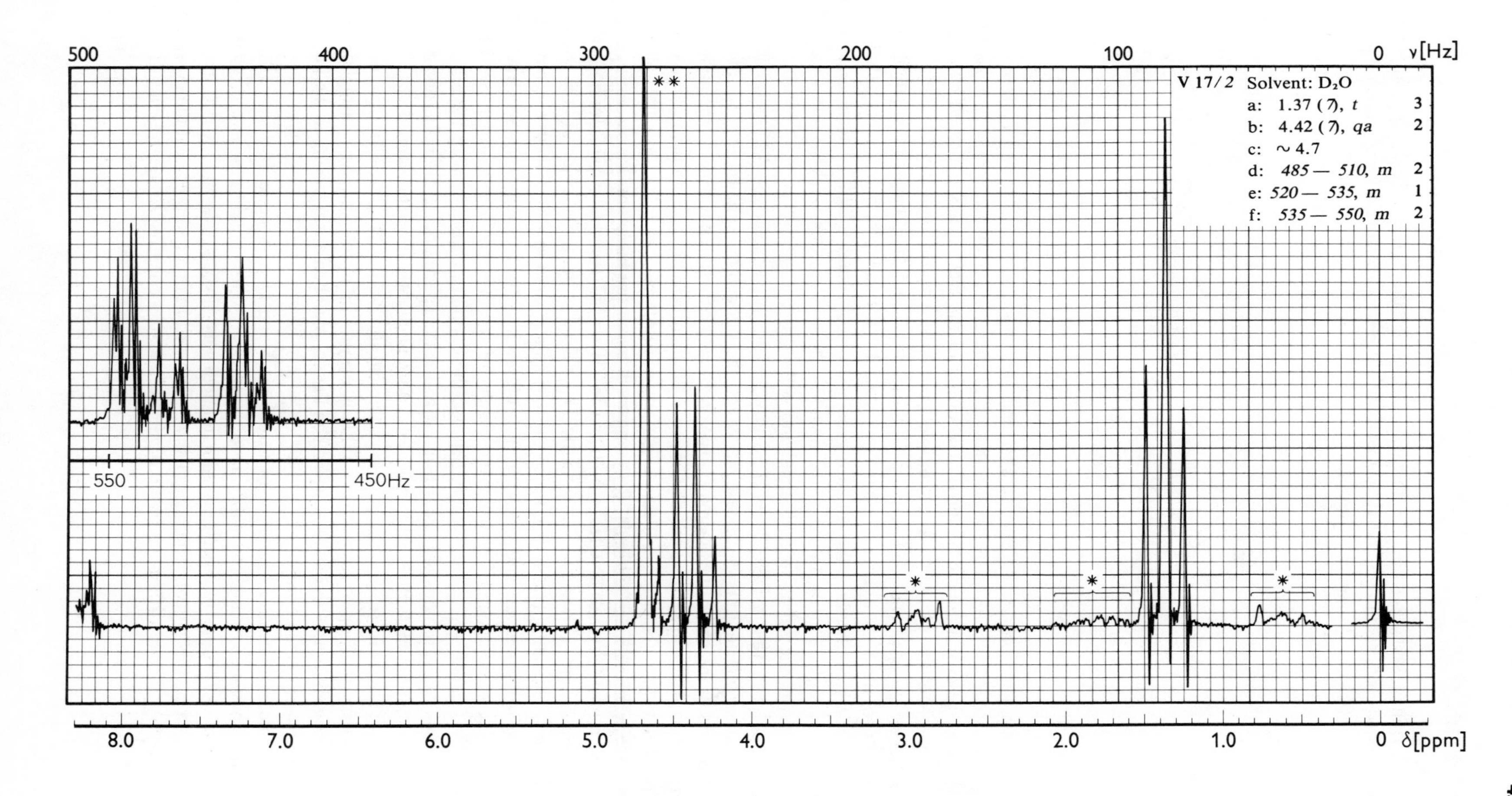
ν[Hz]
500
400
300
200
100
0
V 17/2 Solvent: D_2O
a: 1.37 (7), t 3
b: 4.42 (7), qa 2
c: ~4.7
d: 485 — 510, m 2
e: 520 — 535, m 1
f: 535 — 550, m 2
**
*
550
450Hz
8.0
7.0
6.0
5.0
4.0
3.0
2.0
1.0
0
δ[ppm]

PROBLEM 18[1344,1408]

Determine, without using the integrals, which of the Spectra **18**/*1*, *1a* and **18**/*2*, *2a* corresponds to Structures **I** and **II**, respectively. (Spectra *1a* and *2a* were recorded after the addition of D_2O.) Assign the signals.

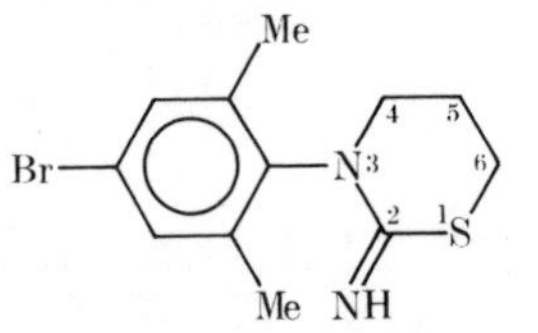

18/I 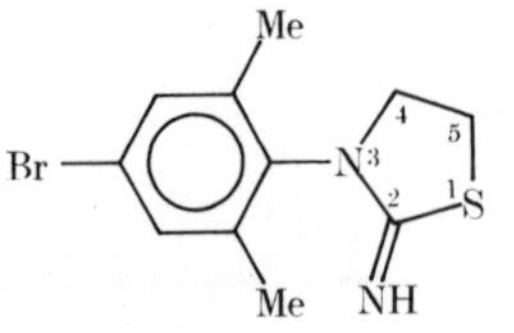 **18/II**

ANSWER

The singlets observed in Spectra **18**/*1* and **18**/*2* at 2.24 and 2.22, at 5.75 and 5.85, and at 7.23 and 7.25 ppm, respectively, belong in the above sequence to the two methyl, the NH, and the two aromatic protons. (The origin of the NH signals is proved by their shift upon the addition of D_2O.) The multiplets between 3 and 4 ppm can be assigned to the methylene groups. The identification of Structures **I** and **II** is based most simply on the multiplet of the 5-methylene group in compound **I**. This multiplet must appear upfields from the other signals because the 4 and 6 groups are deshielded by the adjacent heteroatoms. Such a multiplet cannot, however, be found in the spectra; this weak signal of many lines overlaps with the strong methyl signal and is therefore insignificant.

In Spectrum **18**/*2* the $N–CH_2–CH_2–S$ group yields a characteristic $AA'BB'$ multiplet approximating the limiting A_2X_2 case, in which the outer lines split further and the relative intensities of the inner lines increase. The chemical shifts (the positions of the middle lines) are, in the A_2X_2 approximation: $\delta NCH_2 = 3.80$ ppm; $\delta SCH_2 = 3.55$ ppm; since $\delta NCH > \delta SCH_2$ owing to the stronger electronegativity of nitrogen. $J = 6$ Hz for both triplets. Note that the whole $AA'BB'$ multiplet is symmetrical about the midpoint.

Spectrum **18**/*1* yields $\delta NCH_2 = 3.40$ and $\delta SCH_2 = 3.10$ ppm ($J = 6$ Hz for both triplets). The line system is, however, asymmetrical (the lines in the upfield side of both triplets are stronger), indicating that they correspond to a spin system in which they are coupled with a common third group and not with each other. It is then evident that Spectrum **18**/*1* is assignable to compound **I** and that the quintet of the third methylene group overlaps with the δCH_3 signal. (A quintet occurs because the coupling constants with the *cis* and *trans* NCH_2 and SCH_2 protons are equal.) Investigating the spectra more carefully, one can easily observe that the methyl signal in Spectrum **18**/*1* is slightly more broadened.

Note that with increasing ring size the chemical shifts of NCH_2 and SCH_2 protons decrease by 0.40 and 0.25 ppm, respectively. This is easy to interpret, since the deshielding effect of the nonadjacent heteroatom is weaker in the six-membered ring owing to the insertion of a further methylene group.

It is noted that the assignment of Structures **I** and **II** to Spectra **18**/*1* and **18**/*2* is very simple knowing the integral data, since the intensity corresponding to five protons of the signal at 2.24 ppm in Spectrum **18**/*1* immediately reveals that the CCH_2C multiplet is concealed by this signal. The integrals were omitted in order to point out the possibility of employing a slight difference in the order of spin-spin coupling to identify the structure.

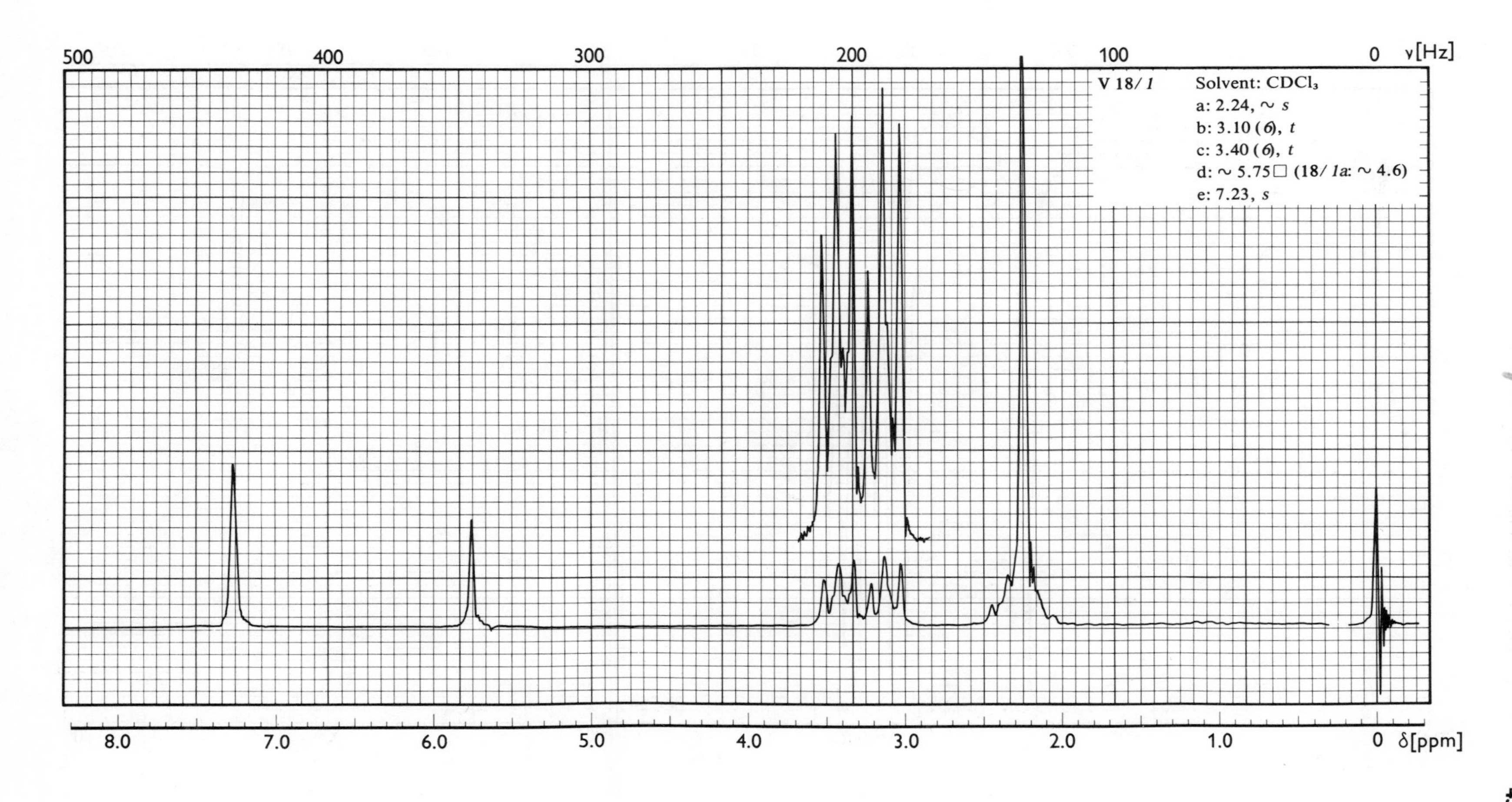
500
400
300
200
100
0
ν[Hz]
V 18/ 1
Solvent: $CDCl_3$
a: 2.24, ~ *s*
b: 3.10 (6), *t*
c: 3.40 (6), *t*
d: ~ 5.75□ (18/ 1a: ~ 4.6)
e: 7.23, *s*
8.0
7.0
6.0
5.0
4.0
3.0
2.0
1.0
0
δ[ppm]

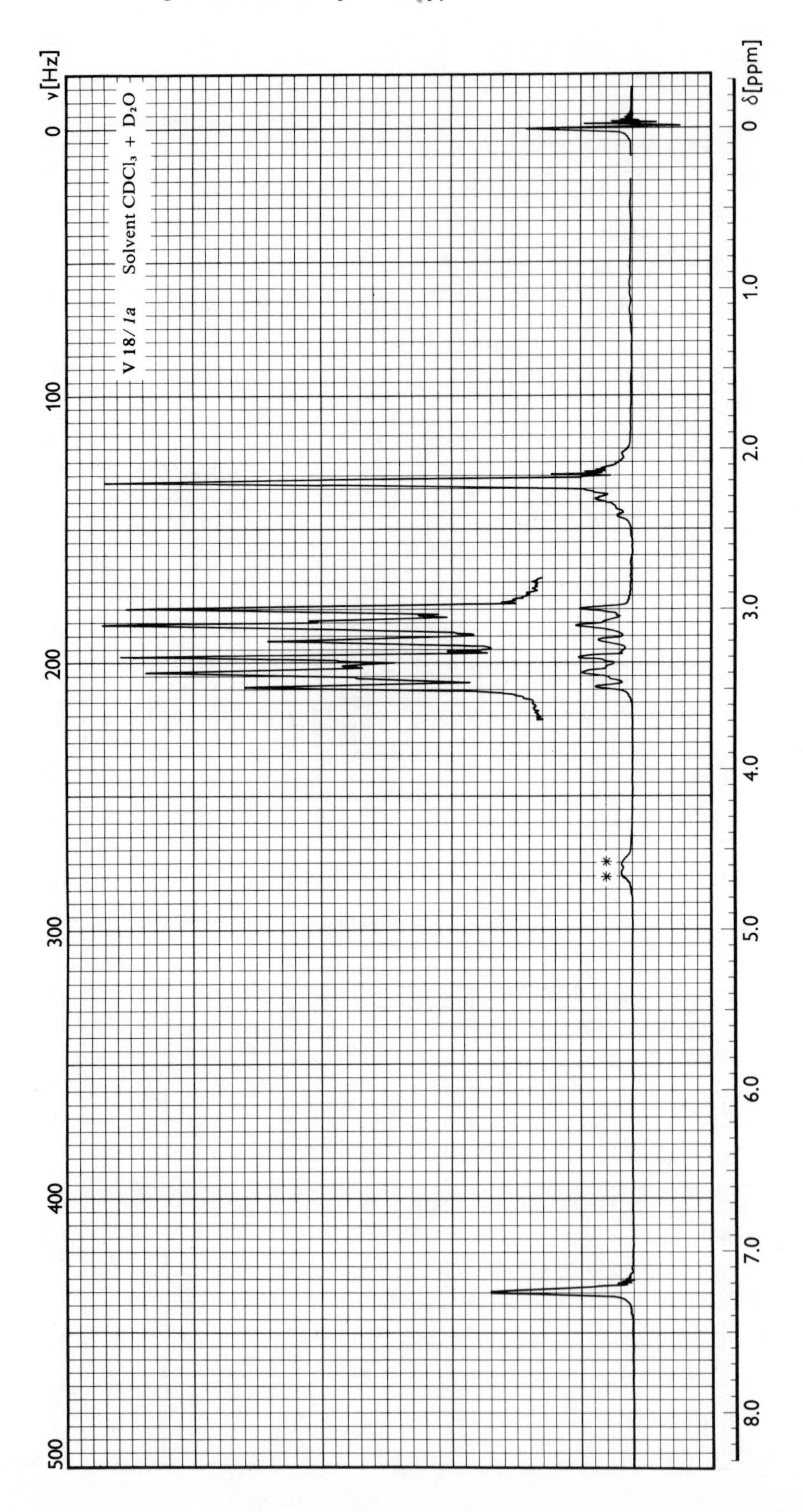
V 18/ 1a
Solvent CDCl3 + D2O
ν[Hz]
0
100
200
300
400
500
δ[ppm]
0
1.0
2.0
3.0
4.0
5.0
6.0
7.0
8.0
**

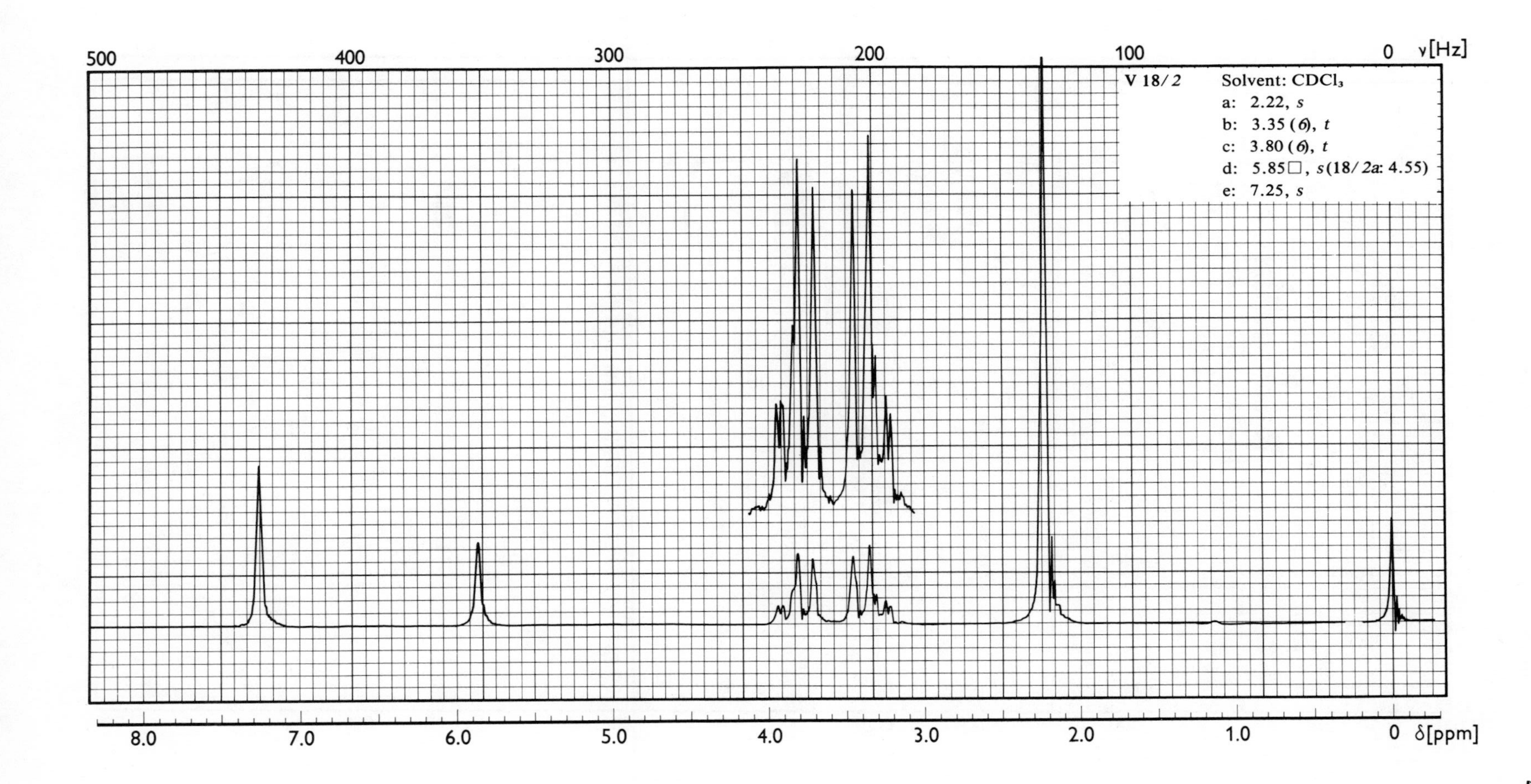

V 18/2
Solvent: $CDCl_3$
a: 2.22, s
b: 3.35 (6), t
c: 3.80 (6), t
d: 5.85□, s(18/2a: 4.55)
e: 7.25, s
500
400
300
200
100
0
ν[Hz]
8.0
7.0
6.0
5.0
4.0
3.0
2.0
1.0
0
δ[ppm]

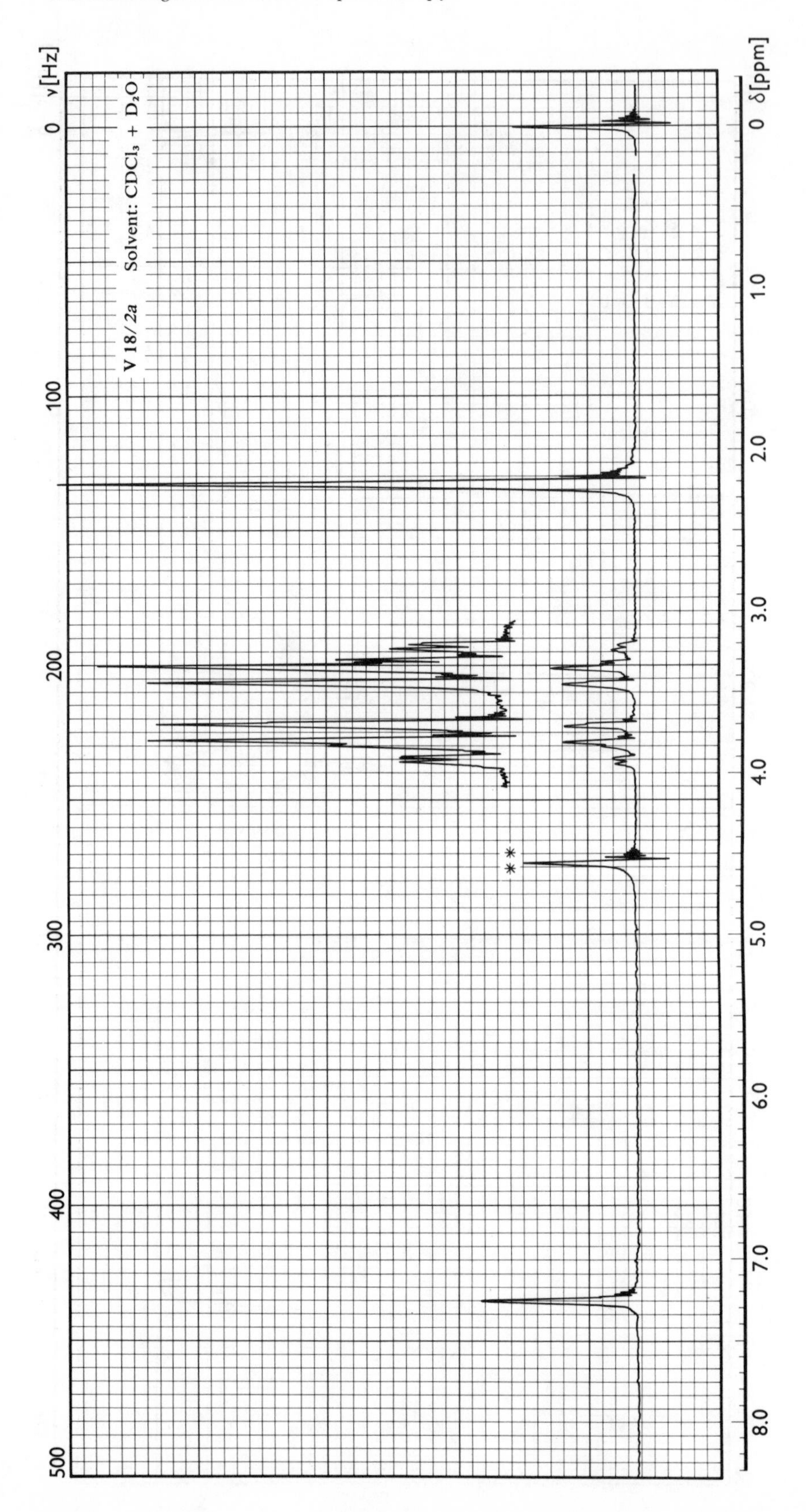
V 18/2a
Solvent: CDCl₃ + D₂O
ν[Hz]
0
100
200
300
400
500
δ[ppm]
0
1.0
2.0
3.0
4.0
5.0
6.0
7.0
8.0
**

PROBLEM **19**[884a]

The compound with Spectrum **19**/*1* was prepared by introducing *t*-butyl group(s) into the tryptophan molecule (**323**). The problem is how many *t*-butyl groups were introduced and to which positions, provided that only carbon atoms were substituted.

ANSWER

Spectrum **19**/*1* contains three signals at 1.40, 1.60, and 1.68 ppm with equal intensities, proving the presence of three *t*-butyl groups. The methylene and methine signals at approximately 3.4 and 4.6 ppm are hardly significant, because they are weak and split.

In the region of aromatic protons two singlet-like broad signals appear at 7.70 and 7.80 ppm, both with 1H intensity. The small distance between the signals suggests C-2,5,7 substitution. Namely, the H-2 and H-4,7 atoms are deshielded by the heteroatom and the heteroring, respectively. Thus, if either of these hydrogens remained in the molecule, the signals arising from the two aromatic protons would be more distant in the spectrum. However, 2,4,7-substitution is incompatible with the spectrum, in which the adjacent H-4,5 ring protons would give an *AB* pattern with $J_{AB} \approx 9$ Hz (*ortho* coupling). Thus, 2,4,6- or 2,5,7-*t*-butyl groups are probable and H-7(4) deshielded by the heteroring gives signal near to H-5(6), due to the opposite effects of the two *vicinal t*-butyl groups in position 5,7 and 4,6, respectively. A substitution on C-4, however, is improbable for steric reasons (adjacent side chains with great steric requirements) and thus the product is compound **I**.

The splitting of approximately 2 Hz due to the coupling of H-4,6 in *meta* position cannot be observed in TFA, because in this solution the lines are broader than 2 Hz. The signals of acidic protons appear, of course, outside the recorded range, together with the absorption of the solvent.

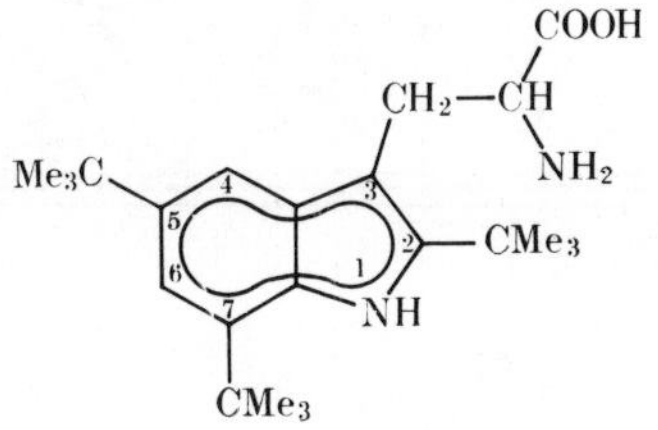

19/I

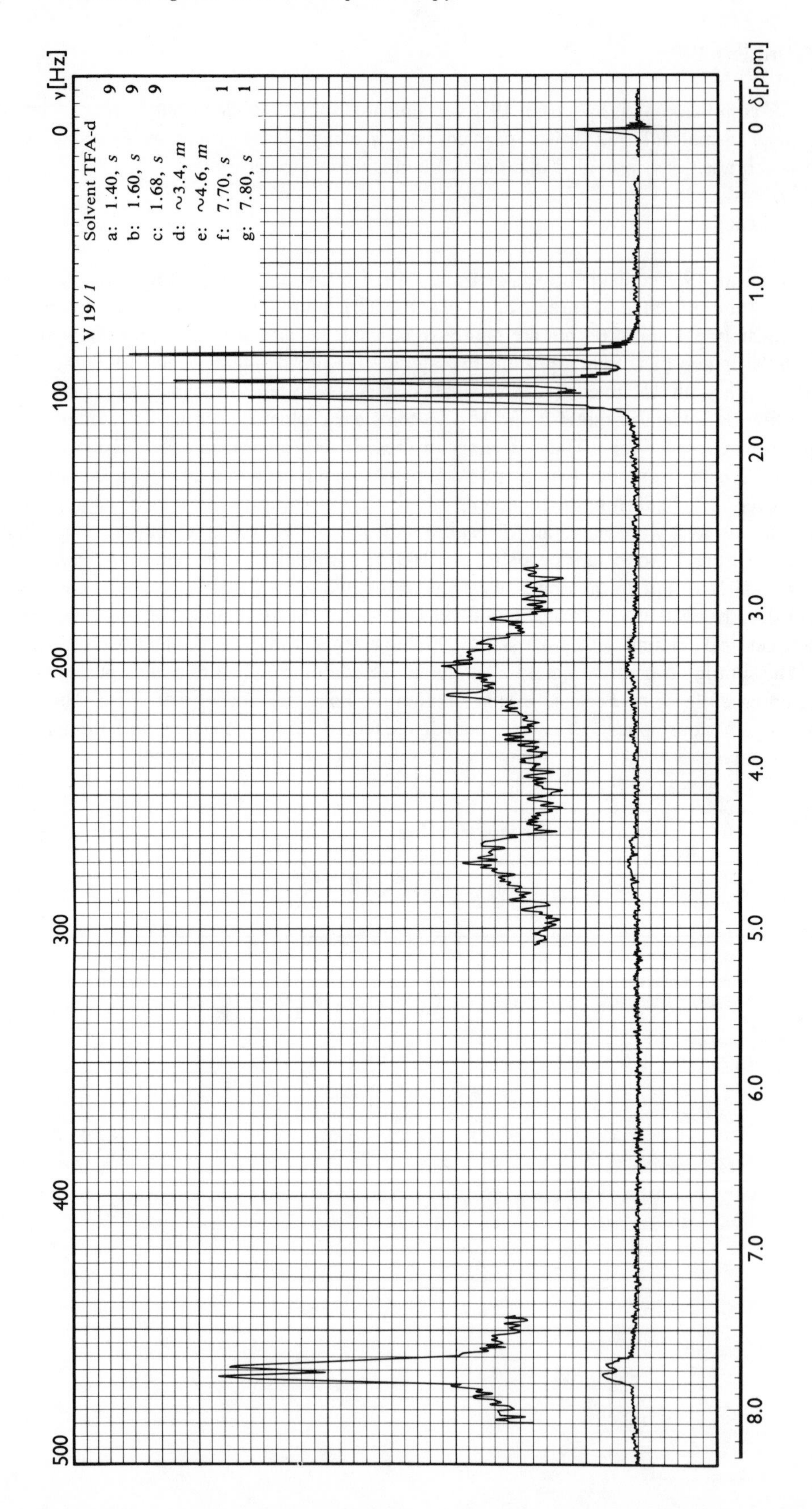
V 19/*1*
Solvent TFA-d
a: 1.40, *s* 9
b: 1.60, *s* 9
c: 1.68, *s* 9
d: ~3.4, *m*
e: ~4.6, *m*
f: 7.70, *s* 1
g: 7.80, *s* 1
ν[Hz]
0
100
200
300
400
500
δ[ppm]
0
1.0
2.0
3.0
4.0
5.0
6.0
7.0
8.0

PROBLEM 20

Assign Structures **I** and **II** to Spectra **20**/*1* and **20**/*2* and interpret the chemical shift differences. Is the spectrum compatible with the information obtained from the IR spectrum that the chelate-type hydrogen bond is stronger in compound **II** (following from the lower νC=O frequency)?

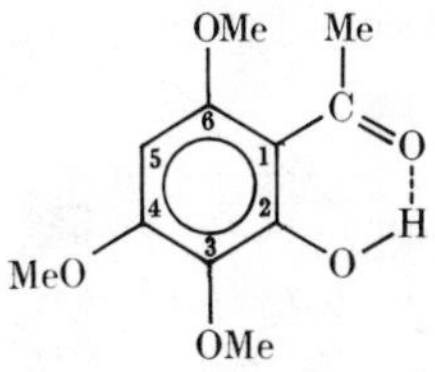

20/I 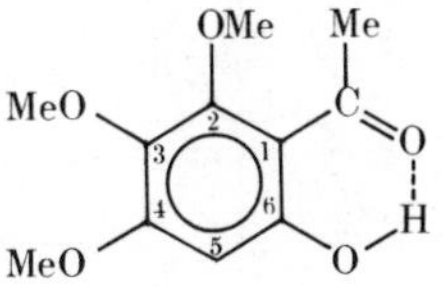 **20/II**

ANSWER

The assignment of the signals in Spectra **20**/*1* and **20**/*2* can be given as follows: δCH_3(Ac): 2.66 and 2.57 ppm; δOCH_3: 3.78, 3.88 and 4.00 and 3.80, 3.88, and 3.93 ppm; δArH: 6.22 and 5.98 ppm; δOH: 13.40 and 13.80 ppm, respectively. The chemical shift of the ring proton is higher in Structure **II** where the proton is located near to the quasiaromatic chelate ring causing paramagnetic shift. Structure **II** can therefore be assigned to Spectrum **20**/*1*. Owing to the delocalization of the π-electrons, namely, ring currents may flow in the chelate ring. Similar to the signal of ring proton, a paramagnetic shift in the acetyl-methyl signal can be expected too. In Spectrum **20**/*1* assigned to Structure **II**, CH_3CO is, indeed, higher (by 0.09 ppm), also supporting the conclusion drawn from the IR spectrum on the strength of chelation.

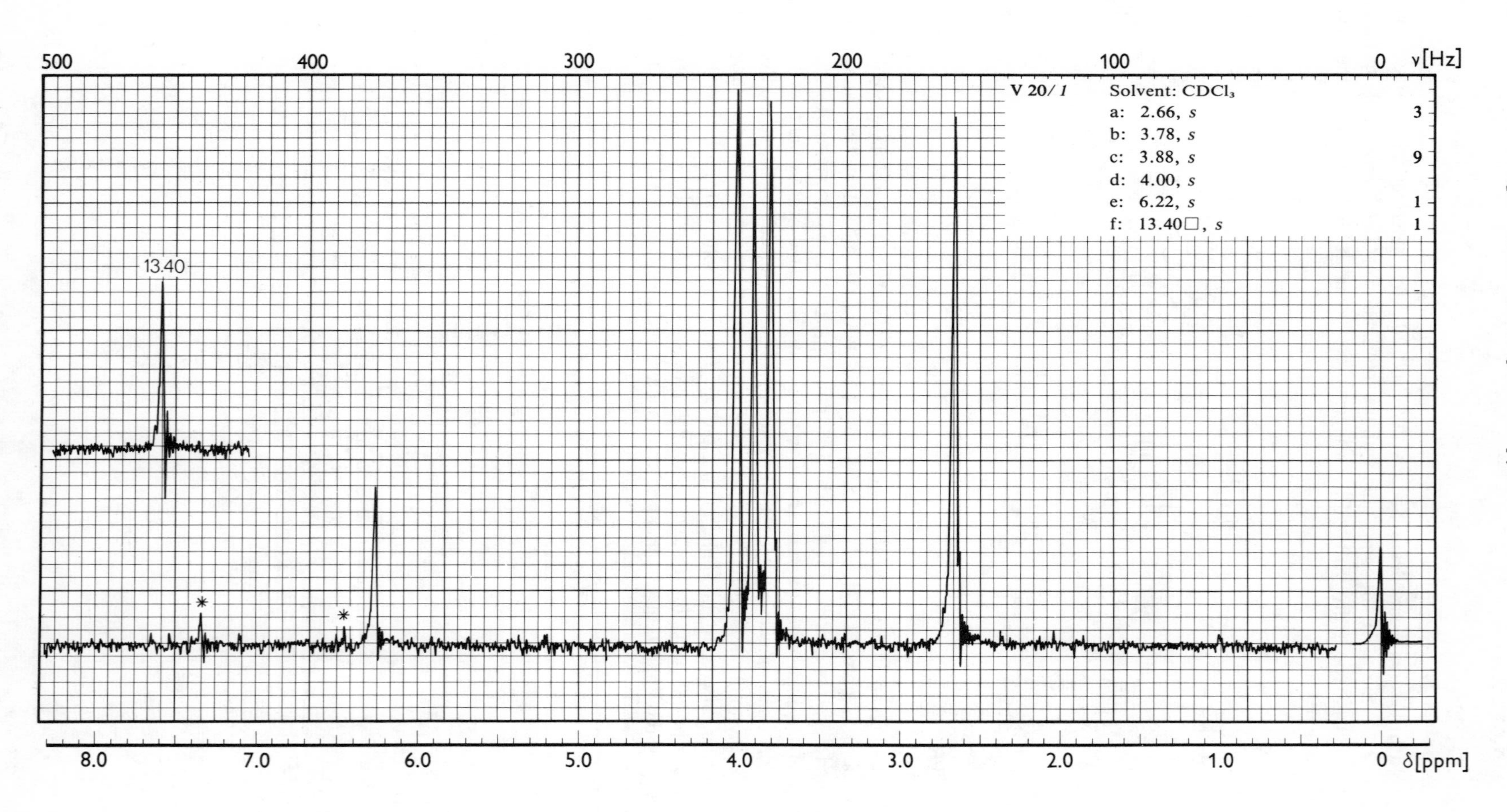
500
400
300
200
100
0 ν[Hz]
V 20/ 1
Solvent: $CDCl_3$
a: 2.66, s 3
b: 3.78, s
c: 3.88, s 9
d: 4.00, s
e: 6.22, s 1
f: 13.40□, s 1
13.40
*
*
8.0
7.0
6.0
5.0
4.0
3.0
2.0
1.0
0 δ[ppm]

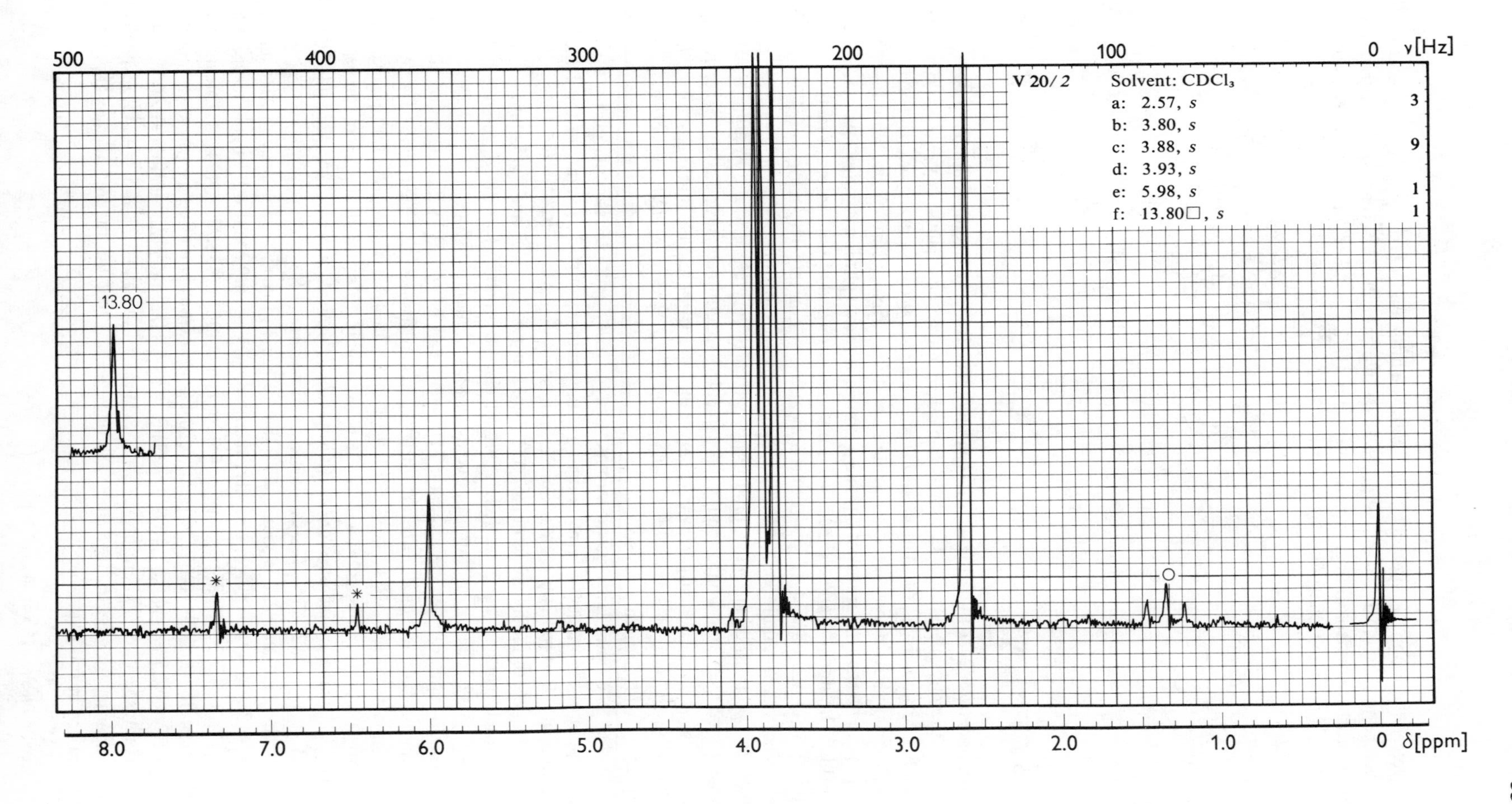

V 20/2
Solvent: $CDCl_3$
a: 2.57, *s* 3
b: 3.80, *s*
c: 3.88, *s* 9
d: 3.93, *s*
e: 5.98, *s* 1
f: 13.80□, *s* 1
13.80
500
400
300
200
100
0
ν[Hz]
8.0
7.0
6.0
5.0
4.0
3.0
2.0
1.0
0
δ[ppm]

PROBLEM **21**

Determine from the spectra of the starting compound (**21**/*1*) and of the product (**21**/*2*) whether the reaction **I** → **II** has proceeded. Interpret the differences. What spin systems give the characteristic line groups in Spectrum **21**/*1*? Calculate all the parameters (chemical shifts and coupling constants) of Spectrum **21**/*2*.

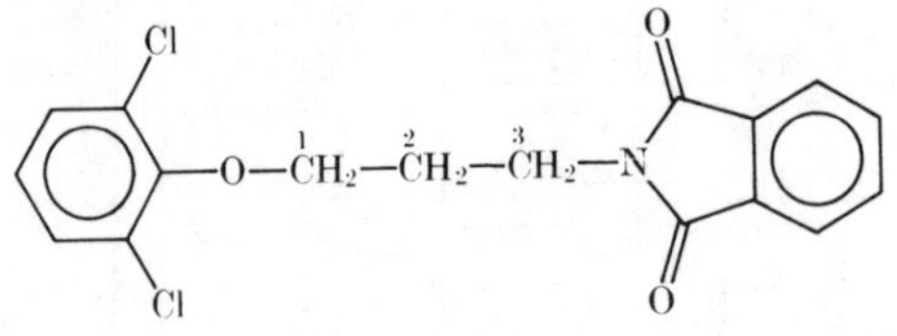

Cl O Cl

—O—$\overset{1}{C}H_2$—$\overset{2}{C}H_2$—$\overset{3}{C}H_2$—N —O—CH_2—CH_2—CH_2—NH_2

Cl O Cl

21/I 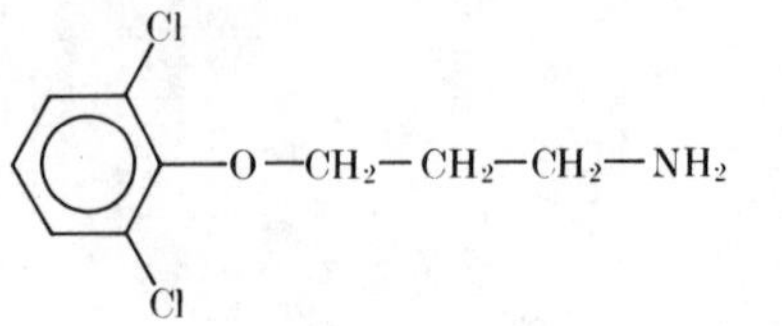 **21/II**

ANSWER

In Spectrum **21**/*1* four signals are separated. The first, with midpoint at 2.25 ppm, belongs to the 2-methylene group. It should consist of nine lines, since $J_{1,2} \neq J_{2,3}$, and thus the number of lines is $(n_1 + 1)(n_2 + 1) = 3 \times 3 = 9$ ($n_1 = n_2 = 2$). However, $J_{1,2}$ and $J_{2,3}$ are only slightly different, and thus the three triplets form a pseudo-quintet with an approximate intensity ratio of 1:3:4:3:1, but a further splitting of the "quintet" is clearly visible. The signal at approximately 4.1 ppm is composed of the triplets of methylene groups 1 and 3 (at 4.03 and 4.13 ppm) and appears as a quartet. The region of 400 to 500 Hz contains the AB_2 multiplet of the protons of the trisubstituted ring and not far apart the $AA'BB'$ spectrum of the phthalimido ring, which is very close to the limiting A_4 case. The analysis of the latter would require an expanded spectrum recorded with higher resolution (as in case of Problem 56). However, even this spectrum reveals that $\delta A \approx \delta B \approx 7.8$ ppm. The intensity ratio of 2:4:3:4 of the line groups is in agreement with the above assignment, therefore with Structure **I**, too.

In the spectrum (**21**/*2*) of the product the $AA'BB'$ multiplet is replaced by a singlet at 1.30 ppm of 2H intensity corresponding to the amino group. The signal of the 2-methylene protons at 1.92 ppm is a regular quintet indicating that $J_{1,2} = J_{3,2} = 6.5$ Hz.

The triplets of methylene groups 1 and 3 are separated at 4.07 and 2.95 ppm, respectively. The upfield shift of the 3-methylene signal can be attributed to the difference in electronegativity between the amino group and the phthalimido substituent.

The AB_2 spectrum of the aromatic hydrogens can be found in the region of 400 to 450 Hz, and the line of the combinational transition can also be identified at 461 Hz. According to the spectrum, $\delta B > \delta A$ and the B lines **5** and **6** are overlapping. As this also pertains to the AB_2 multiplet in Spectrum **21**/*1*, it is preferable to evaluate the spectrum parameters jointly. The line frequencies are as follows: **21**/*1*: 409, 415, 418, 425, 433, 440, and 442 Hz; **21**/*2*: 406, 412, 415, 422, 431, 438, and 440 Hz.

The chemical shift δA is identical to that of the A_3 line, i.e., for Spectra **21**/*1* and **21**/*2*, $\nu A = 418$ and 415 Hz, respectively, therefore, $\delta A = 6.97$ and 6.92 ppm, respectively.* $\nu B = (B_5 + B_7)/2 = (433 + 440)/2 = 436.5$ and $(431 + 438)/2 = 434.5$ Hz, respectively, therefore, $\delta B = 7.28$ and 7.24 ppm, respectively. Consequently, $\Delta\nu AB = 18.5$ and 19.5 Hz, $J_{AB} = (425 + 442 - 409 - 433)/3 = 8.3$ Hz, and $(422 + 440 - 406 - 431)/3 = 8$ Hz, respectively. Therefore, $J/\Delta\nu = 8/19.5 = 0.41$ and $8.3/18.5 = 0.45$, respectively.

Comparing the spectra to Figure 48, one can see that the theoretical ana experimental spectra are in full overlap.

* See also Problem **15** and Volume I, p. 106.

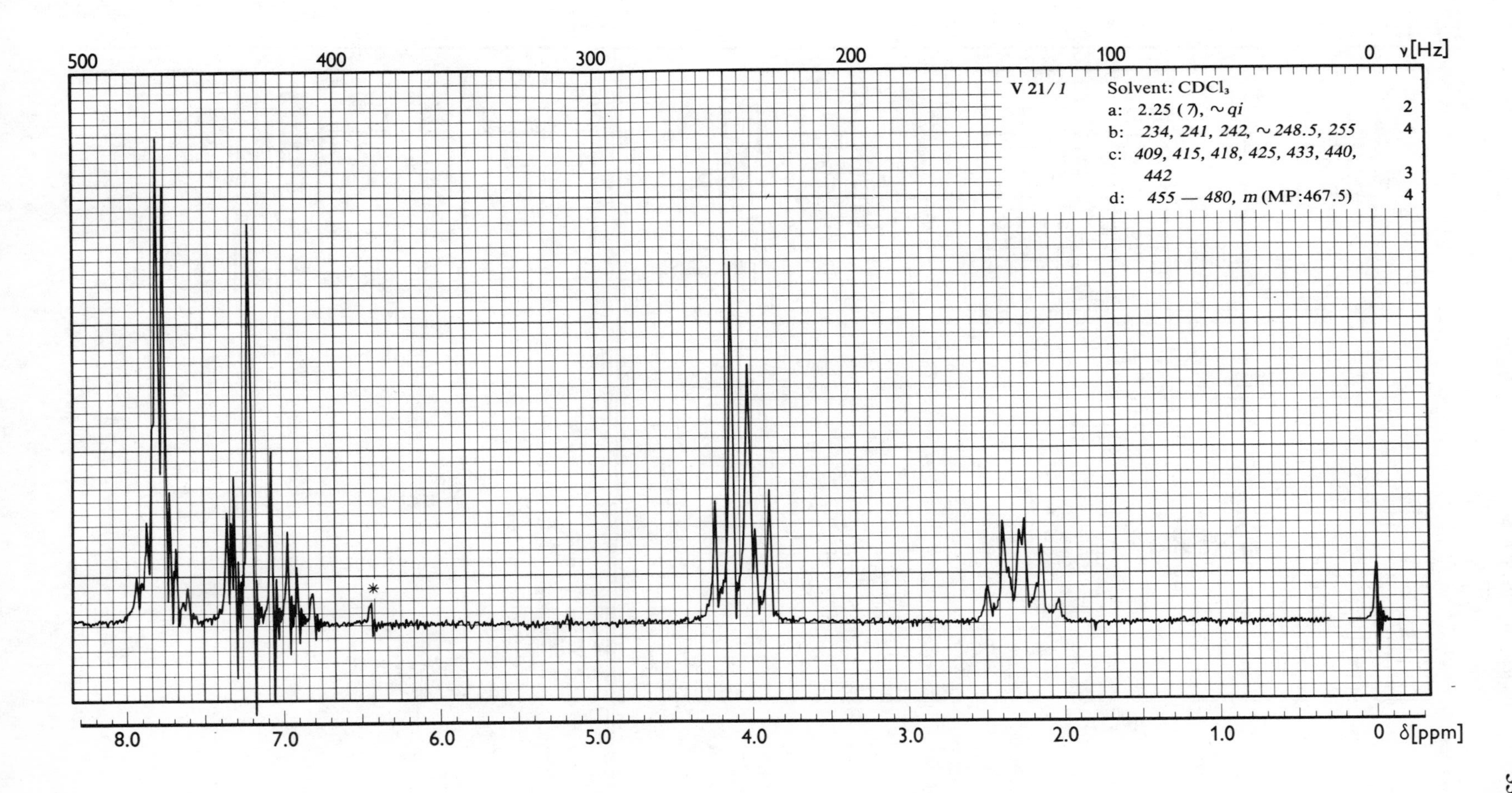

V 21/1
Solvent: CDCl₃
a: 2.25 (7), ~qi 2
b: 234, 241, 242, ~248.5, 255 4
c: 409, 415, 418, 425, 433, 440, 442 3
d: 455 — 480, m (MP:467.5) 4
500
400
300
200
100
0
ν[Hz]
*
8.0
7.0
6.0
5.0
4.0
3.0
2.0
1.0
0
δ[ppm]

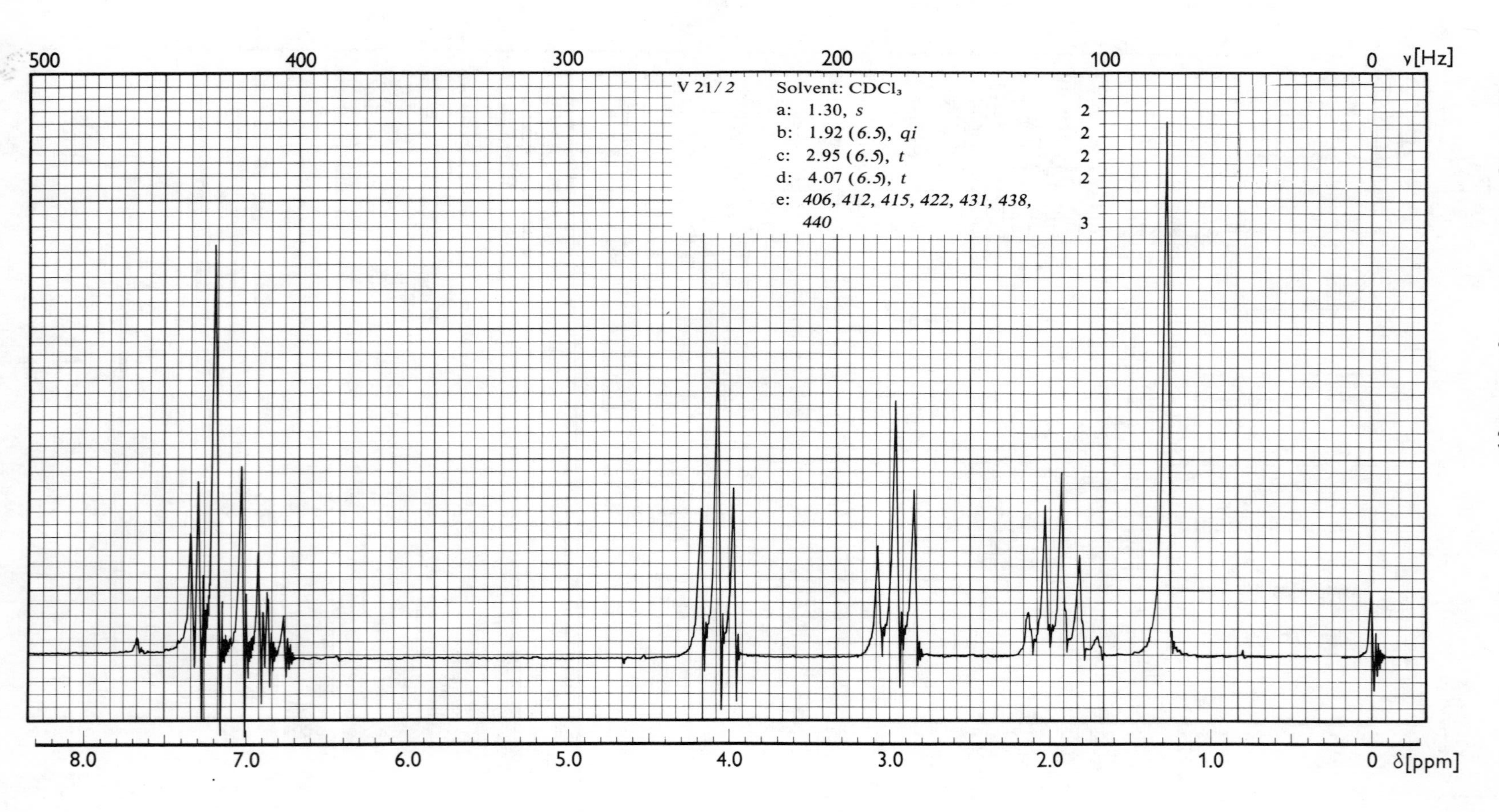
500
400
300
200
100
0 ν[Hz]
V 21/2
Solvent: CDCl₃
a: 1.30, s 2
b: 1.92 (6.5), qi 2
c: 2.95 (6.5), t 2
d: 4.07 (6.5), t 2
e: 406, 412, 415, 422, 431, 438, 440 3
8.0
7.0
6.0
5.0
4.0
3.0
2.0
1.0
0 δ[ppm]

PROBLEM 22

Determine the structure of compound C_4H_5N on the basis of Spectrum **22/*1***.

ANSWER

In Spectrum **22/*1*** the double doublet at 1.89 ppm, corresponding to three protons, proves the presence of a methyl group, and the two groups of signals between 5 and 7 ppm each correspond to one olefinic hydrogen. Therefore, the unknown must be one of compounds **I** to **III**.

H(β)(H)C=C(α)(CN)(CH₃)	H(H₃C)C=C(CN)(H)	H₃C(H)C=C(CN)(H)
22/I	**22/II**	**22/III**

The splitting of the methyl signal is due to the different interactions with the two non-equivalent olefinic hydrogens, which are in *AB*-type interaction, approaching the limiting *AX* case. The parameters of the resulting ABX_3 spin system can be determined in the first-order AMX_3 approximation.

The decision between Structures **I** to **III** can be based on the coupling constant J_{AM} of the olefinic protons which is characteristically different for 1,1-, *Z*-1,2-, and *E*-1,2 derivatives, and $J^E > J^Z > J^{gem}$.* J_{AM} is equal to the distance between the pairs of quartets of olefinic protons, in this case 16 Hz. This value is in the region characteristic of *E*-alkenes, therefore, Structure **II** corresponds to Spectrum **22/*1***. On this basis the assignment of signals, assuming that $\delta A > \delta M$, is straightforward: δCH_3 = 1.89 ppm, 2 × *d*; $J_{MX} \equiv J(H^{Me}, H^{\alpha})$ = 2 Hz; $J_{AX} \equiv J(H^{Me}, H^{\beta})$ = 7 Hz (H^{Me} denotes methyl hydrogen, and H^{α} and H^{β} denote olefinic hydrogens adjacent to the nitrile and methyl groups, respectively. $\delta M \equiv \delta(=CH^{\alpha})$ = 5.38 ppm, 2 × *qa*; $J_{AM} \equiv J(H^{\alpha},H^{\beta})$ = 16 Hz; J_{MX} = 2 Hz, $\delta A \equiv \delta(=CH^{\beta})$ = 6.72 ppm, 2 × *qa*; J_{AM} = 16 Hz; J_{AX} = 7 Hz.

Some further conclusions can also be drawn from the spectrum.

1. The H^{α} and H^{β} signals are far apart, and this, in itself, proves *E*-1,2 disubstitution, since the signals of *Z*-isomers are closer to one another.*
2. $\delta H^{\alpha} < \delta H^{\beta}$, which is due to the anisotropy of the nitrile substituent.**
3. $J_{AM} \gg J_{AX} > J_{MX}$, and the value of $^4J_{MX}$ = 2 Hz is, for a long-range coupling, relatively large due to the interaction transmitted by π-electrons.***

* See Volume II, p. 52.
** See Volume II, p. 62, last paragraph.
***See Volume I, p. 62, last paragraph.

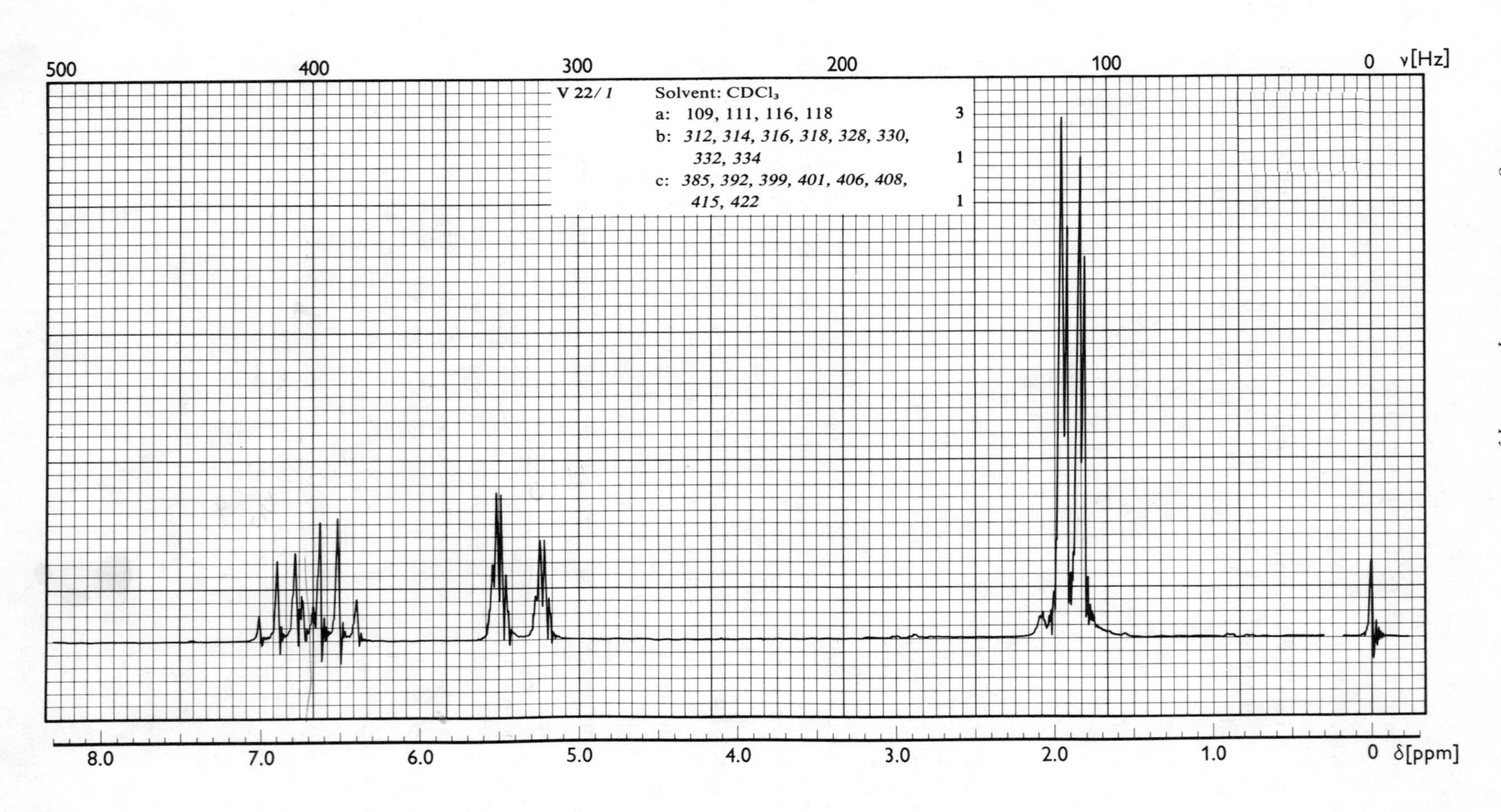
500
400
300
200
100
0 ν[Hz]
V 22/1
Solvent: CDCl₃
a: 109, 111, 116, 118 3
b: 312, 314, 316, 318, 328, 330, 332, 334 1
c: 385, 392, 399, 401, 406, 408, 415, 422 1
8.0
7.0
6.0
5.0
4.0
3.0
2.0
1.0
0 δ[ppm]

PROBLEM 23

Determine the structure of compound C_3H_5–SCN. What kind of spin system is present in the Spectrum **23**/*1*.

ANSWER

The lines of Spectrum **23**/*1* appear in the regions of 240 to 260 and 300 to 380 Hz, suggesting an unsaturated system. At 4.20 ppm a symmetric multiplet of six lines can be found corresponding to two hydrogens. The very complex line system of three protons between 300 and 380 Hz suggests that it can be assigned as the *ABC* multiplet of a vinyl group, also coupled with an α-methylene substituent which gives the multiplet at 4.20 ppm. Accordingly, the sample is allylthiocyanate CH_2=CH–CH_2–SCN, and its protons form an $ABCX_2$ spin system, where *A* is the olefinic hydrogen adjacent to the methylene group, with the highest chemical shift. The signals of the more shielded terminal olefinic hydrogens *B* and *C* are clearly closer to one another. Since the two long-range (allylic) coupling constants are practically the same ($J_{BX} \approx J_{CX}$) and evidently much smaller than J_{AX}, corresponding to a *vicinal* ^3J coupling, the symmetric *X* part of the spectrum is composed of two triplets, at a distance of J_{AX} = 4.5 Hz. Using first-order approximation the schematic structure of the *X* multiplet is depicted in Figure 199.

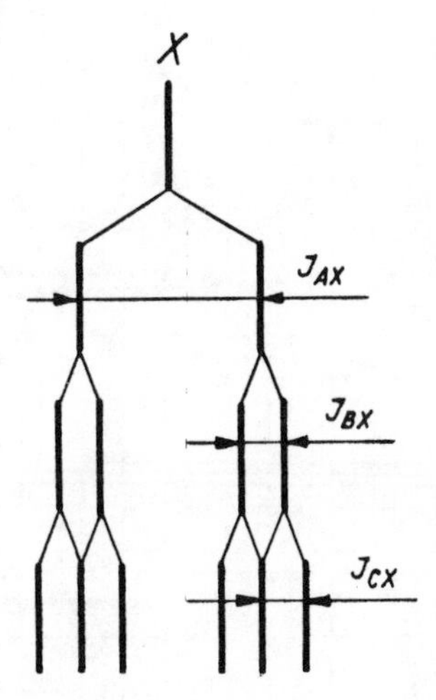

FIGURE 199. The schematic structure of the methylene signal (the *X* part of the $ABCX_2$ multiplet) of allyl-iso-thiocyanate.

The evaluation of chemical shifts in the *ABC* part would require computer methods. Therefore, as the accurate shifts cannot be determined by simple calculation for protons *A*, *B*, and *C*, it is reasonable to give the positions of multiplets in hertz, as is done here and in other similar cases. The value of δ*X* (4.20 ppm), however, can be given as the midpoint of the *X* multiplet, and from the splitting (see Figure 199), the approximate value of J_{BX} and J_{CX} can also be estimated. Actually, however, the exact values of J_{BX} and J_{CX}, supposed to be identical, cannot be determined from the *X* multiplet, only the absolute value of their sum (the distance of the two neighboring lines in the triplets is ($|J_{BX} + J_{CX}|$)/2, like in the analogous case of *ABX* system.* At any rate, these two coupling constants are ≈ 1 Hz, in good agreement with the expected allylic ^4J coupling.**

* Compare Volume I, p. 126.
** Compare Volume I, p. 62, last paragraph.

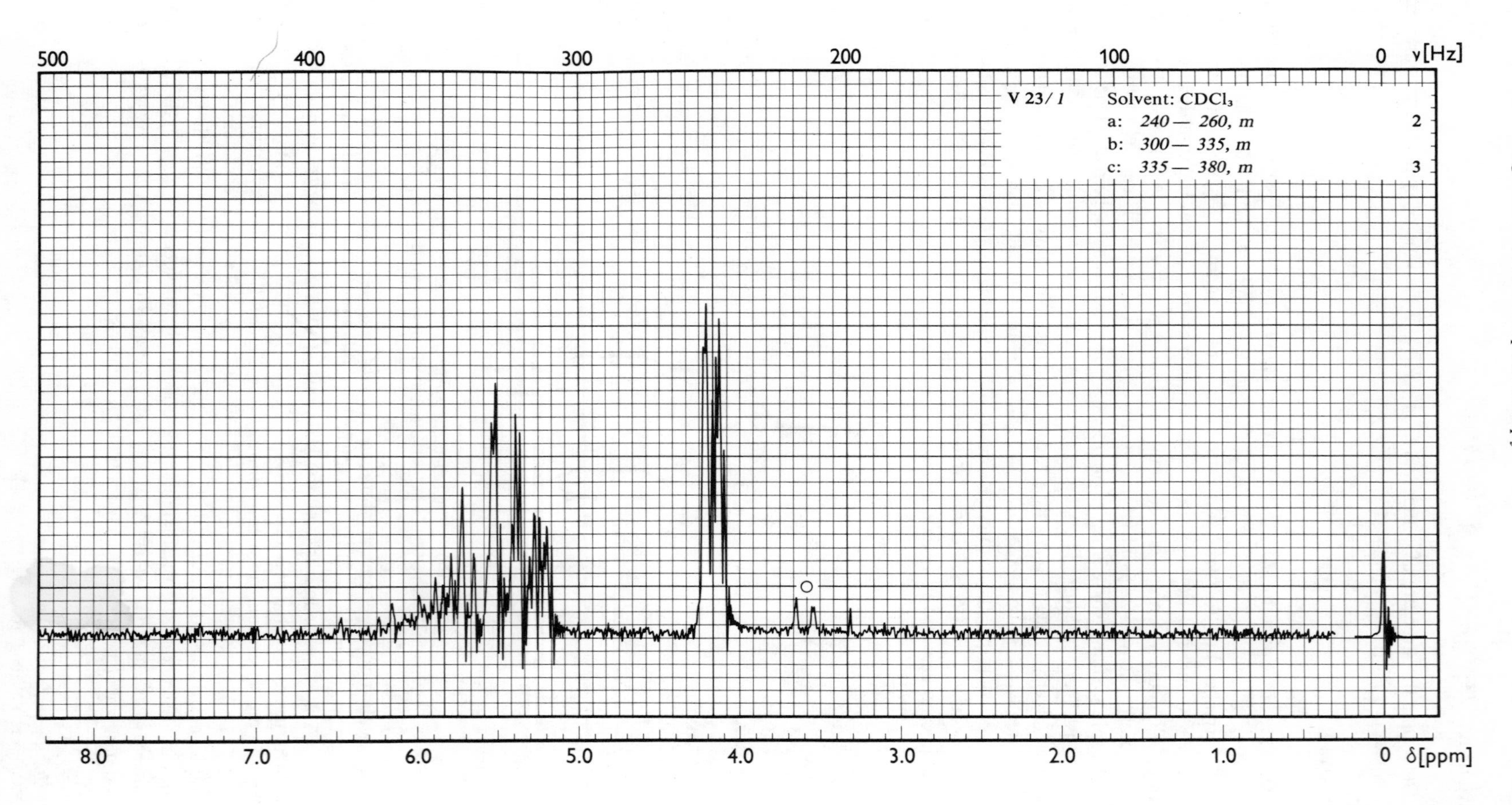
V 23/ 1
Solvent: $CDCl_3$
a: 240 — 260, m 2
b: 300 — 335, m
c: 335 — 380, m 3
500
400
300
200
100
0
ν[Hz]
8.0
7.0
6.0
5.0
4.0
3.0
2.0
1.0
0
δ[ppm]

PROBLEM **24**

Derive the structure of compound $C_4H_6O_2$, interpret its Spectrum **24**/*1*, and calculate all spectrum parameters.

ANSWER

The spectrum consists of a singlet of 3H at 2.08 ppm, and two multiplets of 2H and 1H intensity, respectively, between 260 to 300 and 420 to 450 Hz, characteristic of unsaturated groups. Therefore the molecule contains a methyl and a vinyl group, hence, it is vinyl acetate, CH_2=CH–O–COMe or acrylic acid methyl ester, CH_2=CH–CO–OMe.

The methyl signal at 2.08 ppm contradicts of the latter structure and the strong deshielding of the methine proton (*X*) also supports the former structure, in which the adjacent oxygen accounts for the strong paramagnetic shift of the δX signal.

The vinyl protons give an *AMX* multiplet, where $\delta A \equiv \delta H^E > \delta M \equiv \delta H^Z$ (proton *A*, *cis* to the substituents, is more deshielded in vinyl derivatives).* Consequently, from the *X* part of the spectrum J_{AX} = 14 Hz, J_{MX} = 6.5 Hz, and δX = 7.24 ppm. The magnitudes of J_{AX} and J_{MX} are in good agreement with the coupling constants expected for the *Z* and *E* protons of olefins.**

The assumption $\delta A > \delta M$ is also supported by the *AM* part of the spectrum, in which the distance of the two downfield doublets is J_{AX}. Note that besides the distance of doublets **3** and **4** that of doublets **1** and **3** is also approximately 14 Hz (the lines are numbered in the sequence of decreasing fields). Moreover, not only the distance between doublets **1** and **2** but also between doublets **2** and **3** is nearly equal to J_{MX}, the accurate difference being 8.0 Hz. However, if it is assumed that doublets **1** and **3** belong to nucleus *A*, J_{MX} = 22 Hz would be obtained from the distance of doublets **2** and **4**, which would not only contradict the data obtained from the *X* part of the spectrum, but is also too large for the coupling of any pair of olefinic protons. Similarly, assuming that doublets **2** and **3** belong to nucleus *M*, the values of J_{MX}, which would be then identical with the distance of doublets **1** and **4**, would even be larger 28.5 Hz. Nevertheless, it is also indicated by the relative intensities of the lines that pairs **1** and **2** and **3** and **4** are the corresponding ones, because the lines closer to the *X* part of the spectrum must be stronger. Consequently, J_{AM} = 2.5 Hz and δA = 4.80 and δB = 4.50 ppm. The value of J_{AM} is in the range characteristic of interactions between *geminal* olefinic hydrogens.

* See Volume II, p. 49.

** See Volume II, p. 52.

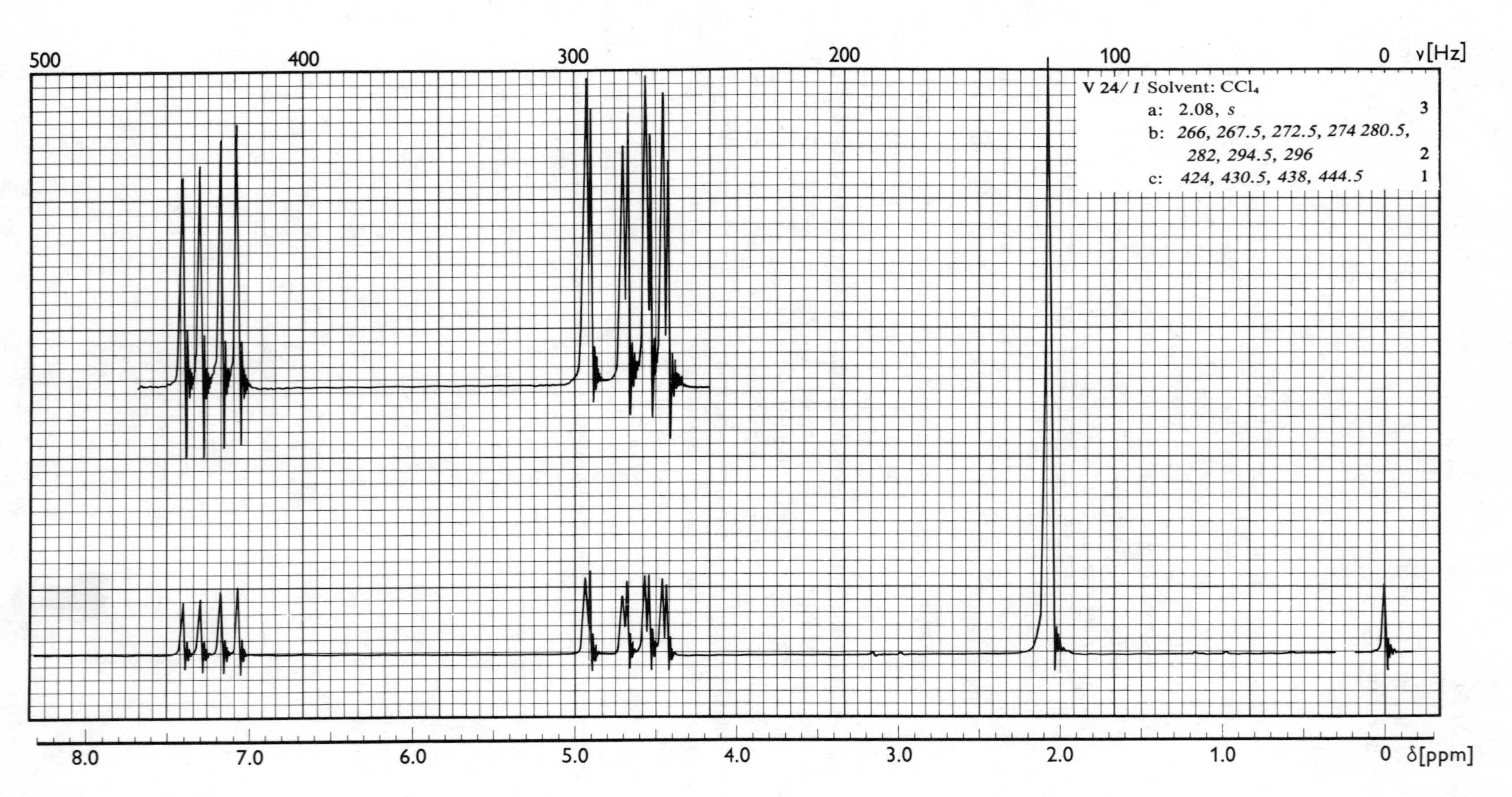
500
400
300
200
100
0 ν[Hz]
V 24/ 1 Solvent: CCl₄
a: 2.08, s 3
b: 266, 267.5, 272.5, 274 280.5, 282, 294.5, 296 2
c: 424, 430.5, 438, 444.5 1
8.0
7.0
6.0
5.0
4.0
3.0
2.0
1.0
0 δ[ppm]

PROBLEM 25

Correlate Structures **I** to **III** with Spectra **25**/*1* to *3* and assign the signals. Spectra **25**/2 and **25**/*2a* were recorded in $CDCl_3$ and D_2O.

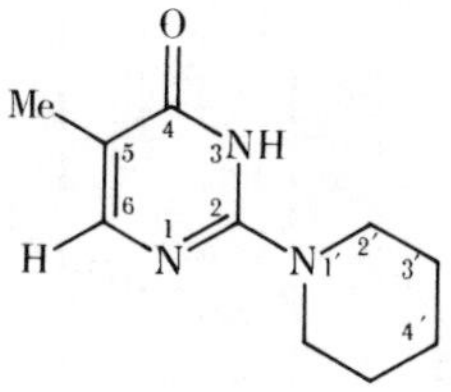

25/I 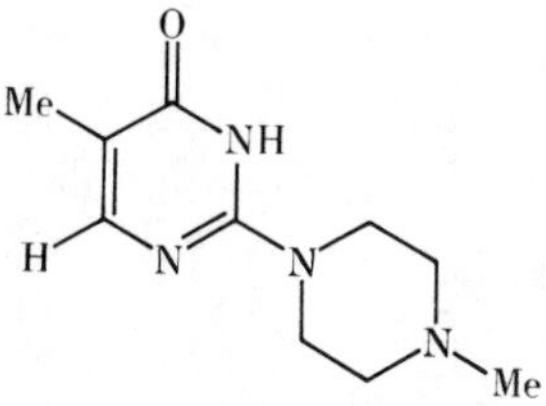 **25/II** 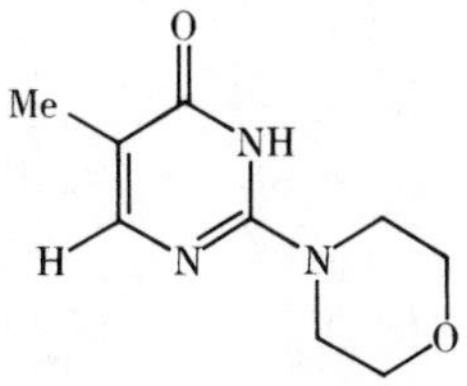 **25/III**

ANSWER

Structures **I** to **III** correspond to Spectra **25**/*1*, **25**/*3*, and **25**/*2*, on the basis of the following arguments. The signal at ~1.65 ppm of 6H intensity in Spectrum **25**/*1* is characteristic of the –C–CH_2–C–type group of the piperidine ring, hence Structure **I**. The *N*-methyl signal of the piperazine ring in compound **II** appears in Spectrum **25**/*3* at 2.35 ppm (singlet of 3H intensity). Consequently, compound **III** must have the spectrum **25**/*2*.

The 5-methyl and H-6 signals are clearly split (to a doublet and a quartet, respectively) by ~1 Hz, owing to the long-range coupling of the methyl and methine protons. They appear in Spectra *1*, *2*, *2a*, and *3* at 1.92, 1.90, 2.00, and 1.93 ppm, respectively, as well as at 7.65, 7.66, 7.60, and 7.66 ppm, respectively.

In the spectrum of compound **I** the δCCH_2 and δNCH_2 signals (at ~1.65 and ~3.7 ppm) broaden owing to a mutual coupling. With compound **II**, the signals of methylene groups 2′,6′ and 3′,5′, respectively, have substantially different chemical shifts (2.50 and 3.79 ppm), owing to the opposite inductive effects of the substituents. Accordingly, both signals split into triplets (J = 5 Hz), and since the electron-withdrawing heteroaromatic ring causes deshielding, the downfield triplet at 3.79 ppm corresponds to the methylene protons adjacent to the nitrogen which is attached to this ring. The NH signal is so diffuse that it cannot be identified, similar to the case of compound **III**. In Spectrum **25**/*1* δNH is ~13.2 ppm; in Spectrum **25**/*2a* the NH signal overlapping with the signal of water content in the solvent appears at ~4.75 ppm.

In the spectrum of the $CDCl_3$ solution of **III**, the NCH_2 and OCH_2 protons have a common singlet at 3.76 ppm, since the heteroaromatic substituent increases the $-I$ effect of the adjacent nitrogen, which effect thus becomes nearly the same as that of the oxygen atom. As a result, all the methylene protons of the morpholine ring become chemically equivalent (accidentally isochronic). In the spectrum of morpholine derivatives substituted on the nitrogen with a not electron-withdrawing group, the methylene hydrogens give generally two well-separated signals, usually triplets. The accidental isochrony can be proved by recording the spectrum in D_2O (**25**/*2a*), in which the *AA′BB′* multiplet is well observable, with a midpoint about 3.87 ppm.

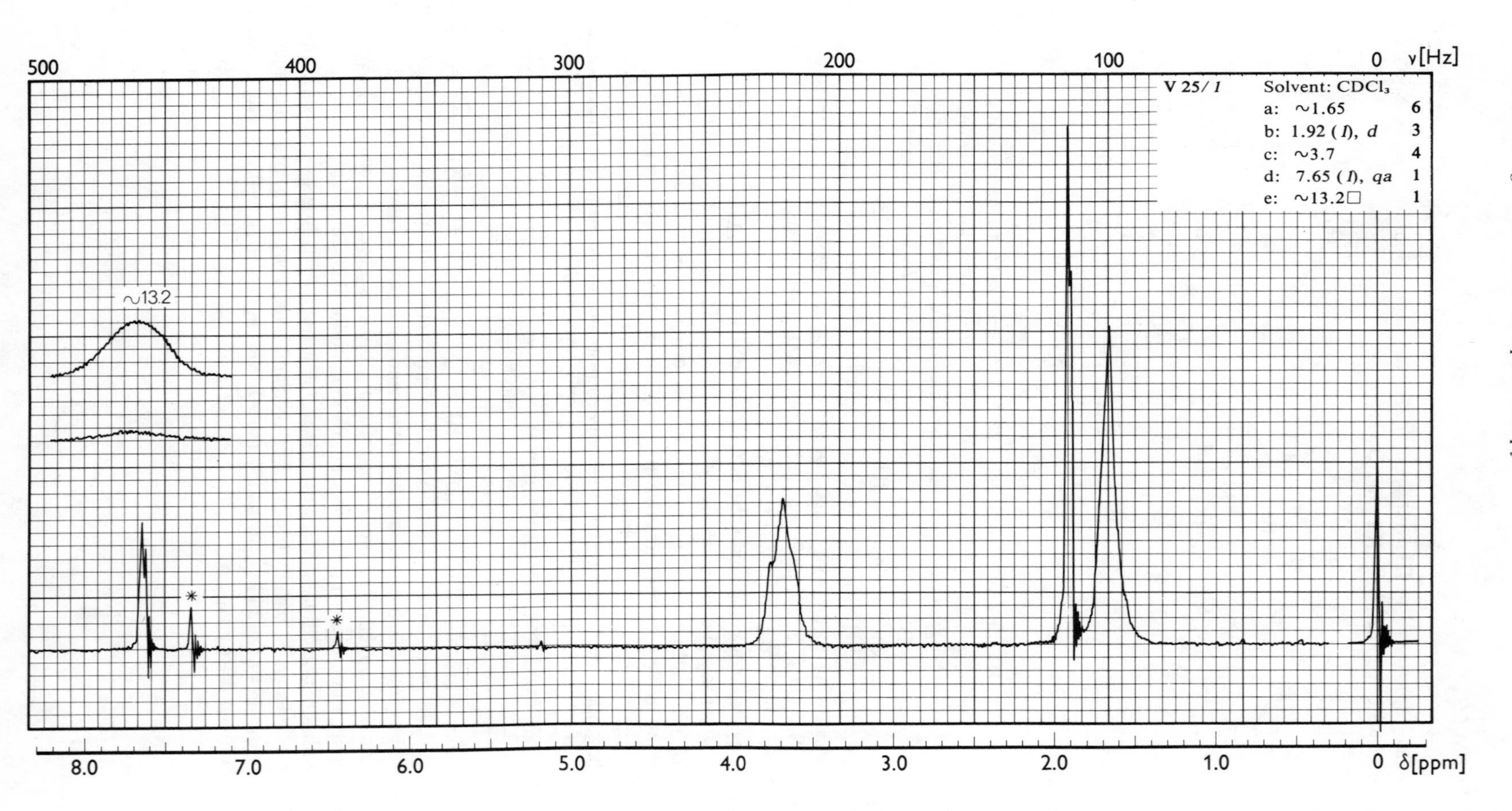
V 25/*1*
Solvent: $CDCl_3$
a: ~1.65 6
b: 1.92 (*I*), *d* 3
c: ~3.7 4
d: 7.65 (*I*), *qa* 1
e: ~13.2□ 1
ν[Hz]
500
400
300
200
100
0
δ[ppm]
8.0
7.0
6.0
5.0
4.0
3.0
2.0
1.0
0
~13.2
*
*

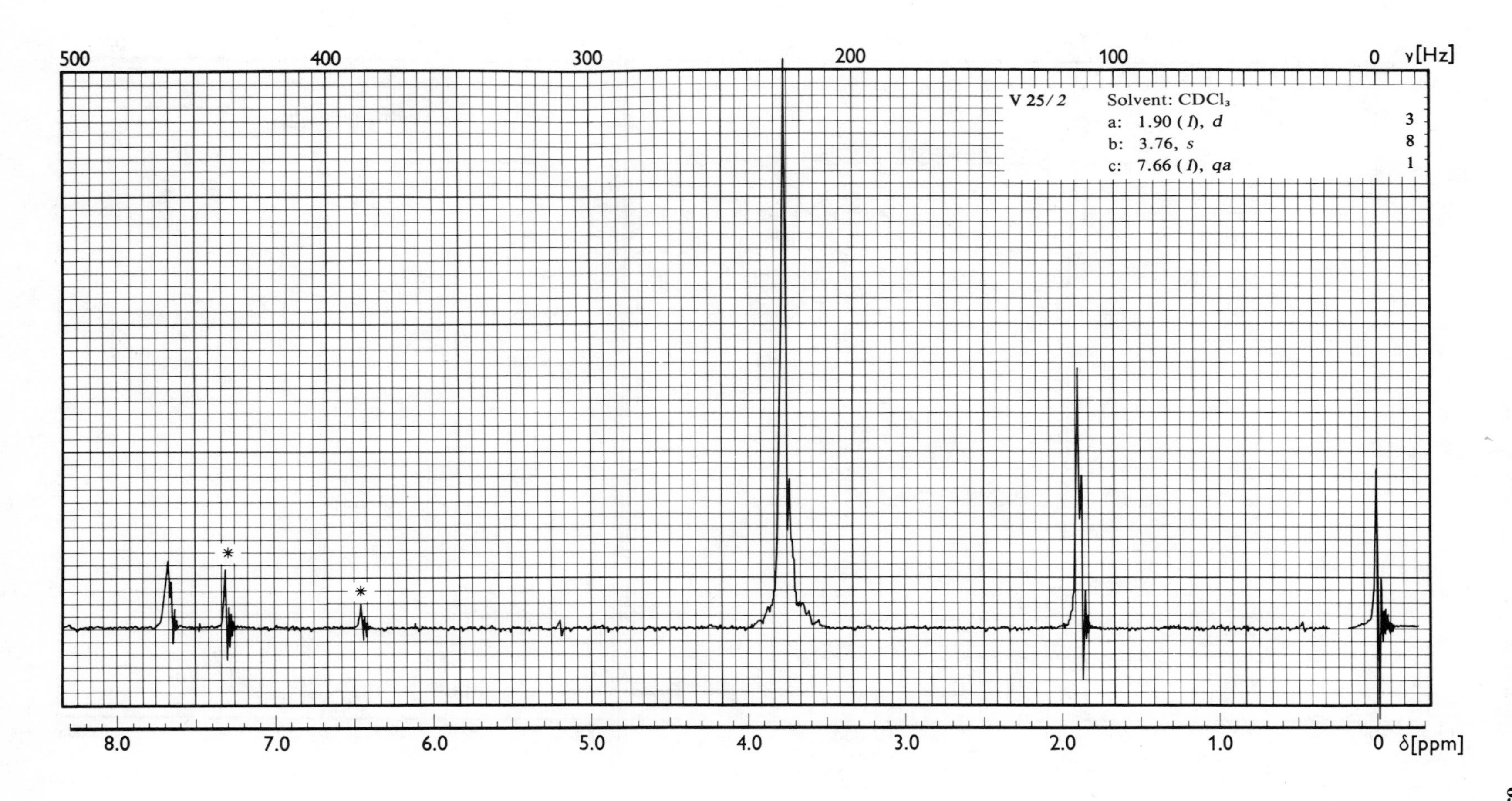
500
400
300
200
100
0
ν[Hz]
V 25/2
Solvent: $CDCl_3$
a: 1.90 (*J*), *d* 3
b: 3.76, *s* 8
c: 7.66 (*J*), *qa* 1
*
*
8.0
7.0
6.0
5.0
4.0
3.0
2.0
1.0
0
δ[ppm]

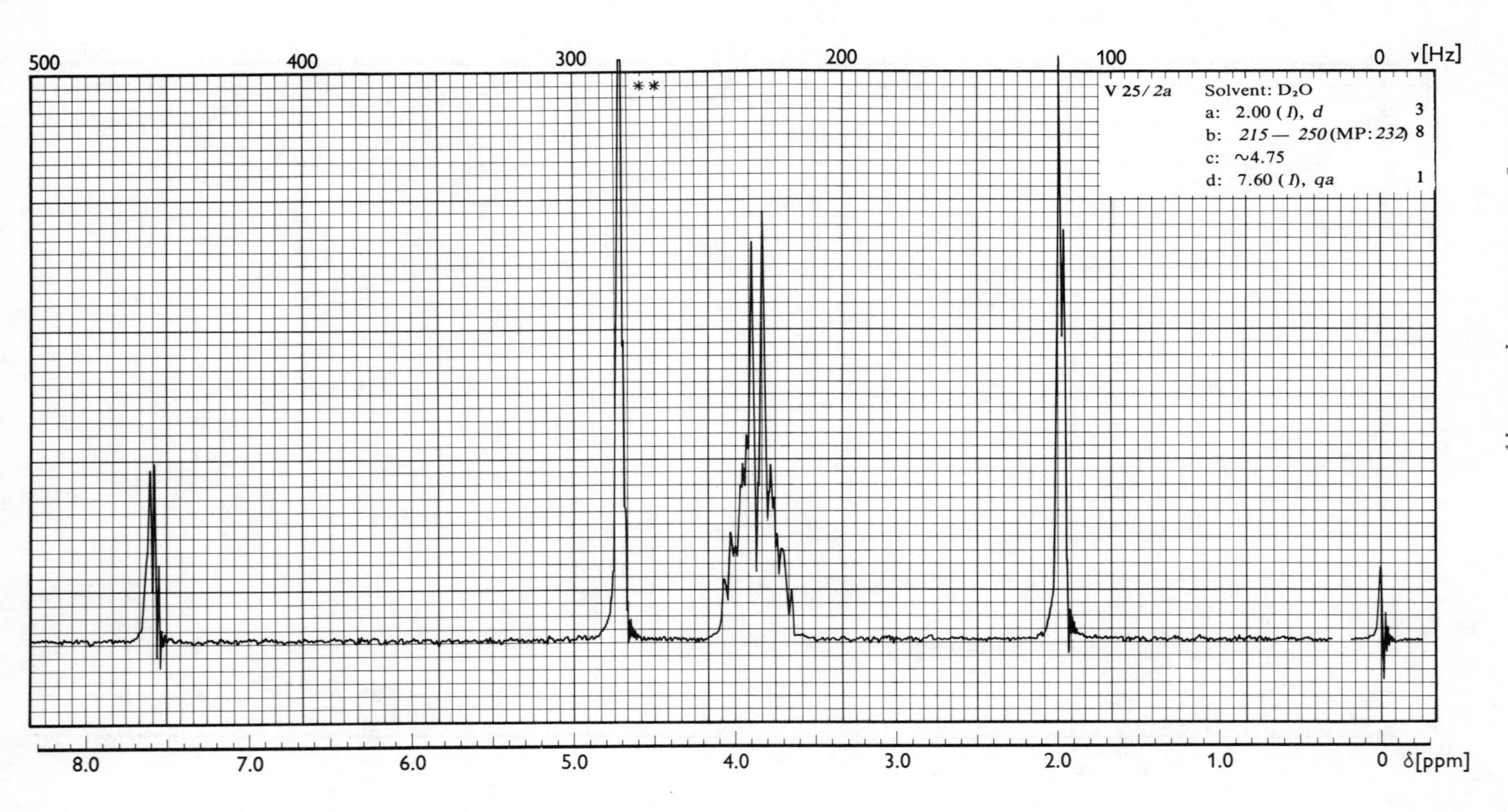
V 25/2a
Solvent: D_2O
a: 2.00 (*I*), *d* 3
b: *215* — *250* (MP: *232*) 8
c: ~4.75
d: 7.60 (*I*), *qa* 1
**
500
400
300
200
100
0 ν[Hz]
8.0
7.0
6.0
5.0
4.0
3.0
2.0
1.0
0 δ[ppm]

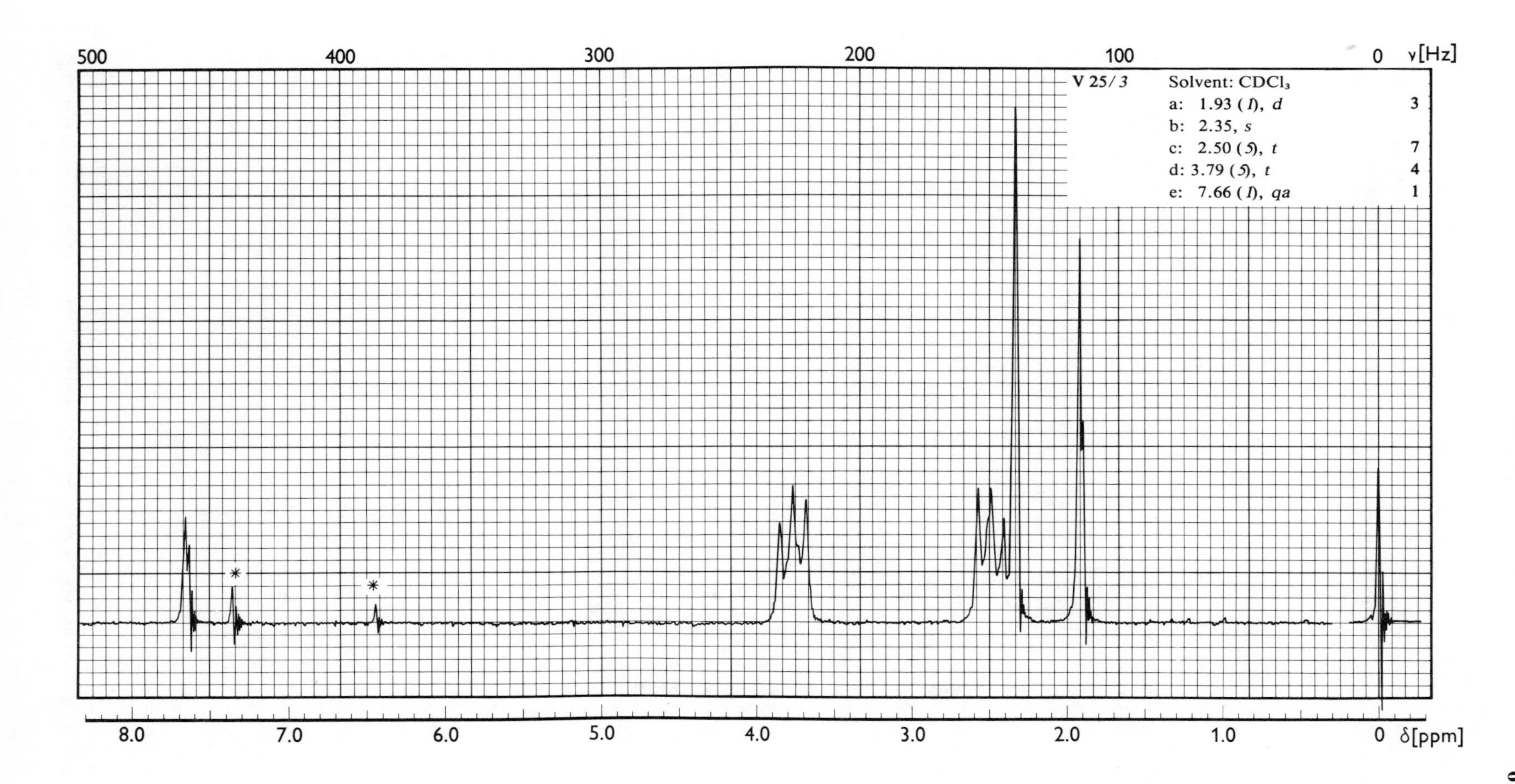

500
400
300
200
100
0
ν[Hz]
V 25/3
Solvent: $CDCl_3$
a: 1.93 (*J*), *d* 3
b: 2.35, *s*
c: 2.50 (*5*), *t* 7
d: 3.79 (*5*), *t* 4
e: 7.66 (*J*), *qa* 1
*
*
8.0
7.0
6.0
5.0
4.0
3.0
2.0
1.0
0
δ[ppm]

PROBLEM **26**

On the basis of Spectra **26**/*1* and **26**/*2* determine the structures of compounds $C_7H_8F_2Si$ and $C_7H_8Cl_2Si$ and decide which spectrum corresponds to which compound.

ANSWER

The spectra contain only two groups of signals of 3:5 intensity, one between 0 and 1 ppm and the other between 7 and 8 ppm. These signals and the empirical formulas limit our choice to structures $PhSiMeCl_2$ and $PhSiMeF_2$.

The upfield signal corresponds to a –Si–Me group, and the multiplet between 7 and 8 ppm corresponds to phenyl hydrogens, in agreement with the integrals. (The signals of –SiH, –CHX, or $-CH_2X$ groups, where X denotes halogen, and of $-CH_2Ph$ or –CHPh groups would appear between 1 and 7 ppm.)

The triplet structure of the methyl signal in Spectrum **26**/*1* is a consequence of an F–H coupling with J(F,H) = 6 Hz. This spectrum corresponds therefore to the fluoro derivatives.

The anisotropy of fluorine causes shielding of the methyl protons (δCH_3 is 0.95 ppm for the chloro and only 0.49 ppm for the fluoro derivative). For the same reason the C part of the AB_2C_2 multiplet of the aromatic protons is better separated with the former (the H_C atoms are more deshielded, while the AB_2 part appears at about the same positions in both spectra). In the chloro derivative, namely, the downfield shift of *ortho* hydrogen signals due to the $-I$ effect of the substituents is compensated by the anisotropy to only a lower degree (the anisotropic shielding effect of fluorine is stronger).

PROBLEM **27**[171]

Compound **I** was reacted with methylamine to yield compound **II**. Are Spectra **27**/*1* and *1a*, recorded after acid addition, compatible with the expected structure? Which of the tautomeric forms is (are) present?

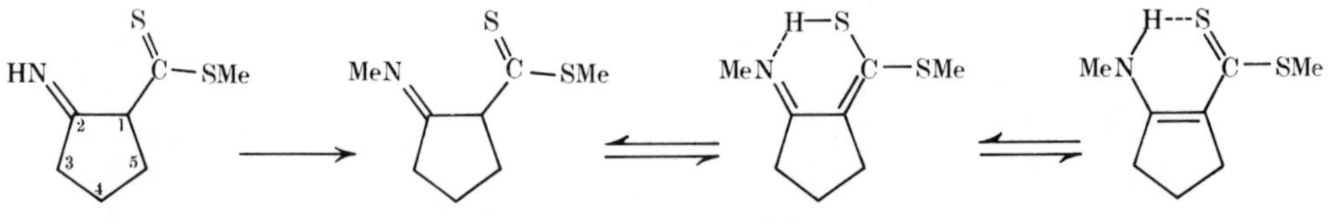

27/I **27/IIa** **27/IIb** **27/IIc**

ANSWER

In Spectrum **27**/*1* a quintet-like multiplet appears at 1.90 ppm with 2H intensity, indicating that it corresponds to the 4-methylene group of tautomeric Structure **IIb** or **IIc**. The spectrum of Structure **IIa** would be more complicated, and the signals in the above range should correspond to 4H (4- and 5-methylene hydrogens). However, the intensity of this signal is about 20% of that of the other signals, which, if this signal corresponded to 4 hydrogens, would require the presence of 20 additional hydrogens.

The singlet at 2.60 ppm of 3H intensity can readily be assigned to the methylmercapto group, and the doublet at 3.10 ppm (J = 5.5 Hz) presumably can be assigned to the N-methyl hydrogens, proving that the expected reaction has proceeded. The lines at 164, 171, and 178 Hz are the overlapping triplets of the 3- and 5-methylene groups, forming four lines; one of them, however, is overlapping with the *S*-methyl signal. This assignment is also confirmed by the relative intensities. From the maxima one obtains $\delta CH_2(3,5)$ = 2.73 and 2.85 ppm. The absorption at 12.3 ppm can be assigned to the chelated NH proton.

There may be two possible reasons why the *N*-methyl signal is split: either a 1:1 mixture of *syn* and *anti* isomers of **IIb** has been formed, or the tautomeric form **IIc** is present, and the splitting is due to a coupling between the NH and methyl protons. In the latter case acid addition will eliminate the splitting, whereas in the former the lines will collapse only at higher temperatures. Since upon the addition of a drop of acid the signal becomes a singlet (see Spectrum **27**/*1a*), there is an NH–CH_3 coupling, which proves that the tautomeric structure is **IIc**.

PROBLEM **28**

Decide between Structures **I** and **II** on the basis of Spectrum **28**/*1*.

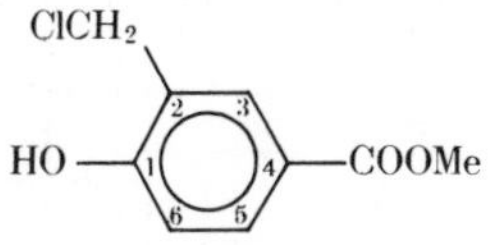

28/I 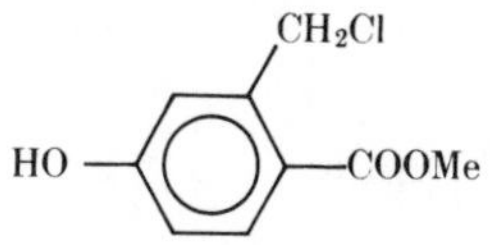 **28/II**

ANSWER

The assignment of the signals is as follows: δCH_3 = 3.85 ppm, *s*; δCH_2 = 4.68 ppm, *s*; δOH = approximately 5.2 ppm, broad signal overlapping with the signal of the water content and the CD_3OH impurity of the solvent. From the eight lines of the *AMX* multiplet of the aromatic protons one obtains: J_{AM} = 8 J_{AX} = 2, and J_{MX} <1 Hz, in good agreement with the values expected for *ortho*, *meta*, and *para* coupling. The further chemical shifts are: δArH_M (H-6) = 6.90, δArH_A (H-5) = 7.84, and δArH_X (H-3) = 7.98 ppm.

According to the deshielding of two ring hydrogens relative to the third, the unknown is compound **I**. In the spectrum of compound **II** the situation would be reverse, because the deshielding of the *ortho* protons is due to the neighboring carbonyl group. Calculating the chemical shift of the aromatic protons with Equation 277, the results are as follows; δArH-3,5(**I**) ≈ δArH-5(**II**) = 7.87 and δArH-6(**I**) ≈ δArH-2,6 (**II**) = 6.87 ppm. Table 39 does not include the substituent constants of the chlormethyl group. Since the singlet signal of the benzylchloride ring protons occurs at 7.32 ppm (in $CDCl_3$ solution),[649] compared to that of the benzene with not more than +0.05-ppm chemical shift, neglecting this substituent effect does evidently not cause any trouble. Thus, in the calculation mentioned above, hydroxy- and ester-substituents have been considered, only. It can be seen that experimental results are in a very good agreement with the data calculated for Structure **I**. The differences for H-3,5,6 are 0.03, 0.03, and 0.11 ppm.

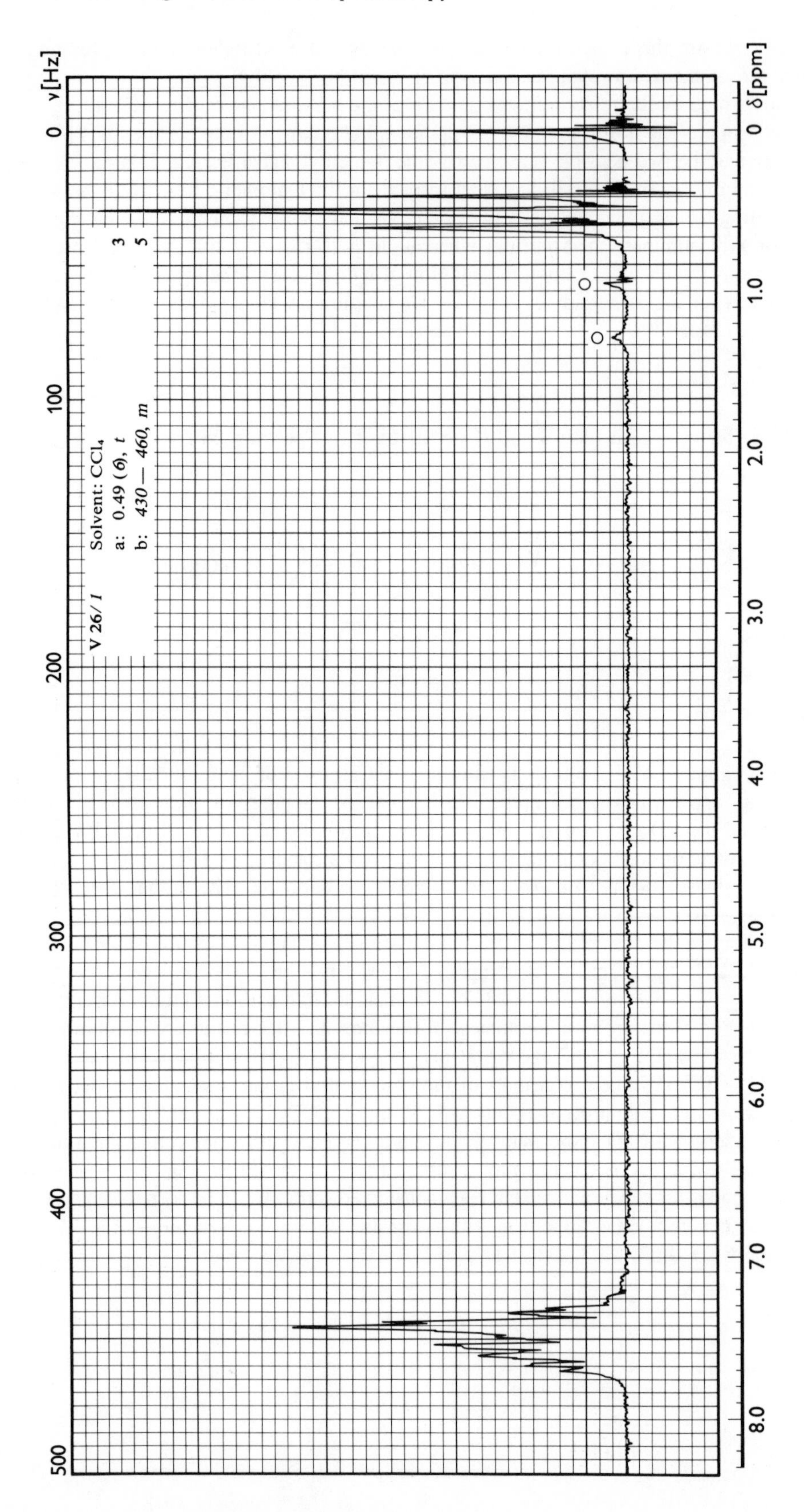
V 26/ 1
Solvent: CCl₄
a: 0.49 (6), t 3
b: 430 — 460, m 5
ν[Hz]
0
100
200
300
400
500
δ[ppm]
0
1.0
2.0
3.0
4.0
5.0
6.0
7.0
8.0

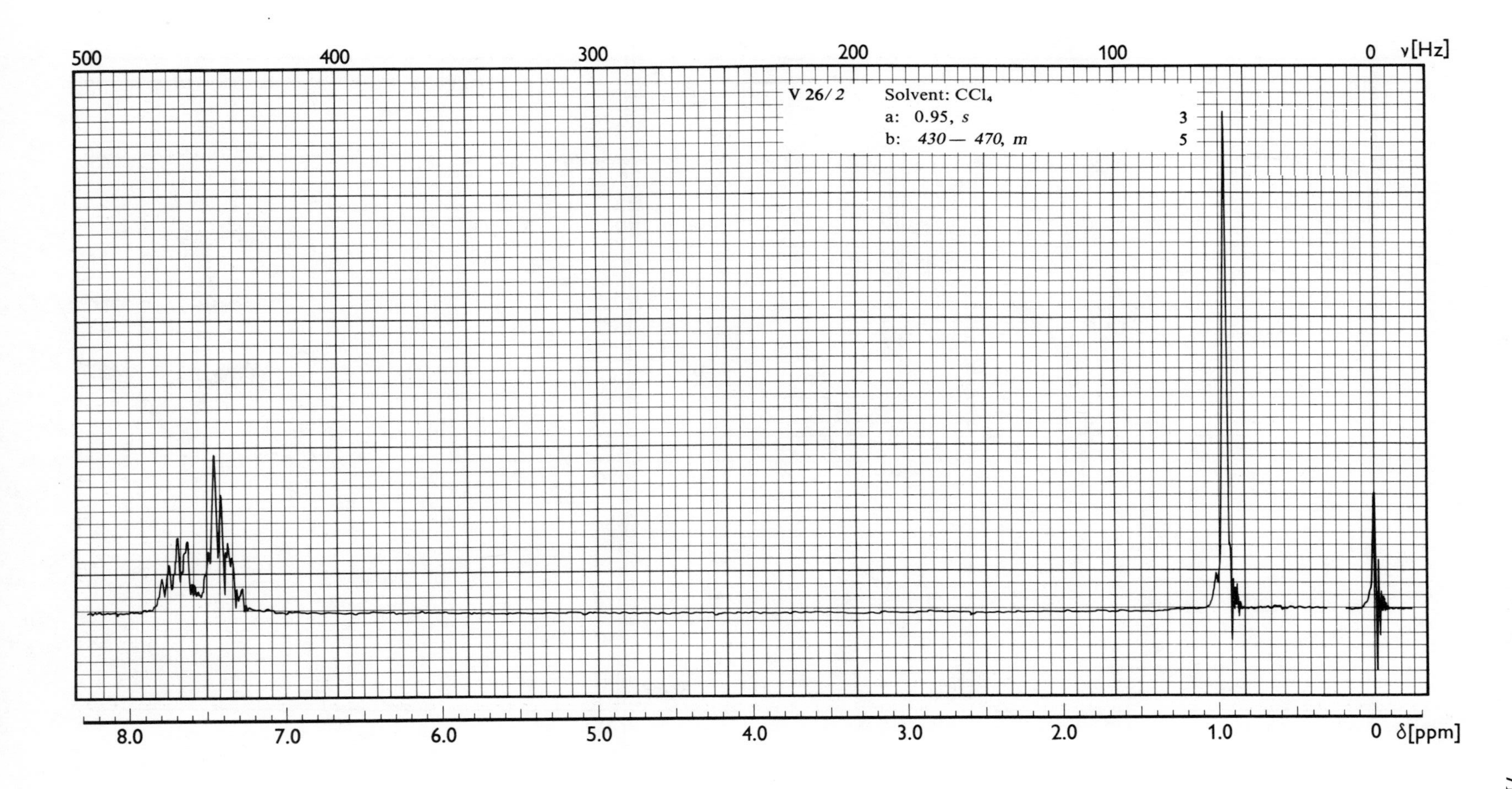
V 26/2
Solvent: CCl_4
a: 0.95, *s* 3
b: *430 — 470, m* 5
ν[Hz]
500
400
300
200
100
0
δ[ppm]
8.0
7.0
6.0
5.0
4.0
3.0
2.0
1.0
0

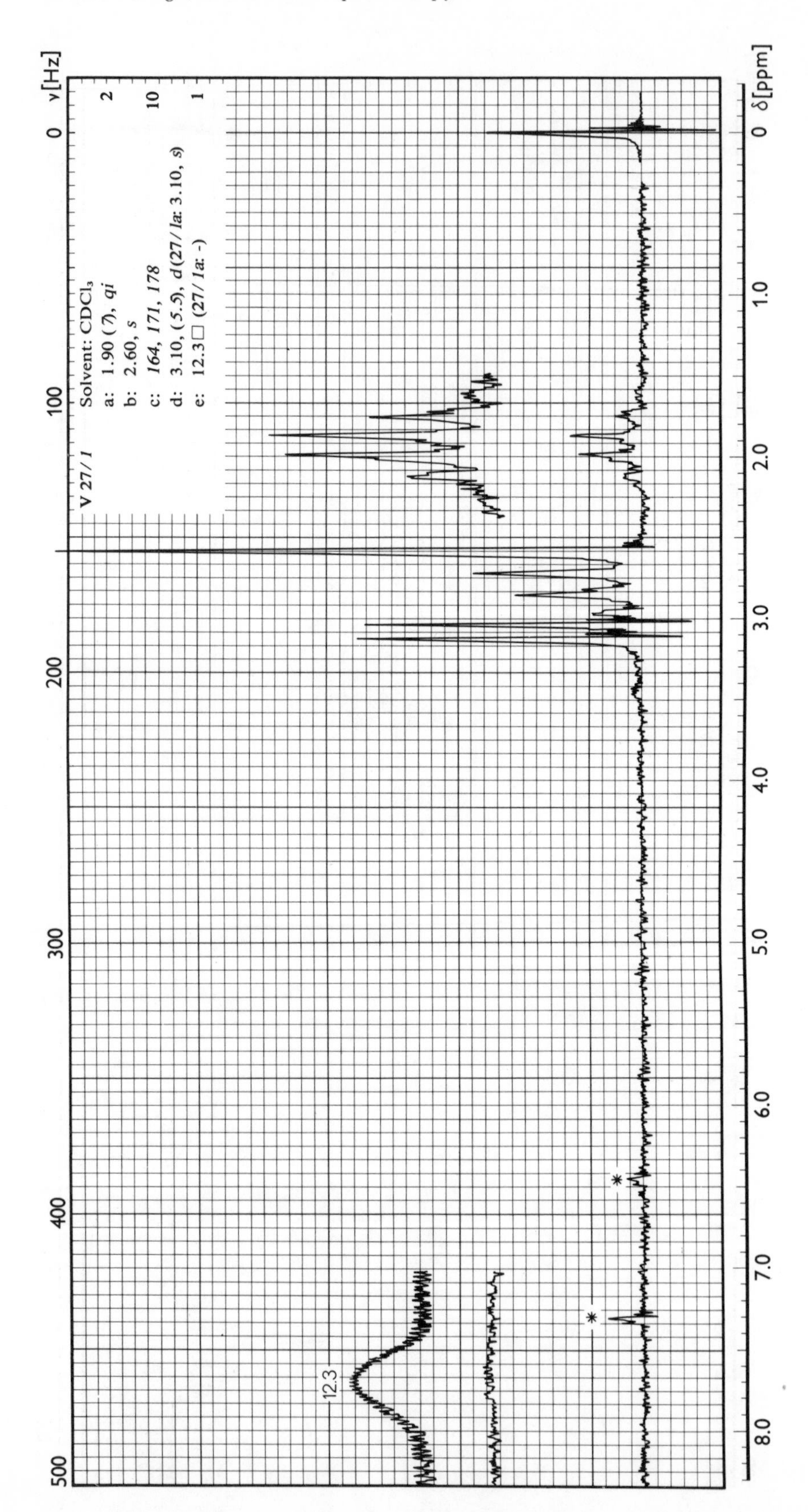
V 27/1
Solvent: $CDCl_3$
a: 1.90 (7), qi 2
b: 2.60, s
c: 164, 171, 178 10
d: 3.10, (5.5), d(27/1a: 3.10, s)
e: 12.3□ (27/1a: -) 1
ν[Hz]
0
100
200
300
400
500
δ[ppm]
1.0
2.0
3.0
4.0
5.0
6.0
7.0
8.0
12.3
*
*

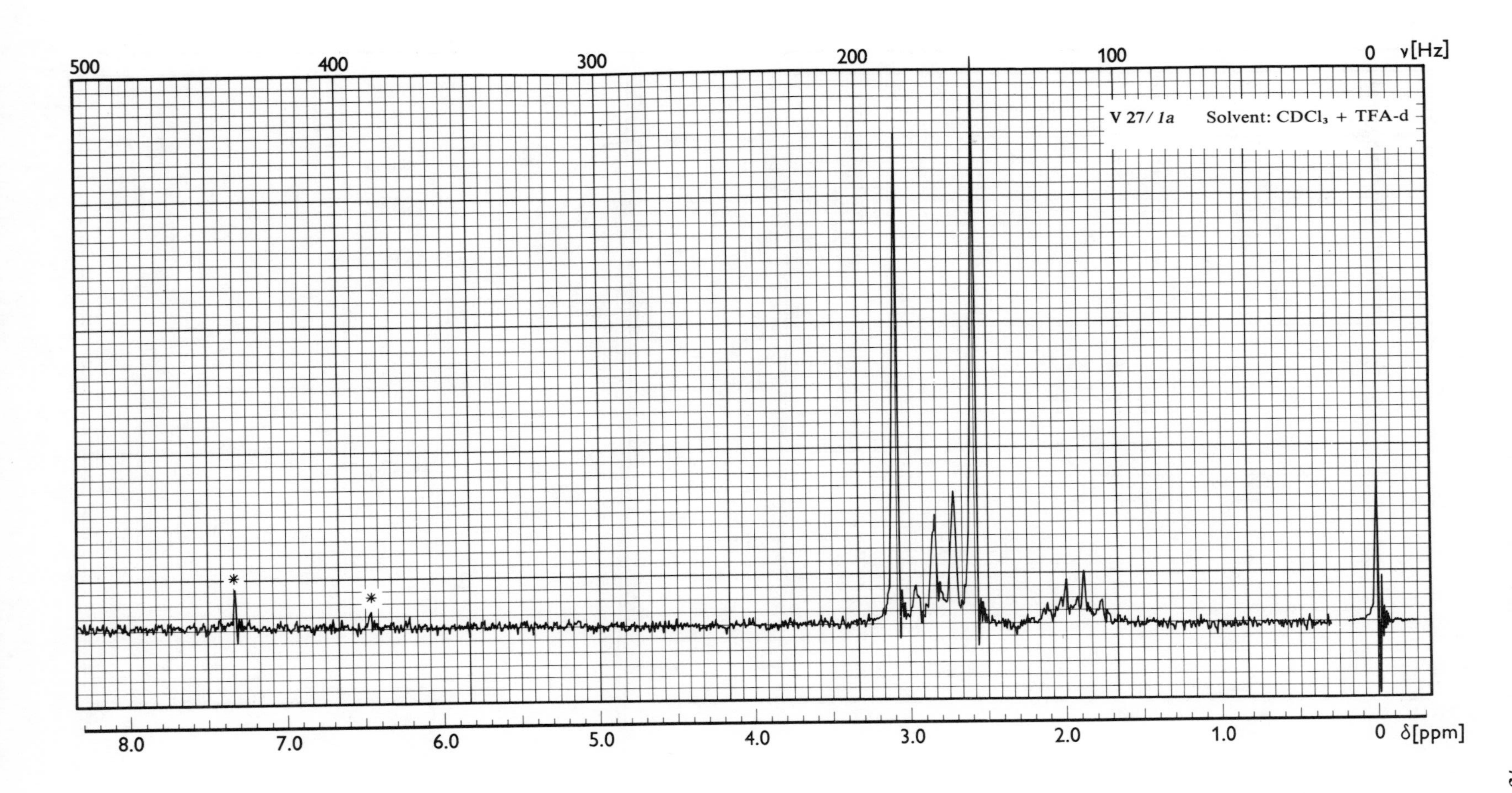

V 27/ 1a
Solvent: CDCl3 + TFA-d
500
400
300
200
100
0
ν[Hz]
8.0
7.0
6.0
5.0
4.0
3.0
2.0
1.0
0
δ[ppm]
*
*

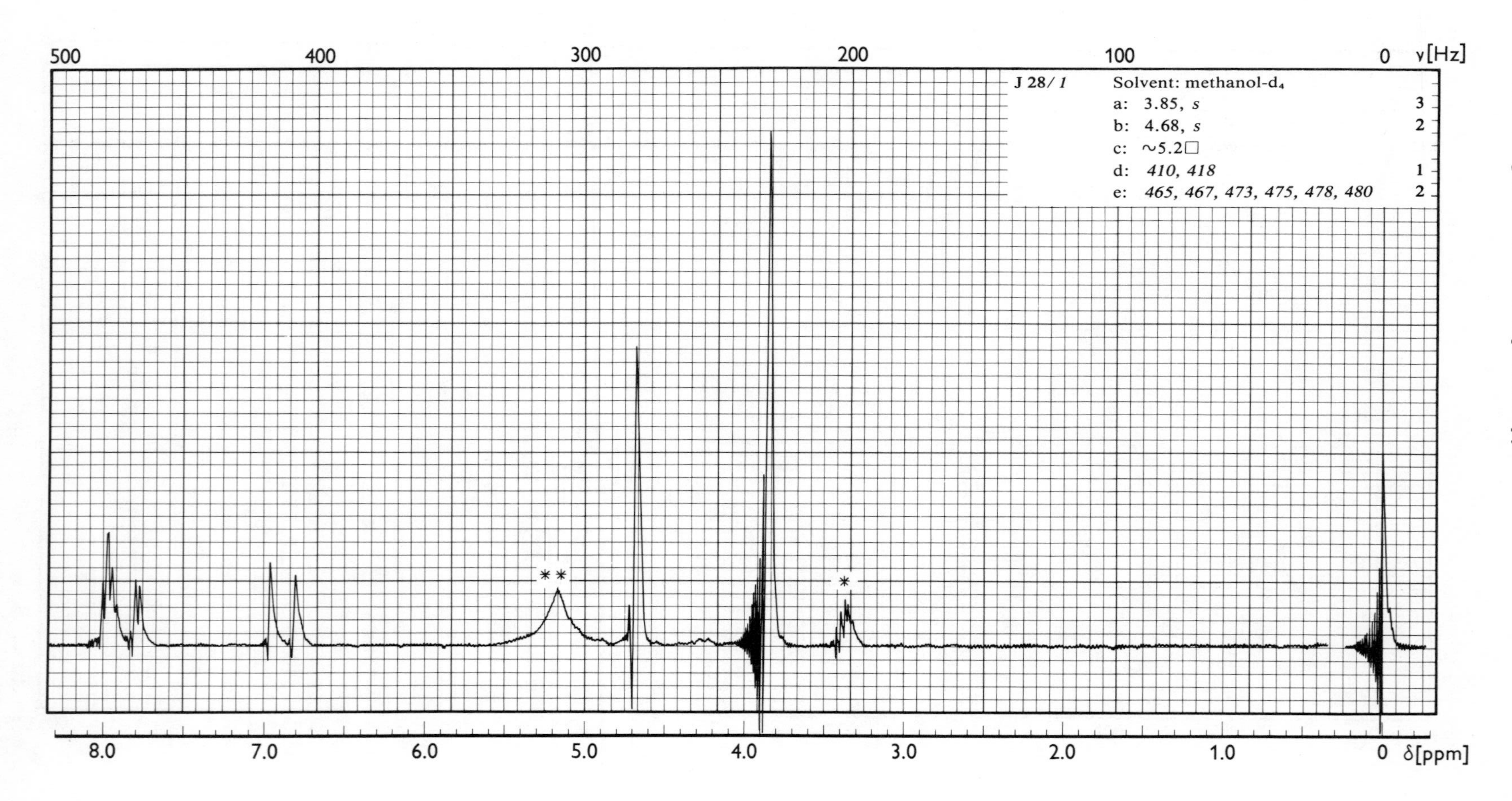
500
400
300
200
100
0 ν[Hz]
J 28/1
Solvent: methanol-d4
a: 3.85, s 3
b: 4.68, s 2
c: ~5.2□
d: 410, 418 1
e: 465, 467, 473, 475, 478, 480 2
**
*
8.0
7.0
6.0
5.0
4.0
3.0
2.0
1.0
0 δ[ppm]

PROBLEM 29[671,1320]

Spectrum **29**/*1* is a compound with the empirical formula $C_3H_6OCl_2$. Derive the structure of the compound. Also determine the structure of the by-product with the empirical formula $C_5H_{10}O_2Cl_2$ on the basis of its Spectrum **29**/*2*.

ANSWER

There is an A_2B_2 multiplet in Spectrum **29**/*1* (see Figures 52 and 53), which must be due to an RCH_2CH_2R' group. In addition to this signal the spectrum shows only a singlet at 5.50 ppm, i.e., the two remaining hydrogens are equivalent. This is also indicated by the intensity ratio 2:1 of the A_2B_2 multiplet and the singlet. It follows from the empirical formula that the singlet arises from a methylene group, hence the structure is $Cl{-}CH_2{-}CH_2{-}O{-}CH_2Cl$ (**I**), which is compatible also with the chemical shifts. From the Shoolery rule (Equation 270), the expected shifts are $\delta OCH_2C = 2.95$, $\delta ClCH_2C = 3.25$, and $\delta OCH_2Cl = 4.95$ ppm. They differ from the measured shifts by approximately 0.5 ppm, all in positive directions.

Spectrum **29**/*2* of the by-product also contains an A_2B_2 multiplet at approximately the same frequencies, but the singlet is shifted considerably upfield, with a half intensity. For this reason the by-product must be $CH_2(OCH_2CH_2Cl)_2$ (**II**).

The measured chemical shift of the singlet (4.75 ppm) differs only by 0.1 ppm from the shift calculated from the Shoolery rule. The presence of two methylene singlets indicates that the two compounds mutually contaminate each other (the main product contains only a minimum amount of the by-product, but the impurity in the latter amounts to about 4.5%, on the basis of the intensity ratio 11:1 of the lines at 4.65 and 5.50 ppm).

When all lines of an A_2B_2 multiplet are separated, the calculation of δA, δB, and J_{AB} is straightforward.* The chemical shifts are given by the position of line **3** or **4** (whichever stronger) with respect to the midpoint of the multiplet, whereas J_{AB} is half the distance between lines **2** and **7**.

In Spectrum **29**/*1* lines **1** and **2** and in Spectrum **29**/*2* lines **3** and **4**, as well, are overlapping. Since, however, line **7** is observable in both spectra, the calculation can be performed from the frequencies to yield: $\delta A = 236/60 = 3.93$ ppm (**I**) and $231/60 = 3.85$ ppm (**II**), $\delta B = 221/60 = 3.68$ ppm (**I**) and $220/60 = 3.67$ ppm (**II**), and finally J_{AB} $(243 - 232)/2 = (225 - 214)/2 = 5.5$ Hz (**I**) and $(239 - 227)/2 = (224 - 212)/2 = 6.0$ Hz (**II**).

* See Volume I, p. 114.

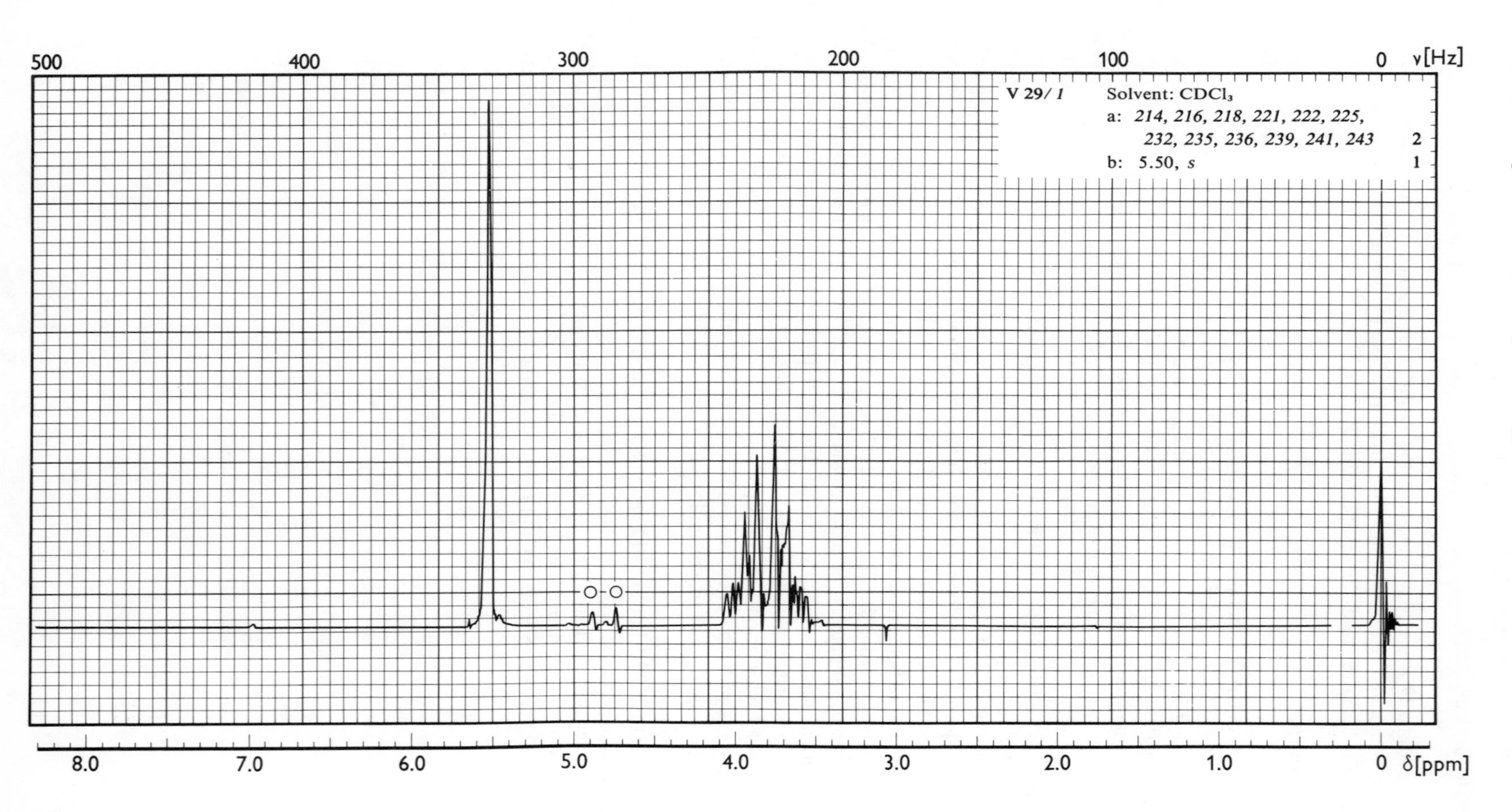
V 29/ 1
Solvent: CDCl3
a: 214, 216, 218, 221, 222, 225, 232, 235, 236, 239, 241, 243 2
b: 5.50, s 1
500
400
300
200
100
0
ν[Hz]
8.0
7.0
6.0
5.0
4.0
3.0
2.0
1.0
0
δ[ppm]

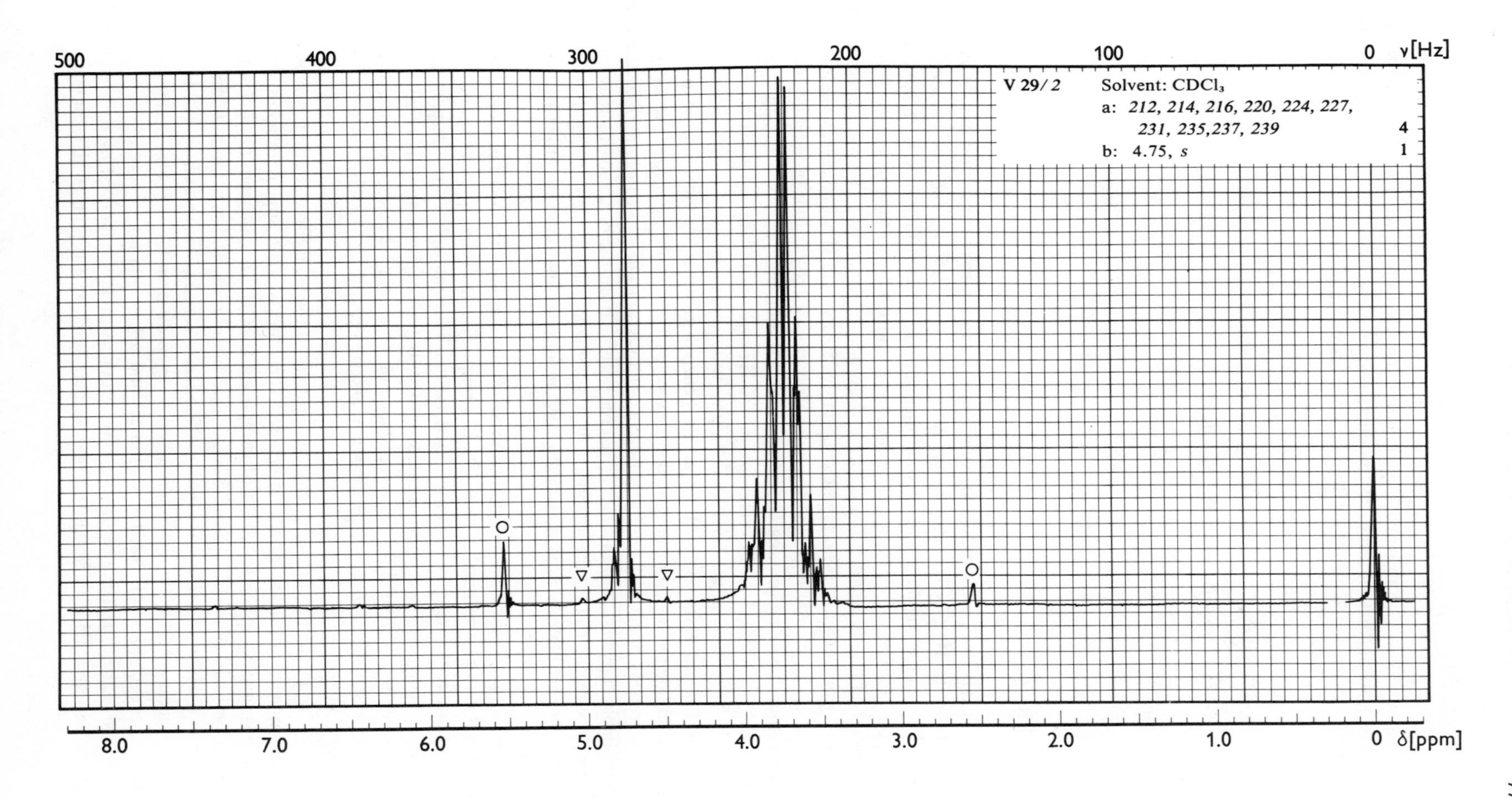
V 29/2
Solvent: CDCl3
a: 212, 214, 216, 220, 224, 227, 231, 235,237, 239 4
b: 4.75, s 1
ν[Hz]
500
400
300
200
100
0
δ[ppm]
8.0
7.0
6.0
5.0
4.0
3.0
2.0
1.0
0

PROBLEM **30**

Compound **I** was brominated to yield derivative **II**. Determine from Spectra **30/*1*** and **30/*2*** of the starting material and the product, respectively, whether the expected product is obtained. Interpret the spectra. (Both compounds contain 1 mol of crystal water.)

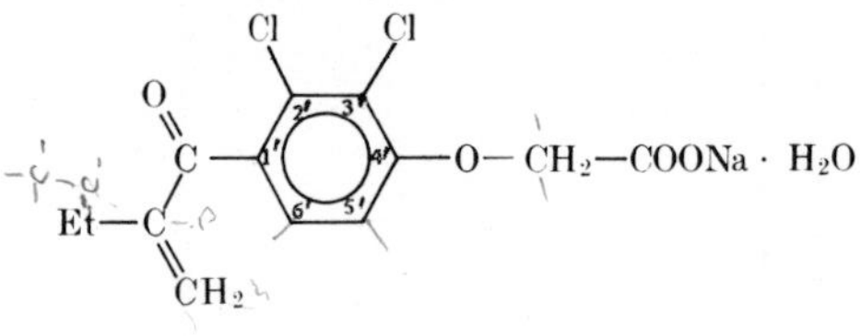

Cl Cl
O
C—(ring)—O—CH_2—COONa · H_2O
Et—C
CH_2

30/I

Cl Cl
O
C—(ring)—O—CH_2—COONa · H_2O
Et—CBr
CH_2Br

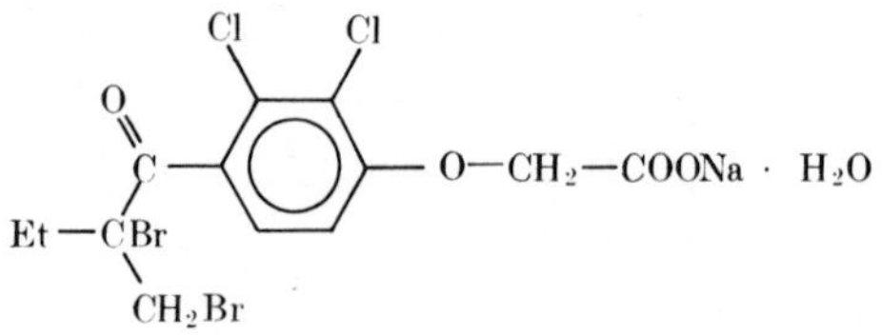

30/II

ANSWER

The signals in Spectrum **30/*1*** of compound **I** can be assigned as follows: δCH_3, *t*: 1.12 ppm (J = 7.5 Hz); δCH_2 (ethyl), *qa*: 2.47 ppm (J = 7.5 Hz, the line broadening is due to the long-range coupling with the olefinic protons); $\delta CH_2(COOH)$, *s*: 4.78 ppm, $\delta(=CH^c)$: 5.60 ppm, $\delta(=CH^t)$: 5.95 ppm. (The indexes *c* and *t* refer to the *cis* and *trans* position of the olefinic protons relative to the carbonyl group, respectively.) The methylene protons of the ethyl group enter a *cisoid* allylic coupling with $^4J \approx 1.5$ Hz. The corresponding *transoid* coupling is, as can be expected,* smaller and causes only a line broadening in the 5.60-ppm signal. The ring protons H-5′,6′ give an *AB* multiplet: J_{AB} = 8 Hz, midpoint: 419.5 Hz, $\Delta\nu AB = \sqrt{31 \cdot 15} \approx 21.6$ Hz, νA = 430 Hz, νB = 409 Hz, $\delta A \equiv \delta$H-6′ = 7.17 ppm, $\delta B \equiv \delta$H-5′ = 6.82 ppm, δOH = 7.9 ppm (water content).

Note that the deshielding effect of the benzoyl group is stronger on the *trans* olefinic proton, although the opposite could be expected (see Table 36) from the assignment of olefinic signals deduced from the difference between *cisoid* and *transoid* allylic couplings. A probable explanation of this fact is that for steric reasons the benzene ring may not become coplanar with the C=O and C=C bonds. If it is perpendicular to the C=C bond, its anisotropy shields the *cis* olefinic proton, which overcompensates the opposite effect of the carbonyl group.

In Spectrum **30/*2*** of the brominated product, the signal of the olefinic protons is replaced by an *AB* multiplet assignable to a $-CH_2Br$ group. The methylene protons of the ethyl group, being diastereotopic, give a complicated (sextet-like) multiplet, the *AB* part of an ABX_3 multiplet. This structure with an intensity distribution of 1:3:4:4:3:1 is obtained from the superposition of the *AB* lines with a distance of about $2J_{A_2X_3}$ between the inner lines of the *AB* components. The outer lines are weak and merge into the baseline or overlap with the stronger lines.

The nonequivalence of the methylene protons in the $-CH_2Br$ and ethyl groups is due to the asymmetric structure of the molecule. The other signals appear in approximately the same positions as in the spectrum of compound **I**: δCH_3, *t*: 1.08 ppm (J = 7.5 Hz), δCH_2 (ethyl), *m*: 2.17 ppm, δCH_2Br: 3.82 and 4.28 ppm (J = 10.5 Hz, the midpoint of the *AB* multiplet: 243 Hz, $\Delta\nu AB = \sqrt{19 \cdot 40}$ = 27.6 Hz, νA = 257 Hz and νB = 229 Hz), δOCH_2, *s*: 4.80 ppm, δArH-6′ = 7.63 and δArH-5′ = 6.80 ppm (J = 9 Hz, midpoint: 433 Hz, $\Delta\nu AB = \sqrt{42 \cdot 60}$ = 50 Hz, νA = 458 Hz and νB = 408 Hz).

Note that the $-I$ effect of the carbonyl group is greatly enhanced by α-bromine substitution, and thus δH-6′ is about 0.5 ppm larger; δOH, *s*: 8.6 ppm (water content).

* See Volume I, p. 63.

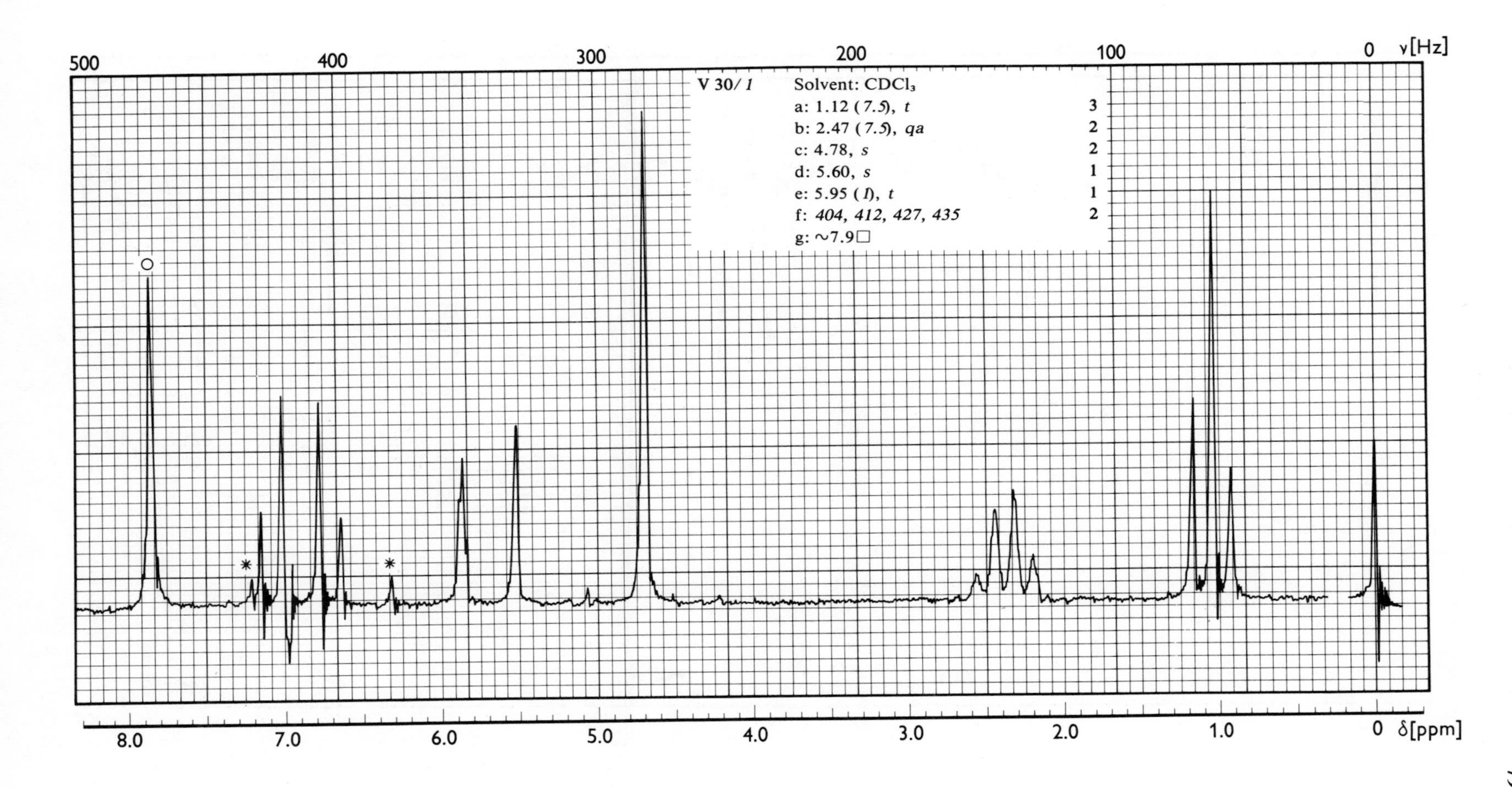

ν[Hz]
500
400
300
200
100
0
V 30/1
Solvent: CDCl₃
a: 1.12 (7.5), t 3
b: 2.47 (7.5), qa 2
c: 4.78, s 2
d: 5.60, s 1
e: 5.95 (1), t 1
f: 404, 412, 427, 435 2
g: ~7.9□
δ[ppm]
8.0
7.0
6.0
5.0
4.0
3.0
2.0
1.0
0

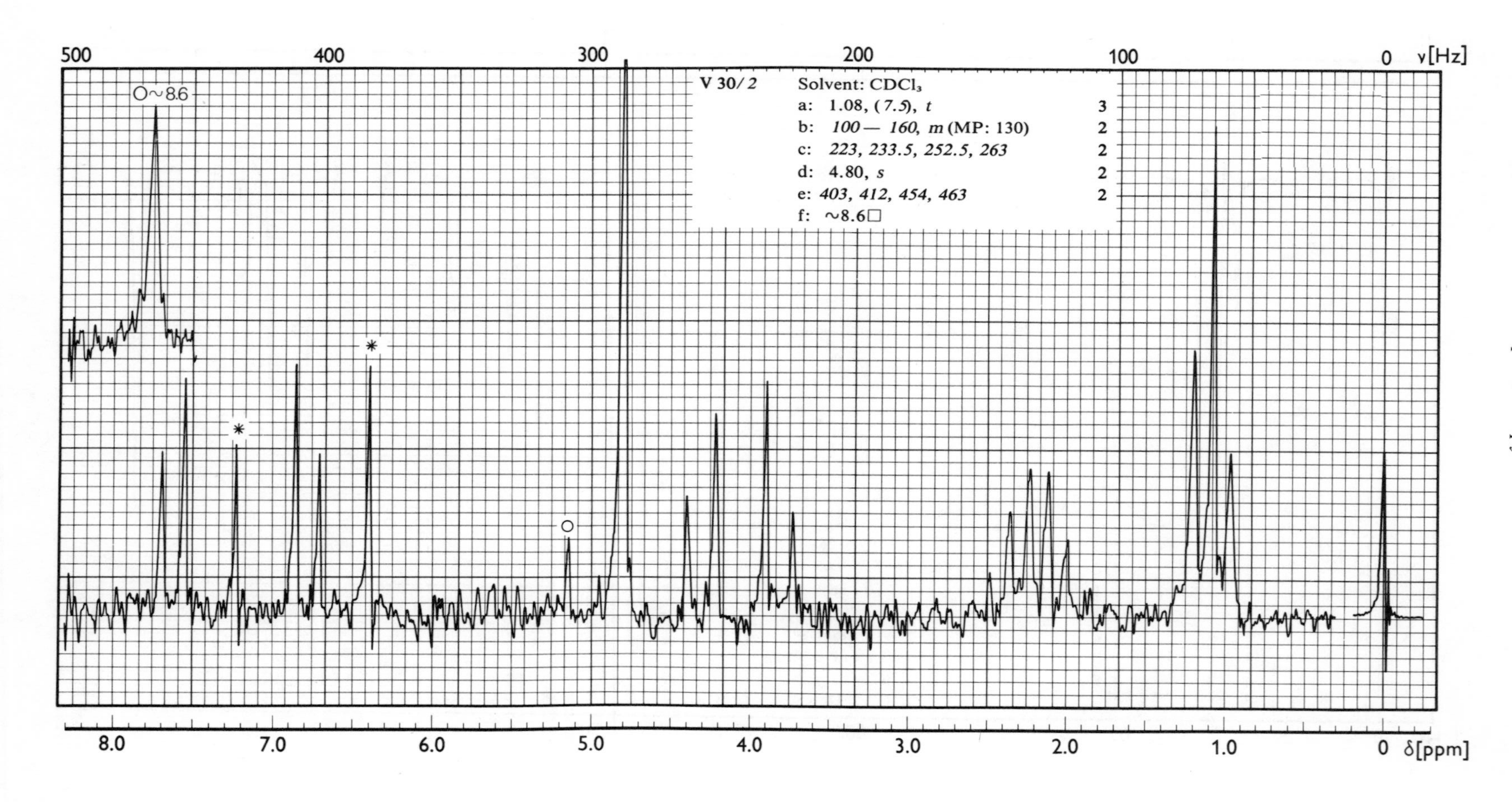

500
400
300
200
100
0 ν[Hz]
V 30/2
Solvent: CDCl₃
a: 1.08, (7.5), t 3
b: 100— 160, m (MP: 130) 2
c: 223, 233.5, 252.5, 263 2
d: 4.80, s 2
e: 403, 412, 454, 463 2
f: ~8.6□
○~8.6
8.0
7.0
6.0
5.0
4.0
3.0
2.0
1.0
0 δ[ppm]

PROBLEM 31

In the reaction of cyclohexanone with hydroquinone performed to prepare compound **I**, two substances were obtained. Decide from Spectra **31**/*1* and **31**/*2* whether either of the products has the expected structure. If so, which spectrum is compatible with the expected product, and what is the other substance?

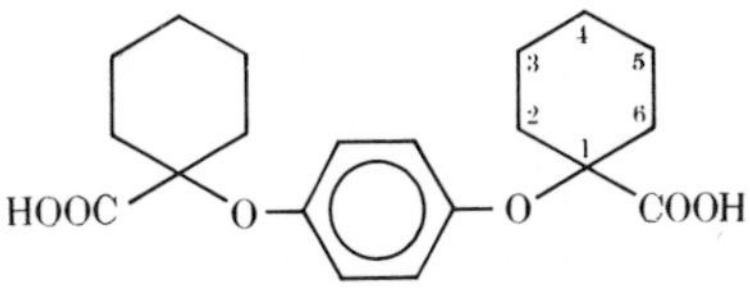

31/I

ANSWER

The broad signals of aliphatic protons appear in both spectra at about 1.5 and 1.6 and at 1.8 and 2.1 ppm, respectively. The relative intensities are 3:2 in both cases. These data indicate that the signals of more shielded protons correspond to 3,4,5-methylene protons, whereas the 2,6 groups are deshielded, giving the downfield signal, proving that both compounds are cyclohexane derivatives. Apart from these maxima, only two other signals can be observed in the spectra. One of them, around 4.65 and 4.85 ppm in the two spectra, is the overlapping signal of the H_2O content of the solvent (D_2O) and the carboxylic hydrogens. The other signal appears in the region characteristic of aromatic hydrogens. In Spectrum **31**/*2* this signal is a singlet at 7.00 ppm and in Spectrum **31**/*1* it is an *AB*-like multiplet with a midpoint at about 399 Hz. The relative intensity of the singlet is 2:5, and that of the *AB* multiplet is 1:5 with respect to the total intensities of the aliphatic protons.

Consequently, Structure **I** is compatible only with Spectrum **31**/*2*, because in this molecule the ratio of aliphatic and aromatic protons is really 5:1 (actually 20:4), and the latter are chemically equivalent, yielding evidently a singlet. In the other compound, the relative number of the aliphatic hydrogens is a half of the above value, and the nonequivalence of the ring protons indicates that the by-product is the monoether analogue **II** of the bis-ether (**I**), and the signal of the phenolic hydrogen overlaps with that of the acidic and solvent protons at 5.65 ppm.

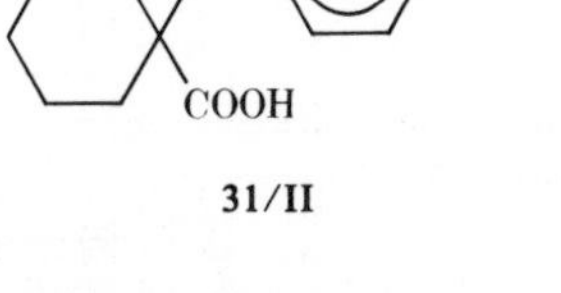

31/II

This assumption is congruent with the *AB*-like form of the aromatic multiplet, since the asymmetric *para*-disubstituted benzene derivatives usually have an $AA'BB'$ spectrum resembling an *AB* pattern. The chemical shifts calculated in the *AB* approximation (see Problem 6) are $\delta A = 6.76$, $\delta B = 6.53$ ppm, and $J_{AB} = 9$ Hz.

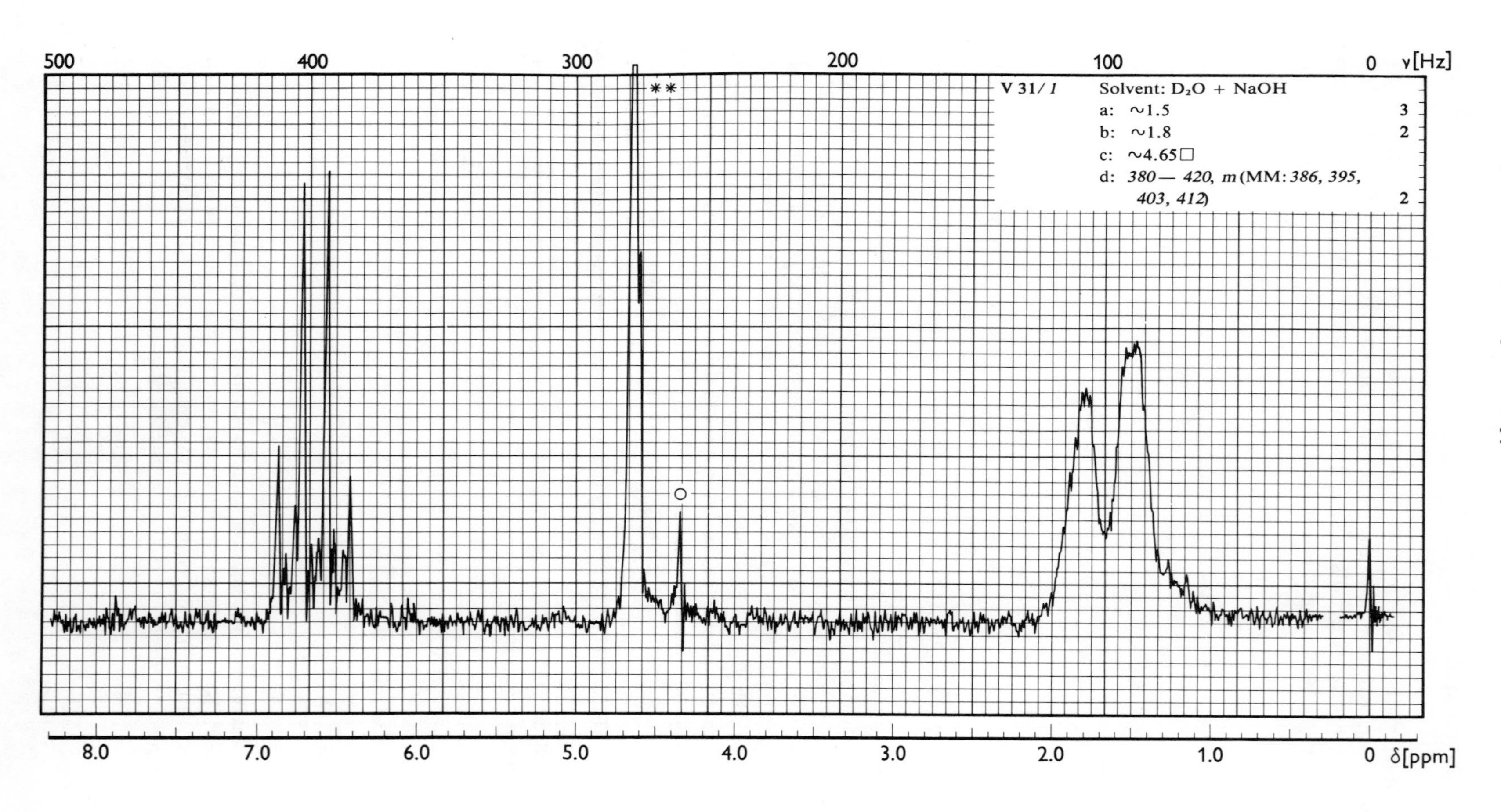

500
400
300
200
100
0 ν[Hz]
**
V 31/1
Solvent: D₂O + NaOH
a: ~1.5 3
b: ~1.8 2
c: ~4.65□
d: 380— 420, m (MM: 386, 395, 403, 412) 2
○
8.0
7.0
6.0
5.0
4.0
3.0
2.0
1.0
0 δ[ppm]

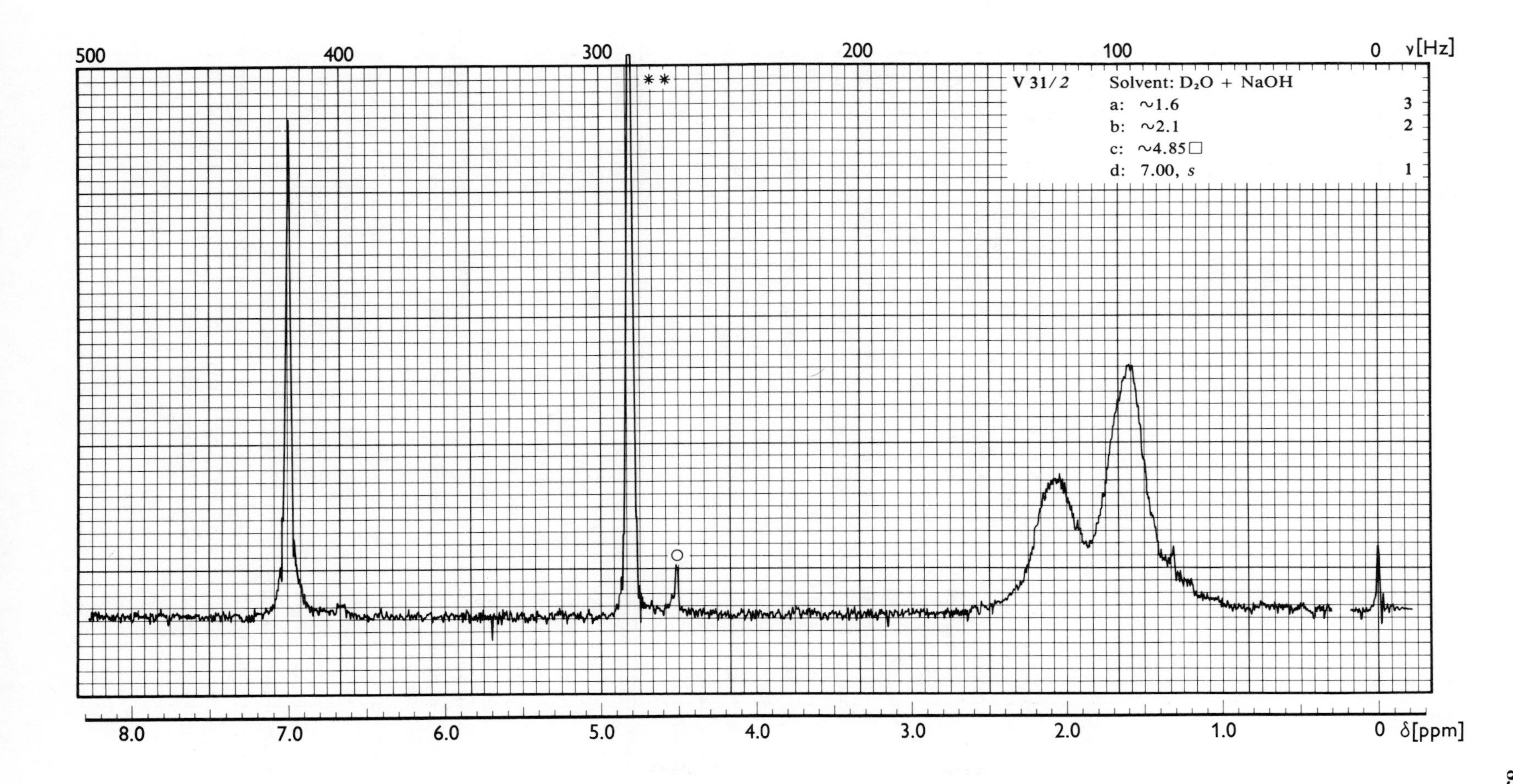
ν[Hz]
500
400
300
200
100
0
**
V 31/2
Solvent: D_2O + NaOH
a: ~1.6 3
b: ~2.1 2
c: ~4.85□
d: 7.00, *s* 1
8.0
7.0
6.0
5.0
4.0
3.0
2.0
1.0
0
δ[ppm]

PROBLEM **32**

Spectra **32**/*1* and **32**/*2* correspond to two of Structures **I** (R = H or Me, R′ = H or OMe). Identify these two compounds and interpret the spectra.

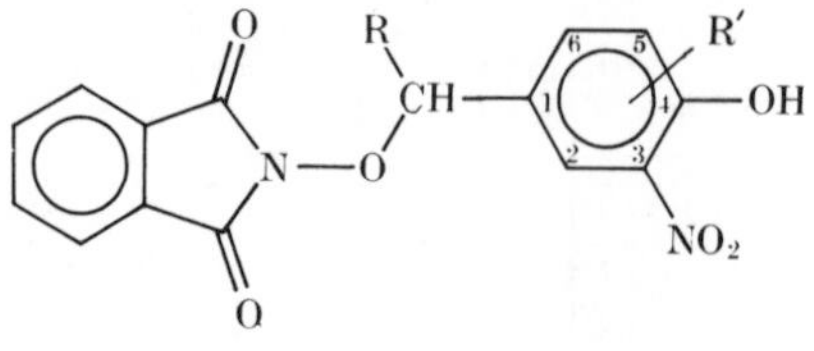

32/I (R, R′ = H or Me)

ANSWER

The CH_3CH group occurring in Structure **I** (R=Me) has presumably an AX_3 spectrum, in which the X_3 doublet can be expected to appear at $\delta X < 2$ ppm. As these line groups are absent in Spectrum **32**/*2*, only Spectrum **32**/*1* may correspond to such a compound. Accordingly, in Spectrum **32**/*1* (also in agreement with the relative intensities), δCH_3, *d*: 1.72 ppm (J = 6.5 Hz); δCH, *qa*: 5.46 ppm (J = 6.5 Hz).

The singlet at 7.88 ppm, in the region of aromatic hydrogens, is due to the phthalimide ring, and the remaining lines (three doublets) form an *AMX* spectrum, the weaker doublet of the *A* part of which is overlapping with the phthalimide signal. From the doublets, J_{AM} = 8.5, J_{AX} = 2, and J_{MX} <1 Hz.

It is clear from the presence of an *AMX* pattern, that Spectrum **32**/*1* is in accordance with Structure **I** (R = Me, R′ = H). The *AM* part of the *AMX* spectrum arises from the adjacent H-5,6 atoms and of course, $J_{2,6} \equiv J^m > J_{2,5} \equiv J^p$. Therefore, the *X* doublet can be assigned to H-2, the chemical shift, δX(H-2) = 8.13 ppm, of which is compatible with the strong paramagnetic shift caused by the adjacent nitro group. The *A* part corresponds, therefore, to H-6.

As J_{AM} is known, the frequencies of the two concealed *A* lines can be determined from the frequencies of two other *A* lines (464 and 466 Hz), to yield 464 + 8.5 = 472.5 and 466 + 8.5 = 474.5 Hz. We obtain A = 7.82 ppm, and thus the doublet at 7.23 ppm belongs to H-5. The shoulders of the lines can be attributed to the small *MX* coupling ($J_{MX} \equiv J_{2,5} \equiv J^p$). The OH signal cannot be identified.

Spectrum **30**/*2* may correspond only to compounds with R = H, and since there is a methoxy signal at 3.93 ppm in the spectrum, R′ = OMe. At 5.22 ppm the methylene signal, and at 7.90 ppm the singlet of the phthalimide protons, can be identified. This assignment is supported by the intensity ratio of 3:2:4 in the above sequence.

The two hydrogens of the tetrasubstituted benzene ring give an *AX* doublet pair from which δA = 7.67, δX = 7.52 ppm, and J_{AX} = 2 Hz. This coupling constant characteristic of *meta* interactions proved the 5-position of the methoxy group and thereby the structure responsible for Spectrum **30**/*2*. It follows from the vicinity of the nitro group that H-2 is strongly deshielded, thus δH-2 $\equiv \delta A > \delta$H-6 $\equiv \delta X$.

The OH signal is insignificant in this case, too. Note that the ring protons of the phthalimide ring represent evidently an *AA′BB′* spin system and the corresponding singlet is due to accidental isochrony in DMSO-d_6 solution, where $\delta A \approx \delta B$.

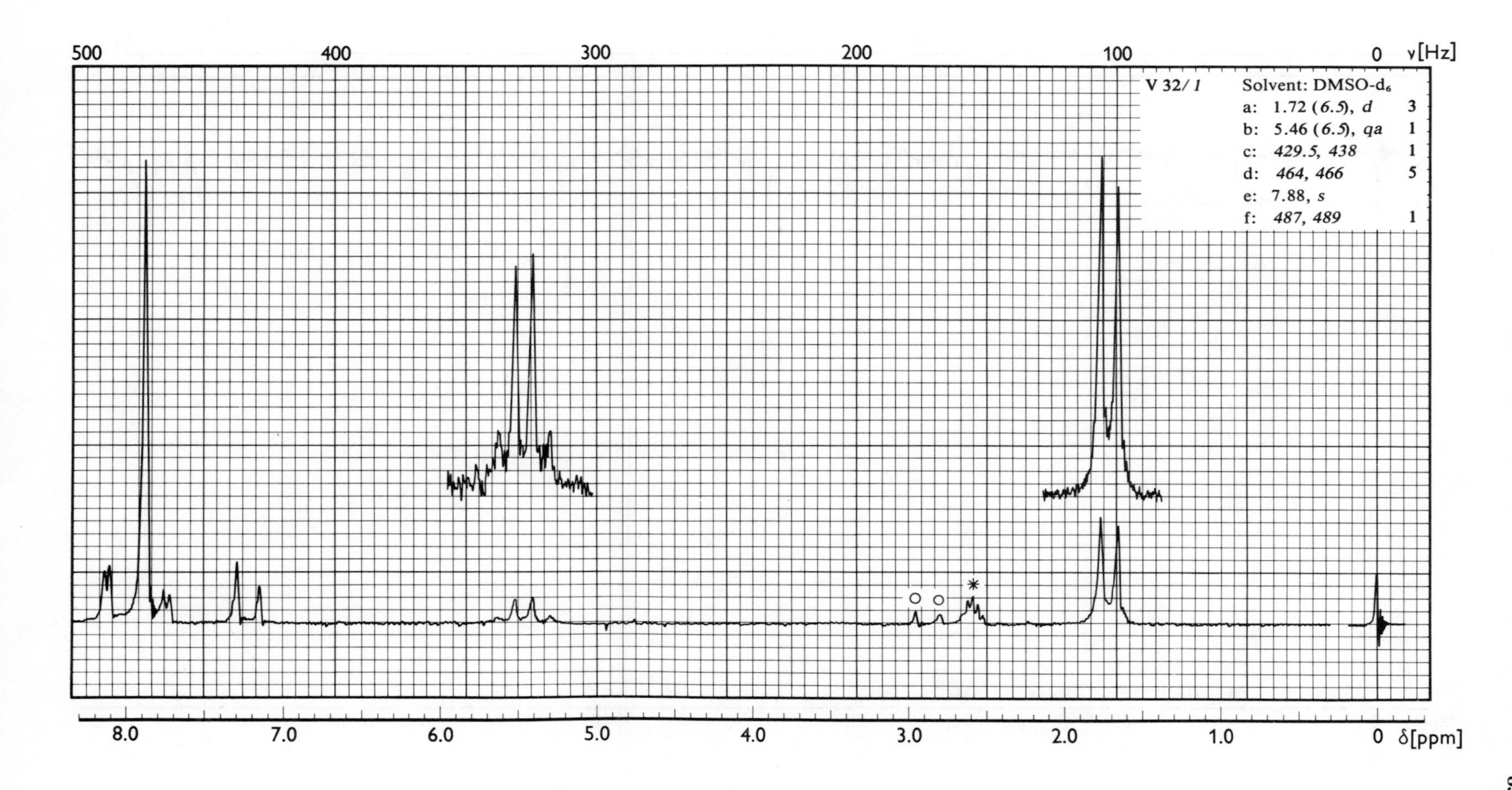
V 32/ 1
Solvent: DMSO-d₆
a: 1.72 (6.5), d 3
b: 5.46 (6.5), qa 1
c: 429.5, 438 1
d: 464, 466 5
e: 7.88, s
f: 487, 489 1
ν[Hz]
500
400
300
200
100
0
δ[ppm]
8.0
7.0
6.0
5.0
4.0
3.0
2.0
1.0
0

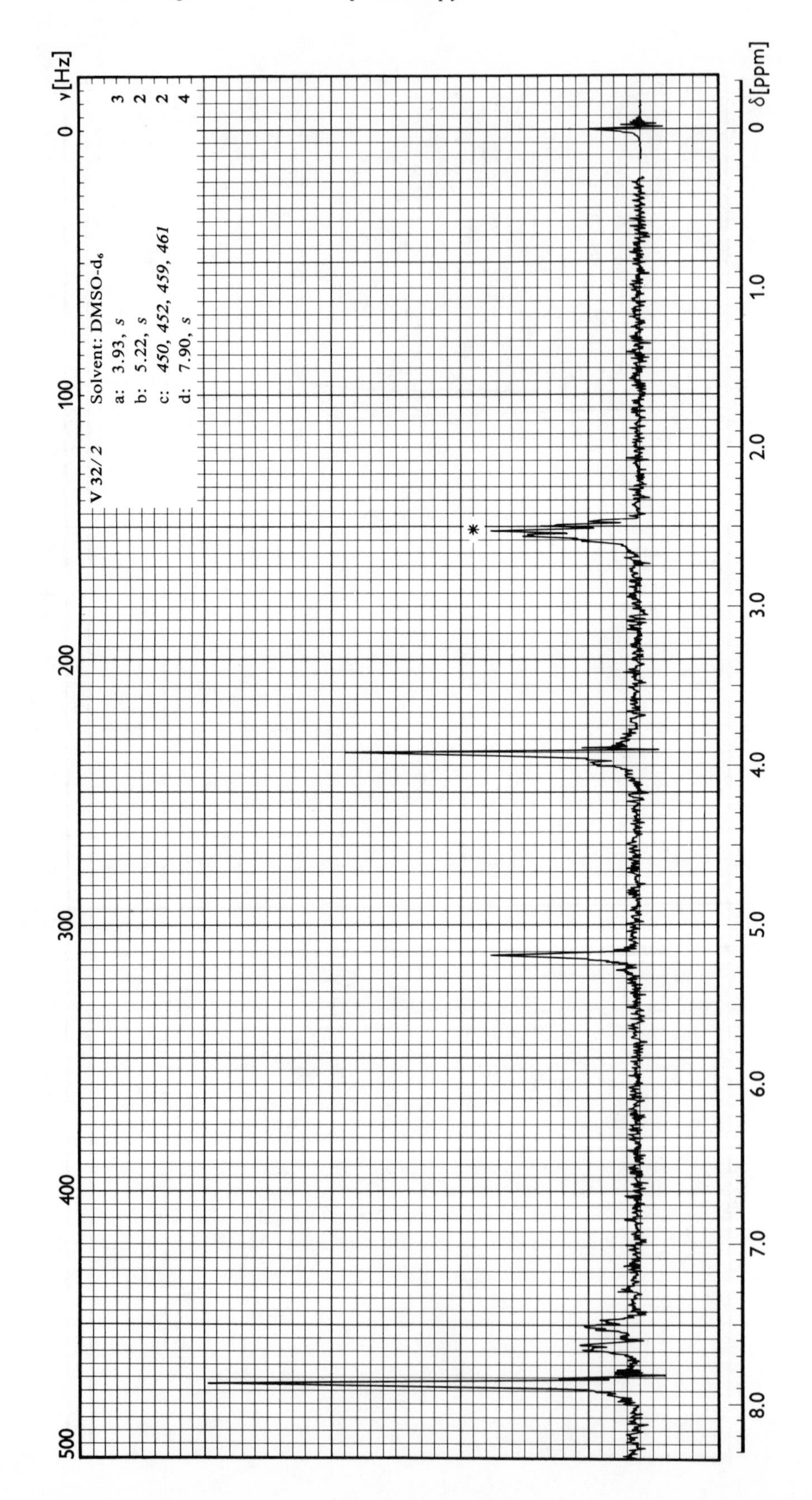
V 32/2
Solvent: DMSO-d6
a: 3.93, s 3
b: 5.22, s 2
c: 450, 452, 459, 461 2
d: 7.90, s 4
ν[Hz]
0
100
200
300
400
500
δ[ppm]
0
1.0
2.0
3.0
4.0
5.0
6.0
7.0
8.0
*

PROBLEM 33[1318,1409]

Assign the signals in Spectrum **33**/*1* of compound **I** and interpret the temperature dependence of the spectrum shown in Figure 200. Calculate the thermodynamical parameter determinable by this temperature dependence.

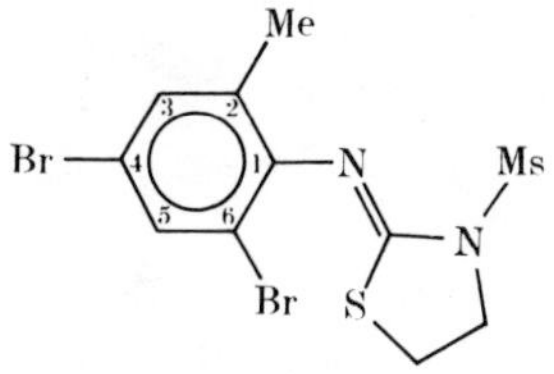

33/I

ANSWER

The assignment is straightforward: $\delta ArCH_3$, *s*: 2.20 ppm, δSCH_2, *t*: 3.28 ppm (J = 6.5 Hz), δSO_2CH_3, *s*: 3.52 ppm, δNCH_2: 4.25 ppm (two triplets at a distance of 4 Hz, J = 6.5 Hz), δArH-3, *d*: 7.30 ppm (J^m = 2 Hz), and δArH-5, *d*: 7.60 ppm (J^m = 2 Hz). The H-5 atom is more deshielded than H-3, due to the vicinity of two bromine atoms causing paramagnetic shift in *ortho* position (compare Table 39).

In the thiazolidine ring, $^3J^c \approx {}^3J^t$ (compare Problem **18**), thus instead of an $AA'BB'$ multiplet, the methylene groups yield an A_2X_2-like triplet pair, and, of course, the upfield one is assigned to the *S*-methylene group (the $-I$ effect of the *N*-acyl group deshields the hydrogens of the *N*-methylene group.

The only unexpected feature of the spectrum is a further splitting, by 4 Hz, of the NCH_2 triplet. As can be seen in Figure 200, it is just this part of the spectrum that varies with the temperature, and at about 62°C it coalesces reversibly into the expected triplet. (*Reversibly* means here that the splitting reappears when the solution is recooled to room temperature.)

The splitting can be attributed to a hindered rotation around the NS-bond. In the rotamers the hydrogens of the NCH_2 group are chemically nonequivalent.* When the rotation becomes free by heating, the difference in shielding of the methylene hydrogens is averaged, and the two triplets collapse again.

From the coalescence temperature of T_c = 62°C = 335 K one can evaluate the activation free enthalpy characteristic of the hindered rotation (rotation barrier). From Equation 288, and assuming that the splitting of triplets is $\Delta\nu$ = 4 Hz, we obtain $\Delta G_{T_c} = 19.13 \cdot T_c (9.97 + \log T_c/\Delta\nu) = 19.13 \cdot 335 = [9.97 + \log (335/4)] = 76.2$ kJ/mol.

* For the hindered rotation of amides, see Volume I, p. 71.

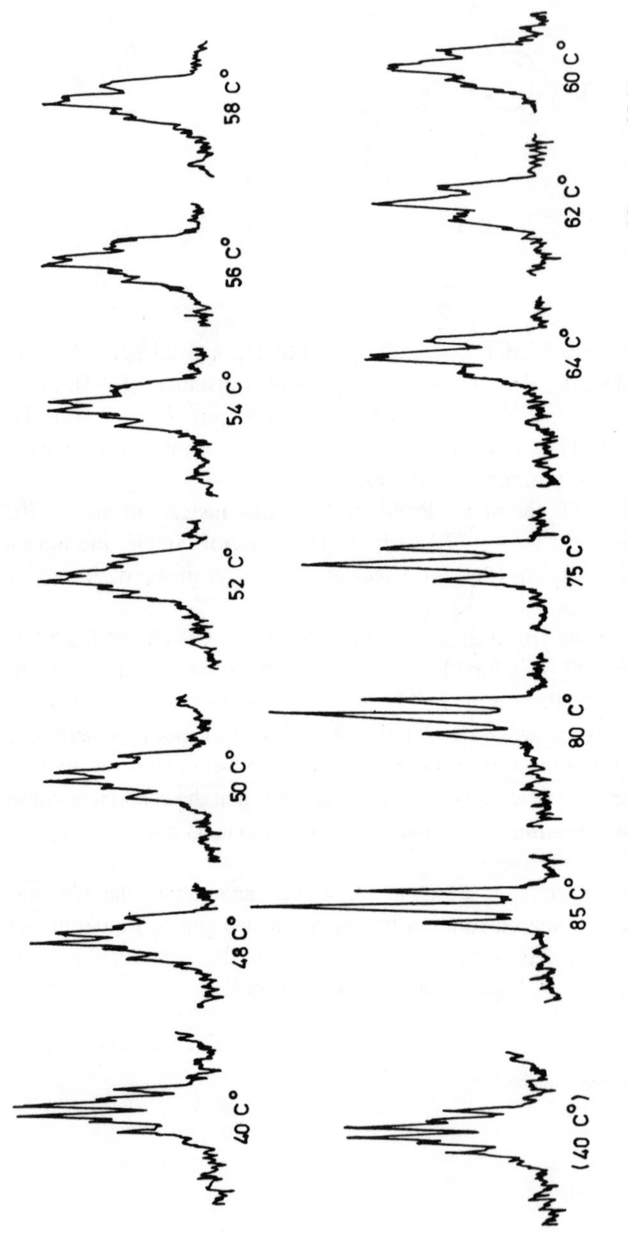

FIGURE 200. The temperature dependence of the double triplet at 3.25 ppm (at room temperature) of Spectrum 33/*1*.

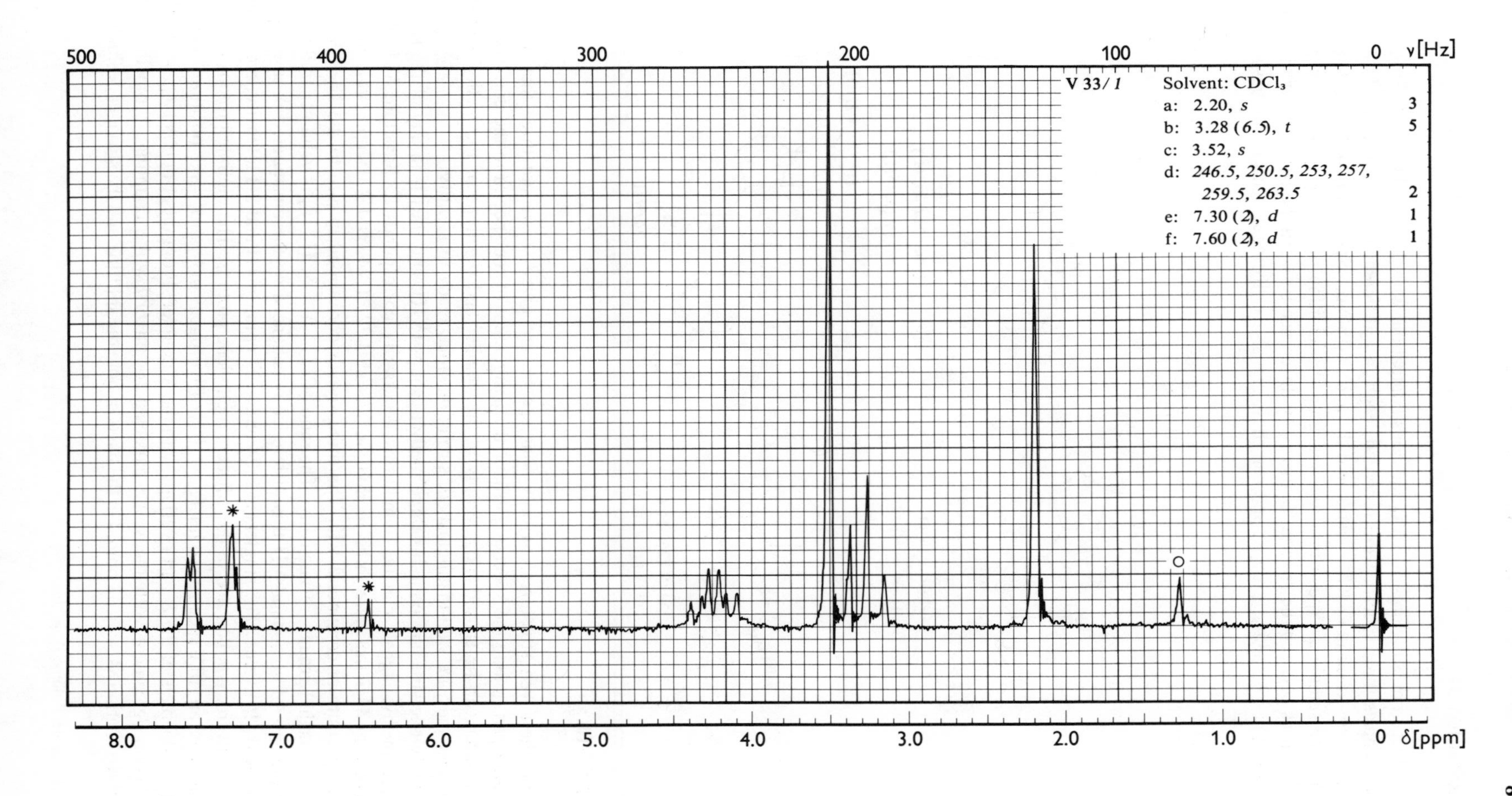

500
400
300
200
100
0
ν[Hz]
V 33/ 1
Solvent: CDCl₃
a: 2.20, s 3
b: 3.28 (6.5), t 5
c: 3.52, s
d: 246.5, 250.5, 253, 257, 259.5, 263.5 2
e: 7.30 (2), d 1
f: 7.60 (2), d 1
*
*
○
8.0
7.0
6.0
5.0
4.0
3.0
2.0
1.0
0
δ[ppm]

PROBLEM 34

Select from Structures **I** to **VII** the ones compatible with Spectra **34**/*1* to *4*. From one of the samples three spectra were recorded, each in a different solvent. Spectra **34**/*1a* and *b* and *c* were taken in $CDCl_3$, acetic acid-d_4, and DMSO-d_6. Spectrum **34**/*2a* was taken from the same sample as Spectrum **34**/*2*, but after the addition of a drop of acid. Interpret the differences between the spectra. Is there any difference between the positions of the signals of 18- and 19- methyl groups, and can these signals be distinguished on this basis?

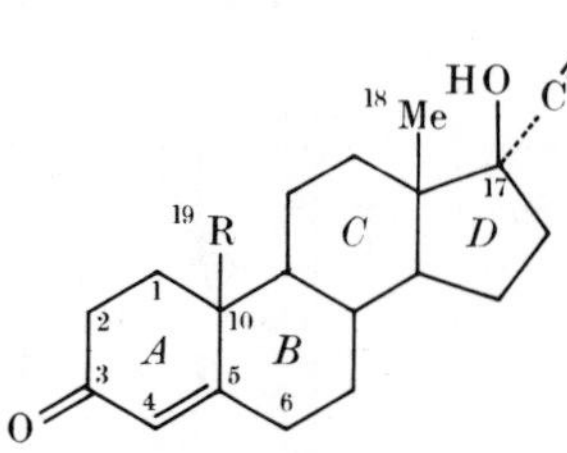

34/I (R = Me)
34/II (R = H)

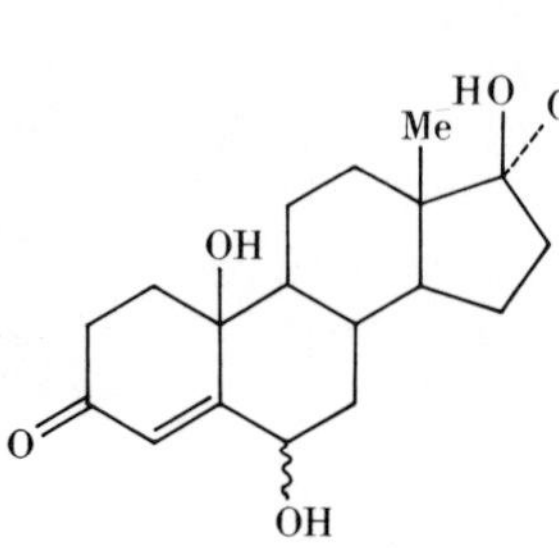

34/III (6βOH$_a$)
34/IV (6αOH$_e$)

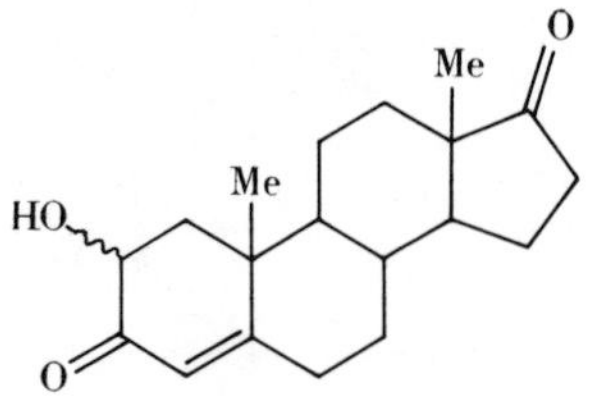

34/V (2βOH$_e$)
34/VI (2αOH$_a$)

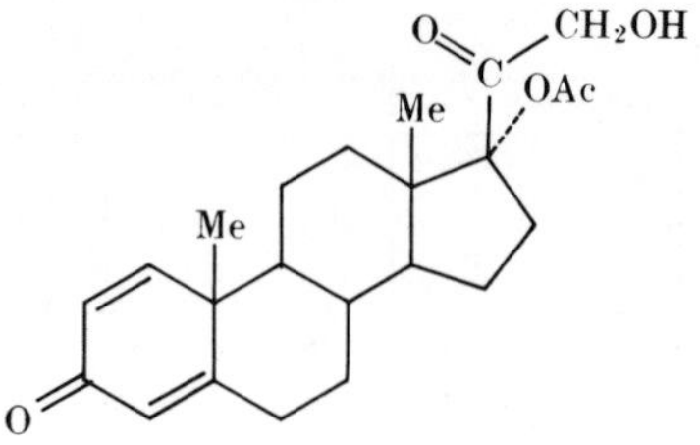

34/VII

ANSWER

Of the seven structures, as only four spectra are given, there are three that have no counterpart among the spectra. Perhaps the assignment of the spectra to the corresponding structures can be started most simply at the number and positions of the methyl signals, because they are well-distinguished, strong maxima, ready to identify.

Only one structure contains three methyl groups (**VII**), and thus if in any of the spectra three methyl signals can be found, this necessarily corresponds to Structure **VII**. As in Spectrum **34**/*3* there are three methyl singlets at 0.72, 1.24, and 2.05 ppm, this must be the spectrum of compound **VII**. The three signals correspond in turn to the 18-, 19-, and acetyl methyl groups, respectively. The ratio $\delta CH_3(19) > \delta CH_3(18)$ is valid generally,* and in the case of compound **VII** this difference is enhanced by the unsaturated A ring raising the $\delta CH_3(19)$ value. The singlets at 3.24 and 4.27 ppm, of 1H and 2H intensity, correspond to the hydroxy proton and to the adjacent methylene protons. The olefinic protons of the A ring have an *ABX* spectrum with $J_{AX} < 1$ Hz (hydrogens in "*para*"-like position), and one of the doublets of the *B* part overlaps with the *X* doublet. From the *A* doublet $J_{AB} = 10$ Hz, in good agreement with the value expected for an *ortho*-like coupling. From the *AMX* approximation, $A = 7.07$ ppm. On the basis of both the splitting and the stronger deshielding,

* Compare Volume II, p. 31.

the A part can be assigned to H-1. In conjugated enones H_β is more deshielded.* Furthermore, J_{BX} = 1.5 Hz and δB = 6.24 ppm. Although the two lines of the B part overlap with the X doublet, from the frequencies of the other two lines and J_{AB}, the chemical shift δB can be calculated accurately (by subtracting $J_{AB}/2$ from the mean frequency of the two B lines). Therefore, we get δB $[(380 + 378.5)/2 - 5]/60 \approx 6.24$ ppm. Finally, $\delta X \approx 6.1$ ppm. Note that the uncharacteristic absorption of the skeletal hydrogens extends from 80 to 170 Hz.

Only two methyl signals (at 0.93 and 1.34 ppm) are present in Spectrum **34**/*4*. Consequently, only one of Structures **I**, **V**, and **VI** can be responsible for it. Due to the unsaturated A ring, the relation $\delta CH_3(19) > \delta CH_3(18)$ holds. In addition, three signals, each corresponding to one hydrogen, appear at higher frequencies. The singlet at 5.83 ppm can be assigned to the olefinic H-4, and the broad maximum at 3.65 ppm can be assigned to the hydroxy proton. Both signals are compatible with any of Structures **I**, **V**, and **VI**. The double doublet at 4.29 ppm, however, excludes the possibility of Structure **I**, because in this compound only the acetylenic hydrogen could give a signal not overlapping with those of the skeletal protons, but this signal would appear upfield (see below) and would not be split.

This signal belongs obviously to H-2 atom, the strong deshielding of which is compatible with the neighborhood of the oxygen (see Tables 26 and 32). The configuration on C-2 can be derived from the magnitude of coupling constants. In compound **V** the hydroxy group is in *equatorial* position, thus H-2 is *axial*. Hence, the magnitudes of couplings with the 1-methylene hydrogens are characteristically different, since one of them corresponds to an *diaxial* coupling, and $J_{aa} > J_{ae} \approx J_{ee}$.** From the frequencies of 247.5, 253.0, 261.0, and 266.5 Hz of the four lines, the two coupling constants are 5.5 Hz and 13.5 Hz (in AMX approximation). The latter, very high value is at the upper limit of the range given for J_{aa} coupling. Thus, it is clear that Spectrum **34**/*4* corresponds to Structure **V**, and Structures **I** and **VI** have no counterparts among the spectra.

The remaining three spectra of unknown origin contain only one methyl signal, and thus their counterparts should be chosen from Structures **II**, **III**, and **IV**.

The assignment of Spectra **34**/*1* and **34**/*2* to Structures **II**, **III**, or **IV** can be based most simply on the fact that Structure **II** yields at most two signals of $\delta > 4$ ppm, whereas compounds **III** and **IV** may give rise to four signals in this region (three hydroxy groups + olefinic hydrogen). On this basis Spectrum **34**/*1* is compatible only with Structure **II**. In the spectrum of the $CDCl_3$ solution (**34**/*1*), the 18-methyl signal appears at 0.93 ppm, the very sharp singlet of the ethynyl proton appears at 2.58 ppm, i.e., in the expected region,*** and the olefinic signal (split by long-range interactions) appears at 5.88 ppm. They hydroxy signal cannot be assigned, and the absorptions of skeletal protons are in the region of 1.0 to 2.5 ppm.

In the spectra of the acetic acid-d_4 and DMSO-d_6 solution (**34**/*1a* and *1b*), the 18-methyl signals (at 0.92 and 0.83 ppm) and the olefinic signals (at 5.95 and 5.80 ppm) have practically unchanged shifts, but the ethynyl signal is shifted considerably, mainly in DMSO-d_6 (to 2.79 and 3.20 ppm, respectively). This significant shift [+0.21 and +0.62 (!) ppm] is characteristic of acetylenic hydrogens.† In the spectrum of DMSO-d_6 solution, as usual, the OH-signal is also readily identified at 5.28 ppm (the signal at 3.38 ppm arises from the water content of the solvent).

* Compare Volume II, p. 52.

** See Volume II, p. 27.

*** See Volume II, p. 46.

† See Volume II, p. 47.

The only remaining problem is to pair Spectrum **62**/*2* with one of Structures **III** and **IV**, i.e., to decide whether the 6-OH group is *equatorial* or *axial*. The signal of the *geminal* H-6 is rather broad and structureless with a half-band width of about 13 Hz. Provided that the coupling constants are similar to those of compound **V** (J_{aa} = 13, J_{ae} = 5.5 Hz), the width of the H-6 signal must be about $J_{aa} + J_{ae} \approx 19$ Hz if H-6 is *axial*. On the other hand, if it is *equatorial*, $J_{ae} + J_{ee} \approx 11$ Hz. Thus, Structure **II** with H-6*e* and OH*g* is proved.

Upon addition of acid (Spectrum **34**/*2a*) to the DMSO-d_6 solution, the signals corresponding to the three hydroxy groups, and, of course, to the ethynyl hydrogen, can be recognized immediately, since the former disappear and the latter undergo a substantial diamagnetic shift. Accordingly the doublet at 5.68 ppm (J = 4 Hz) and the singlets at 4.63 and 5.28 ppm belong to the hydroxy groups, and the signal at 3.27 ppm (after acid addition at 3.10 ppm) belongs to the acetylenic hydrogen. The doublet may, of course, arise only from the 6-hydroxy group (the substituents in positions 10 and 17 are attached to *quaternary* carbons), and the splitting in DMSO-d_6 can be explained by a CH-OH coupling.* Accordingly, the H-6 signal also becomes narrower after acid addition. The true width of the signal is, consequently, only 7 Hz, which finally excludes the possibility of Structure **IV**, i.e., the *equatorial* position of the 6-hydroxy group. Presumably, due to the presence of acetylene group, δOH(17) $<\delta$OH(10), i.e., the maximum at 4.63 ppm corresponds to the former, and the signal at 5.28 ppm corresponds to the latter group. The singlets of the 18-methyl group and of the olefinic H-4 atom appear at 0.83 and 5.80 ppm. The skeletal hydrogens absorb in the region of 60 to 150 Hz. The water content of the solvent yields a signal at 3.3 ppm, which disappears upon acid addition.

* See Volume II, p. 101.

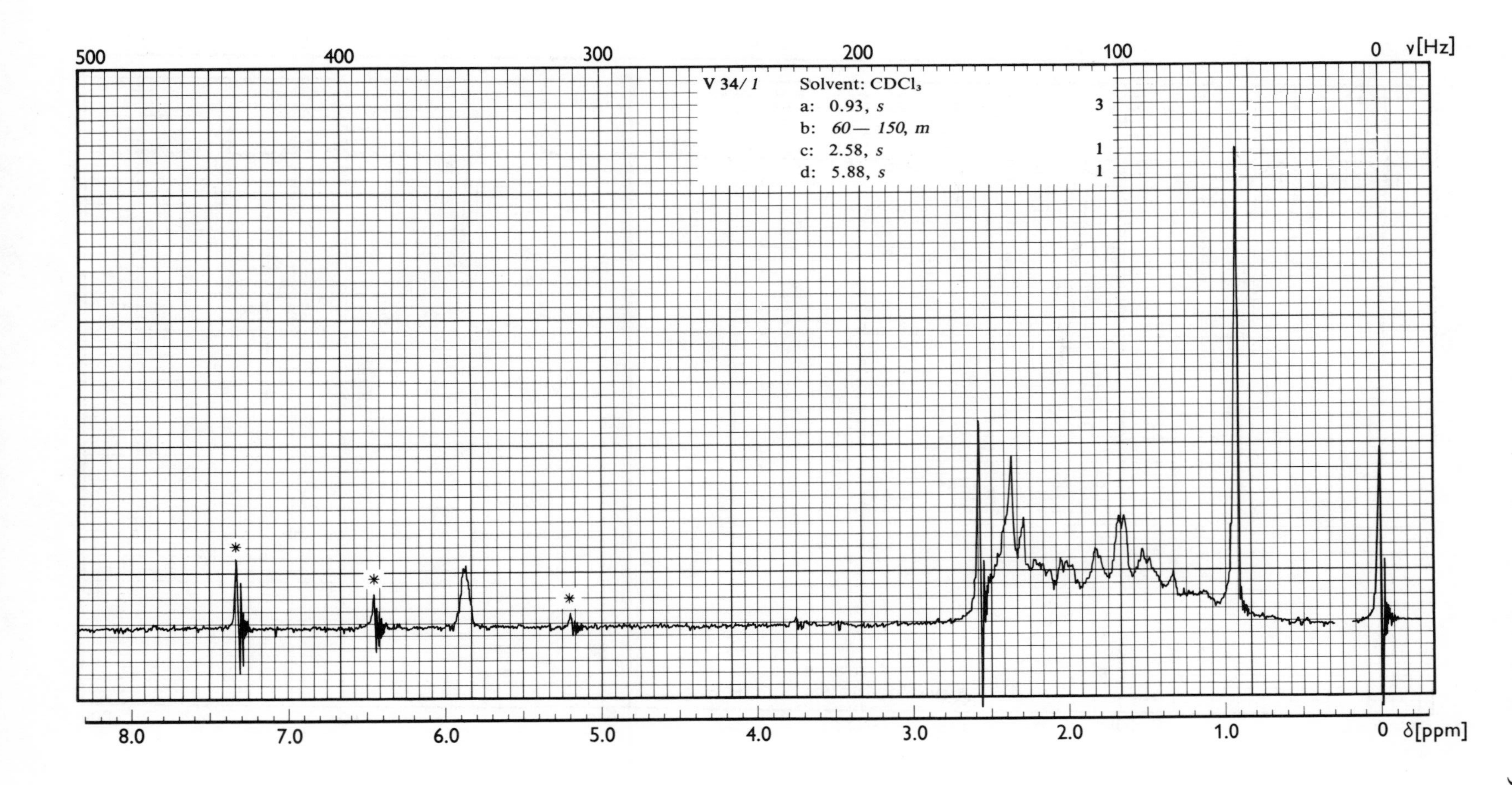
V 34/1
Solvent: $CDCl_3$
a: 0.93, s 3
b: 60— 150, m
c: 2.58, s 1
d: 5.88, s 1
ν[Hz]
500
400
300
200
100
0
δ[ppm]
8.0
7.0
6.0
5.0
4.0
3.0
2.0
1.0
0

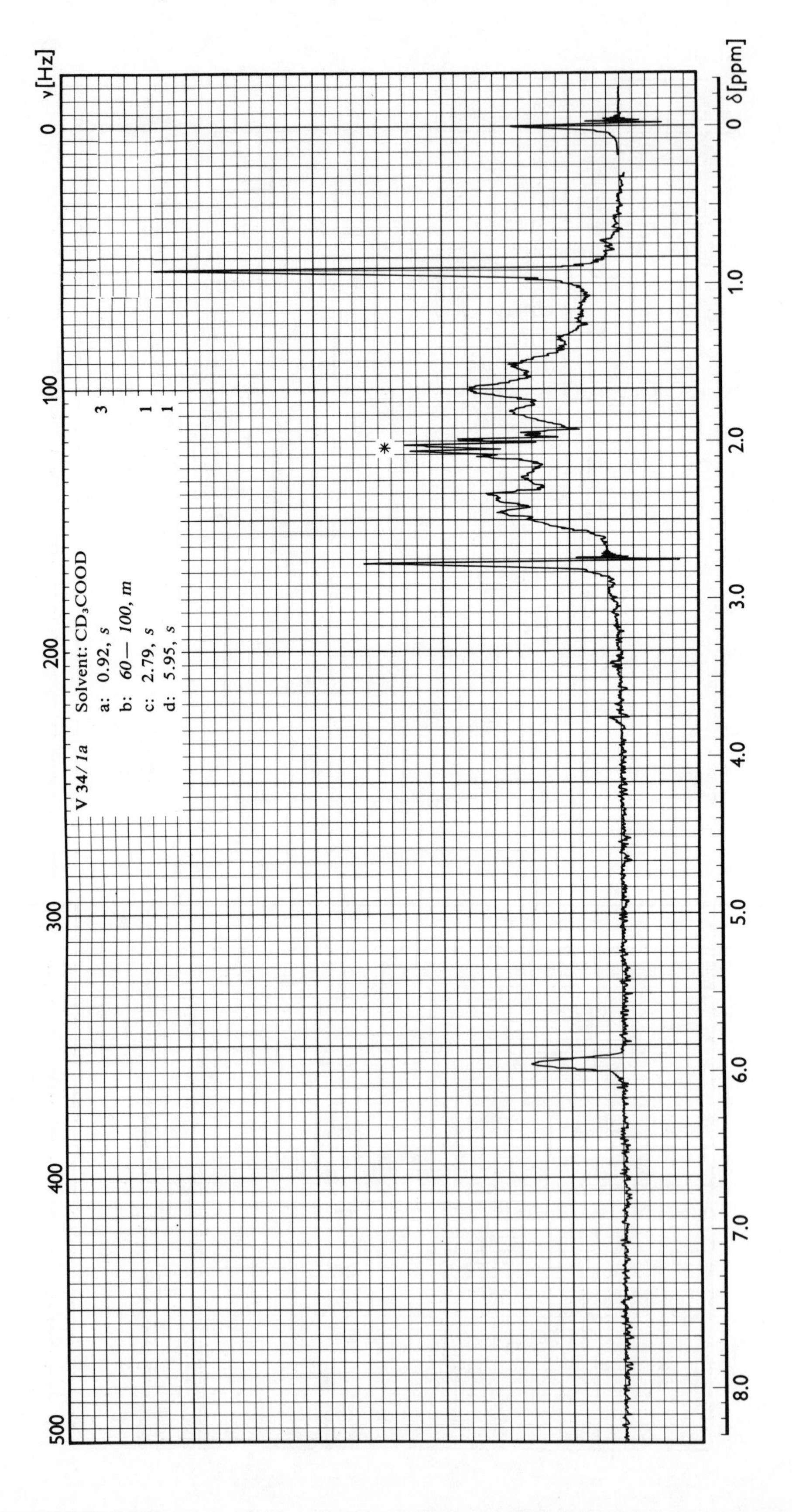
V 34/ 1a
Solvent: CD_3COOD
a: 0.92, s 3
b: 60— 100, m
c: 2.79, s 1
d: 5.95, s 1
*
ν[Hz]
0
100
200
300
400
500
δ[ppm]
0
1.0
2.0
3.0
4.0
5.0
6.0
7.0
8.0

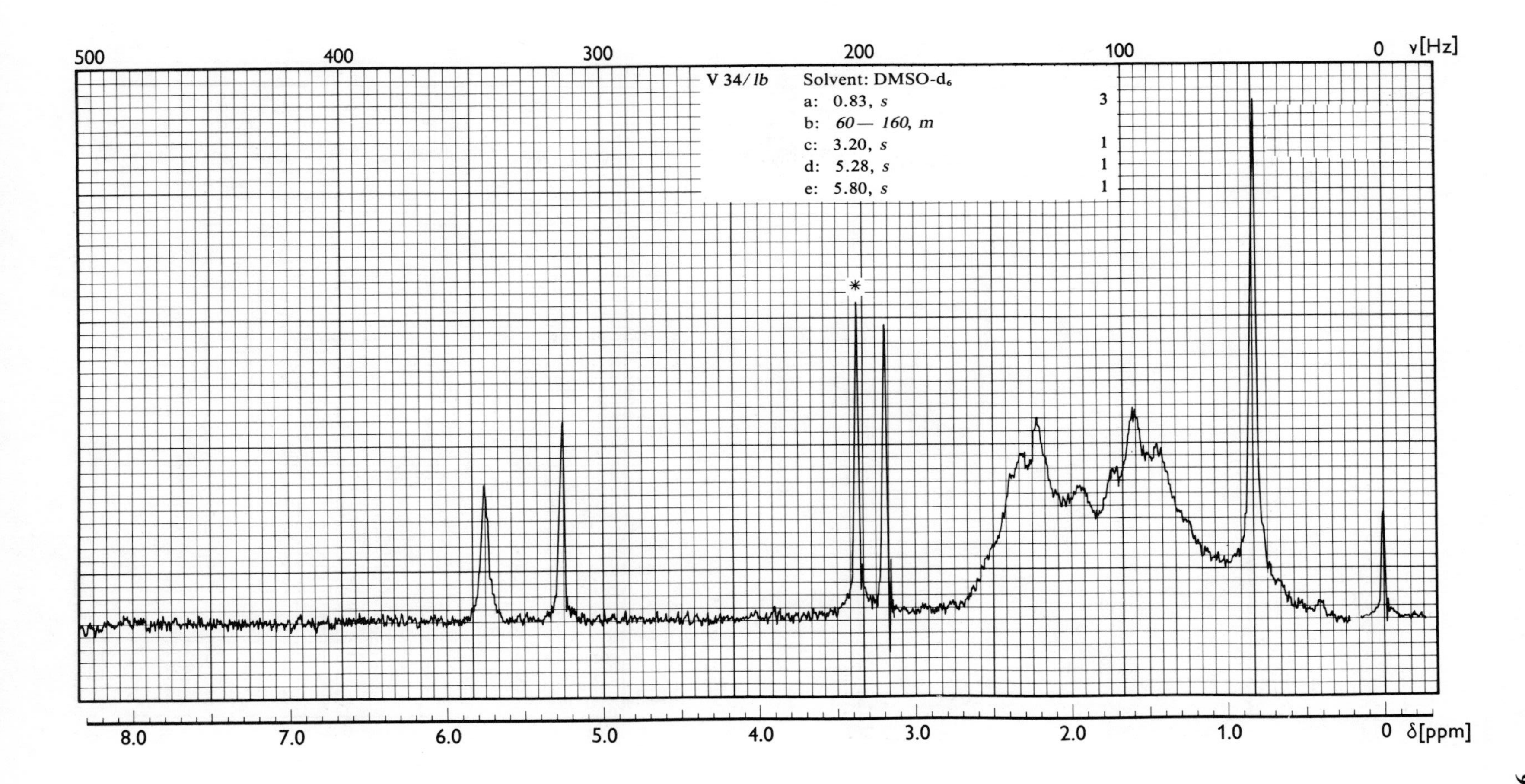

500
400
300
200
100
0
ν[Hz]
V 34/Ib
Solvent: DMSO-d6
a: 0.83, s 3
b: 60 — 160, m
c: 3.20, s 1
d: 5.28, s 1
e: 5.80, s 1
*
8.0
7.0
6.0
5.0
4.0
3.0
2.0
1.0
0
δ[ppm]

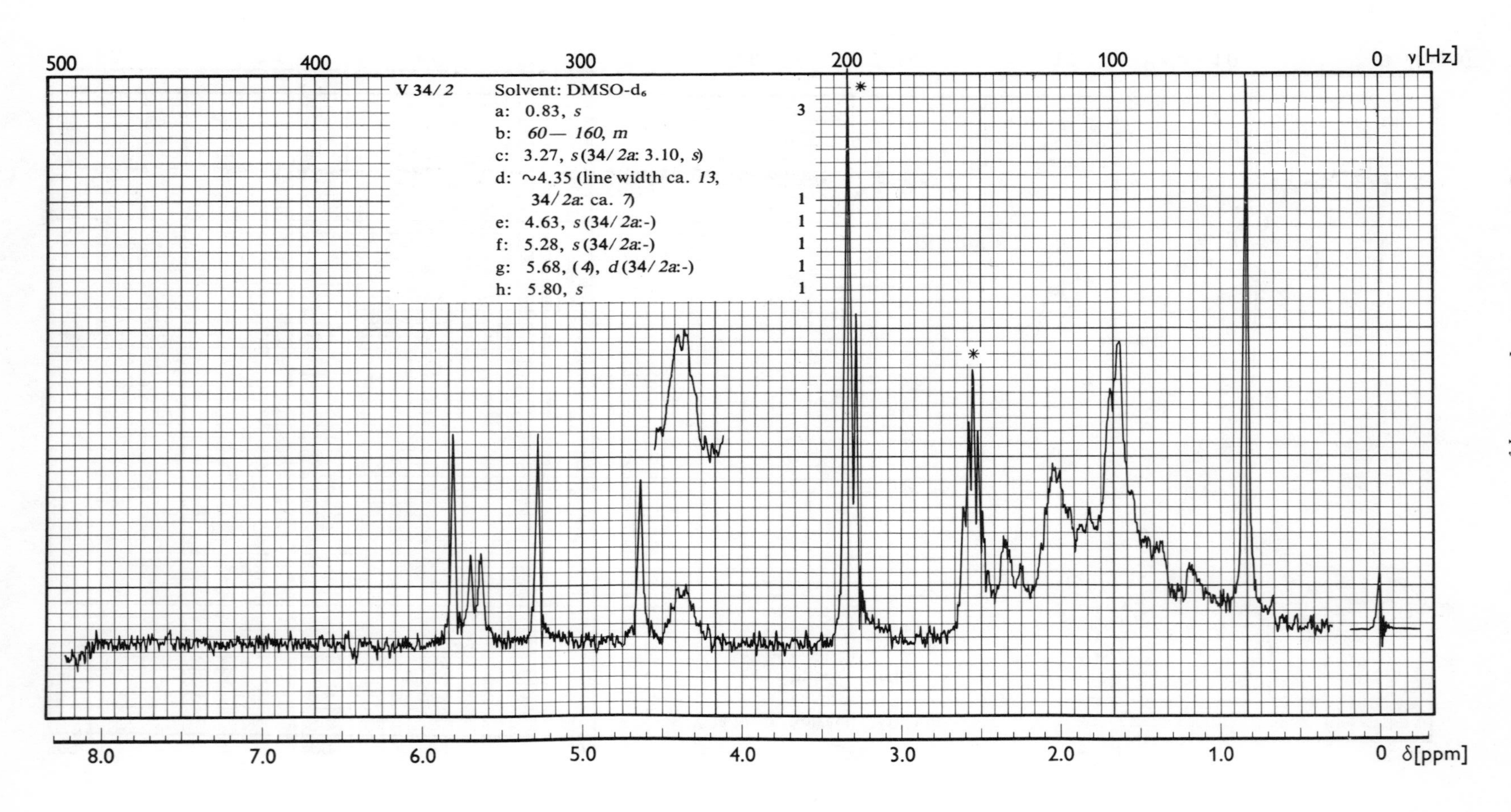
500
400
300
200
100
0 ν[Hz]
V 34/2
Solvent: DMSO-d_6
a: 0.83, s 3
b: 60— 160, m
c: 3.27, s(34/2a: 3.10, s)
d: ~4.35 (line width ca. 13, 34/2a: ca. 7) 1
e: 4.63, s (34/2a:-) 1
f: 5.28, s (34/2a:-) 1
g: 5.68, (4), d (34/2a:-) 1
h: 5.80, s 1
*
*
8.0
7.0
6.0
5.0
4.0
3.0
2.0
1.0
0 δ[ppm]

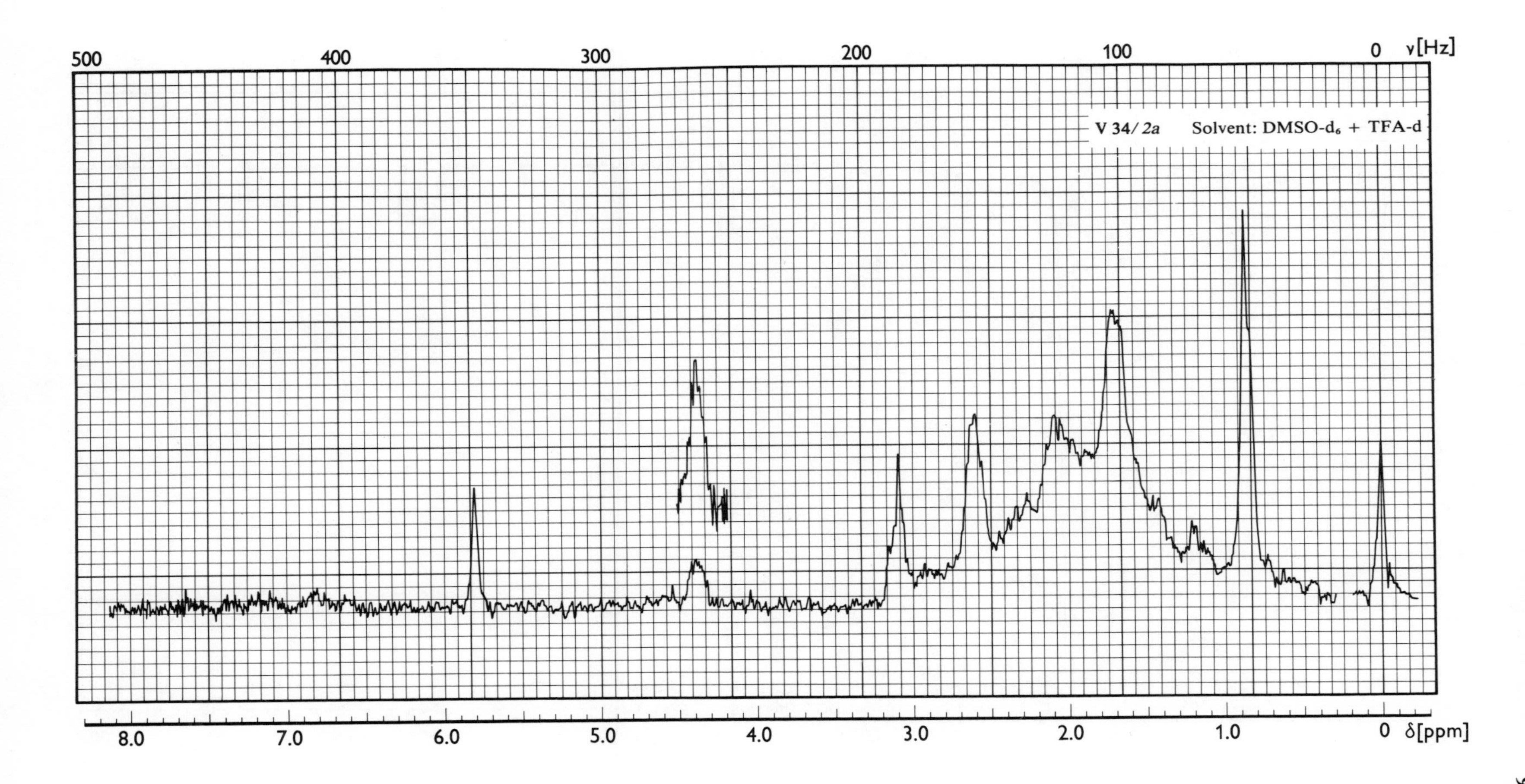
V 34/ 2a
Solvent: DMSO-d6 + TFA-d
ν[Hz]
500
400
300
200
100
0
δ[ppm]
8.0
7.0
6.0
5.0
4.0
3.0
2.0
1.0
0

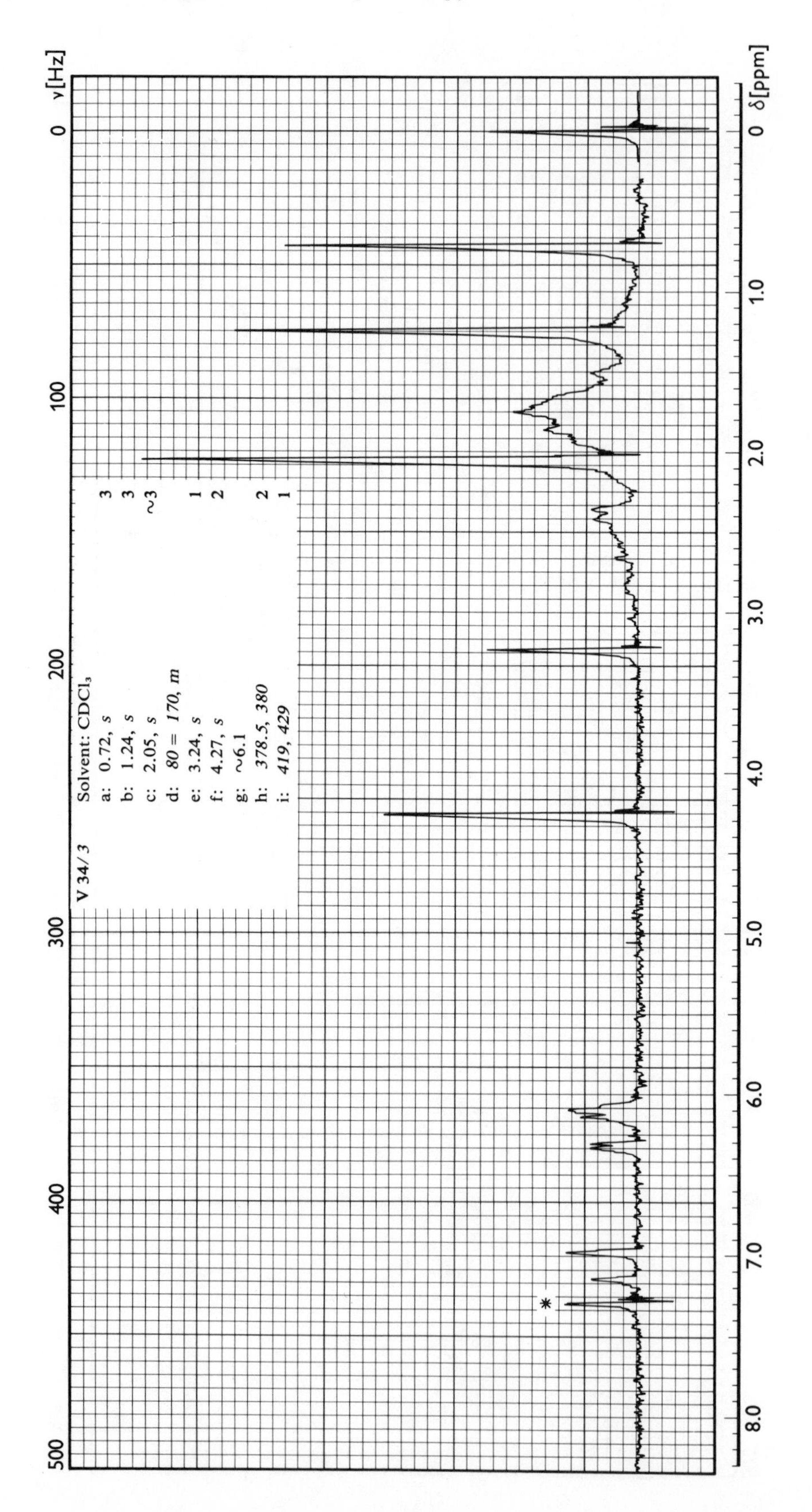

ν[Hz]
0
100
200
300
400
500
δ[ppm]
0
1.0
2.0
3.0
4.0
5.0
6.0
7.0
8.0
V 34/3
Solvent: CDCl₃
a: 0.72, s 3
b: 1.24, s 3
c: 2.05, s ~3
d: 80 = 170, m
e: 3.24, s 1
f: 4.27, s 2
g: ~6.1
h: 378.5, 380 2
i: 419, 429 1
*

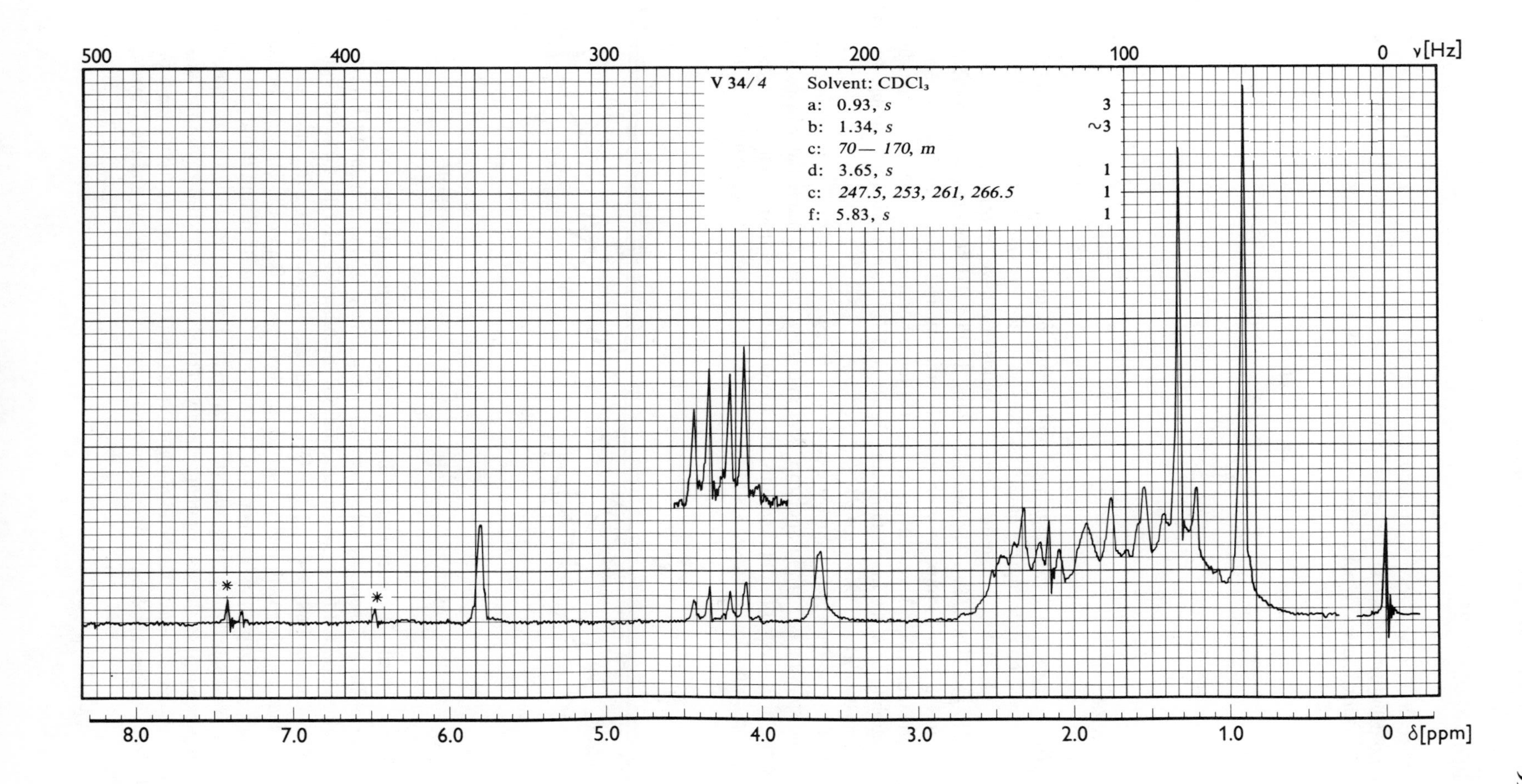
500
400
300
200
100
0 ν[Hz]
V 34/4
Solvent: $CDCl_3$
a: 0.93, s 3
b: 1.34, s ~3
c: 70— 170, m
d: 3.65, s 1
c: 247.5, 253, 261, 266.5 1
f: 5.83, s 1
*
*
8.0
7.0
6.0
5.0
4.0
3.0
2.0
1.0
0 δ[ppm]

PROBLEM 35[120,1318]

Spectra **35/*1*** and **35/*2*** were recorded from the *cis* and *trans* isomers of 1-hydroxy-2-carbomethoxytetraline (**I**). Determine which of the spectra corresponds to the *cis* and which to the *trans* isomer. What are the conformations of the compounds?

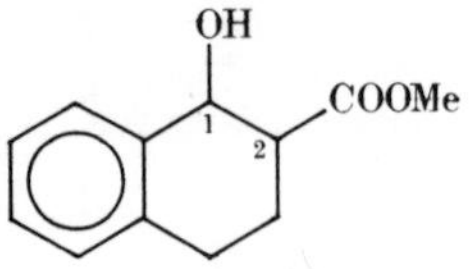

35/I

ANSWER

The isomers may be distinguished on the basis of the chemical shift and splitting of the H-1 signals, considering the empirical rules valid for cyclohexanes: $\delta H_e > \delta H_a$ and $J_{aa} > J_{ae} \approx J_{ee}$.* Both the *cis* and the *trans* compounds may have, in principle, two conformations (**IIa**, **IIb** and **IIIa**, **IIIb**). The formulas show that the H-1 atom in the *cis* isomer may enter only an *e′a* or *a′e* coupling with the only adjacent H-2 atom and that it is *equatorial* in conformation **IIa**, whereas *axial* in **IIb**. On the other hand, in the *trans* isomer there is an *e′e* coupling or in the more stable conformation, **IIIb**, there is an *a′a* coupling. In the latter conformation H-1 is *quasi-axial*.

The H-1 signal appears almost exactly at 5.0 ppm (5.02 and 5.03 ppm) in both spectra, but the splitting is 3 Hz in Spectrum **35/*1*** and 9 Hz in **35/*2***. In the latter case, as indicated by the high value of coupling constant, the coupling is almost certainly of type *a′a*, and thus Spectrum **35/*2*** corresponds to the *trans* isomer, which is conformation **IIIb**.

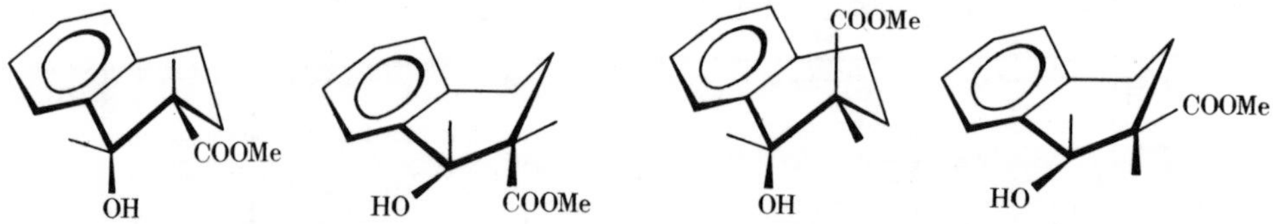

35/II (a, b) **35/III** (a, b)

The identical δH-1 chemical shifts in the spectra suggest that the conformation of the *cis* isomer is **IIb**, and therefore substituent 2 is *axial*. The coupling constant of 3 Hz corresponds to a value expected for an *a′e* interaction. Of the other signals, the methoxy signal appears at 3.75 ppm in both spectra, the OH signal can be found at 3.25 ppm, the methylene multiplets, together with the H-2 signal, are between 110 and 180 Hz, and the lines of the aromatic hydrogens are between 410 and 470 Hz. It is worth noting that in the spectrum of the *trans* isomer the signal of one of the aromatic protons (certainly the one adjacent to the hydroxy group) is shifted paramagnetically. This fact is due probably to the anisotropy of the C–O bond, since this bond, in conformation **IIIb**, is coplanar with the aromatic ring fixed by the intramolecular hydrogen bond between the carbonyl and the 1-hydroxy group.

* See Volume II, p. 27.

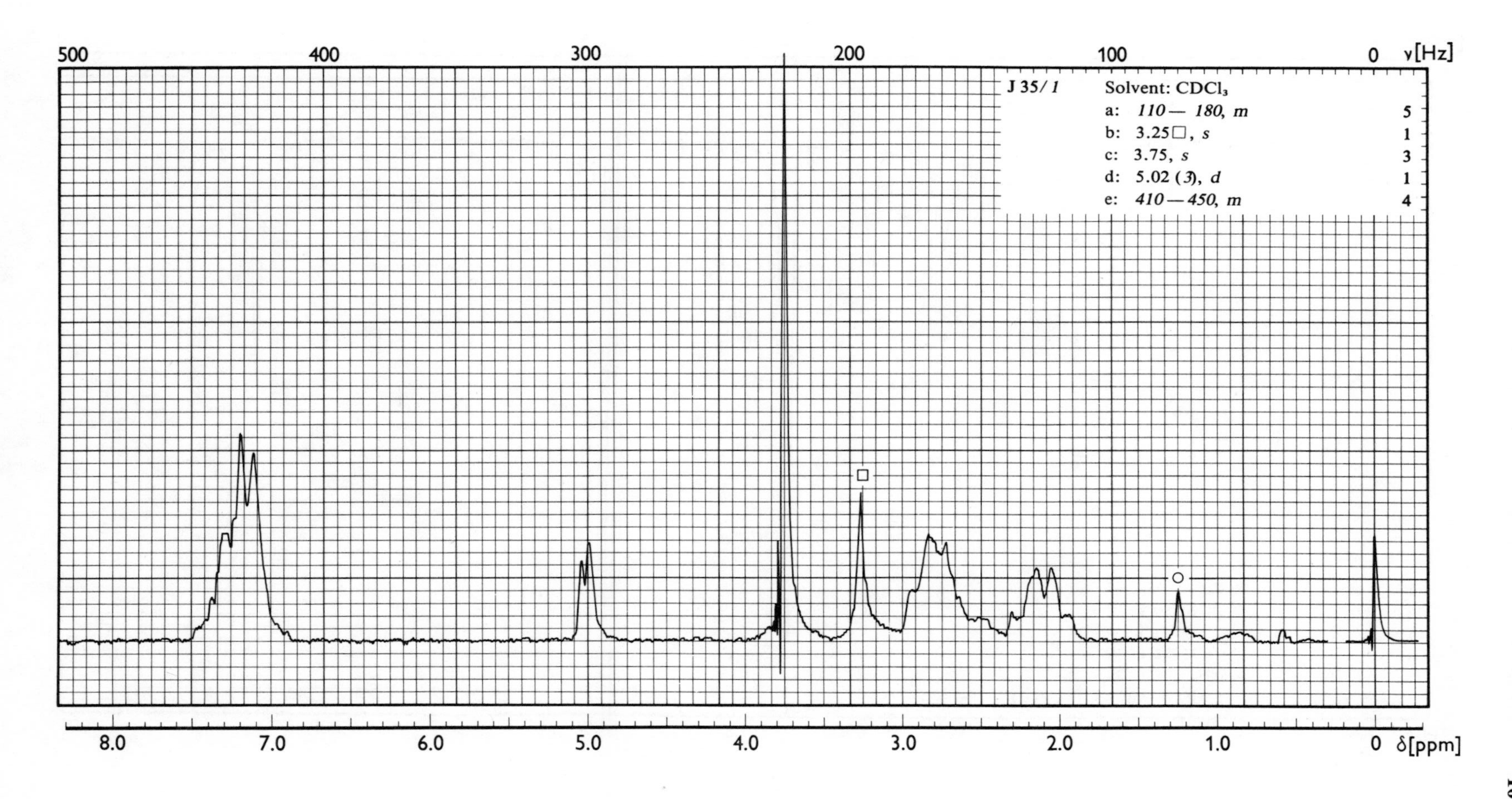

J 35/1
Solvent: $CDCl_3$
a: 110— 180, m 5
b: 3.25□, s 1
c: 3.75, s 3
d: 5.02 (3), d 1
e: 410—450, m 4
500
400
300
200
100
0 ν[Hz]
8.0
7.0
6.0
5.0
4.0
3.0
2.0
1.0
0 δ[ppm]

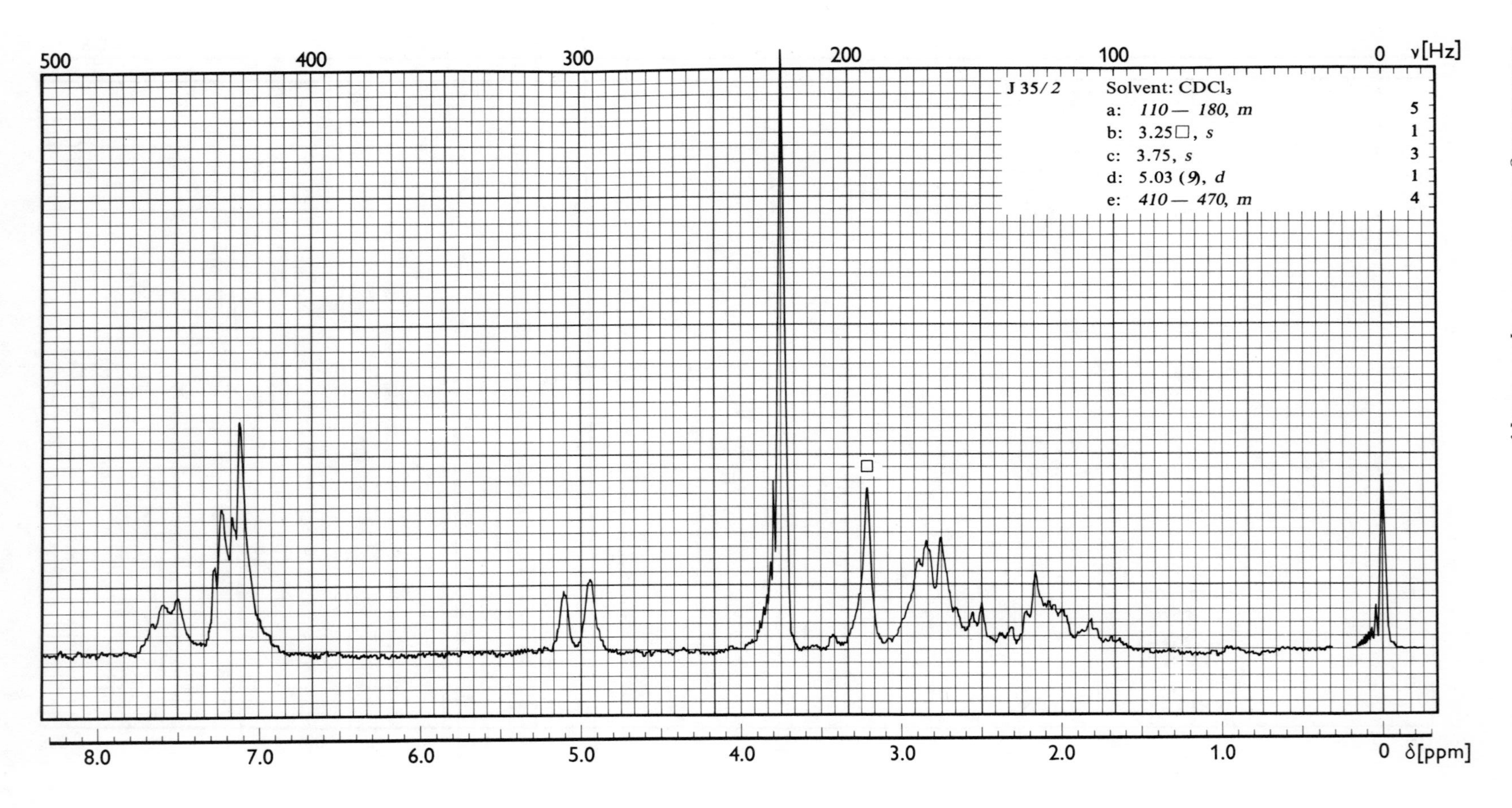
500
400
300
200
100
0 ν[Hz]
J 35/2
Solvent: CDCl₃
a: 110 — 180, m 5
b: 3.25□, s 1
c: 3.75, s 3
d: 5.03 (9), d 1
e: 410 — 470, m 4
□
8.0
7.0
6.0
5.0
4.0
3.0
2.0
1.0
0 δ[ppm]

PROBLEM 36[1318,1341]

In a condensation reaction of ω-bromoacetophenone and *ortho*-nitrobenzaldehyde, the formation of compounds **I** to **III** is conceivable. Actually two products are formed simultaneously; on the basis of Spectra **36**/*1* and **36**/*2*, determine the structures.

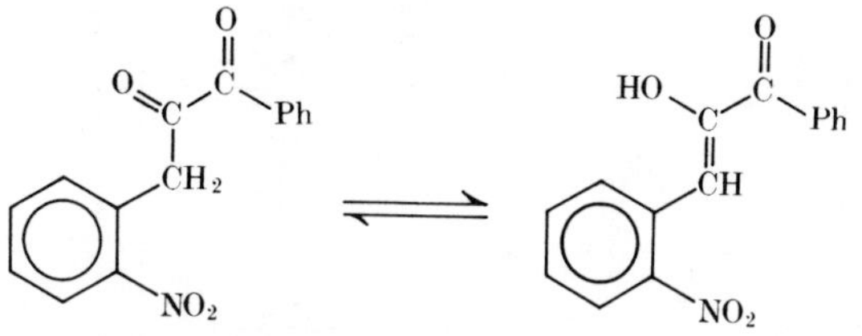

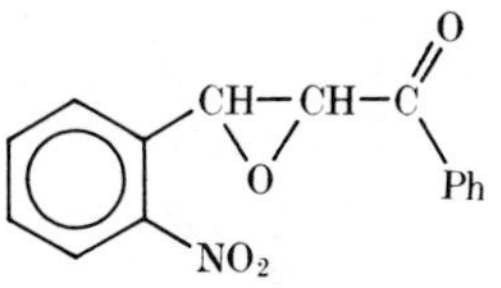

36/I **36/II** **36/III**

ANSWER

Disregarding the multiplet of the aromatic protons between 430 and 500 Hz in both spectra only one group of signals can be found: an *AB* line group in Spectrum **36**/*1* and an *AX* doublet pair in Spectrum **36**/*2*. The corresponding spectrum parameters are as follows: $\delta A = 5.07$ ppm, $\delta B = 4.83$ ppm, $J_{AB} = 4.5$ Hz; and $\delta A = 4.63$ ppm, $\delta = 4.23$ ppm, $J_{AX} = 2$ Hz.

The *AB* and *AX* spectra may be due neither to the methylene groups of compound **I** (which would give a singlet or an *AB* spectrum with J_{AB} about 16 Hz [*geminal* coupling of methylene protons]),* nor to the olefinic proton of compound **II** (the singlet of which would appear at higher frequencies), consequently, we are restricted to Structure **III**.

The two products are the *cis* and *trans* isomers, **IV** and **V** according to the substitution of the oxyrane ring. The coupling constants of the ring protons of 1,2-disubstituted oxiranes are characteristically different.** Therefore, it is also certain that Spectrum **36**/*1* corresponds to the *cis*, and Spectrum **36**/*2* corresponds to the *trans* isomer, because in the former case a coupling constant of 4.6 ± 0.5 Hz and in the latter a value of 2.2 ± 0.5 Hz can be expected on the basis of the literature data.[1187]

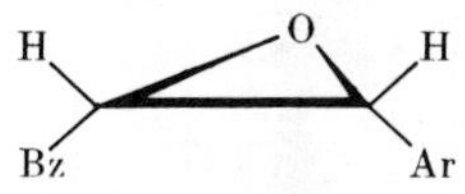

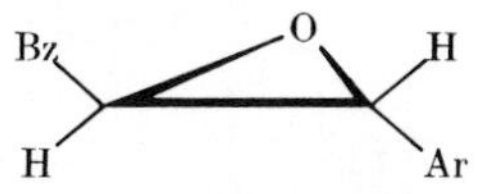

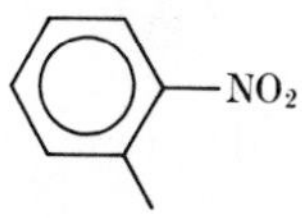

36/IV **36/V**

Note that the change in the ratio $J/\Delta\nu$ causes an transition between the *AB* and *AX* spectra: $J/\Delta\nu\ (cis) = 4.5/14.4 \approx 0.3 > 0.1$: *AB* spectrum, whereas $J/\Delta\nu(trans) = 2/24 = 0.08 < 0.1$: *AX* spectrum. The sample of the *trans* isomer is slightly contaminated with the *cis* isomer.

* Compare Volume I, p. 59.

** See Volume I, last line on p. 14.

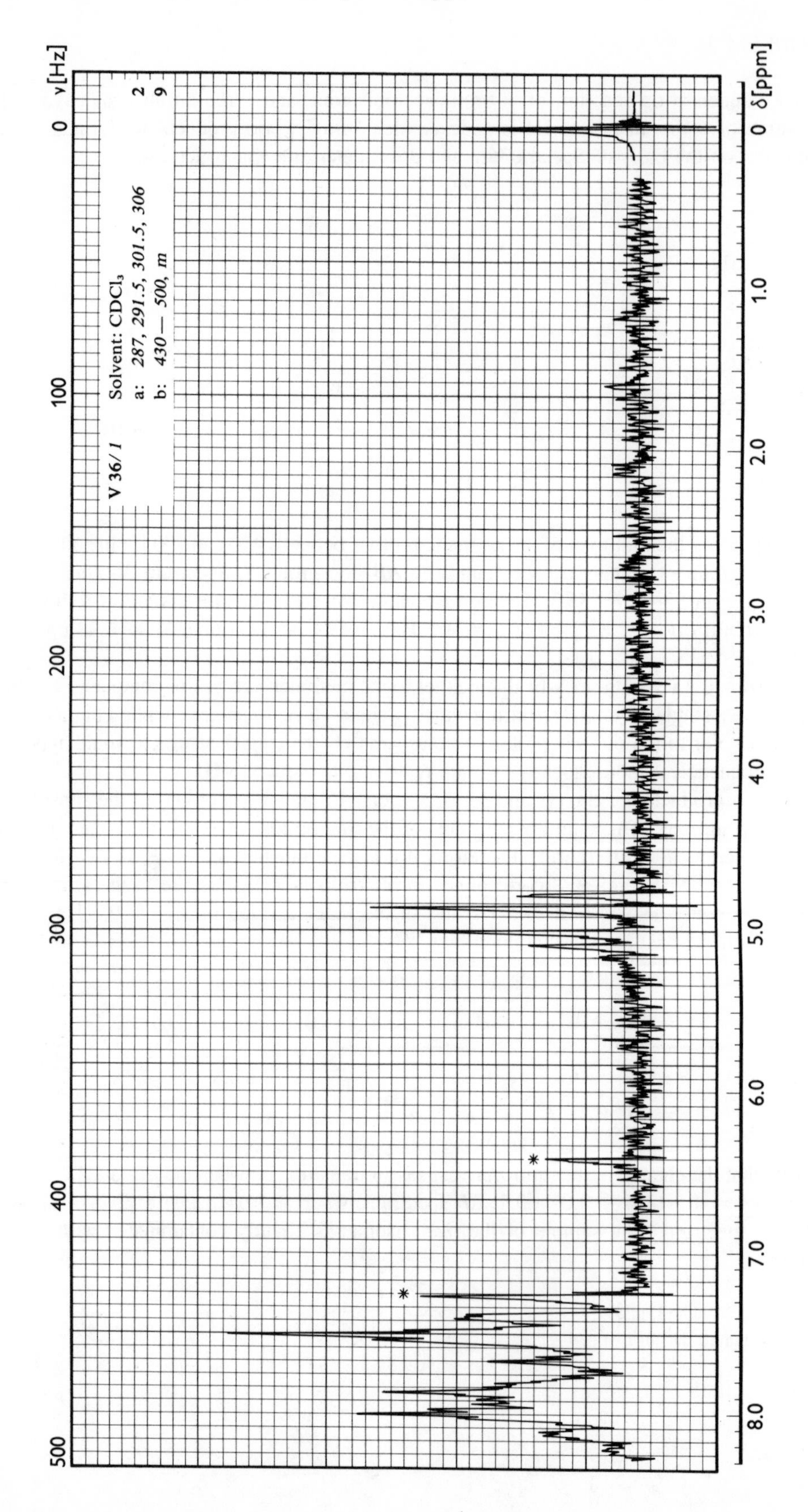

V 36/1
Solvent: $CDCl_3$
a: 287, 291.5, 301.5, 306 2
b: 430— 500, m 9
ν[Hz]
0
100
200
300
400
500
δ[ppm]
0
1.0
2.0
3.0
4.0
5.0
6.0
7.0
8.0

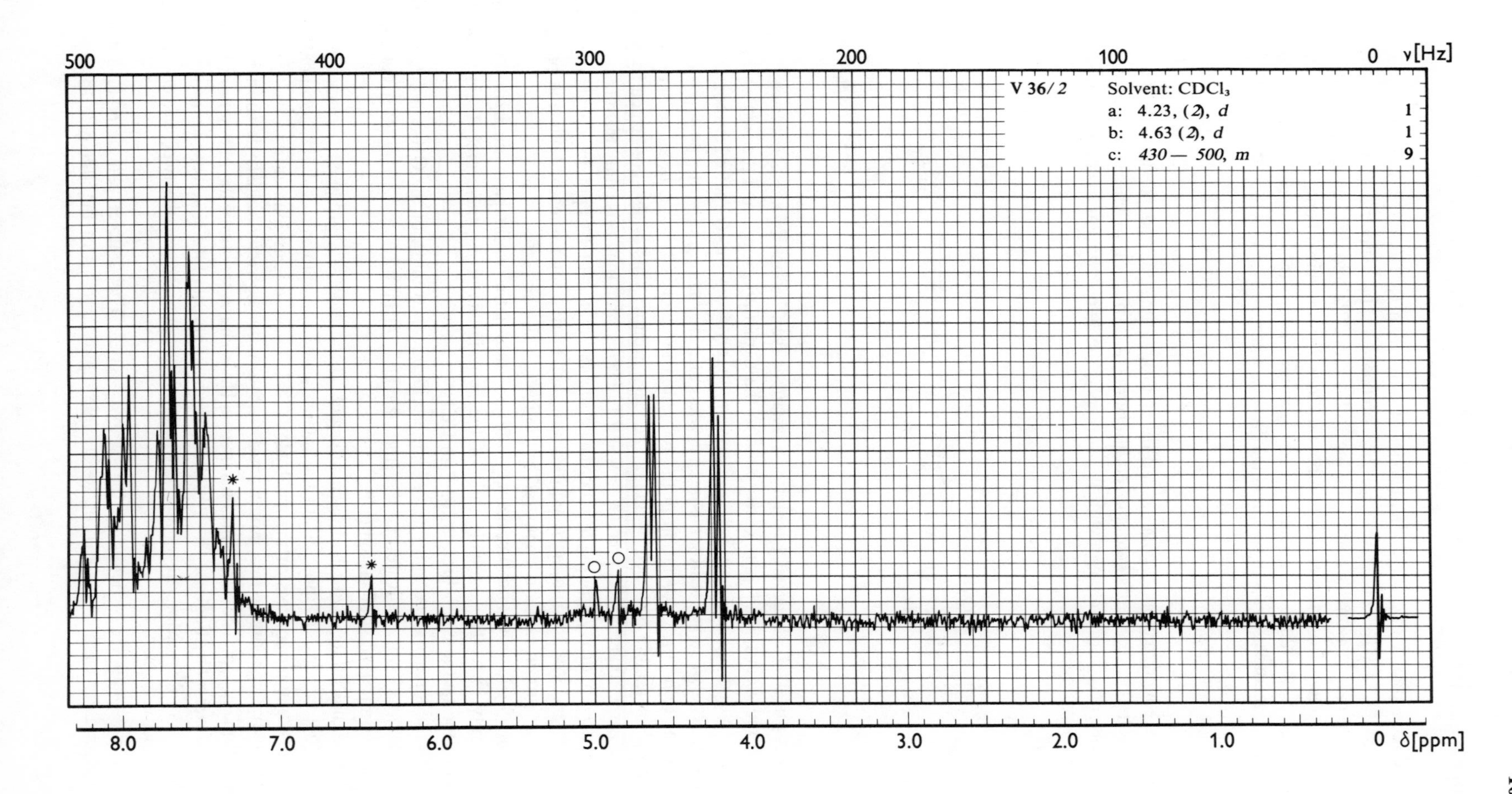
500
400
300
200
100
0 ν[Hz]
V 36/2
Solvent: CDCl₃
a: 4.23, (2), d 1
b: 4.63 (2), d 1
c: 430 — 500, m 9
8.0
7.0
6.0
5.0
4.0
3.0
2.0
1.0
0 δ[ppm]

PROBLEM 37[390]

By reduction with $LiAlH_4$ of malonester derivative RCH(COOR′) an aldehyde-ester of type ROOC–CHR–CHO was to be prepared. The spectrum of the product is **37**/*1*. The pair of isomers corresponding to Spectra **37**/*2* and **37**/*3* have been obtained by a simple reaction from the above compound. Identify the substituents R and R′. Derive the structures of the three compounds, and interpret the changes observable in Spectra **37**/*1a* and **37**/*1b*, recorded in acetone-d_6 and DMSO-d_6, whereas **37**/*1* was recorded in CCl_4 solution.

ANSWER

As the reduction product contains a β-dioxo group, the solvent dependence of the spectrum indicates the presence of a keto-enol equilibrium (**Ia** ⇌ **b** ⇌ **c**), in which the relative amounts of tautomers **a** and **b** change with the solvent. The interpretation of the spectra should be based on this fact.

R–CH(COOR′)(CHO) ⇌ R–C(C(=O)OR′)=CH–O–H (chelated) ⇌ R–C(C(=O)OR′)=C(H)–OH

37/Ia **37/Ib** **37/Ic**

Spectrum **37**/*1* of the CCl_4 solution contains seven groups of signals, of which the six lines in the vicinity of 1 ppm correspond to the *C*-methyl group, and the two or three very closely spaced lines around 3.8 ppm arise probably from the *O*-methyl protons. This assignment is also supported by the intensity ratio (2:1). Accordingly, the presence of two *C*-methyl and one methoxy group in the reduction product is certain, in agreement with the expected invariable presence of one of the ester groups. This is also confirmed by the two doublets at 9.70 and 11.55 ppm (J = 4 and 13 Hz), which may arise only from the aldehyde and chelated hydroxy groups. In accordance, the doublet at 11.55 ppm disappears upon D_2O addition, and due to tautomeric equilibrium the relative intensities of both line pairs give noninteger proton ratios if compared to the methyl signals.

The aldehyde proton occurs, namely, only in tautomer **a**, while the enolic hydroxy proton occurs only in **b**. The intensity ratio of the two signals gives, therefore, directly the **a**:**b** ratio of the tautomers, which is ~3:2 in CCl_4.

Both the intensity and splitting, which are identical to those of the OH-doublet, and the large chemical shift indicate that the line pair at 7.08 ppm (J = 13 Hz) must correspond to the olefinic proton of tautomer **b**.

The signal of the *meso* hydrogen adjacent to the aldehyde group in tautomer **a** is the double doublet at 3.05 ppm. Both the intensity and the splitting of 4 Hz, identical to the corresponding parameters of the aldehyde signal, prove this assignment.

The two neighbors of this proton must therefore be the *vicinal* aldehyde substituent and the methine group. The coupling constant of 8 Hz is characteristic of the interactions between saturated *vicinal* hydrogens. This completes the assignment of the signals and proves structure **I** (R = *i*Pr, R = Me). Accordingly, the combined intensity of the two adjacent methine hydrogens is 1 + 3/5.

These explain the structure of the methyl signal. The doublet pair of equal intensity corresponds to each of the two diastereotopic methyl groups of the aldehyde form **a**, and the enol form gives rise to the third line pair. The spacing of two line pairs is, accordingly, the same (approximately 7 Hz), wherefrom the three chemical shifts are 1.01, 1.04, and 1.11 ppm. The complex structure of the methine signal of the isopropyl group is due to similar reasons: in this signal the septet corresponding to the enol form **b** and the octet arising from the aldehyde form **a** (the coupling constants with the hydrogens of the aldehyde and methyl groups are approximately the same) are superimposed.

The splitting of the methoxy signal is also due to the simultaneous presence of the above tautomers. The stronger line at 3.78 ppm corresponds to tautomer **a**, whereas the weaker at 3.81 ppm corresponds to the **b** form.

Spectrum **37**/*1a* is quite analogous to Spectrum **37**/*1*, the only difference occurs in the relative intensities. Consequently, only the ratio of the tautomers, has changed, actually in favor of **a** (**a**:**b** = 3:1).

In Spectrum **37**/*1b* of the DMSO-d_6 solution, however, more essential changes can be observed. The *C*-methyl, methoxy, and methine signals have become more complicated, and two new broad signals appear around 7.65 and 10.40 ppm.

Both these signals and their doublet counterparts, assigned to the enol form, at 7.25 and 11.48 ppm disappear upon D_2O addition, and the intensities of the signals of the aldehyde form **a** increase. It is obvious to assume that the new signals correspond to the *trans (E)* enol (**Ic**), which is incapable of chelation and which participates in the tautomeric equilibrium to an observable extent only in DMSO-d_6 solution. The evidence is as follows:

1. The vinyl proton is more deshielded than in **Ib** (R = *i*Pr, R′ = Me), due to the fact that in **c** form the carbonyl is in *cis* position with respect to the olefinic hydrogen, whereas in the *cis* enol the situation is reverse. A carbonyl group has, however, a stronger deshielding effect on the *cis vicinal* olefinic hydrogens than on that of the *trans* ones (see Table 36).
2. There is no splitting attributable to a =CH–OH coupling, whereas it occurs with the **a** form (of 13 Hz). The reason is that in the *E* enol **Ic** a fast intermolecular exchange of hydroxy protons may proceed, which averages the spin states, whereas the chelate structure formed in the *Z* enol retards these exchange processes.* The relative shielding of the hydroxy proton is also due to the weaker association.**
3. The disappearance of hydroxy and ethyl signals belonging to form **Ic** upon D_2O addition also confirms the *E* enol structure. Namely, the tautomeric equilibrium in the solution of increased polarity is shifted towards the oxo-form **Ia**, and the proton exchange becomes faster causing the disappearance of the hydroxyl signals.

From the relative intensities of the aldehyde proton and the two olefinic protons, the **a**:**b**:**c** ratio is about **2**:**2**:**3**.

In the very similar Spectra **37**/*2* and **37**/*3*, the signals of the isopropyl group are invariably present. The doublet and septet structure of the methyl and methine signals, respectively, indicate that these derivatives do not take part in tautomeric equilibrium. In accordance, the signal of acidic protons is absent. The singlets of 1H intensity at 6.25 and 7.10 ppm, respectively, suggest that the samples are derivatives of the enol forms **b** and **c**, thus they are *Z-E* isomers. The isomer containing the more shielded olefinic hydrogen is probably the derivative of **Ib**, in which the former and the carbonyl are in *trans* position.

* See Volume II, p. 102.
** See Volume II, p. 101.

Since both spectra (**37**/*2* and **37**/*3*) contain two methoxy signals, each corresponding to three hydrogens, a methoxy substituent must also be present in addition to the methyl ester group. As there is no other signal in the spectra, Spectrum **37**/*2* belongs to Structure **IIa** and Spectrum **37**/*3* belongs to Structure **IIb**. This assignment following from the relative shielding of the olefinic protons is also supported by further facts.

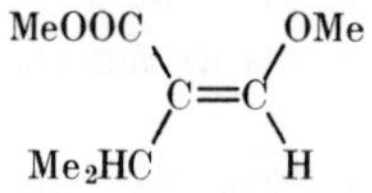

```
MeOOC       OMe          MeOOC       H
     \     /                  \     /
      C=C                      C=C
     /     \                  /     \
Me2HC       H            Me2HC       OMe

    37/IIa                   37/IIb
```

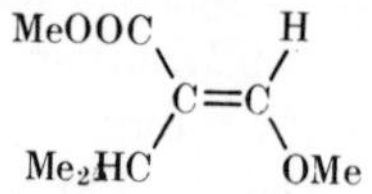

The relative ''height'' of the olefinic signal is much lower in Spectrum **37**/*2* assigned to Structure **IIa** derivable from the *Z* enol. The weak splitting which causes this reduced height is clearly visible as a shoulder, at the low frequency side of the signal. The splitting must be due to a long-range coupling with the methine proton of the isopropyl group, which also broadens the lines of the methine septet. Since a *cisoid* allyl coupling is always stronger than a *transoid* coupling,* the significant long-range coupling is a further proof for Structure **IIa**.

* See Volume I, p. 63.

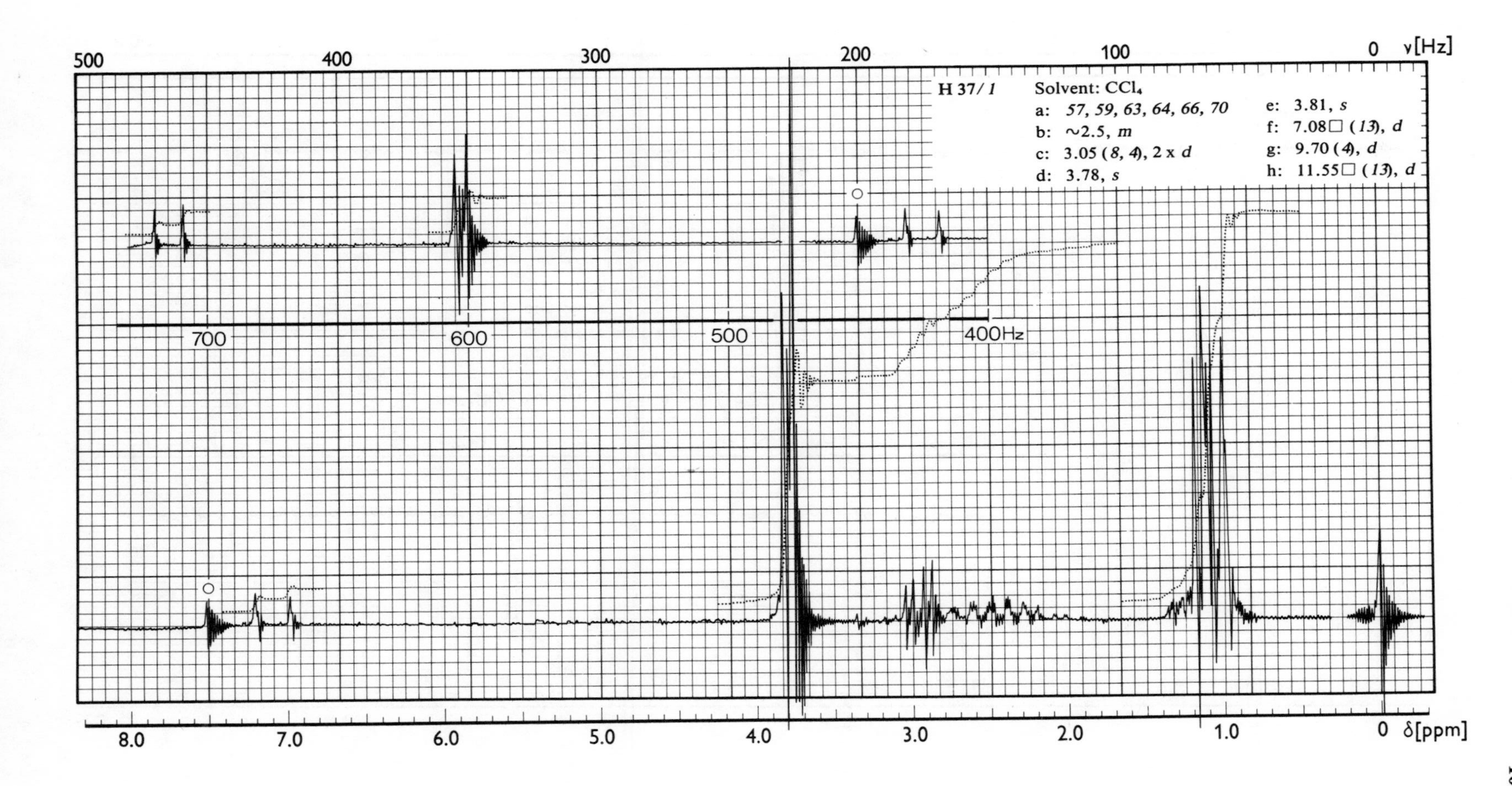

H 37/1
Solvent: CCl4
a: 57, 59, 63, 64, 66, 70
b: ~2.5, m
c: 3.05 (8, 4), 2 x d
d: 3.78, s
e: 3.81, s
f: 7.08□ (13), d
g: 9.70 (4), d
h: 11.55□ (13), d
ν[Hz]
δ[ppm]
700
600
500
400Hz

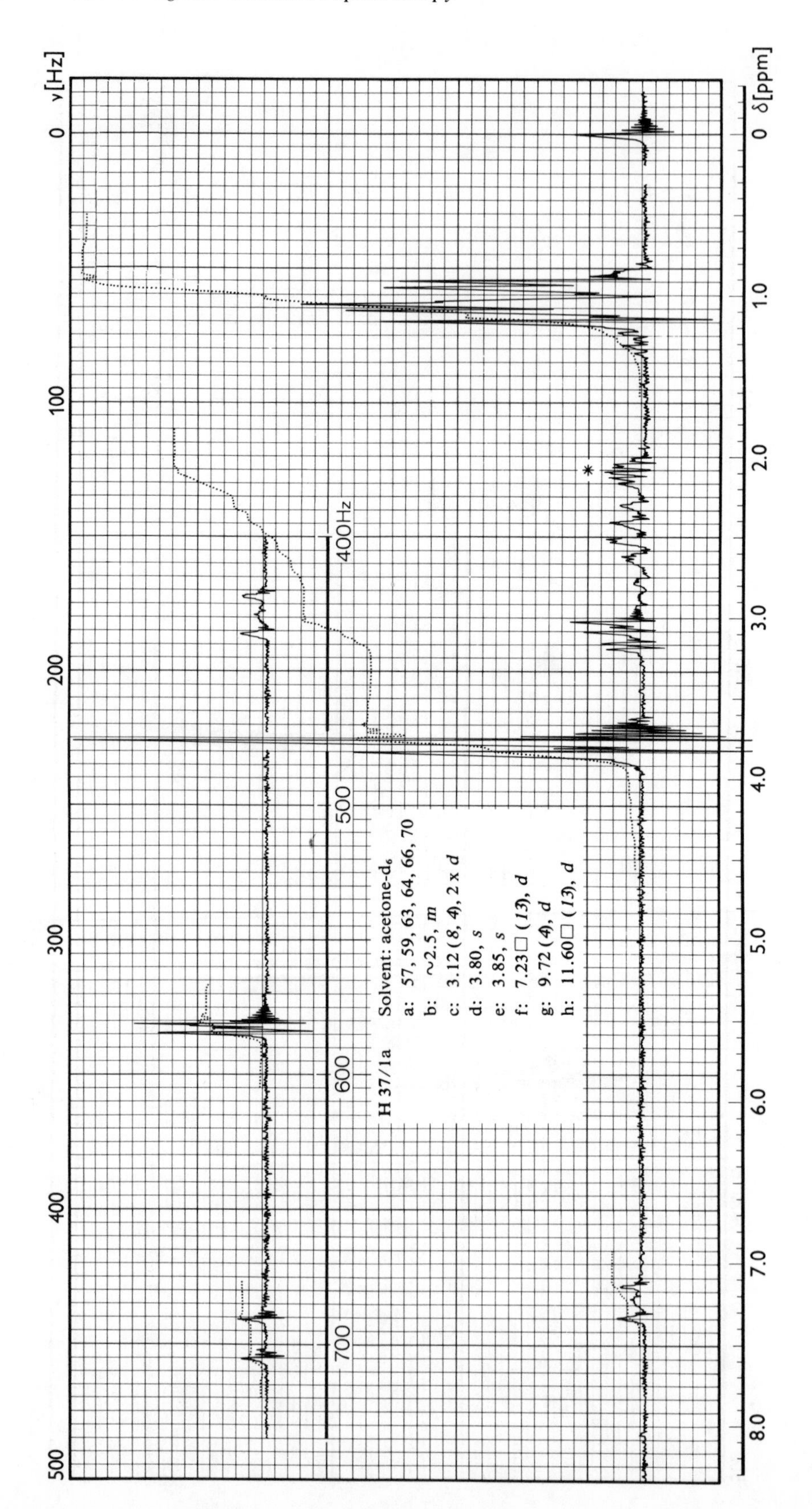
ν[Hz]
0
100
200
300
400
500
400Hz
500
600
700
δ[ppm]
0
1.0
2.0
3.0
4.0
5.0
6.0
7.0
8.0
H 37/1a
Solvent: acetone-d6
a: 57, 59, 63, 64, 66, 70
b: ~2.5, m
c: 3.12 (8, 4), 2 x d
d: 3.80, s
e: 3.85, s
f: 7.23□ (13), d
g: 9.72 (4), d
h: 11.60□ (13), d

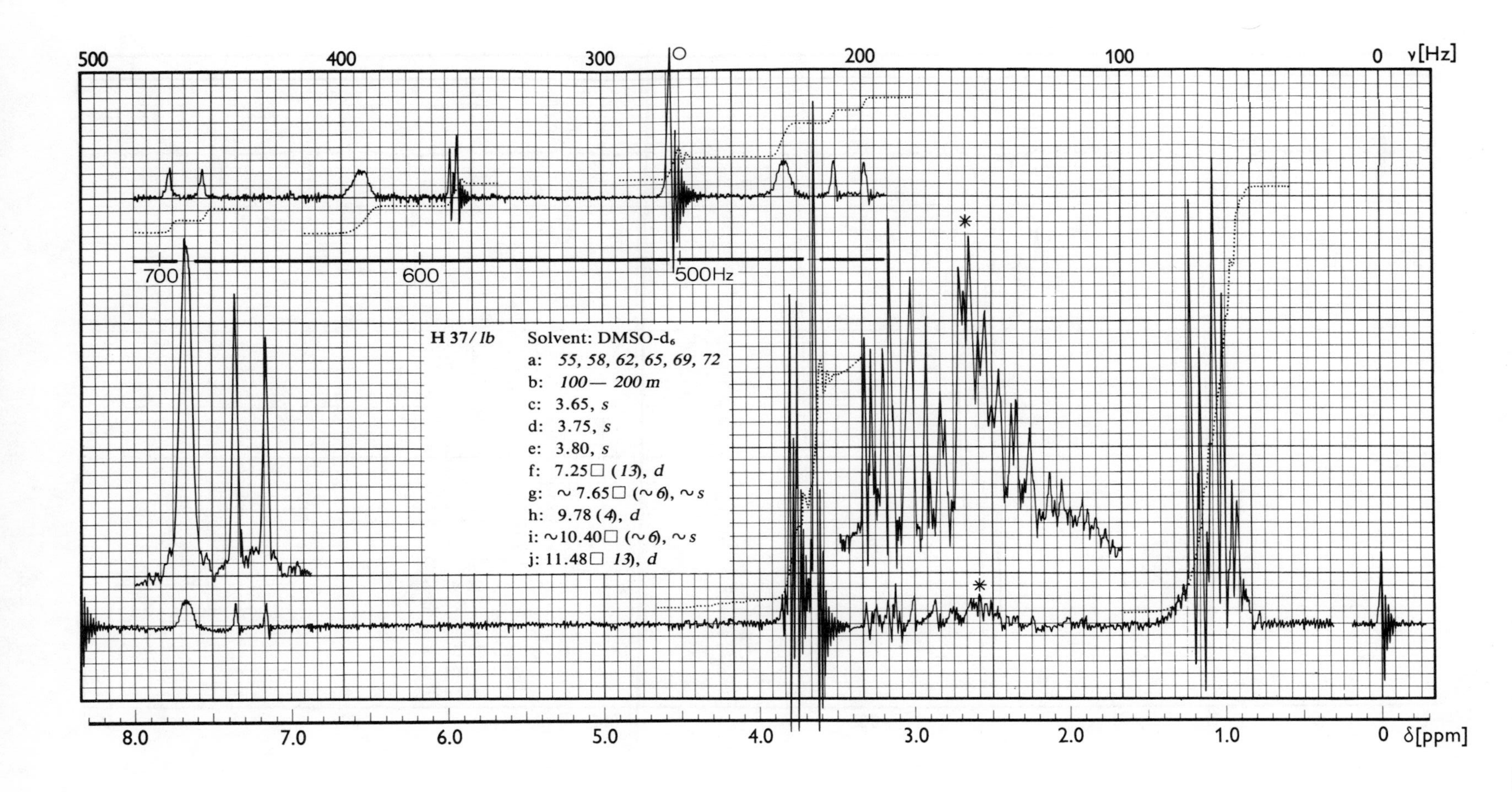

500
400
300
200
100
0
ν[Hz]
700
600
500Hz
H 37/ lb
Solvent: DMSO-d₆
a: 55, 58, 62, 65, 69, 72
b: 100— 200 m
c: 3.65, s
d: 3.75, s
e: 3.80, s
f: 7.25□ (13), d
g: ~ 7.65□ (~6), ~s
h: 9.78 (4), d
i: ~10.40□ (~6), ~s
j: 11.48□ 13), d
8.0
7.0
6.0
5.0
4.0
3.0
2.0
1.0
0
δ[ppm]

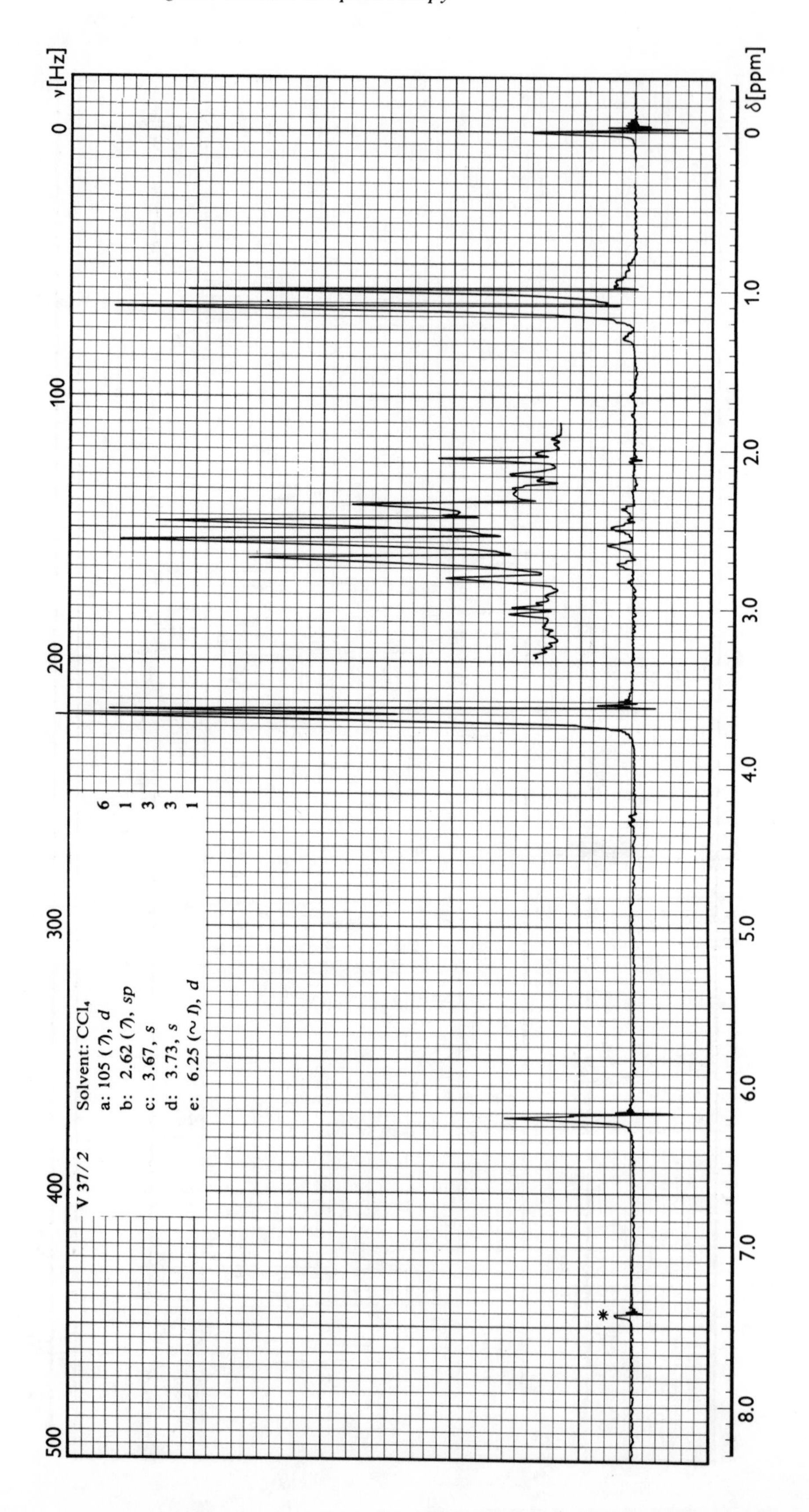

ν[Hz]
0
100
200
300
400
500
V 37/2
Solvent: CCl$_4$
a: 105 (7), d 6
b: 2.62 (7), sp 1
c: 3.67, s 3
d: 3.73, s 3
e: 6.25 (~1), d 1
*
δ[ppm]
0
1.0
2.0
3.0
4.0
5.0
6.0
7.0
8.0

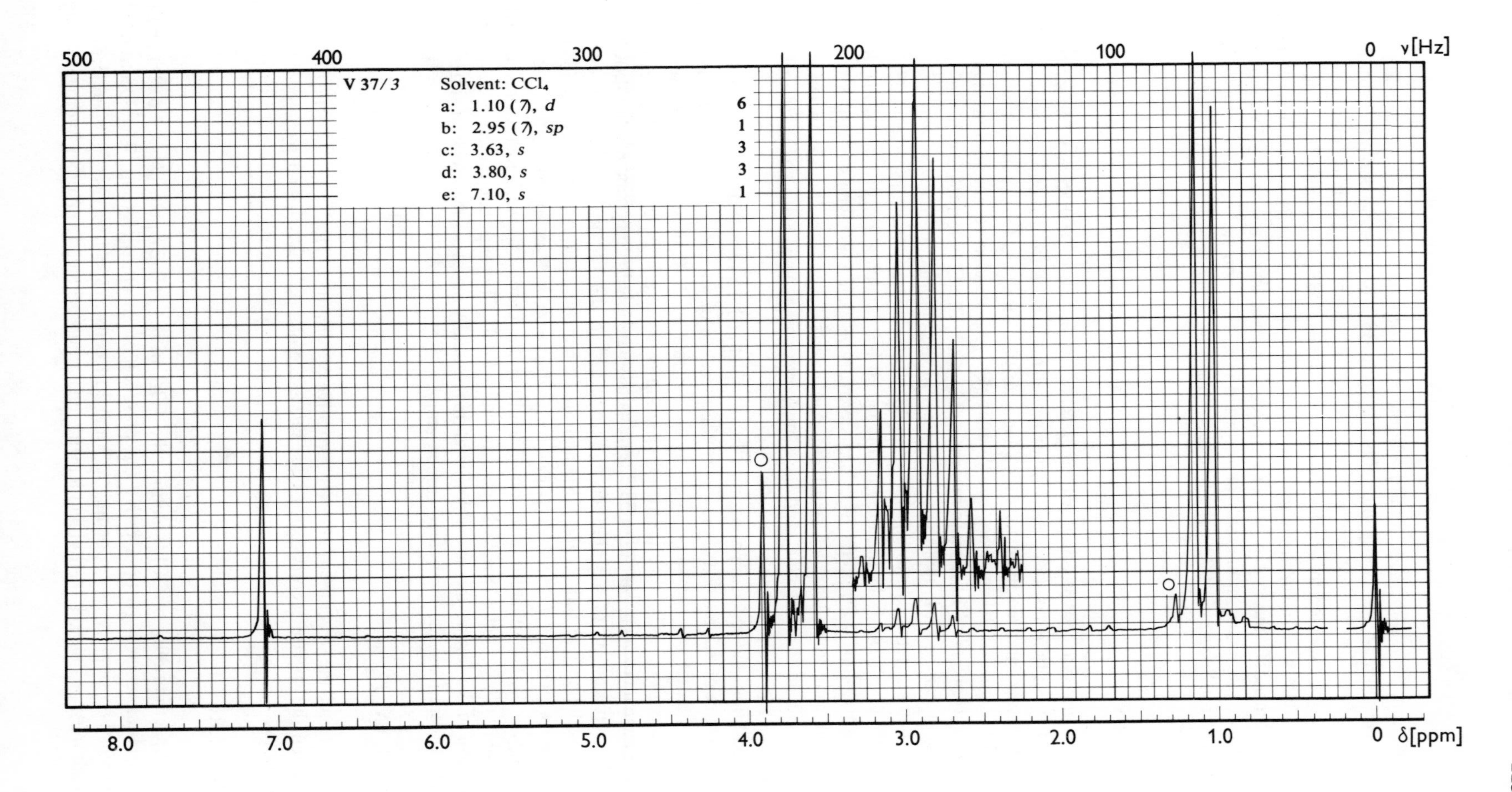

V 37/3
Solvent: CCl_4
a: 1.10 (7), d 6
b: 2.95 (7), sp 1
c: 3.63, s 3
d: 3.80, s 3
e: 7.10, s 1
ν[Hz]
500
400
300
200
100
0
δ[ppm]
8.0
7.0
6.0
5.0
4.0
3.0
2.0
1.0
0

PROBLEM 38

Spectrum **38**/*1* is the spectrum of a liquid with formula $C_3H_8Cl_2Si$. What is the structure of the compound? The compound with Spectrum **38**/*2* is a simple derivative of this liquid. Determine the structure of the latter and decide whether the sample is homogeneous. As a signal can be expected to occur very close to that of TMS, the standard must not be added to the sample without preliminary investigations (see Problem **7**). Since it was found that in Spectrum **38**/*1* the signal of most shielded protons is far enough (0.55 ppm) from the TMS line, the spectrum can also be recorded in $CDCl_3$ containing TMS. This condition is not valid for Spectrum **38**/*2*, which was recorded in a $CDCl_3$ solution containing no TMS.

ANSWER

Spectrum **38**/*1* contains only two singlets, of 3:1 intensity, at 0.55 and 2.98 ppm. On the basis of its upfield position, the former signal must arise from methyl groups attached to silicon atoms, and the intensity ratios indicate that there are two such methyl groups. As the other signal of 2H intensity is also a singlet, it must also be due to a methylene group attached to silicon. Thus, the structure is $Me_2SiCl–CH_2Cl$. Deshielding of the methylene hydrogens can be attributed to the $-I$ effect of the chlorine atom.

The two singlets constituting Spectrum **38**/*1* can also be found in Spectrum **38**/*2* (at 0.55 and 2.90 ppm), but both lines are accompanied by a much stronger line on the downfield sides (at 0.20 and 2.70 ppm). The intensity ratio of these two signals is also 3:1, and they are about nine times stronger than their counterparts. Therefore, the Me_2SiCH_2Cl group is still present in the molecule, and the sample is contaminated with about 10% of the first substance.

In addition, the spectrum contains a 1:2:1 triplet at 1.15 (J = 8 Hz) and a 1:3:3:1 quartet at 3.70 ppm of 3:2 intensity, indicating the presence of an ethyl group, taking into account that the quartet has the same intensity as the line at 2.70 ppm. The deshielding of the methylene group indicates that the ethyl group is not attached directly to the silicon atom, but through an oxygen. Spectrum **38**/*2* corresponds therefore to compound $Me_2Si(OEt)CH_2Cl$.

The replacement of the chlorine atom attached to the silicon by an ethoxy group also explains the upfield shift of the SiMe and $SiCH_2$ signals, relative to the first compound. As the chlorine with a $-I$ effect is replaced by a substituent with an opposite effect, the electron density on the silicon atom, and consequently on the protons as well, increases. This effect is weaker on the methylene hydrogens of the $-SiCH_2Cl$ group than on the $-SiMe$ groups because in the former the two chlorine atoms mutually suppress the electron shifts towards one of them from the methylene carbon atom.

PROBLEM 39

Determine the structure of an ester derivative of formula $C_6H_{10}O_3$. Explain the differences between the spectra recorded in CCl_4 (**39**/*1*), acetone-d_6 (**39**/*1a*), and pyridine-d_5 (**39**/*1b*) solutions.

ANSWER

In Spectrum **39**/*1* of CCl_4 solution, the 1:2:1 triplet at 1.27 ppm (J = 7.5 Hz) and the 1:3:3:1 quartet at 4.16 ppm prove the presence of an ethoxy group. The spectrum also contains two singlets corresponding to 3 and 2 protons, respectively, at 2.20 and 3.37 ppm, which must be due to a methyl and a methylene group isolated from other hydrogens. These features (see Table 102), taking also into account the ester group, indicate the presence of a second carbonyl group, and thus the derivative must be acetoacetic ester MeCO–CH_2–COOEt.

This structure also explains the origin of the two unassigned weak singlets. The molecule, being a β-dioxo derivative, may form a keto-enol equilibrium; the two weak signals can therefore be assigned to the enol form Me–C(OH)=CH–COOEt. The signal at 1.91 ppm corresponds to the methyl group adjacent to the unsaturated carbon, whereas at 4.93 ppm the olefinic proton absorbs.[953]

Table 102
THE CHEMICAL SHIFTS (PPM) OF THE KETO AND ENOL FORMS OF ACETYL ACETIC ETHYL ESTER IN $CDCl_3$, ACETONE-d_6, AND PYRIDINE-d_5 SOLUTIONS

Spectrum	39/*1*	39/*1a*	39/*1b*
Solvent	CCl_4	acetone-d_6	pyridine-d_5
Assignment			
δCH_3(Et), *t*[a]	1.27	1.22	1.18
δCH_3(Ac), *s*	2.20[b]	2.21[b]	2.21[b]
	1.91[c]	1.93[c]	—
δCH_2CO, *s*[b]	3.37[d]	3.48[d]	3.60[e]
δCH_2(Et), *qa*	4.16	4.14	4.17
δ(=CH), *s*[c]	4.93	5.02	—

[a] *J* = 7.5 Hz.
[b] Keto.
[c] Enol.
[d] Sharp.
[e] Broad.

The relative intensity of the methylene quartet of the ethyl group with respect to the methylene singlet gives the relative proportions of the keto and enol forms. The same ratio can also be obtained from the relative intensities of the methyl singlets or from the intensity ratio corresponding to one hydrogen and the intensity of the olefinic signal. The intensity corresponding to one proton can be determined, e.g., from the signals of the ethyl group.

The proportion of the enol form is negligible in pyridine-d_5 solution (Table 102), since the above two signals do not appear in Spectrum **39**/*1b*. In CCl_4 and acetone-d_6 solutions, the keto-enol equilibrium is shifted towards the enol form (see Spectra **39**/*1* and *1a*), but still the diketo form is predominant. According to the integrals, the ratio of keto-enol forms in the two solutions are 95:5 and 87:13, respectively.

The methylene signal is rather broad in Spectrum **39**/*1b*, and its shape is similar to that of acidic protons (hydroxy, amino, etc.). This shape is clearly due to the fact that in the alkaline medium the methylene protons are in exchange between the molecules.*

* Compare Section 3.4.1.1.

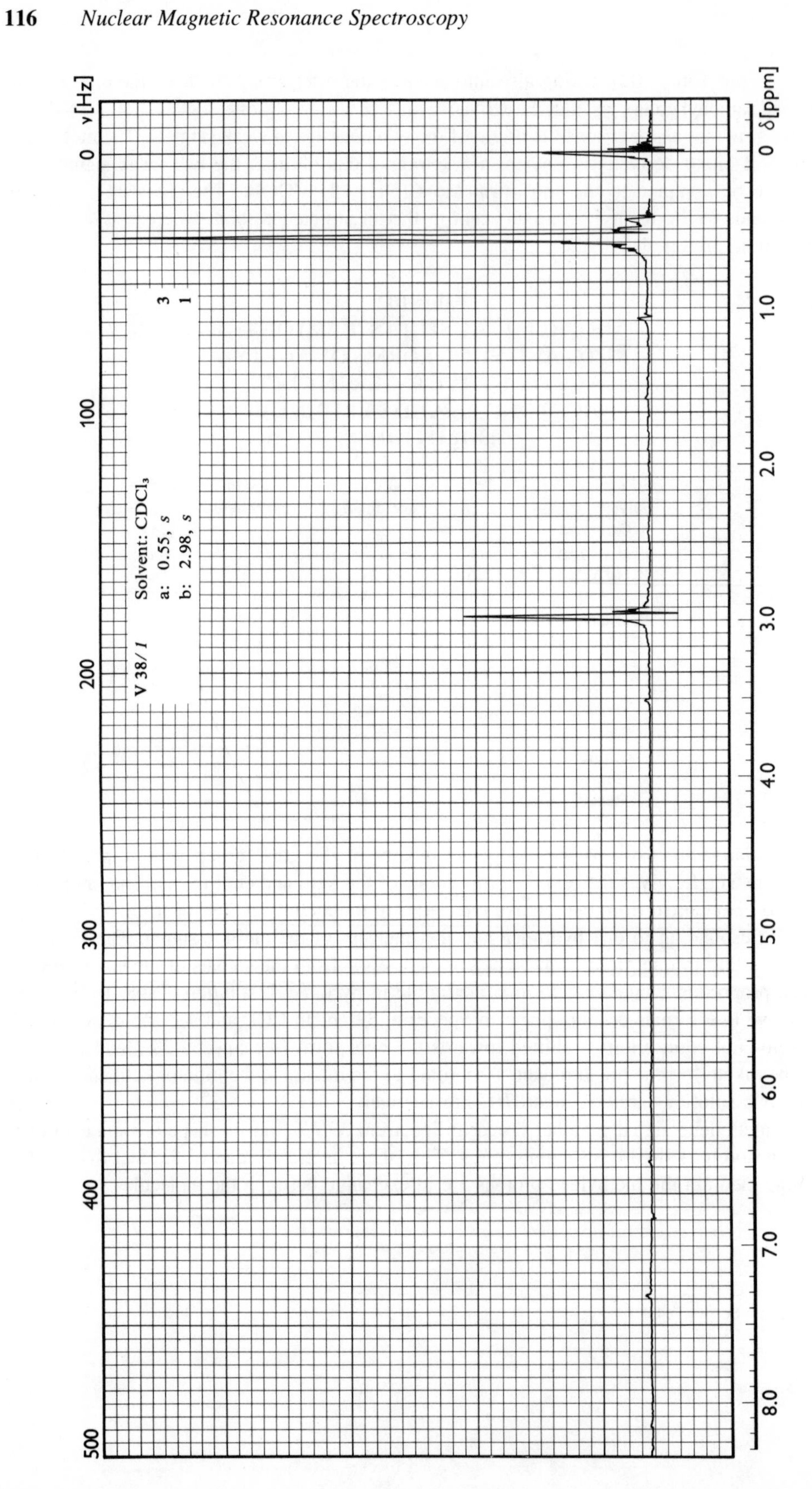
V 38/ 1
Solvent: $CDCl_3$
a: 0.55, s 3
b: 2.98, s 1
ν[Hz]
0
100
200
300
400
500
δ[ppm]
0
1.0
2.0
3.0
4.0
5.0
6.0
7.0
8.0

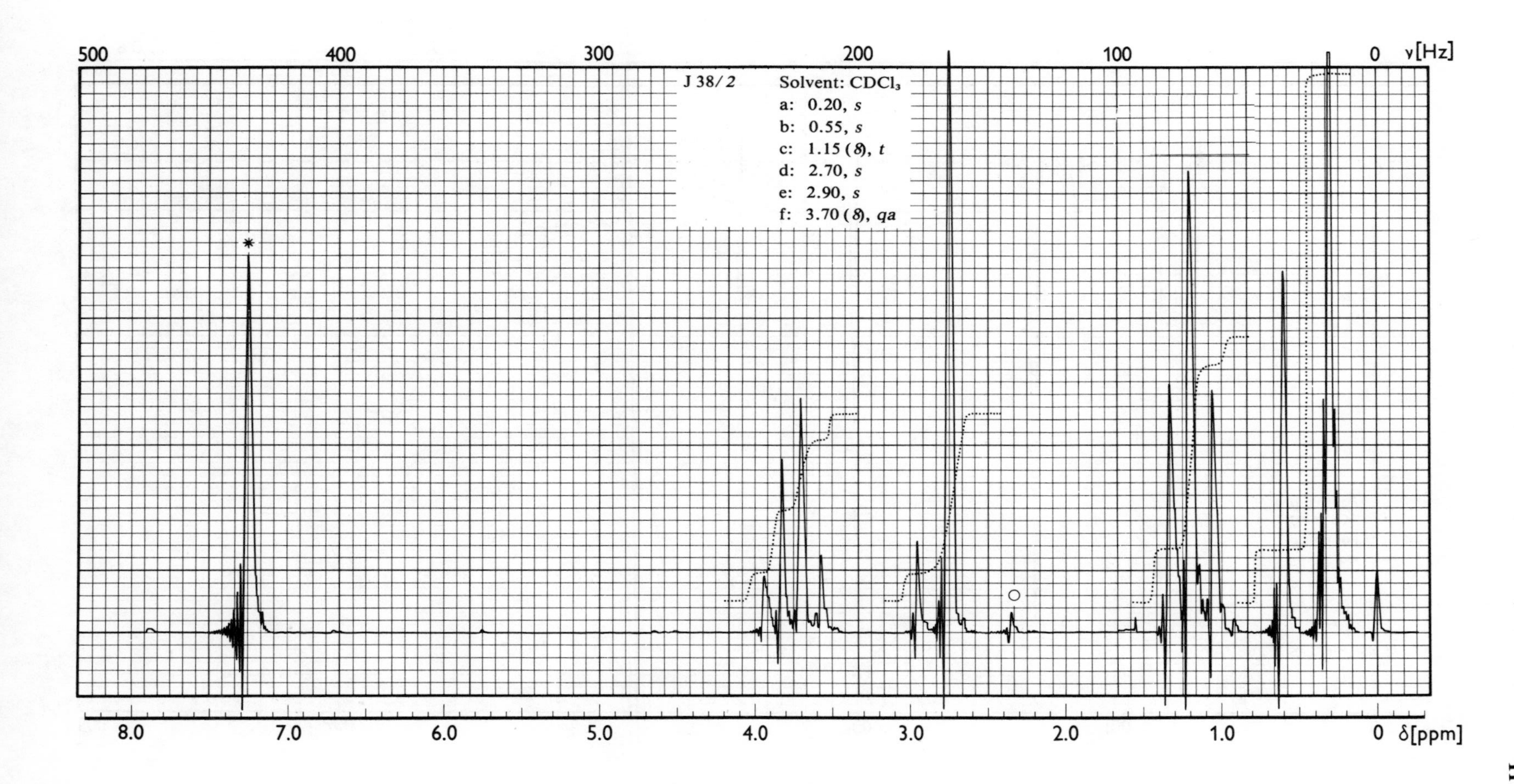
ν[Hz]
500
400
300
200
100
0
J 38/2
Solvent: $CDCl_3$
a: 0.20, *s*
b: 0.55, *s*
c: 1.15 (8), *t*
d: 2.70, *s*
e: 2.90, *s*
f: 3.70 (8), *qa*
δ[ppm]
8.0
7.0
6.0
5.0
4.0
3.0
2.0
1.0
0

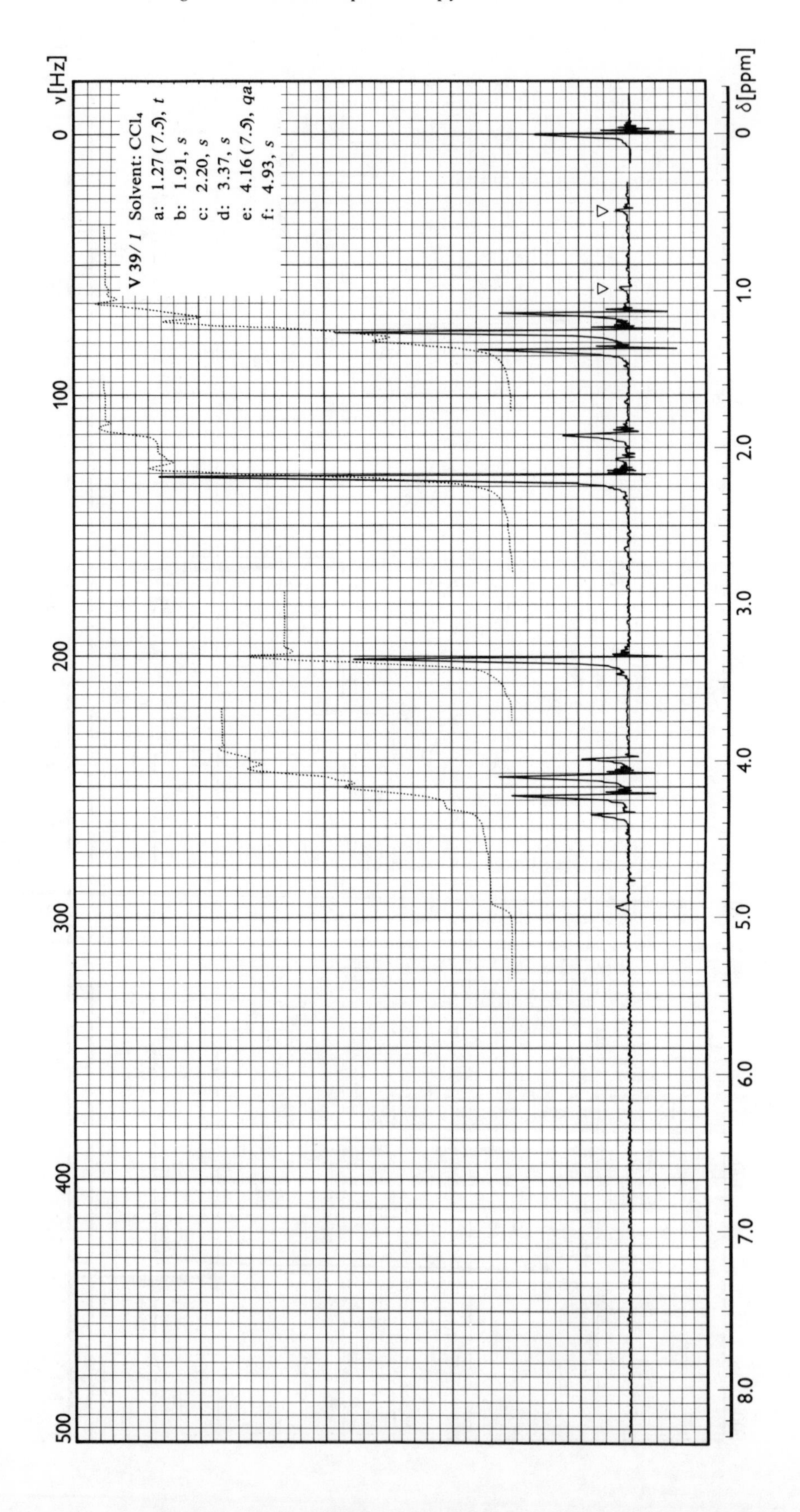
V 39/1 Solvent: CCl₄
a: 1.27 (7.5), t
b: 1.91, s
c: 2.20, s
d: 3.37, s
e: 4.16 (7.5), qa
f: 4.93, s
ν[Hz]
0
100
200
300
400
500
δ[ppm]
0
1.0
2.0
3.0
4.0
5.0
6.0
7.0
8.0

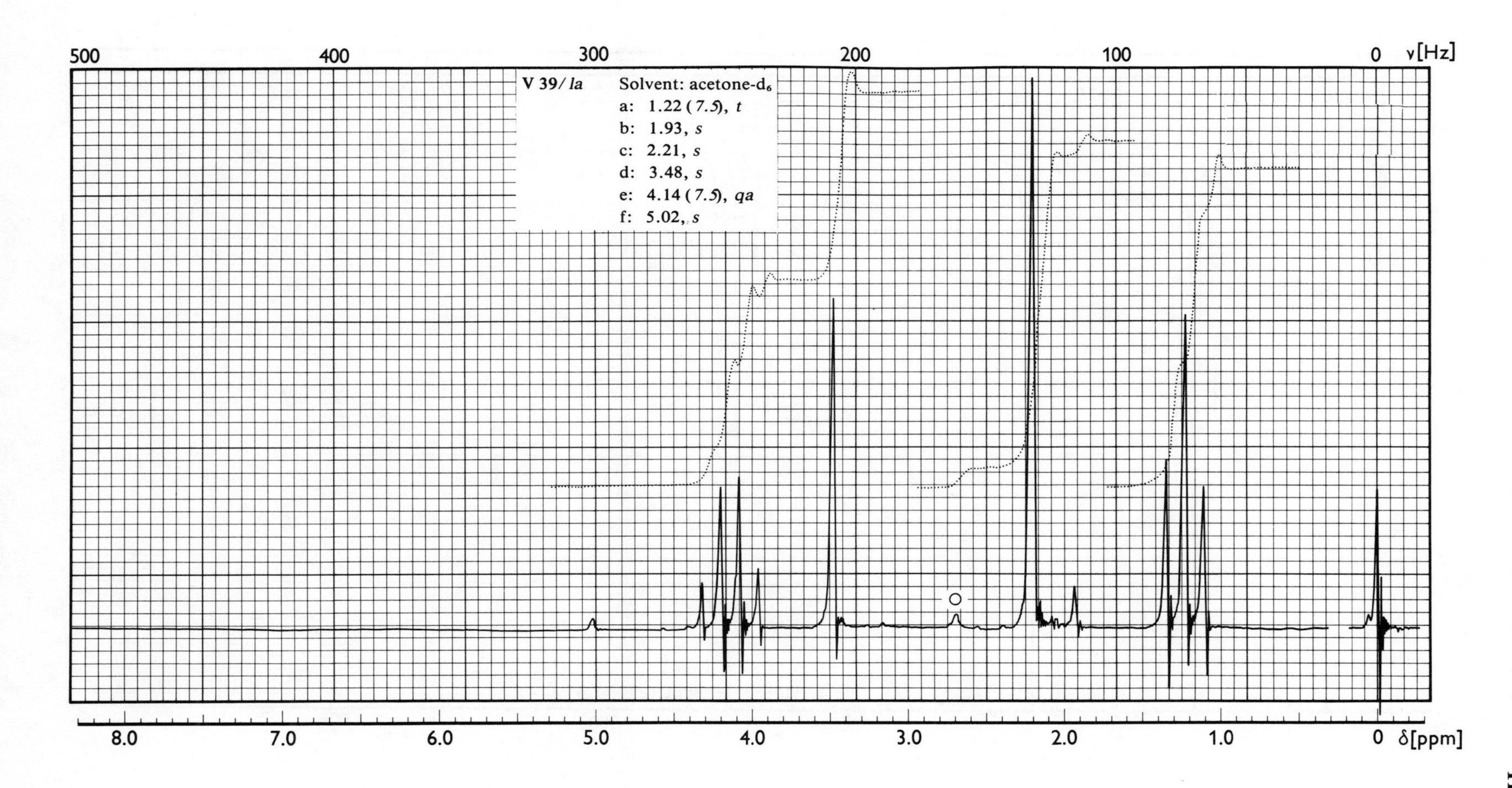
500
400
300
200
100
0 ν[Hz]
V 39/ Ia
Solvent: acetone-d6
a: 1.22 (7.5), t
b: 1.93, s
c: 2.21, s
d: 3.48, s
e: 4.14 (7.5), qa
f: 5.02, s
8.0
7.0
6.0
5.0
4.0
3.0
2.0
1.0
0 δ[ppm]

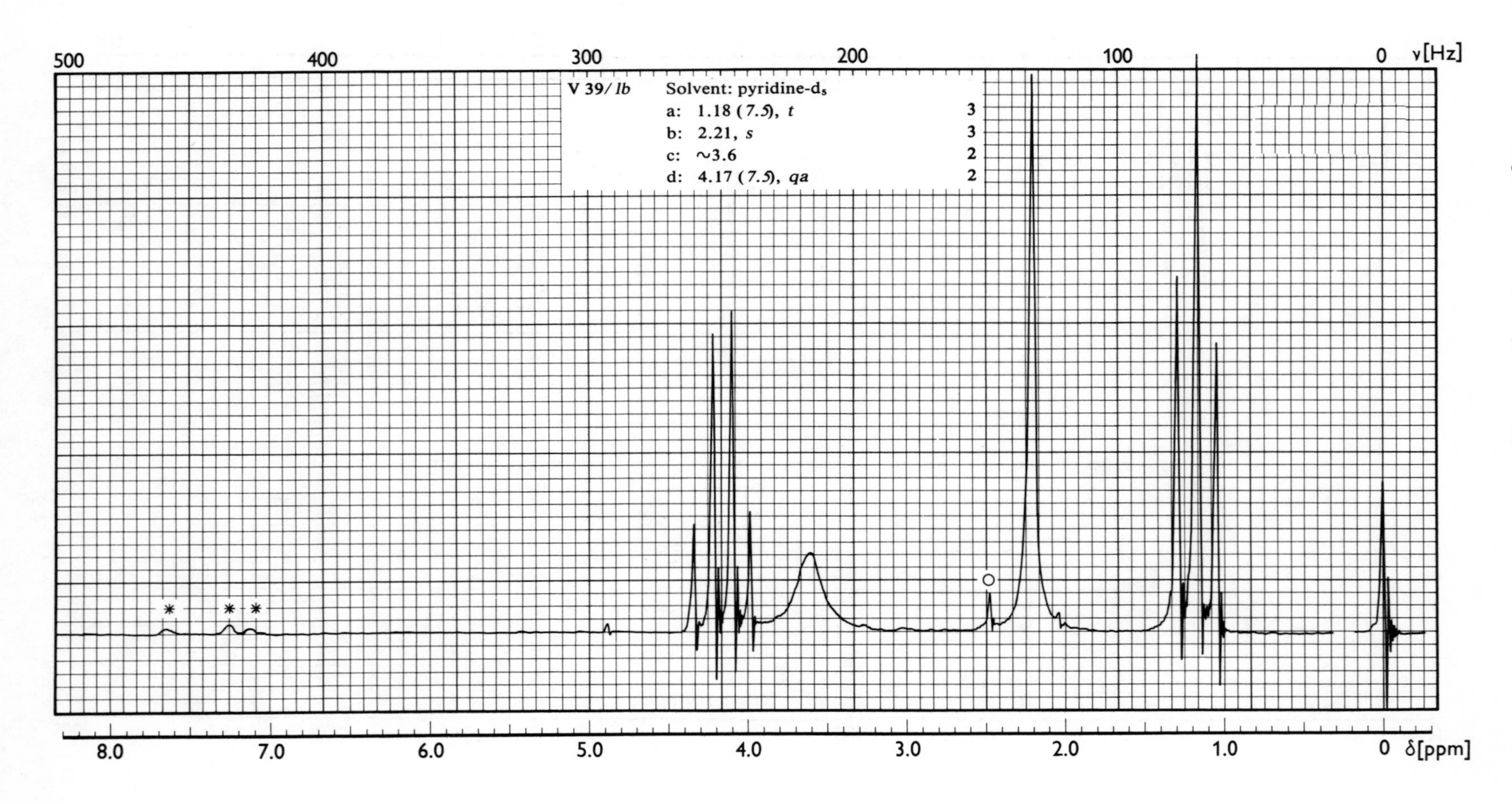

V 39/1b
Solvent: pyridine-d$_5$
a: 1.18 (7.5), t 3
b: 2.21, s 3
c: ~3.6 2
d: 4.17 (7.5), qa 2
ν[Hz]
500
400
300
200
100
0
δ[ppm]
8.0
7.0
6.0
5.0
4.0
3.0
2.0
1.0
0

PROBLEM **40**[840]

Correlate Spectra **40**/*1* and **40**/*2* with Structures **I** and **II**. Assign the signals.

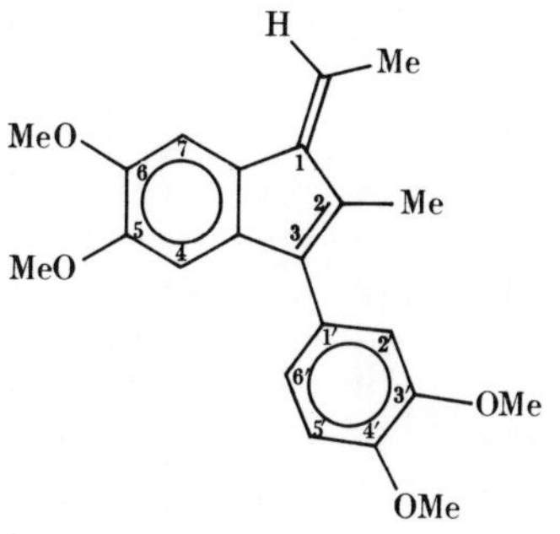

40/I 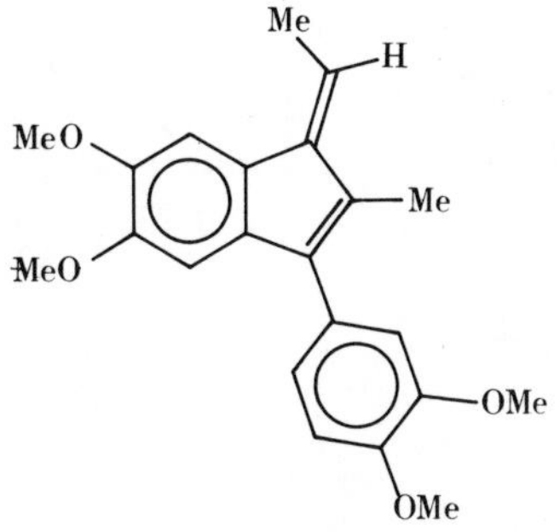 **40/II**

ANSWER

In the *Z* and *E* isomers the shieldings of the methine and methyl hydrogens of the =$CHCH_3$ group are different. It can be expected that in compound **I** the methine, whereas in compound **II** the methyl protons will be deshielded by the anisotropy of the coplanar aromatic ring.*

The assignment of Spectra **40**/*1* and **40**/*2* is as follows: $\delta CH_3(2)$, *s*: 2.10 and 2.31 ppm, $\delta CH_3(1)$, *d*(J = 8 Hz): 2.35 and 2.27 ppm, δOCH_3, *s*: 3.85, 3.90, 3.95 and 3.83, 3.91, 3.93, and 3.95 ppm, δCH, *qa*(J = 8 Hz): 6.40 and 6.68 ppm, δArH-4, *s*: 6.82 and 6.73 ppm, δArH (the 3 protons of the tri-substituted ring): 6.98 and 6.97 ppm (singlet-like signal), δArH-7, *s*: 7.38 and 7.15 ppm, respectively.

On the basis of the assignment, Structure **I** corresponds to Spectrum **40**/*2*, and Structure **II** corresponds to Spectrum **40**/*1*, because $\Delta_{I,II}\ \delta CH_3 = 2.27 - 2.35 = -0.08$ ppm, and $\Delta_{I,II}\ \delta CH = 6.68 - 6.40 = +0.28$ ppm.

The methine quartet in Spectrum **40**/*2* overlaps with the signal of one of the aromatic protons, and one of the lines of the methyl doublet also coalesces with the 2-methyl singlet. Nevertheless, the line positions can be determined accurately, because the two outer lines and one of the inner lines of the methine quartet are separated, and thus the magnitude of the coupling constant can be determined from the spectrum (which is the same for the two isomers, within the experimental error).

The aromatic protons can also be assigned, because the shifts in the two isomers are different. The larger difference in chemical shift between the two isomers can be expected for H-7. Therefore, the signal of the most deshielded proton can be assigned to H-7, and the less shifted signal, of 1H intensity, must belong to H-4. The veratryl protons form an *ABC* spin system, close to the limiting A_3 case, and their chemical shifts are practically the same for the isomers. The H-4 and H-7 signals are singlets because of the weak J^p coupling (the splitting is insignificant). Note that the signal of the 2-methyl group is also shifted by 0.19 ppm in a direction opposite to the shift of the ethylidine-methyl doublet. This fact can also be utilized in distinguishing between analogous isomeric pairs.

* Compare Volume I, p. 35.

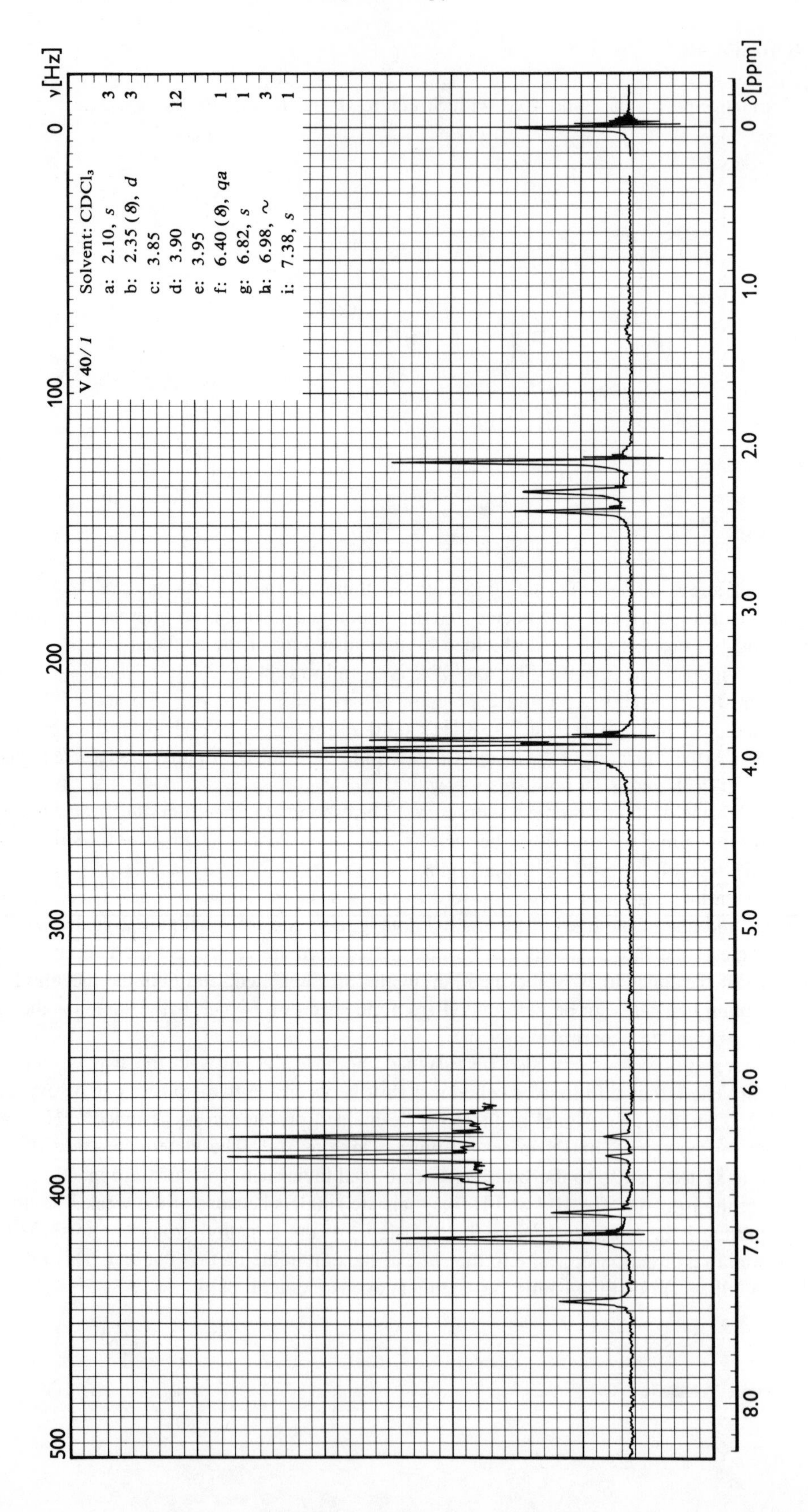
V 40/1
Solvent: CDCl₃
a: 2.10, s 3
b: 2.35 (8), d 3
c: 3.85
d: 3.90 12
e: 3.95
f: 6.40 (8), qa 1
g: 6.82, s 1
h: 6.98, ~ 3
i: 7.38, s 1
ν[Hz]
0
100
200
300
400
500
δ[ppm]
0
1.0
2.0
3.0
4.0
5.0
6.0
7.0
8.0

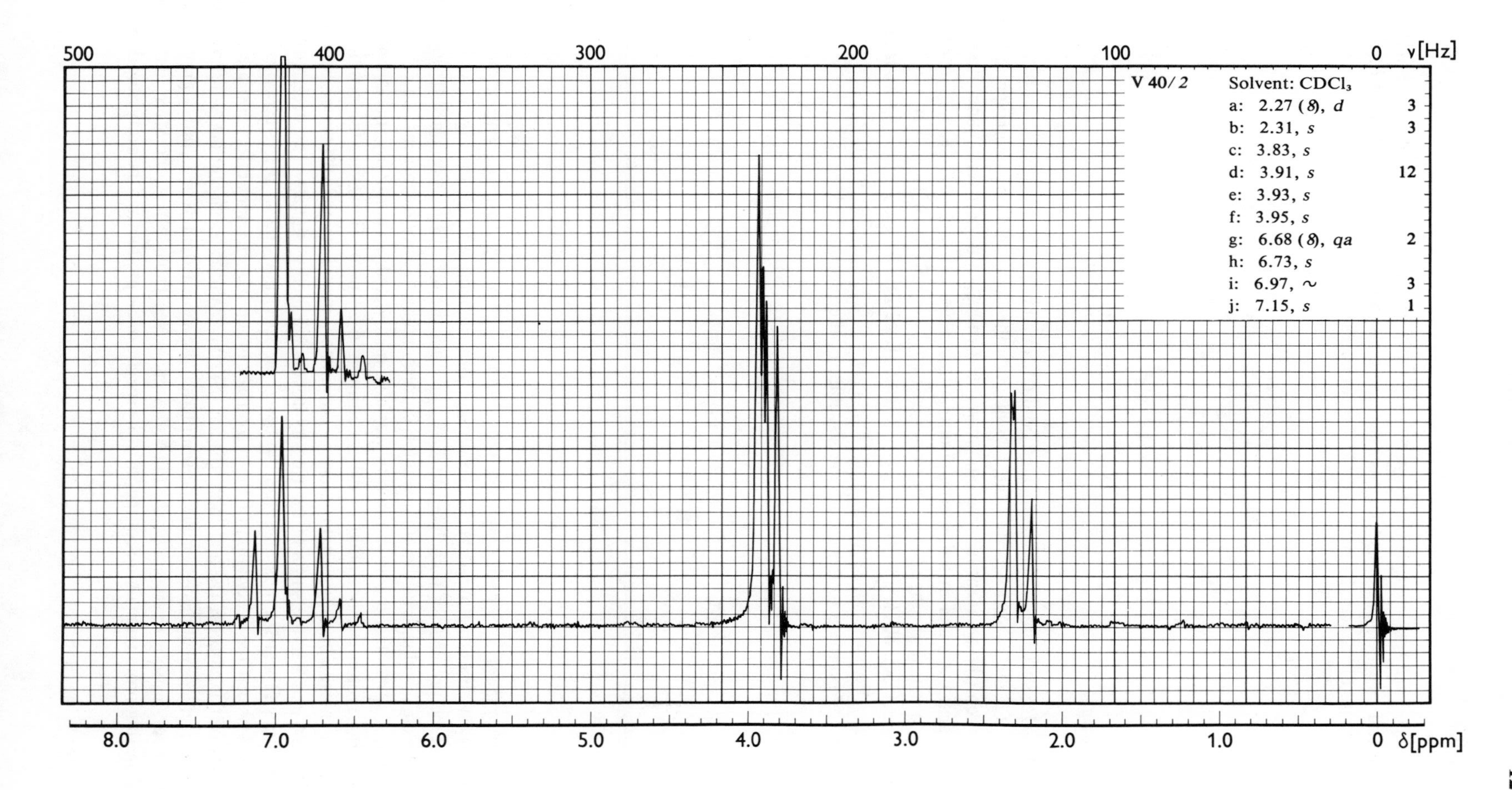
500
400
300
200
100
0 ν[Hz]
V 40/2
Solvent: $CDCl_3$
a: 2.27 (8), *d* 3
b: 2.31, *s* 3
c: 3.83, *s*
d: 3.91, *s* 12
e: 3.93, *s*
f: 3.95, *s*
g: 6.68 (8), *qa* 2
h: 6.73, *s*
i: 6.97, ~ 3
j: 7.15, *s* 1
8.0
7.0
6.0
5.0
4.0
3.0
2.0
1.0
0 δ[ppm]

PROBLEM 41[1343]

Compound **I** was methylated. How many methyl groups were incorporated into the molecule and in which positions? To answer these questions the Spectrum **41**/*1* of the reaction product and of its acetyl derivative (**41**/*2*) can be used.

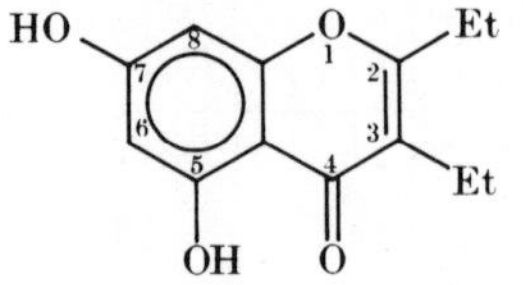

41/I

ANSWER

The **41**/*1* spectrum of the methylated product contains the A_2X_3 multiplets of the two ethyl groups. Of the two triplets and the two quartets the downfield ones correspond to the 2-ethyl group (the adjacent heteroatom causes deshielding). In the overlapping triplets, lines **1, 2, 4,** and **3, 5, 6** and in the overlapping quartets, lines **1, 2, 4, 6** and **3, 5, 7, 8,** respectively, belong to the same ethyl group. For both cases J_{AX} = 7.5 Hz. δCH_3 (3-Et) = 1.11, δCH_3 (2-Et) = 1.30, δCH_2 (3-Et) = 2.49, and δCH_2 (2-Et) = 2.66 ppm. Apart from these multiplets, the spectrum contains four singlets at 2.05, 3.83, 6.24, and 13.0 ppm, of 3:3:1:1 intensity. According to these data the structure of the product is **II** or **III**.

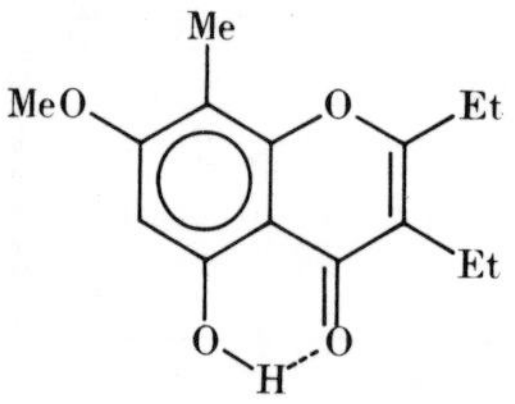

41/II 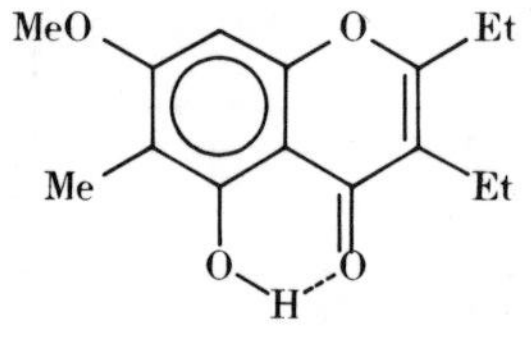 **41/III**

The signal at 3.83 ppm may arise only from a methoxy group, i.e., one of the hydroxy groups has been methylated. As the strong paramagnetic shift of the signal at 13.0 ppm proves the presence of a chelated hydroxy group,* the methoxy group must be in position 7. The chemical shift of the other singlet, corresponding to three hydrogens, i.e., to a methyl group, suggests that the aromatic ring was also methylated. Accordingly, the signal of only one ring proton can be found in the spectrum at 6.24 ppm. However, the position of the methyl substituent is uncertain; there is presumably no substantial difference between positions 6 and 8 in the shifts of the ring hydrogen and methyl signals. The decision between Structures **II** and **III** must therefore be based on Spectrum **41**/*2* of the *O*-acetyl derivative. It is clear that the acetylation of the 5-hydroxy group has different effects on the chemical shifts of the adjacent (6) and more distant (8) ring hydrogens; the former will presumably be deshielded, and the latter will be shielded to the same extent. The hydroxy group causes, however, a strong diamagnetic shift in both signals (see Table 39). If, therefore, the positions of the ring-proton signals are compared in the spectra of the product and the acetylated derivative,

* See Volume II, p. 101.

a strong paramagnetic shift (about +0.6 ppm) can be expected for Structure **II**, and a much lower shift (about +0.2 ppm) can be expected for Structure **III**.

In Spectrum **41**/*2* of the acetyl derivative, the signals of the ethyl, methyl, and methoxy groups appear practically unchanged; the signal of the chelated hydroxy group is, of course, replaced by the acetoxy singlet with a treble intensity. This signal of the ring proton is, however, strongly shifted (by +0.46 ppm). The assignment is as follows: δCH_3 (3-Et), *t*: 1.09 ppm (J = 7.5 Hz), δCH_3 (2-Et), *t*: 1.31 ppm (J = 7.5 Hz), δCH_3 (ring-Me), *s*: 2.10 ppm, δCH_3 (acetyl), *s*: 2.49 ppm, δCH_2 (3-Et), *qa*: 2.51 ppm, δCH_2 (2-Et), *qa*: 2.68 ppm, δOCH_3, *s*: 3.92 ppm, and δArH, *s*: 6.70 ppm. Although one of the lines of both methylene quartets overlap with the methyl signal of the acetyl group, the coupling constant extracted from the remaining three lines and the ethyl triplets enables to deduce their frequencies. The strong paramagnetic shift in δArH with respect to Spectrum **41**/*1* proves the 6-position of the ring hydrogen, and thus a 8-methyl substituent, and therefore, Structure **II**.

PROBLEM **42**

Determine the structure of the glycine derivative $R_2N^+H–CH_2–COR + Cl^-$ from its Spectrum **42**/*1* recorded in D_2O solution. Interpret the changes upon acidification (**42**/*1a*), and those observed at varying temperature (Figure 201) in Spectrum **42**/*1b* of a DMSO-d_6 solution.

ANSWER

Apart from the methylene singlet at 5.03 ppm and a signal in its neighborhood, which is shifted upon acid addition, Spectrum **42**/*1* contains three groups of signals of 12:5:4 intensity. The acid-sensitive signal, which shifts from 4.9 to 5.4 ppm (see Spectrum **42**/*1a*), is the overlapping absorption of the acidic proton (N^+H) and the water content of the solvent. Of the signal groups, the first four lines between 85 and 105 Hz correspond either to four, chemically different methyl groups, attached to *tertiary* carbons, or to two pairs of pairwise equivalent methyl groups of type $–CHCH_3$. In the latter case, a methine multiplet must also appear, consisting of at least four (with $–CHCH_3$ groups) or seven [with $–CH(CH_3)_2$ group] lines. Since the spectrum contains a multiplet in the range of 225 to 275 Hz, overlapping with a sharp, strong singlet at 4.10 ppm, there must be two pairs of methyl groups. The singlet, according to its shift, arises from a methoxy group, the multiplet corresponds therefore to two hydrogens. The coupling constant of the CH–CH_3 interaction is ~7 Hz,* and thus the lines at 89.5 and 96 and at 93.5 and 100 Hz, respectively, correspond to each other. From the frequencies δCH_3 = 1.55 and 1.61 ppm, respectively, the lines of the multiplet must therefore have a distance of 6.5 Hz. In addition to the strong singlet at 4.10 ppm, seven lines can be identified between 225 and 275 Hz, where the outer lines are at a distance of 6 × 6.5 = 39 Hz. Since the relative intensities correspond to a 1:6:15:20:15:6:1 septet, the presence of two isopropyl groups, with equivalent methine protons, must be assumed (AX_6 spin system). Should two pairs of equivalent $–CHCH_3$ groups occur in the molecule, their two quartets, even in the case of overlapping lines, could not yield a symmetric septet. Assuming that their two outer lines overlap, although seven lines would occur, their intensity ratio would be 1:3:3:2:3:3:1. Consequently, R = $–CH(CH_3)_2$. From the middle line, δCH = 4.15 ppm.

Naturally, $J(CH_3,CH)$ = 6.5 Hz. The multiplet having the strongest lines at 431, 440,

* Compare Volume I, Table 5 and Volume II, p. 7.

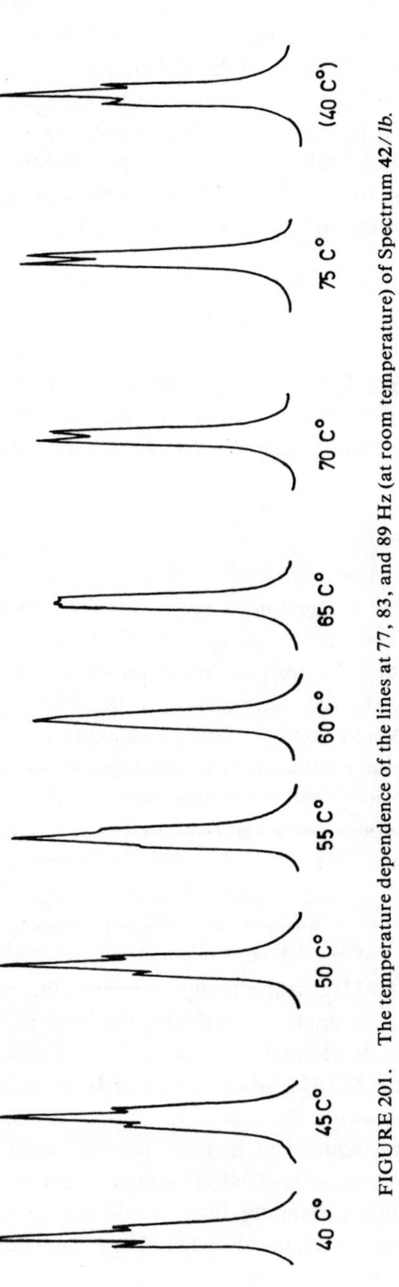

FIGURE 201. The temperature dependence of the lines at 77, 83, and 89 Hz (at room temperature) of Spectrum **42**/*Ib*.

491, and 500 Hz together with the methoxy singlet could be assigned to the R′ substituent. The four strong lines surrounded by weaker lines belong clearly to the *AA′BB′* multiplet of aromatic protons, and the spin system is close to the limiting *AA′XX′* case. The *AB* (or *AX*) pattern indicates the presence of a *para*-disubstituted system (in which $J_{AB} \equiv J^o \approx 9$ Hz). The large difference between δ*A* and δ*B* suggests that the substituents are strongly different in character. This is the case if R′ is a *para*-anysyl group, and the structure of the unknown is **I**.

$$\text{MeO}-\text{C}_6\text{H}_4-\text{CO}-\text{CH}_2-\overset{\oplus}{\text{N}}\text{H}(\text{CHMe}_2)_2 \quad \text{Cl}^{\ominus}$$

42/I

The spectrum of the DMSO-d_6 solution (**42**/*1b*) is practically identical to **42**/*1*, only the solvent effect causes some variations in certain chemical shifts. In addition, the signals are broadened in the strongly polar solvent. The shift is, of course, the strongest for the signal of the acidic protons ($N^+H + H_2O$): it is shifted to 5.25 ppm ($\Delta\delta = +0.35$ ppm) and broadened. One line in each of the two methyl doublets is now overlapping and thus a pseudo-triplet can be observed ($\delta CH_3 = 1.35$ and 1.44 ppm; $\Delta\delta = -0.20$ and -0.17 ppm). The methine septet overlaps with the methoxy signal at 3.97 ppm ($\Delta\delta = -0.13$ ppm), and the methylene signal is at 4.62 ppm. The strong diamagnetic shift (-0.41 ppm) can be attributed to the anisotropic shielding of the S=O bond of the solvent molecules, which are attached upon solvation primarily to the N^+H group. In the *AB* approximation (see Problem **6**), the shifts of the aromatic protons are δ*A* = 8.25 and 7.26 ppm and δ*B* = 8.30 and 7.37 ppm (in D_2O and in DMSO-d_6, respectively), i.e., the shift differences are +0.05 and +0.11 ppm, where the downfield signals correspond, of course, to the ring hydrogens *ortho* to the carbonyl group.

The last problem is why the methyl signal is split, although the methyl groups are equivalent in Structure **I**. Owing to the two bulky isopropyl groups, there is a high barrier for the rotation about the quaternary nitrogen, and this must be the reason of the splitting. This interpretation is supported by the high-temperature spectrum in which the splitting is absent, and only one doublet can be observed see (Figure 201). Upon recooling the sample, the original inequivalence reappears. From the magnitude of the splitting (6 Hz) and the coalescence temperature ($T_c = 68°C$), the free enthalpy of activation, characteristic of the rotational barrier, can be determined as follows:* $\Delta G\ddagger = 19.13 \cdot 341[9.97 + \log(341/6)] = 76.5$ kJ/mol.

* Compare Equation 288.

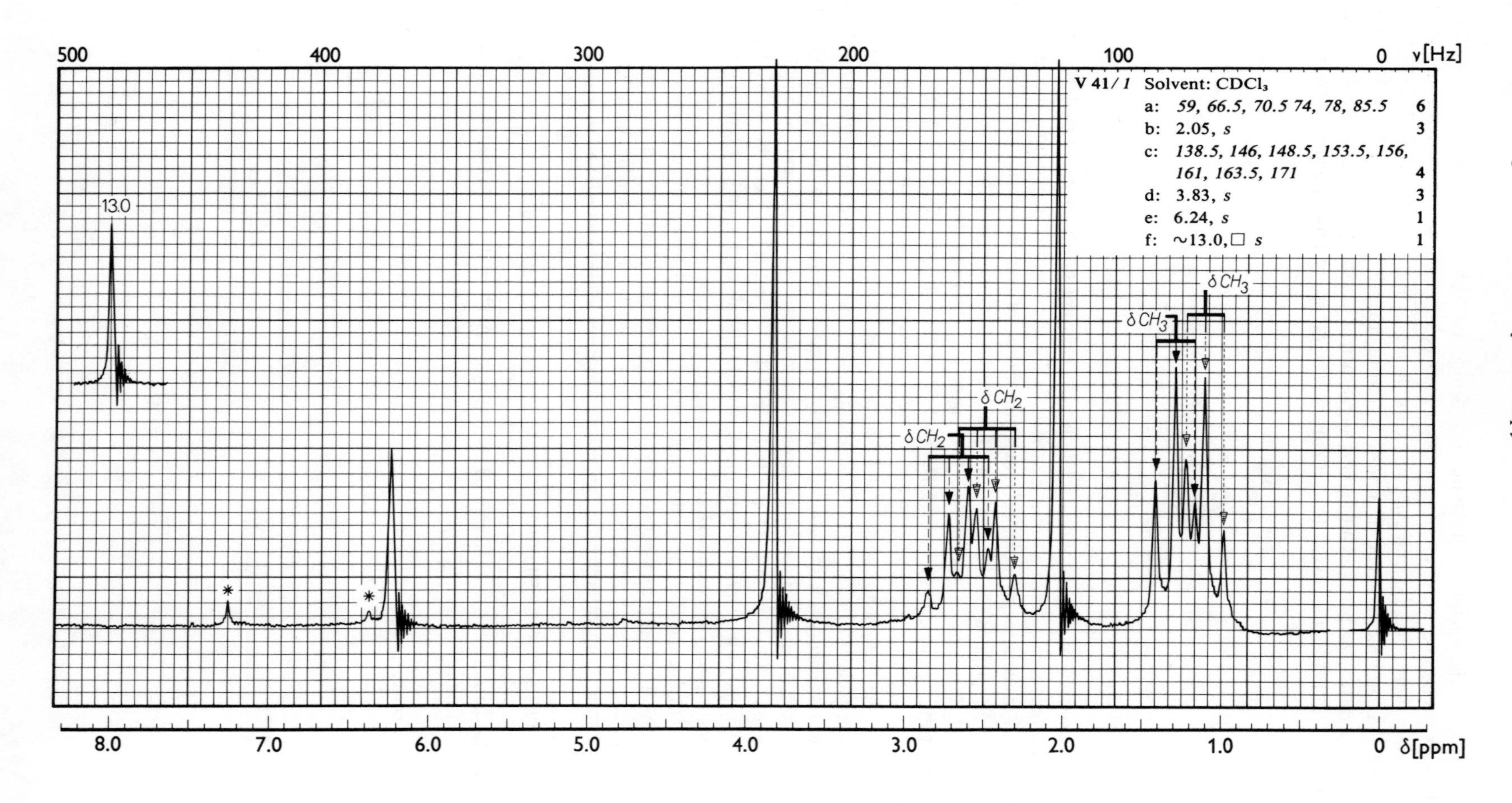
500
400
300
200
100
0 ν[Hz]
V 41/1 Solvent: $CDCl_3$
a: 59, 66.5, 70.5 74, 78, 85.5 6
b: 2.05, s 3
c: 138.5, 146, 148.5, 153.5, 156, 161, 163.5, 171 4
d: 3.83, s 3
e: 6.24, s 1
f: ~13.0, □ s 1
13.0
δCH_2
δCH_2
δCH_3
δCH_3
8.0
7.0
6.0
5.0
4.0
3.0
2.0
1.0
0 δ[ppm]

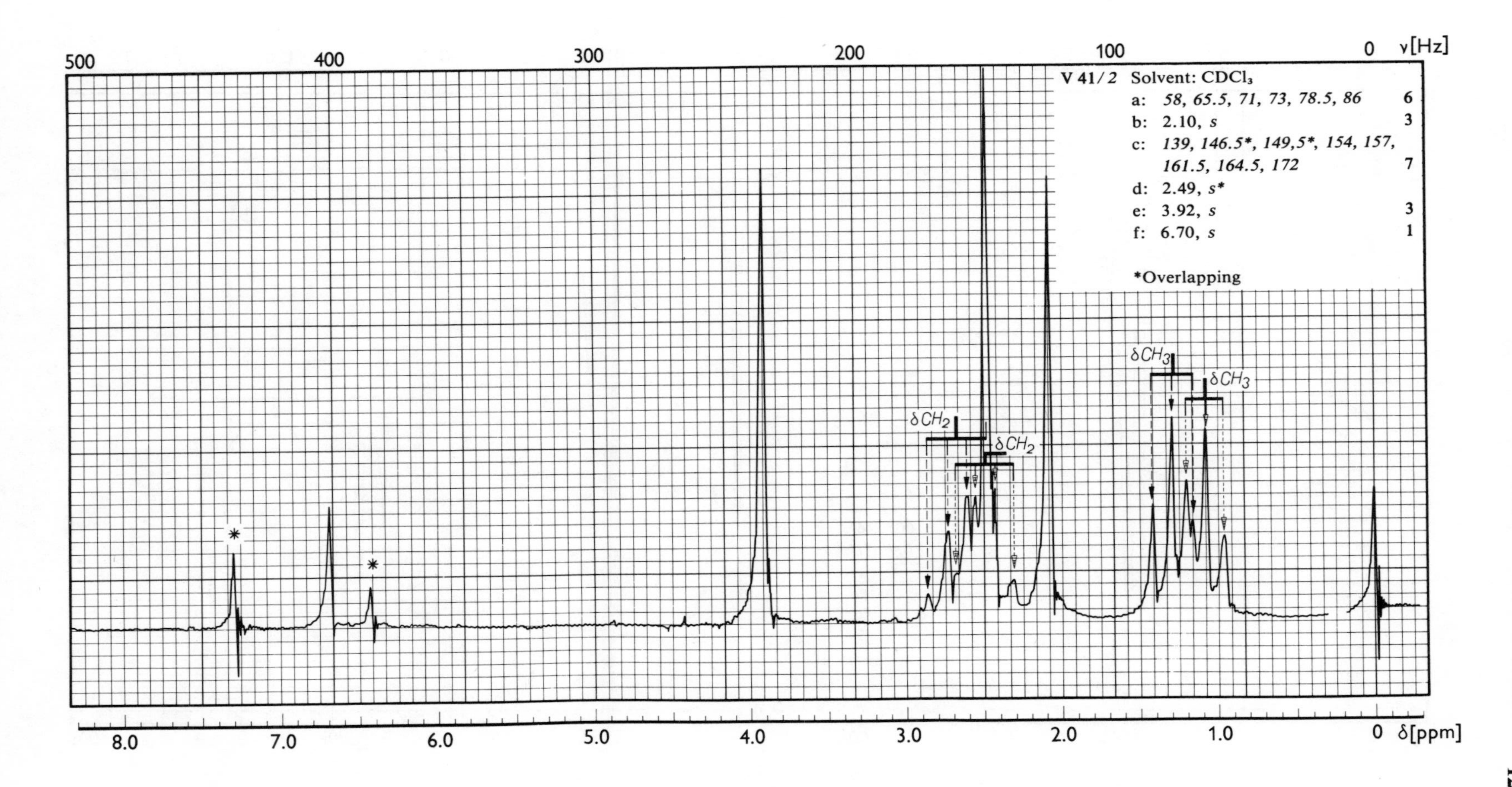
V 41/2 Solvent: $CDCl_3$
a: 58, 65.5, 71, 73, 78.5, 86 6
b: 2.10, s 3
c: 139, 146.5*, 149,5*, 154, 157, 161.5, 164.5, 172 7
d: 2.49, s*
e: 3.92, s 3
f: 6.70, s 1
*Overlapping
δCH3
δCH3
δCH2
δCH2
*
*
500
400
300
200
100
0
ν[Hz]
8.0
7.0
6.0
5.0
4.0
3.0
2.0
1.0
0
δ[ppm]

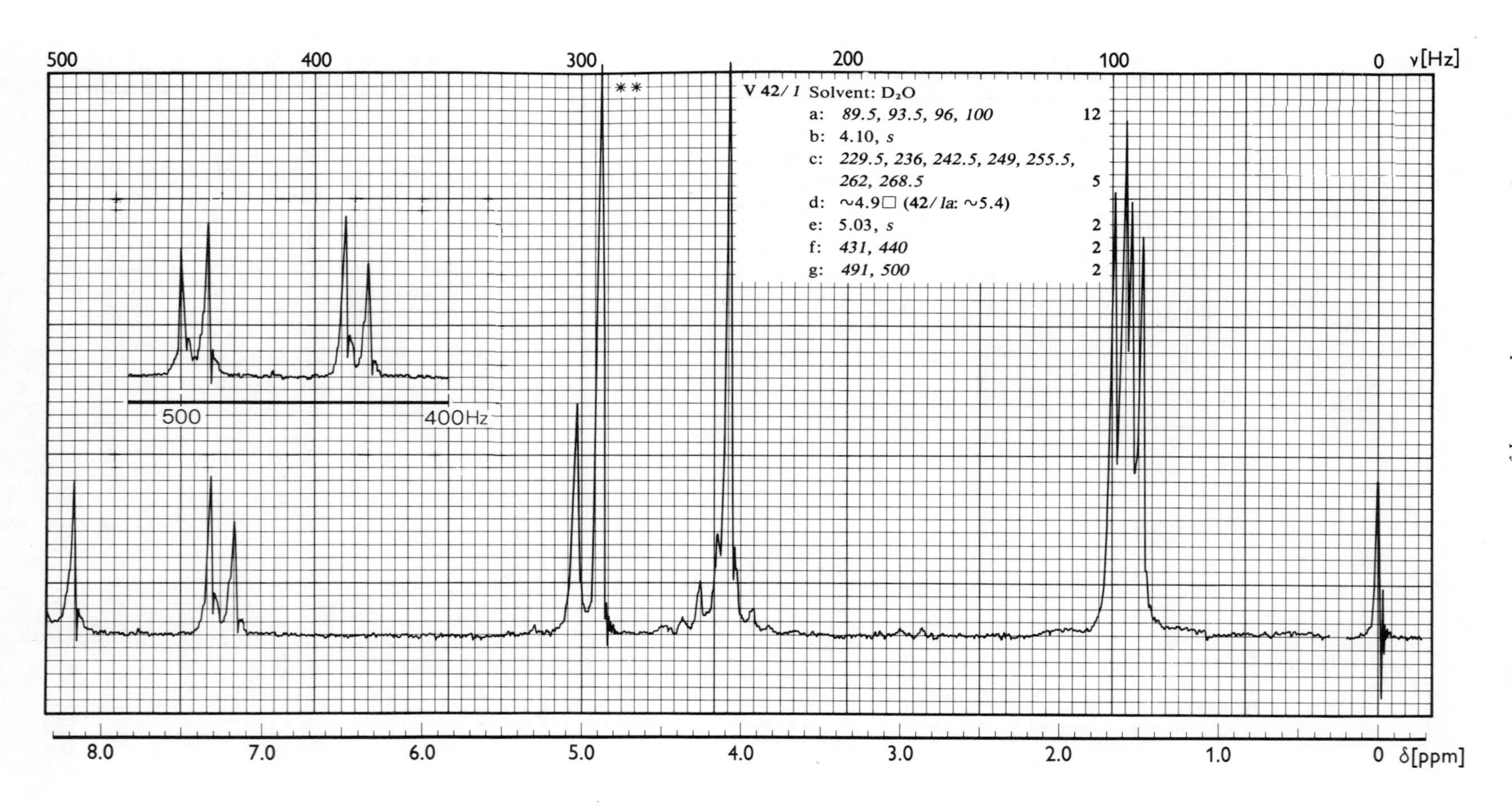
V 42/1 Solvent: D_2O
a: 89.5, 93.5, 96, 100 12
b: 4.10, s
c: 229.5, 236, 242.5, 249, 255.5, 262, 268.5 5
d: ~4.9□ (42/1a: ~5.4)
e: 5.03, s 2
f: 431, 440 2
g: 491, 500 2
**
500 400Hz
ν[Hz] 500 400 300 200 100 0
δ[ppm] 8.0 7.0 6.0 5.0 4.0 3.0 2.0 1.0 0

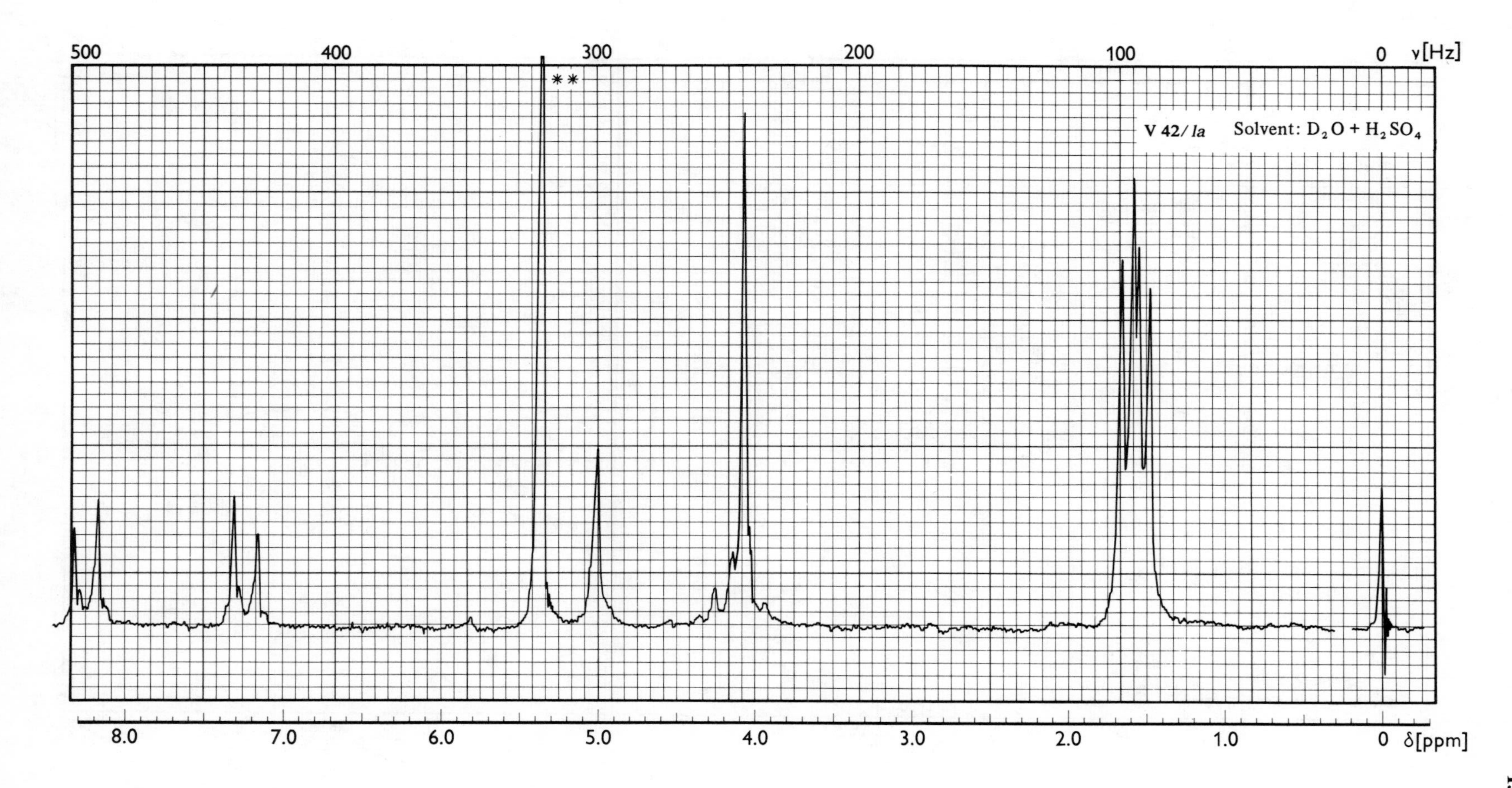
500
400
300
200
100
0
ν[Hz]
**
V 42/Ia
Solvent: $D_2O + H_2SO_4$
8.0
7.0
6.0
5.0
4.0
3.0
2.0
1.0
0
δ[ppm]

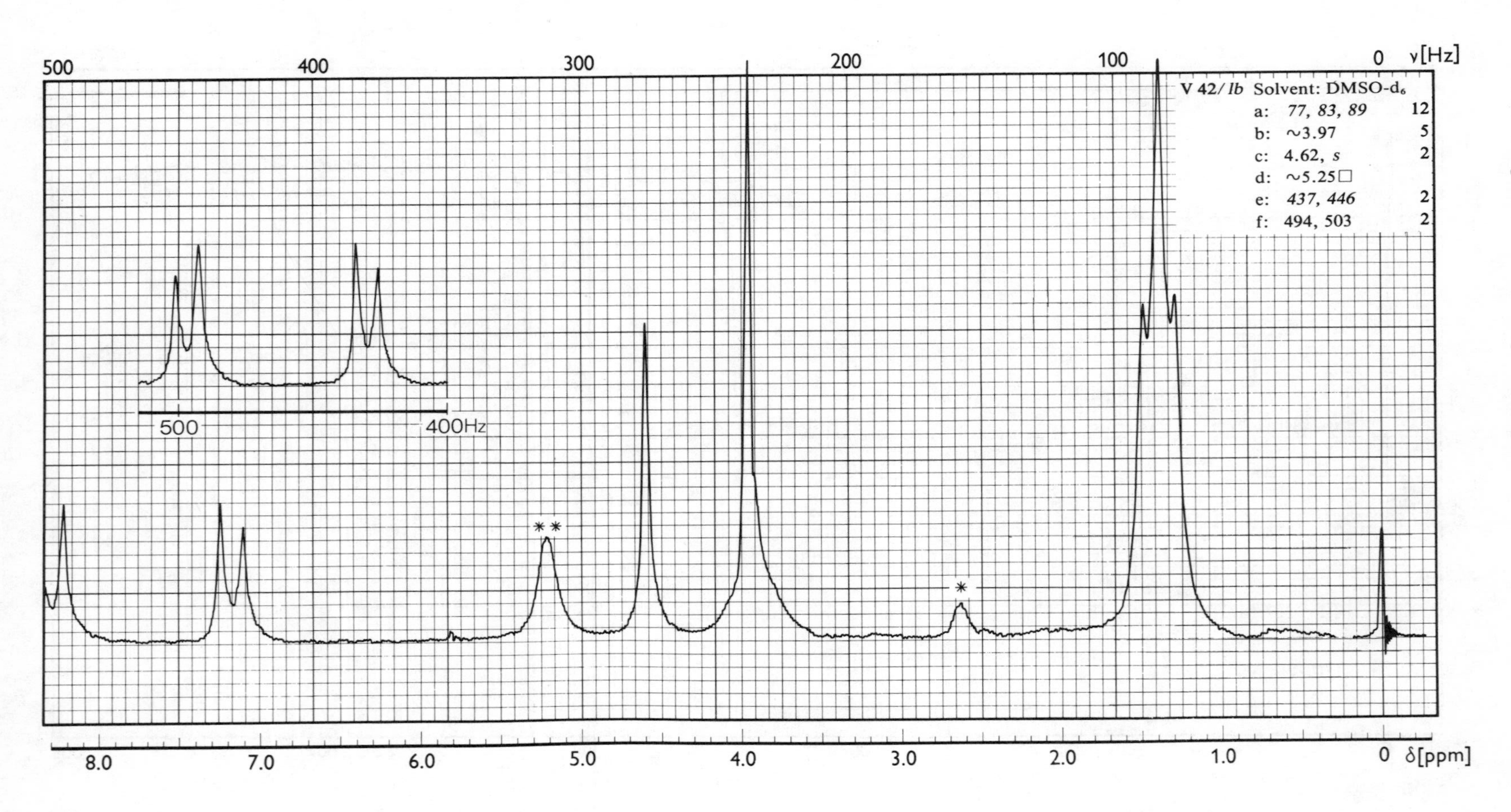
V 42/Ib Solvent: DMSO-d₆
a: 77, 83, 89 12
b: ~3.97 5
c: 4.62, s 2
d: ~5.25□
e: 437, 446 2
f: 494, 503 2
ν[Hz]
500
400
300
200
100
0
δ[ppm]
8.0
7.0
6.0
5.0
4.0
3.0
2.0
1.0
0
**
*
500
400Hz

PROBLEM 43[684,804]

On the basis of Spectrum **43**/*1* choose between isomers **I** and **II**, and assign the signals. Both isomers may be present theoretically in two conformations 8(**a** and **b**), but for steric reasons only the given structures with *equatorial* phenyl substituent must be considered.

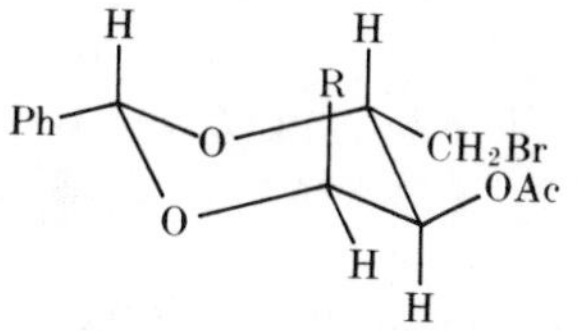

43/I

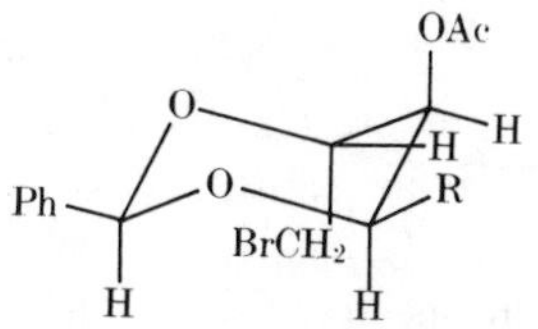

43/II

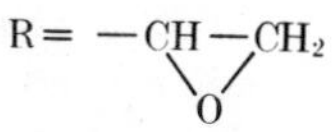

ANSWER

Isomers **I** and **II** can be distinguished by the chemical shift and splitting of the methine signal of the –CHOAc group. This methine hydrogen is *axial* in **I**, and it enters a *diaxial* coupling, presumably larger than 4 Hz, with one of the adjacent methine protons. In isomer **II**, the *equatorial* hydrogen is more deshielded (δCH > 5 ppm), and since the type of coupling with the two adjacent protons may be only *ee* or *ea*, the splitting may not be higher than 4 Hz.

The methine signal of the –CHOAc group appears in the spectrum at 5.10 ppm, and it is, in the first approximation, a triplet with $J = 2$ Hz. Accordingly, Spectrum **43**/*1* must correspond to compound **II**. The assignment of the further signals is δCH_3CO, *s*; 2.15 ppm, δCH_2 (oxyrane ring), *d*($J = 3$ Hz); 2.88 ppm, δCH (oxyrane ring), *m*: ~3.2 ppm, δCH_2Br, *d*($J = 7$ Hz): 3.73 ppm, δH-4, *m*: ~3.95 ppm, δH-2, *m*: ~4.35 ppm, δCHPh, *s*: 5.80 ppm, and νArH, *m*: 430 to 460 Hz.

Only the assignment of the methine protons adjacent to the three-membered ring and to the bromomethyl group is problematic. The above assignment has been chosen because for cyclohexane derivatives $\delta H_e > \delta H_a$. This assignment is also confirmed by the splitting patterns; the signal of the *equatorial* hydrogen is a triplet, and that of the *axial* proton closely resembles a pair of doublets. (The former proton is adjacent to a methylene, whereas the latter is adjacent to two methine groups, one of which causes much smaller — *ea* — splitting.)

PROBLEM 44

Correlate Spectra **44**/*1* and **44**/*2* with isomers **I** and **II**. Determine the value of *n* and the substituent R. Derive which of the possible tautomeric forms of compound **I** is present. To answer the questions, the spectra recorded after D_2O addition (**44**/*1a* and **44**/*2a*) may also be used.

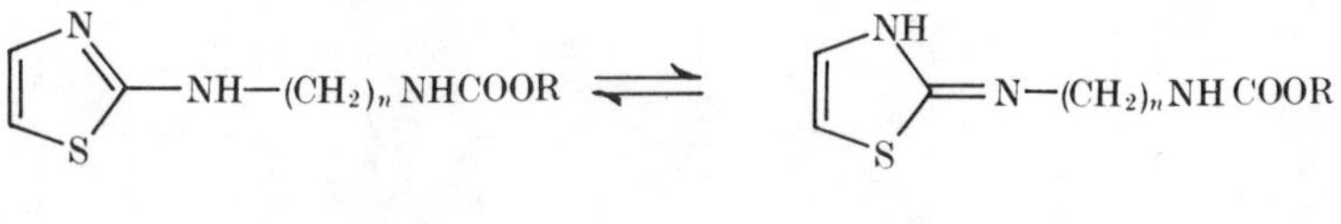

44/Ia

44/Ib

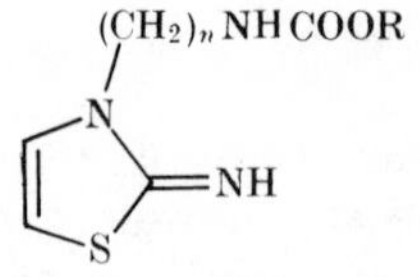

$(CH_2)_n$NHCOOR

=NH

N S

44/II

ANSWER

Since Spectra **44**/*1* and **44**/*2* contain a symmetric triplet at 1.23 ppm (J = 7 Hz) and a corresponding quartet at 4.12 ppm, substituent R is an ethyl group and the above signals constitute its A_2X_3 spectrum.

In the region $\delta < 5$ ppm in Spectrum **44**/*1*, a triplet and a quartet-like multiplet of 2-2H intensity can be observed between 195 and 240 Hz, overlapping in part with the ethyl quartet, and a doublet at 3.45 ppm (J = 3 Hz) of 4H intensity in Spectrum **44**/*2*. Therefore, the molecule contains two further methylene groups, i.e., n = 2. The two hydrogens of the heteroring give, in both cases, an AX spectrum with δA = 6.44 and 7.10 ppm, δB = 5.78 and 6.46 ppm, and J_{AX} = 5 and 3.5 Hz, respectively.

In the aromatic Structure **Ia**, the delocalized π-electrons induce a local field causing a substantial deshielding of the ring protons. Consequently, the spectrum in which the AX lines have higher frequencies will belong to isomer **I**, present in tautomeric form **Ia**. On this basis Spectrum **44**/*2* can be assigned to Structure **Ia** (R = Et, n = 2) and Spectrum **44**/*1* can be assigned to Structure **II** (R = Et, N = 2).

In Spectrum **44**/*1* of compound **II** the imino group has a relatively sharp signal at 5.65 ppm, and the broad absorption of the urethane–NH group, showing a perceptible triplet splitting, can be found at approximately 6.15 ppm. The relative shifts can be interpreted by the fact that the imino–NH group is more basic and forms weaker hydrogen bonds than the other NH group, and thus its chemical shift is smaller. This assignment is proved by the triplet-like structure of the other signal, because the urethane–NH proton is adjacent to a methylene group, which latter may cause a triplet splitting through the NH–CH_2 coupling. After addition of D_2O, the quartet-like multiplet at 3.45 ppm of one of the methylene groups collapses into a line group resembling a triplet, and both NH signals disappear simultaneously, being replaced by an overlapping NH–OH signal at approximately 4.6 ppm (see Spectrum **44**/*1a*). These facts can also be used to assign the methylene signals. The signal at 3.45 ppm must arise from the methylene group adjacent to the urethane group, whereas the one around 3.85 ppm, with an approximate triplet structure, is due to the methylene hydrogens adjacent to the ring nitrogen. The latter protons give an A_2B_2 spectrum, very close to the limiting A_2X_2 case, explaining why the signal resembles a triplet.

According to Spectrum **44**/*2* of isomer **I** the methylene groups are practically isochronous, and the splitting (3 Hz) is due only to the NH–CH_2 coupling. This conclusion is proved by Spectrum **79**/*2a* (recorded after D_2O addition), in which the methylene signal collapses into a singlet, and the broad NH signals, appearing originally at 5.85 and approximately 6.5 ppm, disappear, in favor of a broad common maximum at about 4.6 ppm. The isochrony of the methylene groups is in accordance with Structure **Ia**, because the substituents of the two groups are, quite similar, whereas they are widely different in the case of **Ib** or **II**.

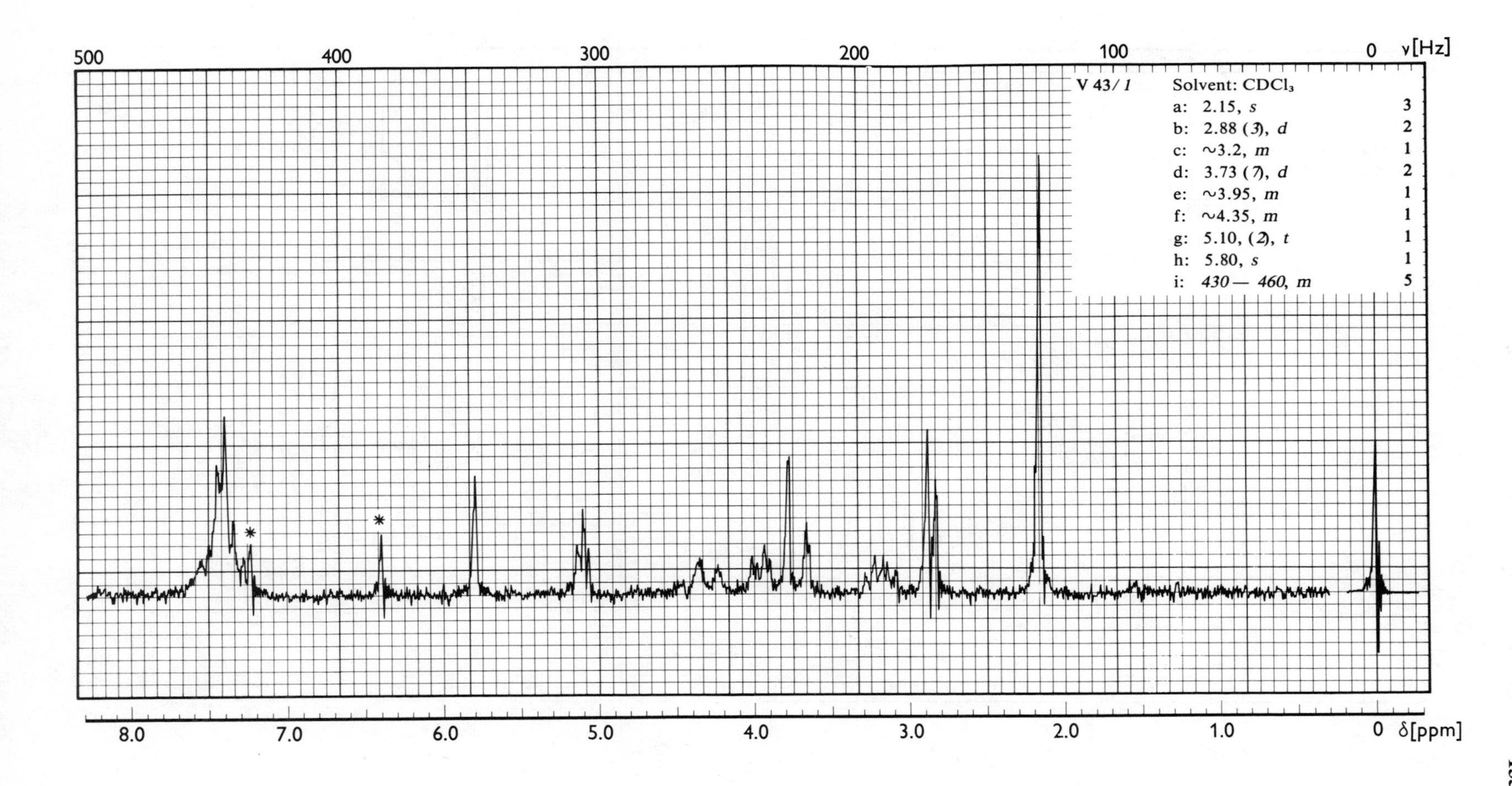

500
400
300
200
100
0 ν[Hz]
V 43/1
Solvent: CDCl₃
a: 2.15, s 3
b: 2.88 (3), d 2
c: ~3.2, m 1
d: 3.73 (7), d 2
e: ~3.95, m 1
f: ~4.35, m 1
g: 5.10, (2), t 1
h: 5.80, s 1
i: 430— 460, m 5
*
*
8.0
7.0
6.0
5.0
4.0
3.0
2.0
1.0
0 δ[ppm]

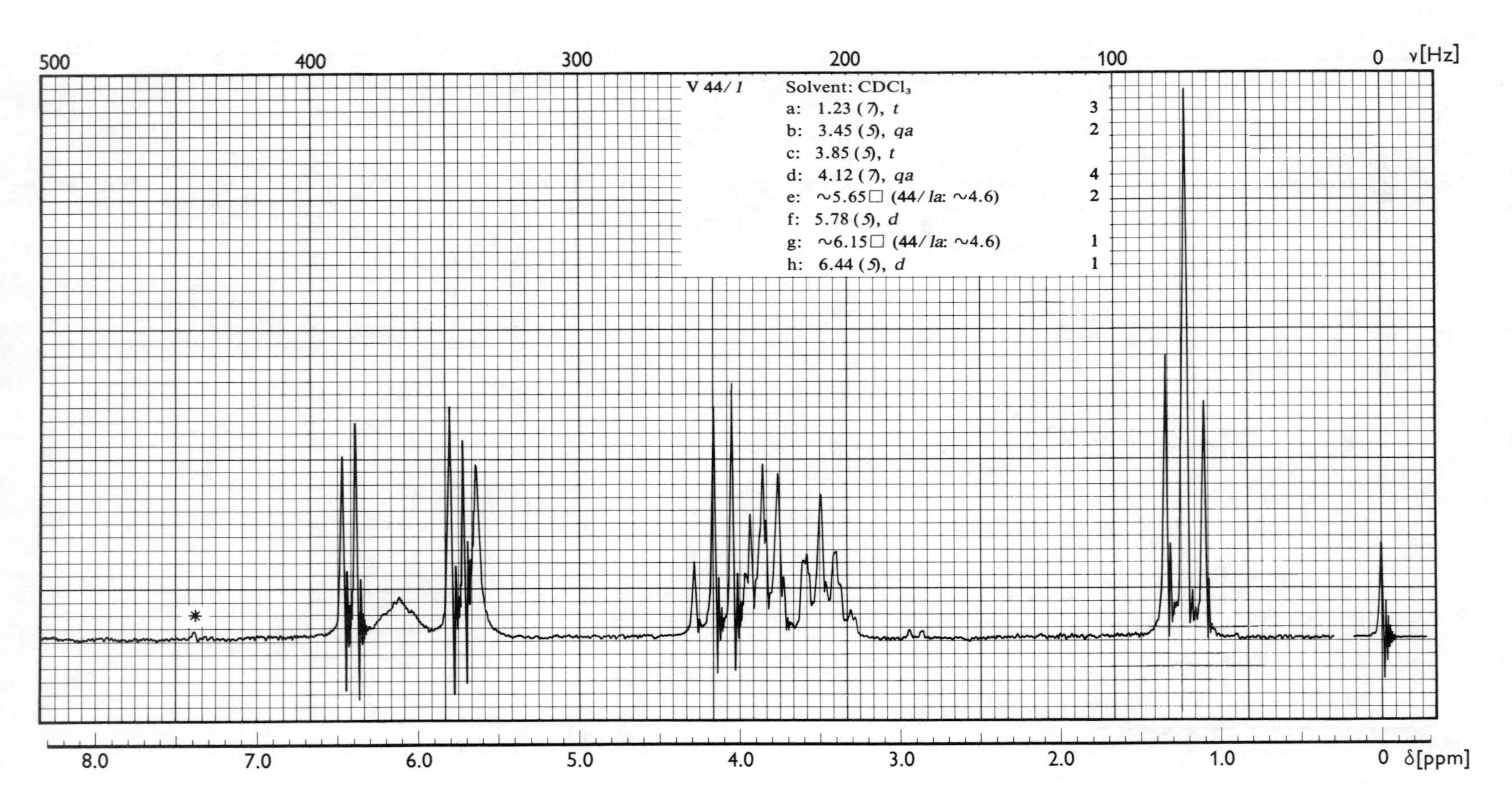

500
400
300
200
100
0
ν[Hz]
V 44/1
Solvent: $CDCl_3$
a: 1.23 (7), t 3
b: 3.45 (5), qa 2
c: 3.85 (5), t
d: 4.12 (7), qa 4
e: ~5.65□ (44/1a: ~4.6) 2
f: 5.78 (5), d
g: ~6.15□ (44/1a: ~4.6) 1
h: 6.44 (5), d 1
*
8.0
7.0
6.0
5.0
4.0
3.0
2.0
1.0
0
δ[ppm]

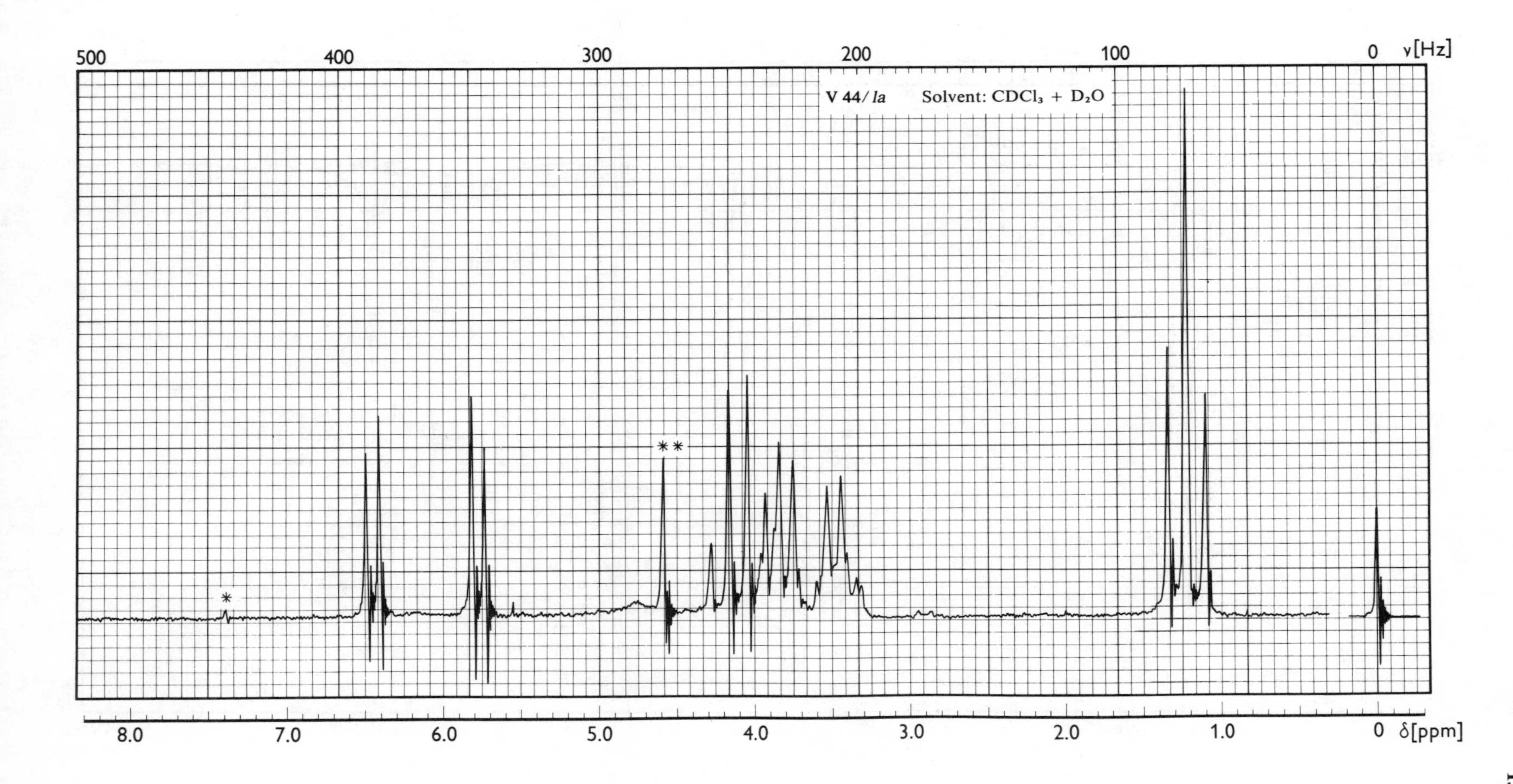

500
400
300
200
100
0
ν[Hz]
V 44/Ia
Solvent: CDCl₃ + D₂O
*
**
8.0
7.0
6.0
5.0
4.0
3.0
2.0
1.0
0
δ[ppm]

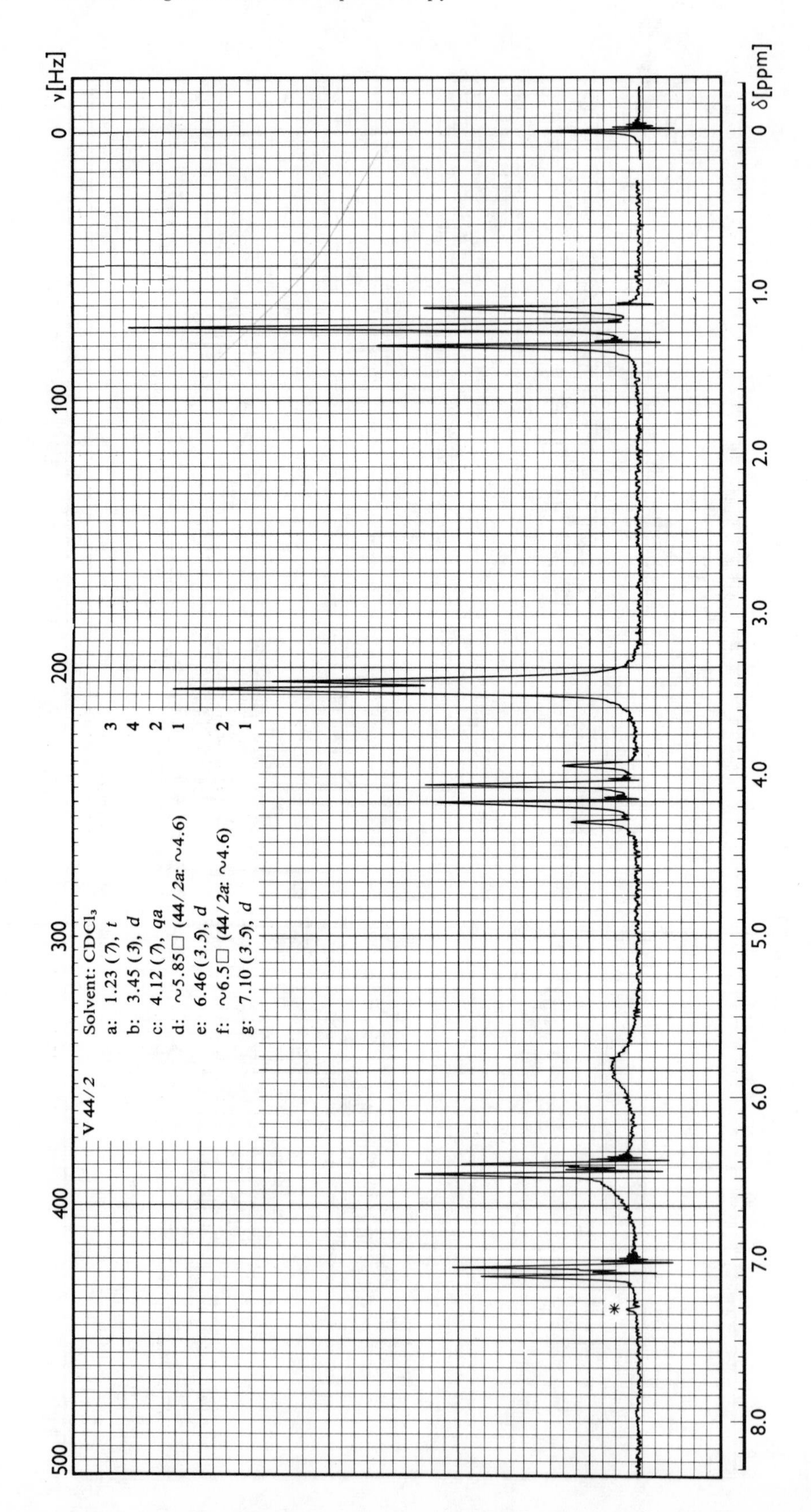
ν[Hz]
0
100
200
300
400
500
δ[ppm]
0
1.0
2.0
3.0
4.0
5.0
6.0
7.0
8.0
V 44/2
Solvent: CDCl₃
a: 1.23 (7), t 3
b: 3.45 (3), d 4
c: 4.12 (7), qa 2
d: ~5.85□ (44/2a: ~4.6) 1
e: 6.46 (3.5), d
f: ~6.5□ (44/2a: ~4.6) 2
g: 7.10 (3.5), d 1
*

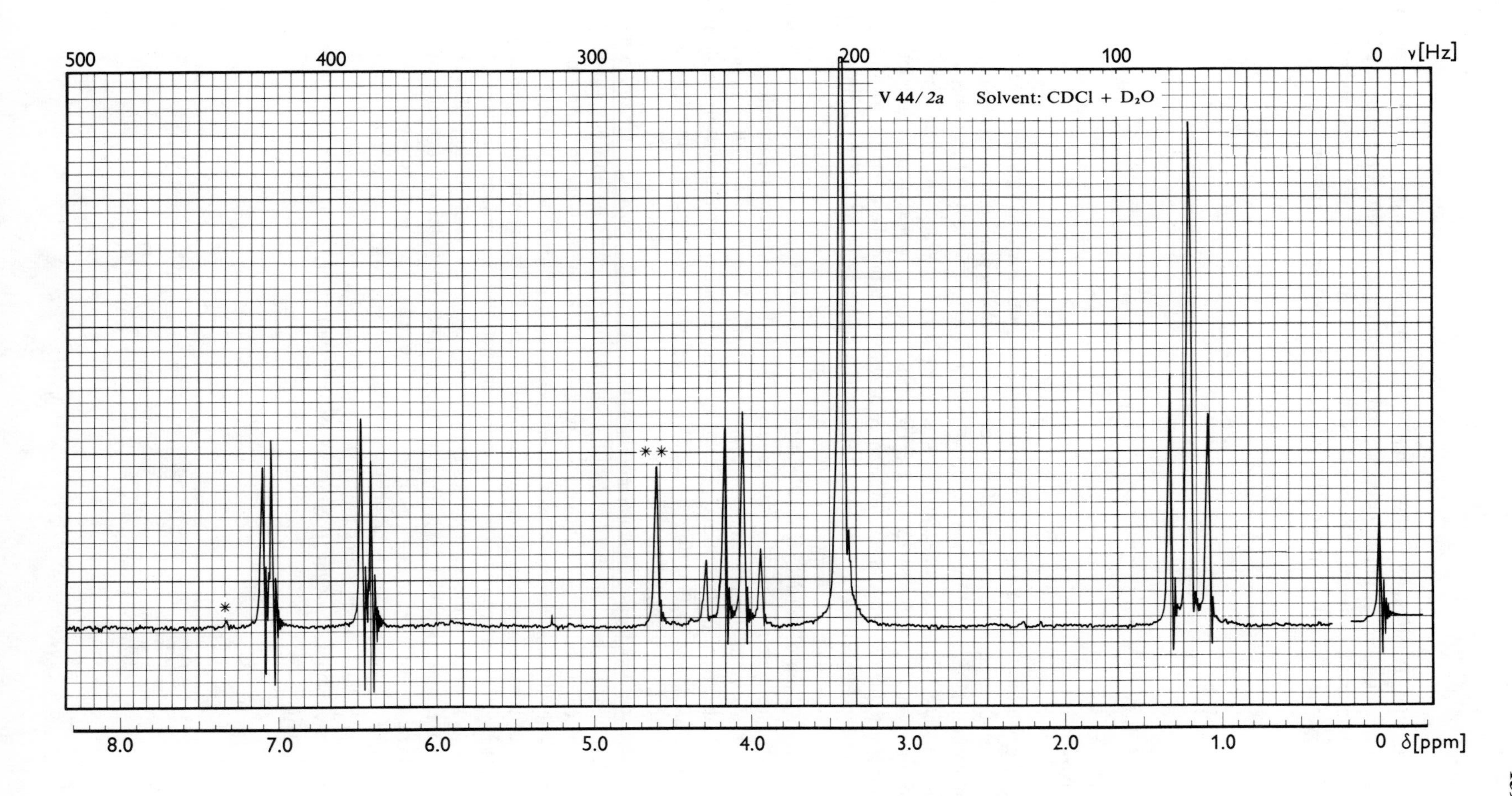
500
400
300
200
100
0 ν[Hz]
V 44/2a Solvent: CDCl + D$_2$O
*
**
8.0
7.0
6.0
5.0
4.0
3.0
2.0
1.0
0 δ[ppm]

PROBLEM **45**[119,1320,1322]

Determine the difference between the two substances both characterized by Structure **I**, but having different chemical, physical, and spectroscopic properties (see Spectra **45**/*1* and **45**/*2*). Interpret the spectra, calculate the determinable spectrum parameters, and explain on this basis the above differences. The spectra are independent of temperature.

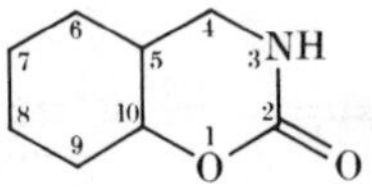

45/I

ANSWER

The two spectra are quite similar and both are compatible with Structure **I**. The methylene protons of the saturated ring and H-5 give rise to an overlapping, uncharacteristic absorption in the upfield region of the spectra between 70 to 130 and 50 to 150 Hz, respectively. The 4-methylene hydrogens adjacent to the nitrogen are deshielded and their signal appears separately (170 to 220 and 155 to 215 Hz, respectively), as well as the multiplet of H-10 adjacent to the oxygen and thus even more downfield shifted (between 255 to 280 and 210 to 250 Hz, respectively). Finally, a broad NH signal appears at larger chemical shifts at ~7.0 and ~6.95 ppm, respectively (urethane group, e.g., see Figure 154).

The structure of the NCH_2 multiplet (12 lines) reveals that H-4*a*, 4*e*,5,10 represent an *ABMX* spin system. The *M* and *X* signals split further owing to the couplings with the adjacent methylene hydrogens of the cyclohexane ring. The values δA, δB, δX, and J_{AB} can be evaluated from the *AB* and *X* parts of the *ABMX* multiplets, respectively, and in first-order *AMX* approximation (for H-4*a*,*e*,5) the approximate value of J_{AX} and J_{BX} can also be derived from the width of the *X* signal.

Before calculating these parameters, however, the difference between the structures must be determined. The close similarity of the spectra and the common Structure **I** indicate that the two compounds may be different in configuration or possibly in conformation. As the spectra are different independently of temperature, the former is the case, i.e., the two compounds must be a pair of *cis* and *trans* isomers. The temperature invariance indicates that the systems are conformationally homogeneous, i.e., the majority of the molecules are in the most stable conformation. With the *trans* isomer, the configuration also determines the conformation, in which both rings are in chair form (**II**). In this conformation the oxygen atom and the 4-methylene group) in *trans* position relative to the cyclohexane ring are both *equatorial*. (Note that the bond angles are slightly distorted by the incorporated carbamoyl group, and thus the positions are in fact *pseudo-axial* and *pseudo-equatorial*, like in cyclohexene, which, however, need not be considered further, since it has no major effect on the spectra.)

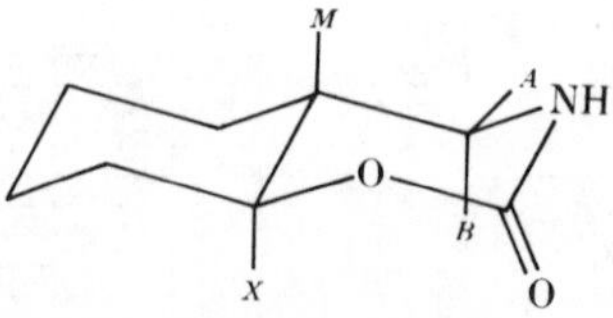

45/II

The *cis* isomer may assume two such arrangements in which the rings have chair conformations. In one of them (**III**) the 4-methylene group is *equatorial* and the oxygen atom is *axial* with respect to the cyclohexane ring, whereas in the other conformer (**IV**) the situation is reversed.

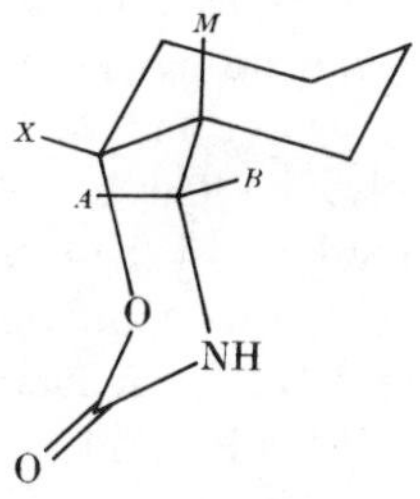

45/III 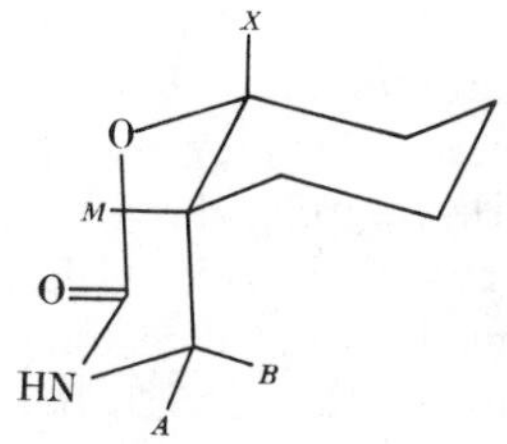 **45/IV**

A comparison of Structures **II, III**, and **IV** reveals that the answer can be based on the relative magnitudes of the coupling constants and chemical shifts of the *AMBX* spin systems. Since $\delta H_e > \delta H_a$ for cyclohexanes (see Table 31), δM_a(**II**) $\approx \delta M_a$(**III**) $< \delta M_e$(**IV**) and δX_a(**II**) $\approx \delta X_a$(**IV**) $< \delta X_e$(**III**). Moreover, because $J_{aa} > J_{ae} \approx J_{ee}$ for the cyclohexanes, J_{MX}(**II**) $> J_{MX}$(**III**) $\approx J_{MX}$(**IV**) and J_{BM}(**II**) $\approx J_{BM}$(**IV**) $> J_{BM}$(**III**), since J_{MX} in Structure **II** and J_{BM} in Structures **II** and **IV** correspond to *diaxial* couplings, whereas the other interactions are *equatorial-axial*.

If, therefore, we can determine δM, δX, J_{MX}, and J_{BX}, it will be easy to distinguish Structures **II**, **III**, and **IV**. The value of δM is inaccessible, since the signal overlaps with absorption of the cyclohexane protons. J_{MX} cannot be obtained either, from the coalesced X multiplet, but it can be estimated from the width of this signal. This parameter, together with J_{BX} and δX is sufficient to correlate the structures with the spectra.

The value of δX is given by the midpoint of the X multiplet. Hence, in Spectrum **45**/*1*, $\delta X = 4.45$ ppm, and in Spectrum **45**/*2*, $\delta X = 3.87$ ppm. As there is a substantial difference between these values, the X proton must be *equatorial* in one of the structures and *axial* in the other, i.e., Spectrum **45**/*1* corresponds to the *cis* isomer in conformation **III**, whereas Spectrum **45**/*2* is compatible with the *trans* structure. Structure **IV** is excluded unambiguously by the widely different magnitudes of J_{MX} in the two spectra, too. As mentioned above, the relative magnitude of J_{MX} can also be deduced from the X part, by determining the half band width of this signal. This is 8 Hz in Spectrum **45**/*1* and 25 Hz in Spectrum **45**/*2*. Consequently, it is an independent evidence of the postulated structures.

To return to the AB part, the 4*e* and 4*a* methylene hydrogens are nonequivalent, and due to J_{AM} and J_{BM} couplings, they give 8 lines. As mentioned, however, the AB part of the spectra consists of 12 lines. This further splitting is due to long-range coupling with the X proton, which is attributed to another, although longer, "path" of interaction through the heteroatoms.* By calculating the midpoints of the doublets and quartets, the multiplet can be reduced into a simple AB quartet, from which δA, δB, and J_{AB} can be evaluated. Since in Spectrum **45**/*1* of the *cis* isomer **III** $\delta A_a < \delta B_e$, the B lines split into doublets (200, 204, 212, and 216 Hz), whereas the A lines form quartets (173, 175, 177, 179, 185, 187, 189, and 191 Hz). The doublet splitting is $J_{BM} \approx 4$ Hz, and the value of J_{AM} is also about 4 Hz. The long-range coupling is $J_{AX} \approx 2$ Hz, in the first-order approximation. The lines of the

* See Volume I, p. 67.

AB quartet appear therefore at (200 + 204)/2 = 202 Hz and 214 Hz, (175 + 177)/2 = 176 and 188 Hz. From these frequencies δA = 3.06, δB = 3.43 ppm and J_{AB} = 12 Hz.

In Spectrum **45**/*2* of the *trans* isomer (**II**), the *AB* multiplet is shifted upfield. (Owing to a different position of the heteroring, the bonds of the methylene hydrogens are approximately perpendicular to the plane of the cyclohexane ring, and, similar to the *axial* ring hydrogens, they are subjected to the anisotropic effect of the C–C bonds, causing diamagnetic shift. In the *cis* isomer the bonds of the methylene hydrogens are parallel to the plane of the ring, similar to the *equatorial* protons.) In this case $\delta A_e > \delta B_a$, and the long-range coupling acts again between the *A* and *X* hydrogens. Accordingly, the lines at 190, 192, 195, 197, 202, 204, 207, and 209 Hz comprise the *A* part, and the lines of the *B* part appear at 164, 175, 176, and 187 Hz. These frequencies yield in the first-order approximation $J_{AM} \approx 5$ Hz, $J_{BM} \approx 11$ Hz, and, for the long-range coupling, $J_{AX} \approx 2$ Hz. By reducing the multiplet into an *AB* quartet (with lines at 194.5, 205.5, 169.5, and 181.5 Hz), δA = 3.30, δB = 2.95 ppm, and J_{AB} = 12 Hz. J_{BM} is much higher (11 Hz) for the *trans* isomer (**II**) than for the **III** *cis* isomer (4 Hz), as expected, which supports further the proposed steric structures.

PROBLEM 46[1188,1189,1340]

Correlate isomeric Structures **I** and **II** (R = Me) with Spectra **46**/*1* and **46**/*2* and then determine the tautomeric structure of the *N*-unsubstituted derivative (R = H) from its spectrum recorded in D_2O solution before and after acidification, respectively (**46**/*3* and **46**/*3a*).

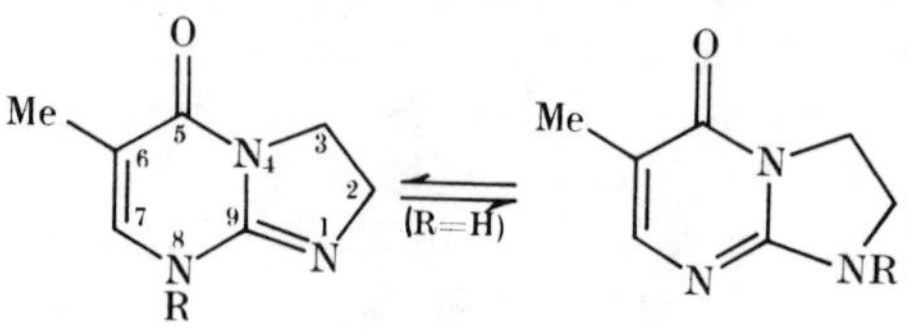

46/I **46/II**

ANSWER

There must be substantial differences between the spectra corresponding to Structures **I** and **II** (R = Me):

1. The olefinic proton in Structure **II** is presumably deshielded, owing to the two double bonds in the six-membered ring.
2. An effect of the same direction, but much smaller in magnitude, is also conceivable for the signal of the *C*-methyl group.
3. The *N*-methyl signal can be expected to undergo an opposite and significant shift.
4. The chemical shift difference of the methylene groups is larger for compound **II**, and thus its methylene multiplet must be closer to the first-order A_2X_2 case than that of isomer **I**. As usual in five-membered saturated ring systems with two heteroatoms, the methylene proton pairs are isogamic and the $AA'BB'$ spin system is simplified into an A_2B_2 case (compare Problems **18** and **33**).

The assignment of the spectra is as follows: $\delta CH_3(6)$, *d* = 1.83 (**I**) and 1.92 (**II**) ppm; δNCH_3, *s*: 3.30 (**I**) and 2.98 ppm (**II**); δCH_2: 3.95 ppm, singlet-like signal (**I**) and 3.63 and

4.13 ppm from the $AA'BB'$ multiplet by A_2B_2 approximation J_{AB} = 9 Hz (**II**): H-7, *qa*: 6.70 (**I**) and 7.48 ppm (**II**).

The doublet splitting of the *C*-methyl signal and the quartet structure of the H-7 signal can be attributed to an allylic coupling (^4J = 1.5 Hz) which is a frequent phenomenon transmitted by the π-electrons with unsaturated CH_3–C=CH groups mainly in cyclic systems.* The lines of the quartet are not separated, but the half-width of the signal (4.5 Hz) is exactly the anticipated value. The sample corresponding to Spectrum **46**/*1* was slightly contaminated; the signal at 2.80 ppm is due to its impurity.

Since all the 14 lines of the A_2B_2 multiplet are separated, the chemical shifts and the coupling constant can be calculated directly. As line **4** from the midpoint is stronger than line **3**, the former yields δ*A* and δ*B* and J_{AB} is one half of the distance between lines **2** and **7**, from either part of the symmetric multiplet (see Problem **29**). The spectra indicate that, on the basis of all presumed differences, Structure **1** corresponds to Spectrum **46**/*1* and **II** corresponds to Spectrum **46**/*2*.

It can be seen from the spectra that δH-7 (**I**) < δH-7(**II**) and δNCH_3(**I**) > δNCH_3(**II**). In Spectrum **46**/*1* of compound **I**, the methylene multiplet is much closer to a singlet structure, whereas in Spectrum **46**/*2*, i.e., in Structure **II**, ΔδAB is significantly larger. The relation δCCH_3(**I**) < δCCH_3(**II**) also holds, but the difference is in the order of experimental errors.

On the above basis it is already easy to solve the problem of tautomeric structures by means of Spectrum **46**/*3*, in which the broad, singlet-like methyl and δH-7 signals can be found at 1.88 and 7.48 ppm, respectively, and the methylene groups have an $AA'BB'$ multiplet. Both data support tautomeric Structure **II** (R = H), in accordance with the expectation (more extensive conjugation in **II**).

The data from the $AA'BB'$ multiplet, calculated in the A_2B_2 approximation, are as follows: δ*A* ≈ 4.17, δ*B* ≈ 3.75 ppm, and J_{AB} = 9 Hz. The downfield signal (δH_A) clearly belongs to the 3-methylene group adjacent to the amide nitrogen, like in Spectrum **46**/*2*.

The NH signal appears at approximately 4.7 ppm, overlapping with the signal of the light isotope content of the solvent. This assignment is also proved by Spectrum **46**/*3a* recorded after acidification, in which this signal shifts to 5.35 ppm. The $AA'BB'$ multiplet also undergoes a substantial shift (δ*A* ≈ 4.12 and δ*B* ≈ 4.35 ppm), due presumably to protonation. Note that in Spectrum **46**/*3a* the splitting of the methyl and methine signals can be clearly seen, which cannot be observed in the original spectrum. This difference arises from a better adjustment of the field homogeneity during the measurement, which results in higher resolution. This phenomenon well demonstrates that even a slightly worse operation of the spectrometer (also compare the shapes of the TMS signals in Spectra **46**/*3* and **46**/*3a*) may cause important information to be lost.

* See Volume I, p. 64.

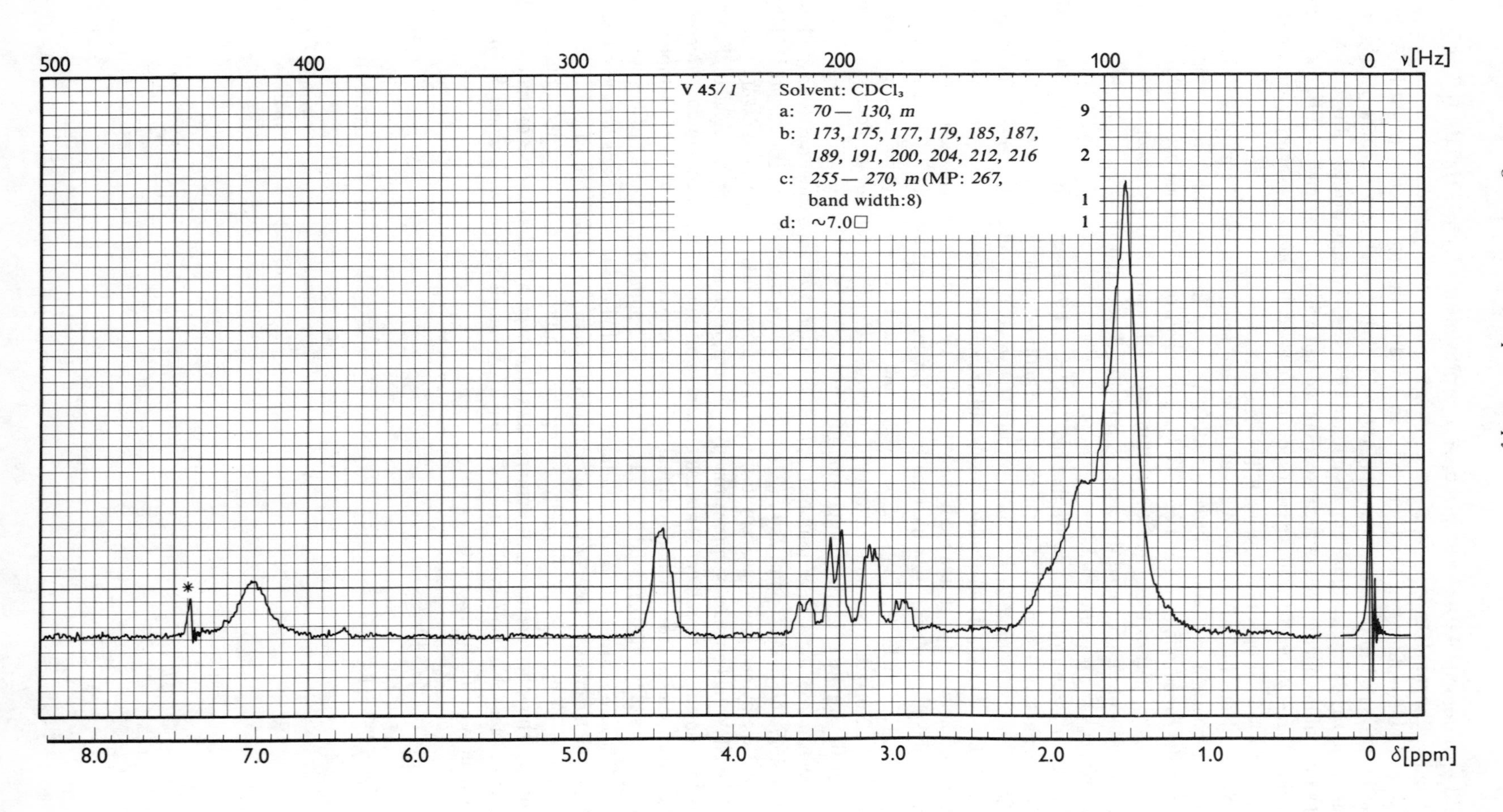
V 45/1
Solvent: $CDCl_3$
a: 70 — 130, m 9
b: 173, 175, 177, 179, 185, 187, 189, 191, 200, 204, 212, 216 2
c: 255 — 270, m (MP: 267, band width:8) 1
d: ~7.0□ 1
500
400
300
200
100
0
ν[Hz]
*
8.0
7.0
6.0
5.0
4.0
3.0
2.0
1.0
0
δ[ppm]

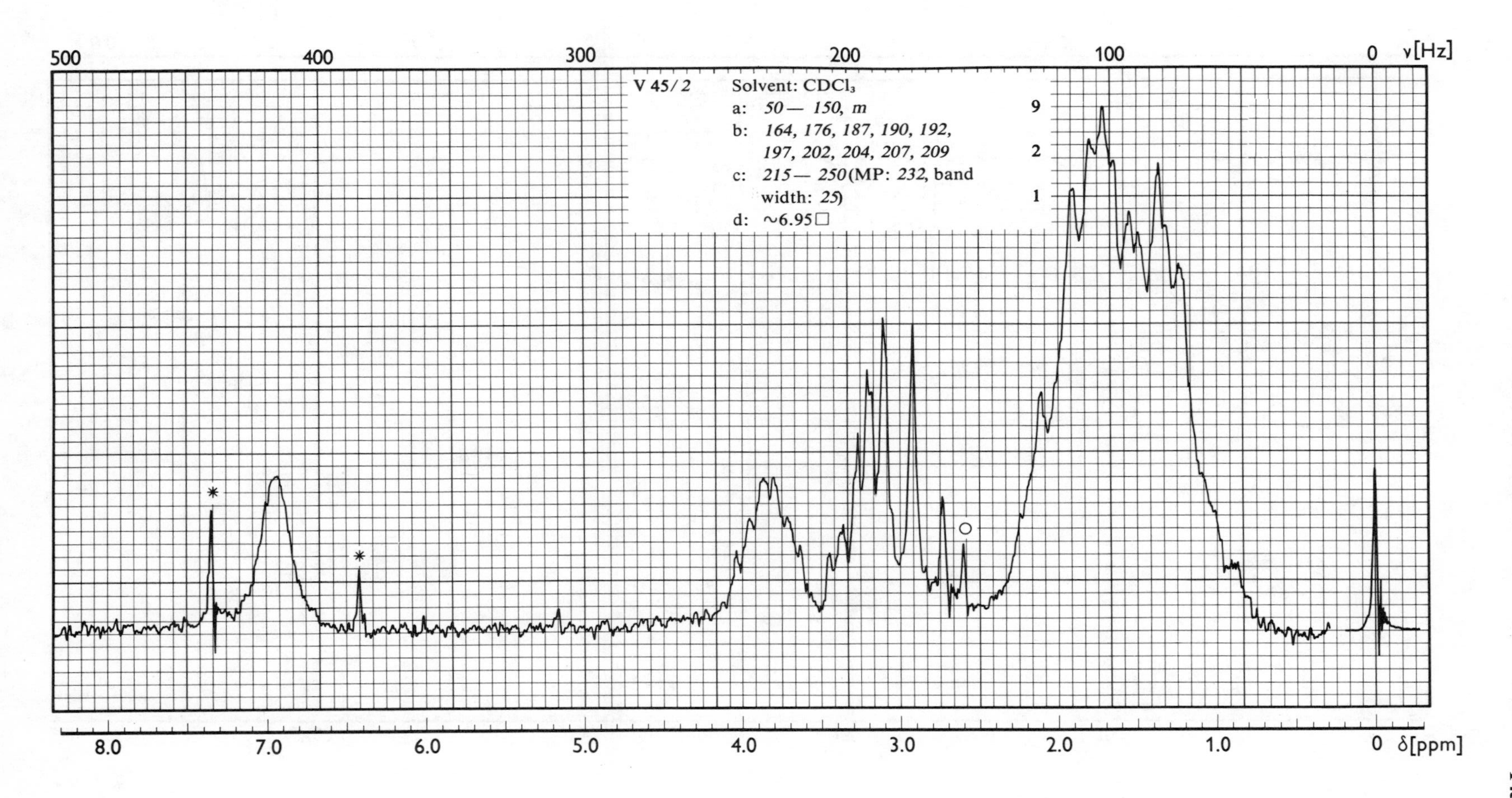
500
400
300
200
100
0 ν[Hz]
V 45/2
Solvent: $CDCl_3$
a: 50 — 150, m 9
b: 164, 176, 187, 190, 192, 197, 202, 204, 207, 209 2
c: 215 — 250 (MP: 232, band width: 25) 1
d: ~6.95□
8.0
7.0
6.0
5.0
4.0
3.0
2.0
1.0
0 δ[ppm]

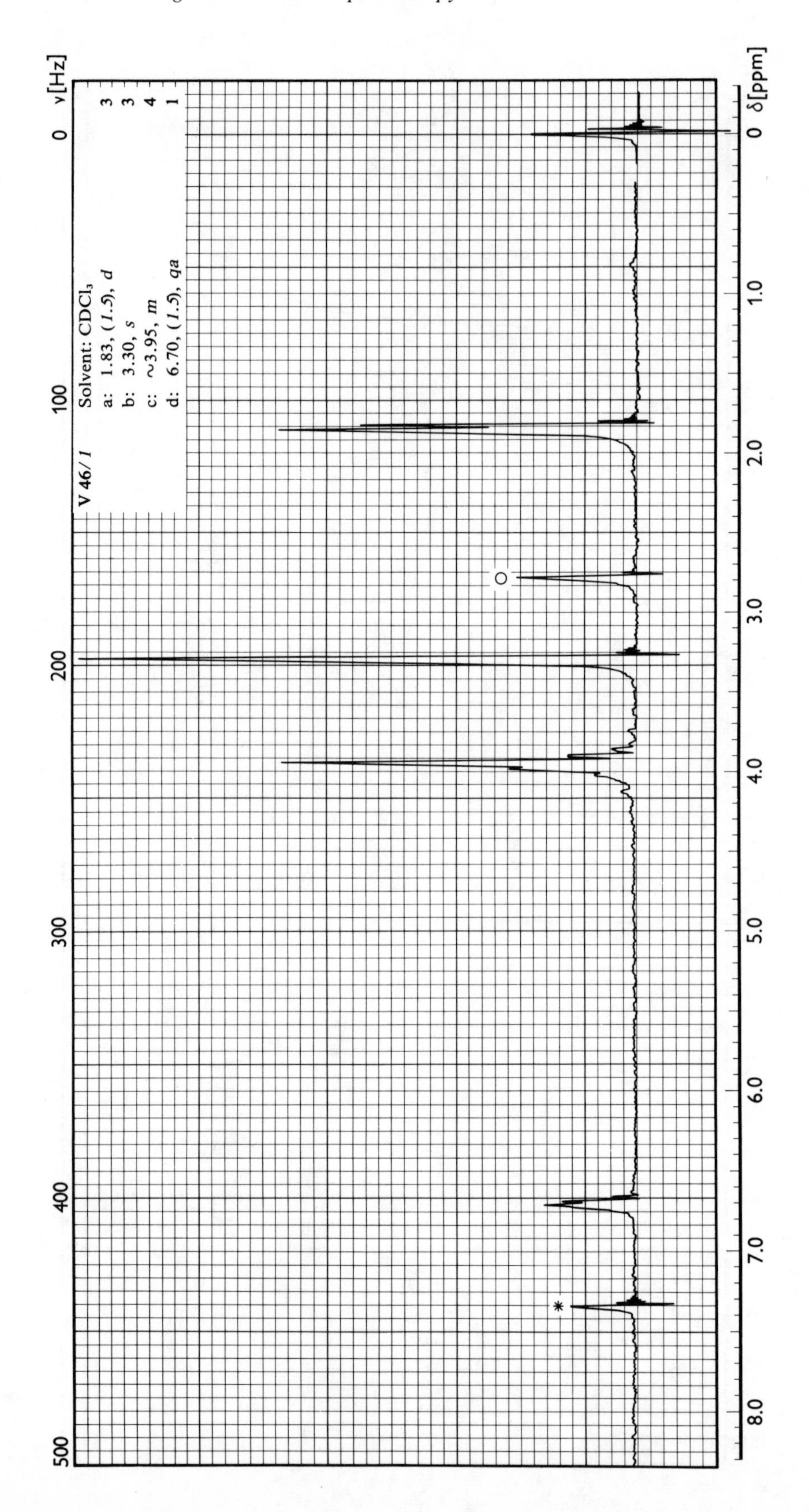
V 46/1
Solvent: $CDCl_3$
a: 1.83, (1.5), d 3
b: 3.30, s 3
c: ~3.95, m 4
d: 6.70, (1.5), qa 1
ν[Hz]
0
100
200
300
400
500
δ[ppm]
0
1.0
2.0
3.0
4.0
5.0
6.0
7.0
8.0

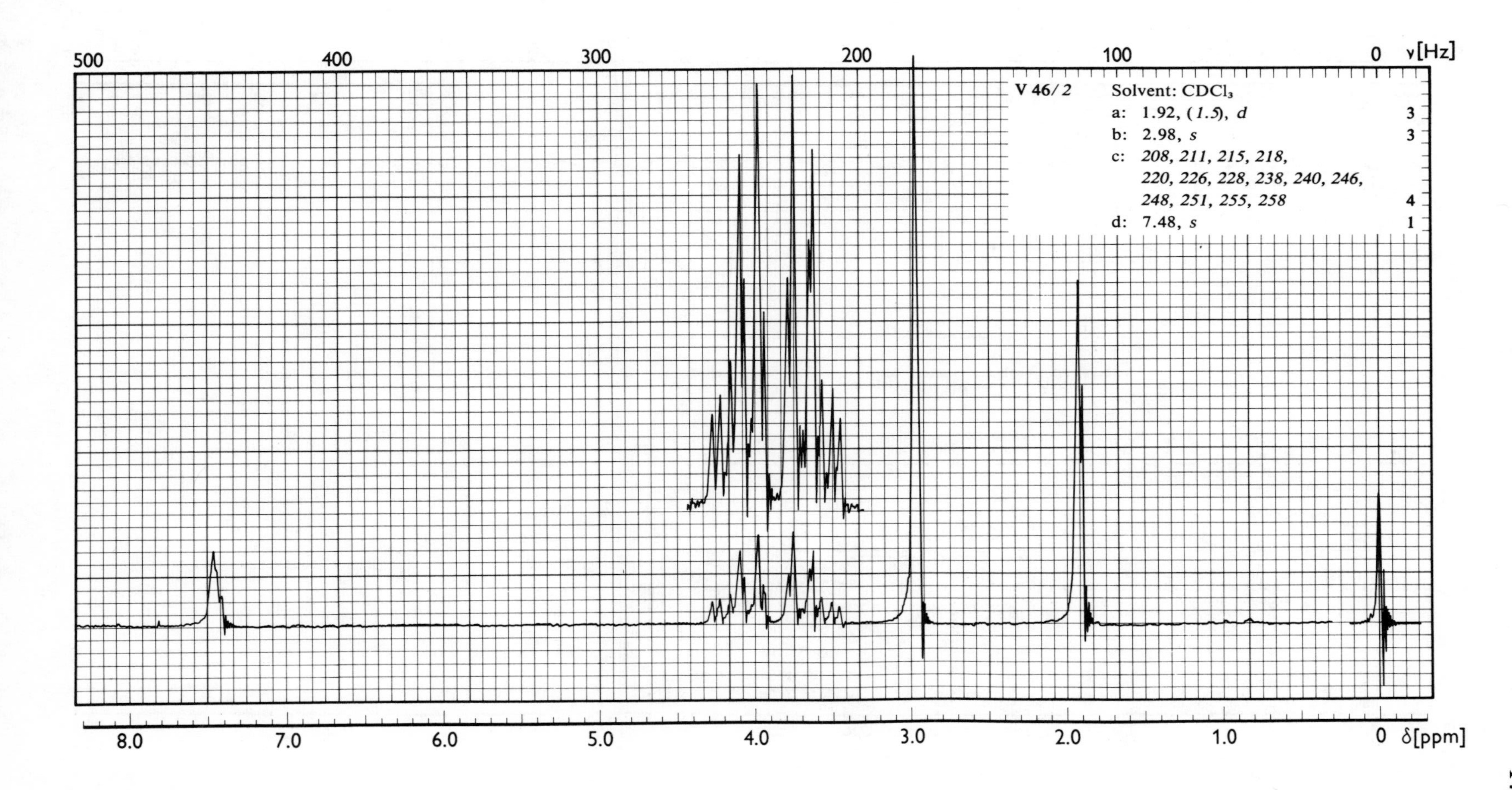

500
400
300
200
100
0 ν[Hz]
V 46/ 2
Solvent: CDCl₃
a: 1.92, (1.5), d 3
b: 2.98, s 3
c: 208, 211, 215, 218,
220, 226, 228, 238, 240, 246,
248, 251, 255, 258 4
d: 7.48, s 1
8.0
7.0
6.0
5.0
4.0
3.0
2.0
1.0
0 δ[ppm]

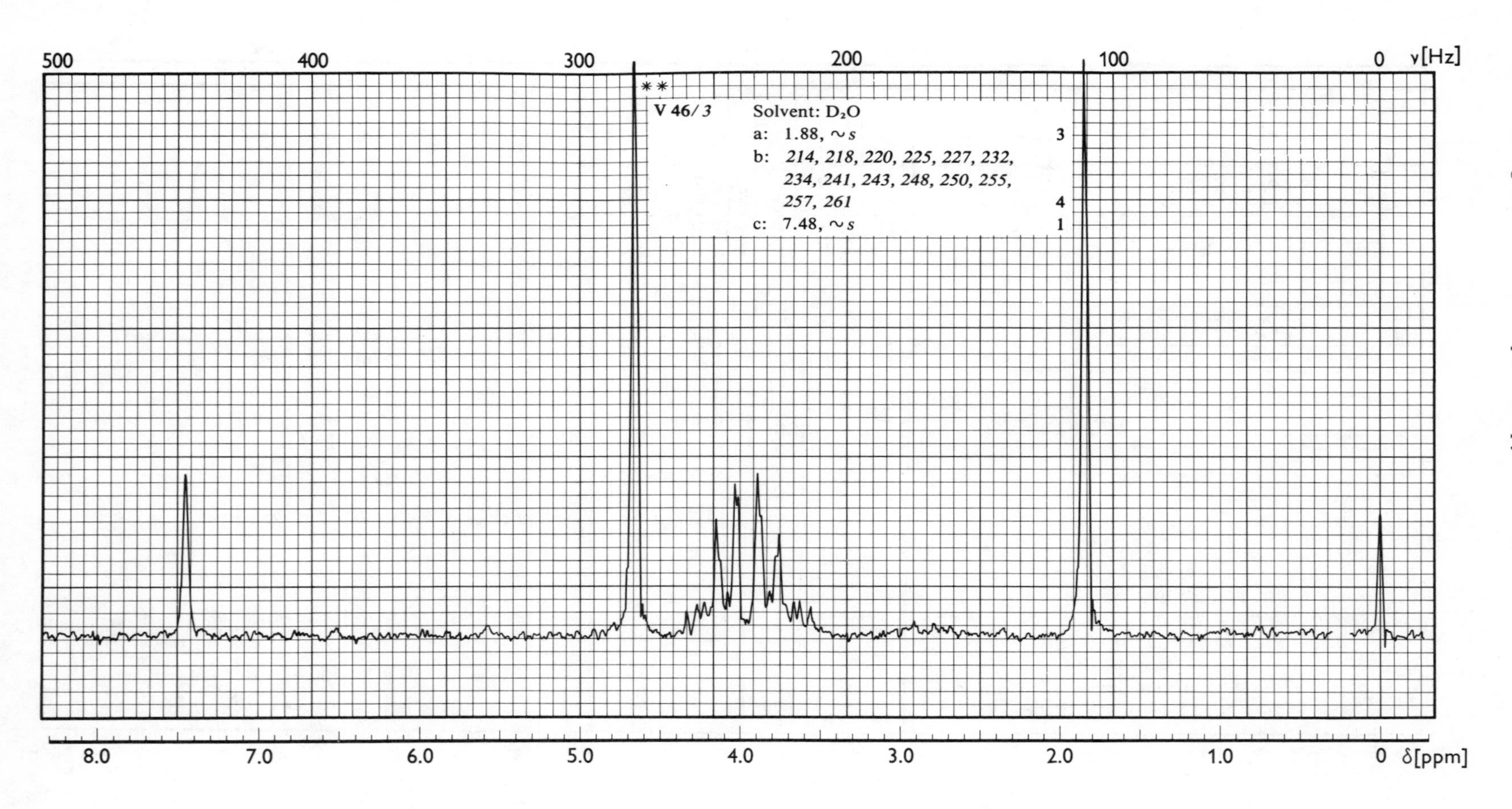

500
400
300
200
100
0 ν[Hz]
* *
V 46/3
Solvent: D2O
a: 1.88, ~s 3
b: 214, 218, 220, 225, 227, 232,
234, 241, 243, 248, 250, 255,
257, 261 4
c: 7.48, ~s 1
8.0
7.0
6.0
5.0
4.0
3.0
2.0
1.0
0 δ[ppm]

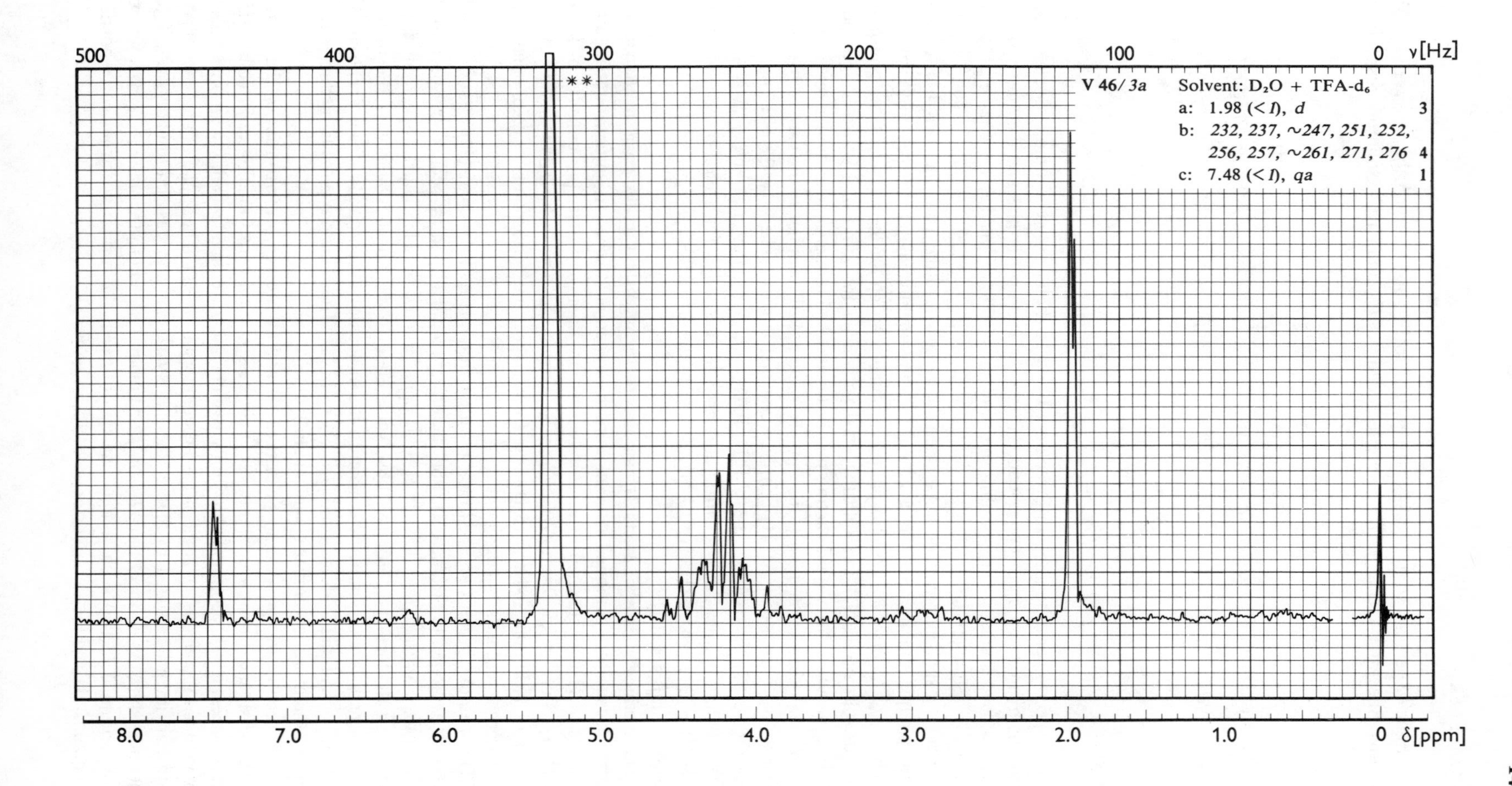

V 46/3a
Solvent: D_2O + TFA-d_6
a: 1.98 (< *I*), *d* 3
b: *232, 237, ~247, 251, 252, 256, 257, ~261, 271, 276* 4
c: 7.48 (< *I*), *qa* 1
500
400
300
200
100
0
ν[Hz]
8.0
7.0
6.0
5.0
4.0
3.0
2.0
1.0
0
δ[ppm]

PROBLEM 47

Correlate Spectra **47**/*1* to *4* with Structures **I** to **IV**, and interpret the observed splitting patterns. Spectrum **47**/*3* and **47**/*3a* were recorded in CCl_4 and $CDCl_3$ solution, respectively.

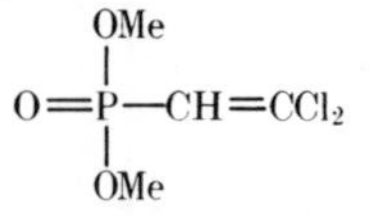

OMe	OMe $\quad CCl_3$	OEt	OEt
$O{=}P{-}CH{=}CCl_2$	$O{=}P{-}O{-}CH$	$O{=}P{-}CH_2CN$	$O{=}P{-}CH_2COOEt$
OMe	OMe $\quad$ OH	OEt	OEt
47/I	**47/II**	**47/III**	**47/IV**

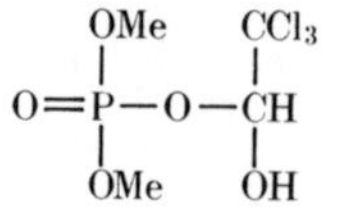

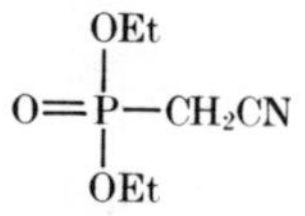

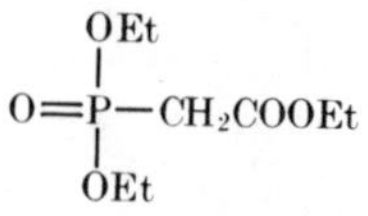

ANSWER

The spectra corresponding to Structures **I** and **II** should be very simple, because, apart from the hydroxy group of compound **II**, only the methoxy groups and methine hydrogens may produce signals. The splitting and chemical shift of the latter signal enable us to distinguish Structure **I** from **II**.

There is a spin-spin interaction between the methine hydrogen and the phosphorus nucleus with spin quantum number $I = 1/2$, and thus the methine signal is expected to split into a doublet. (Of course, in the phosphorus resonance spectrum the signal of the phosphorus nucleus split similarly.) The splitting must be smaller for compound **I** than for compound **II**. In Structure **I** the two interacting nuclei are attached to the same carbon atom, and since this carbon is unsaturated (sp^2), the P–C–H angle is 120°. Similar to the case of the *geminal* olefinic hydrogens, the ^2J(P,H) coupling constant is small when the P–C–H angle is 120° (see Figure 33). In contrast, the ^3J(P,H) interaction occurring in Structure **II** is analogous to a coupling between the *vicinal* hydrogens which is larger for proton-proton interaction: ^2J(H,H) ≈ 2 Hz and ^3J(H,H) ≈ 7 Hz. Similarly the ^2J(P,H) < ^3J(P,H) relation is expected for Structures **I** and **II**.

The methine signals of **I** and **II** can also be distinguished on the basis of the chemical shifts, because in compound **I** the methine hydrogen is attached to an unsaturated carbon, whereas in **II** it is attached to a saturated carbon atom, and thus, presumably, δCH(**I**) > δCH(**II**). Since in Spectrum **47**/*1* δCH = 4.92 ppm and J = 12.5 Hz, but in Spectrum **47**/*2* δCH = 7.00 ppm and J = 5 Hz, the former spectrum corresponds to compound **II** and the latter corresponds to compound **I**.

The methoxy protons are in similar position with respect to the phosphorus atom than the methine hydrogen in Structure **II**, and therefore a similar splitting occurs. In Spectrum **47**/*2* from the corresponding doublet at 3.90 ppm, ^3J(P,H) = 11.5 Hz. In Spectrum **47**/*1*, a further small doublet splitting is observable, due to the nonequivalence of the methoxy groups (there is an asymmetric carbon in the molecule). Accordingly, from Spectrum **47**/*1* δOCH_3 = 3.91 and 3.95 ppm and ^3J(P,H) = 11 Hz. The *P*-methoxy and *C*-methoxy signals appear, therefore, at similar chemical shifts.

The spectra corresponding to Structures **III** and **IV** are very easy to find on the basis of the ethyl signals. It is clear, moreover, that the spectrum corresponding to Structure **III** is simpler, because this compound has only two chemically equivalent ethoxy groups, whereas compound **IV** contains, in addition, a carbethoxy substituent. The signal of the latter appears with half intensity related to that of the ethoxy signals and has, presumably, a different chemical shift. As in Spectrum **47**/*4* only one methyl triplet (1.40 ppm, J = 7 Hz), whereas in Spectrum **47**/*3* two such signals can be observed, the former spectrum corresponds to compound **III** and the latter corresponds to Structure **IV**.

The methylene group of Structure **III** gives a doublet in Spectrum **47**/*4*, due to the ^2J(P,H) coupling. Thus, δPCH_2 = 2.99 ppm and ^2J (P,H) = 21 Hz. It can be seen that a decrease in the P–C–H angle to approximately 109° involves, indeed, a considerable increase in the ^2J(P,H) coupling constant, similar to the case of H–H couplings (see Table 5).

The methylene quartets of the ethyl groups split further due to a coupling with the phosphorus nucleus, to make the multiplet consist of $(2n_HI_H + 1)(2n_PI_P + 1) = 4 \times 2 = 8$ lines, since $n_H = 3$ (methyl group) and $n_P = 1$. The ^3J(P,H) coupling constant can be determined from the spectrum as shown schematically in Figure 202. Accordingly, ^3J(P,H) = 9 Hz and ^3J(H,H) = 7 Hz, furthermore δCH_2(ethyl) = 4.27 ppm.

The Spectrum **47**/*3* corresponding to Structure **IV**, the signal of the two identical ethoxy groups, and that of the carbethoxy group appear separately, as expected. Thus, of the six methyl lines, the ones **1**, **3**, and **5** correspond to the methyl hydrogens of the carbethoxy group, and the lines **2**, **4**, and **6** correspond to those of the two *P*-ethoxy groups. The chemical shifts are 1.29 and 1.33 ppm, respectively, and the coupling constant is 7 Hz in both cases.

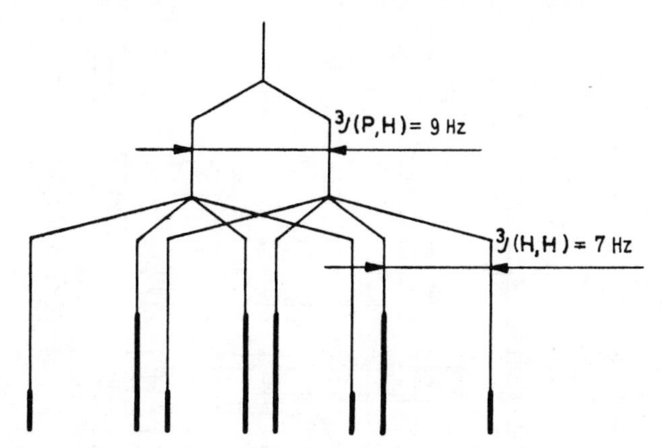

FIGURE 202. The schematic structure of the ethoxy methylene signal of compound 47/III.

The next two lines belong, also in this case, to the isolated methylene group, for which δPCH_2 = 2.86 ppm and ^2J(P,H) = 21.5 Hz. In the region of 230 to 270 Hz, four lines may be assigned to the carbethoxy group, and eight may be assigned to the two $POCH_2$ groups, because the distant phosphorus atom and the carbethoxymethylene protons do not interact. In Spectrum **47**/*3*, however, only 7 of the expected 12 lines are resolved, and further 2 can be recognized as shoulders. The quintet-like multiplet does not even reveal unambiguously which lines of the $POCH_2$ multiplet (appearing approximately as a quintet) overlap with the carbethoxy quartet. It is more convenient to investigate the $CDCl_3$ solution of the sample, since in Spectrum **47**/*3a* all the 12 lines appear (line **4** as a shoulder). This well illustrates the advantages of varying solvents, which makes it easier to determine spectrum parameters.

The structures of the methyl and isolated methylene signals are essentially unchanged in comparison with Spectrum **47**/*3*, except for small changes in the shifts: δCH_3(COOEt), *t* (*J* = 7 Hz): 1.29 ppm; δCH_3(POEt), *t* (*J* = 7 Hz): 1.33 ppm; and δPCH_2, *d* [^2J(P,H) = 21.5 Hz]: 2.99 ppm.

In Spectrum **47**/*4*, lines **2**, **5**, **8**, and **11** of the multiplet between 230 and 270 Hz (see Figure 202) comprise the quartet of the $COOCH_2$ group. The frequencies yield δ$POCH_2$ = 4.17 ppm. δ$COOCH_2$ = 4.18 ppm, and ^3J(H,H) = 7 Hz (COOEt). Projecting now these data onto Spectrum **47**/*3*, we can discover that the quartet overlaps with lines **2** to **5** of the "quintet", and thus δ$COOCH_2$ ≈ 4.18 ppm, and δ$POCH_2$ ≈ 4.12 ppm for CCl_4 solution. The coupling constants do not, of course, vary with the solvent.

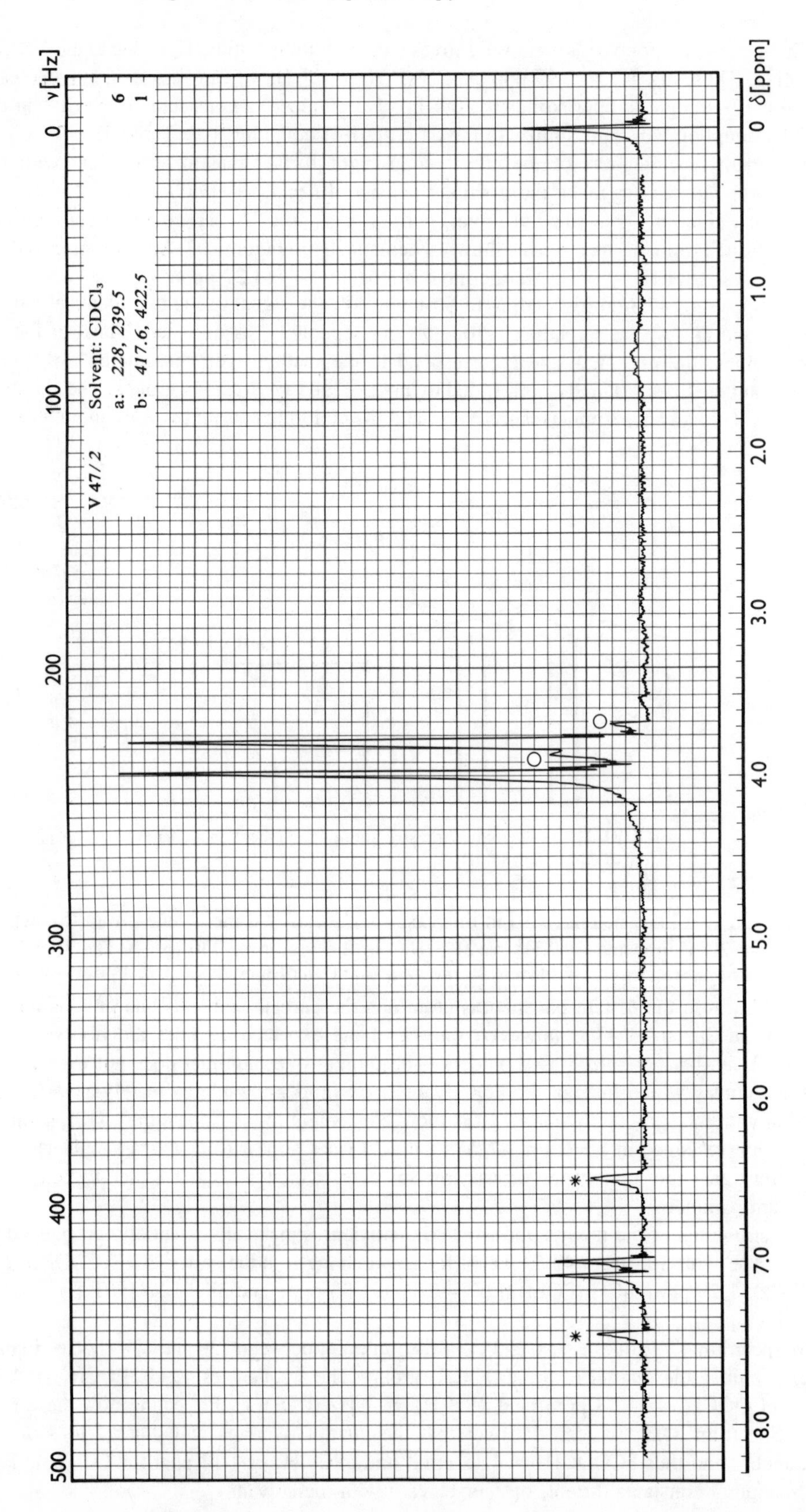
V 47/2
Solvent: $CDCl_3$
a: 228, 239.5 6
b: 417.6, 422.5 1
ν[Hz]
0
100
200
300
400
500
δ[ppm]
0
1.0
2.0
3.0
4.0
5.0
6.0
7.0
8.0

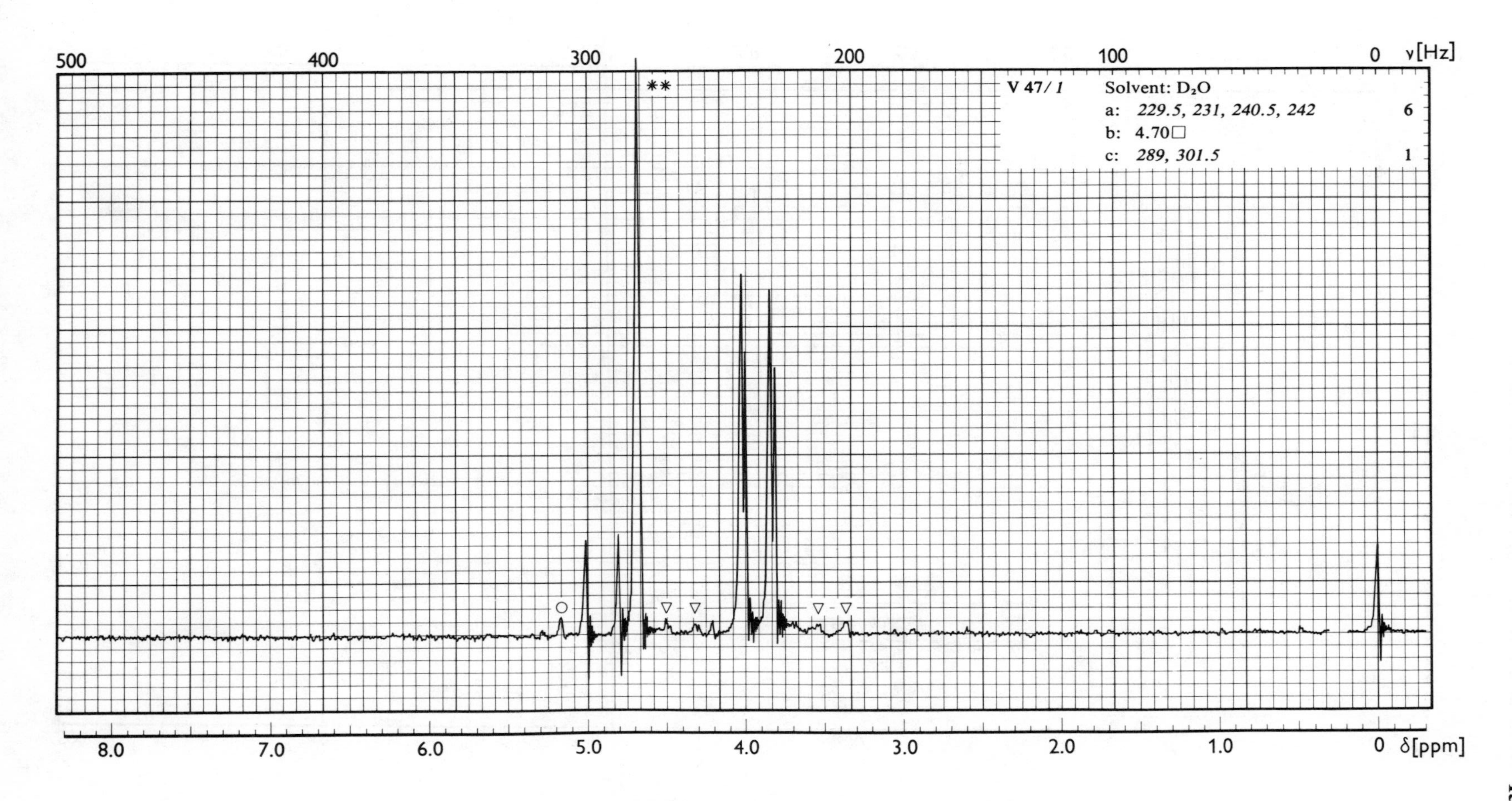
V 47/1
Solvent: D_2O
a: 229.5, 231, 240.5, 242 6
b: 4.70□
c: 289, 301.5 1
ν[Hz]
500
400
300
200
100
0
δ[ppm]
8.0
7.0
6.0
5.0
4.0
3.0
2.0
1.0
0
**

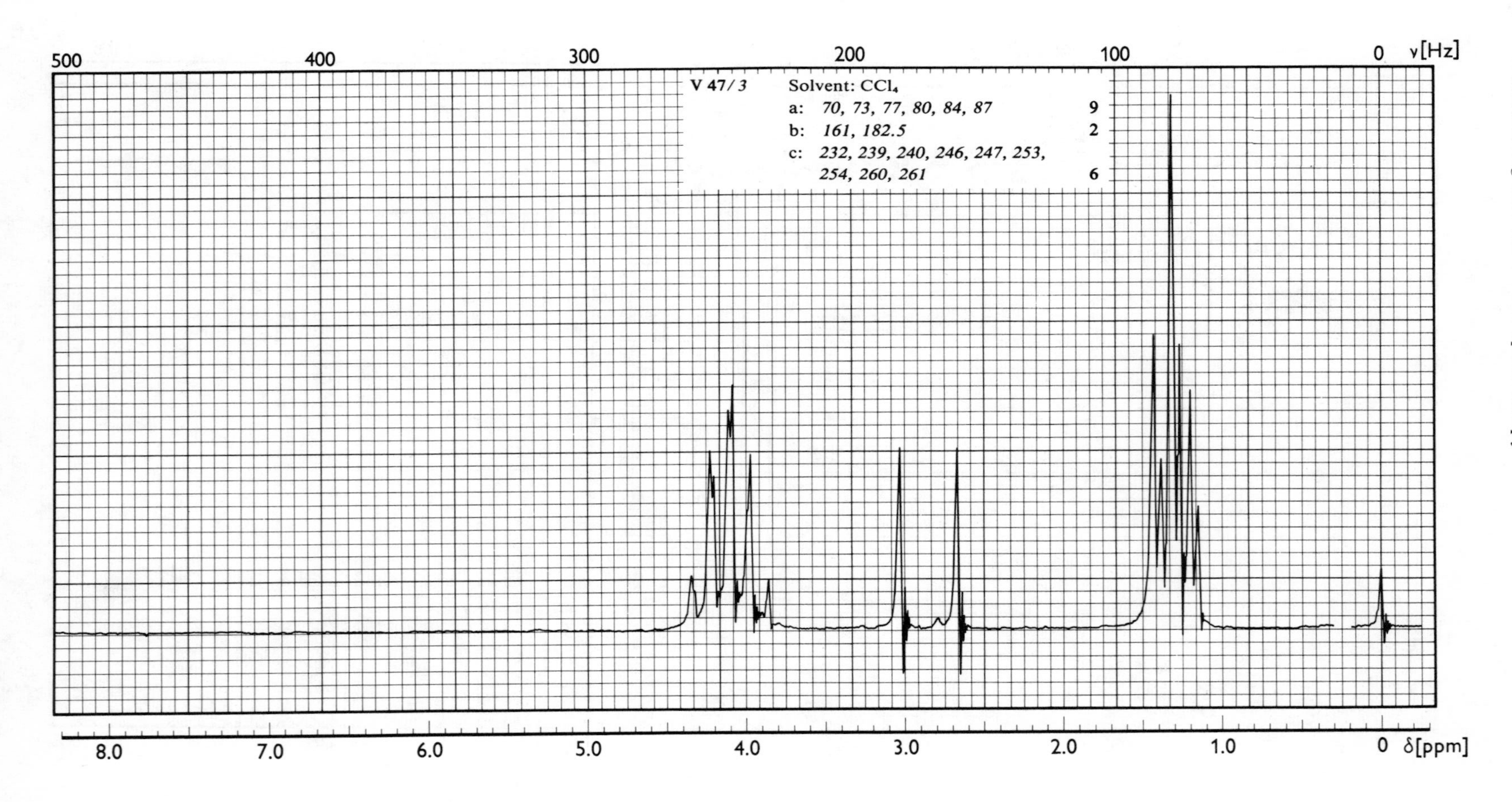
500
400
300
200
100
0 ν[Hz]
V 47/3
Solvent: CCl_4
a: 70, 73, 77, 80, 84, 87 9
b: 161, 182.5 2
c: 232, 239, 240, 246, 247, 253, 254, 260, 261 6
8.0
7.0
6.0
5.0
4.0
3.0
2.0
1.0
0 δ[ppm]

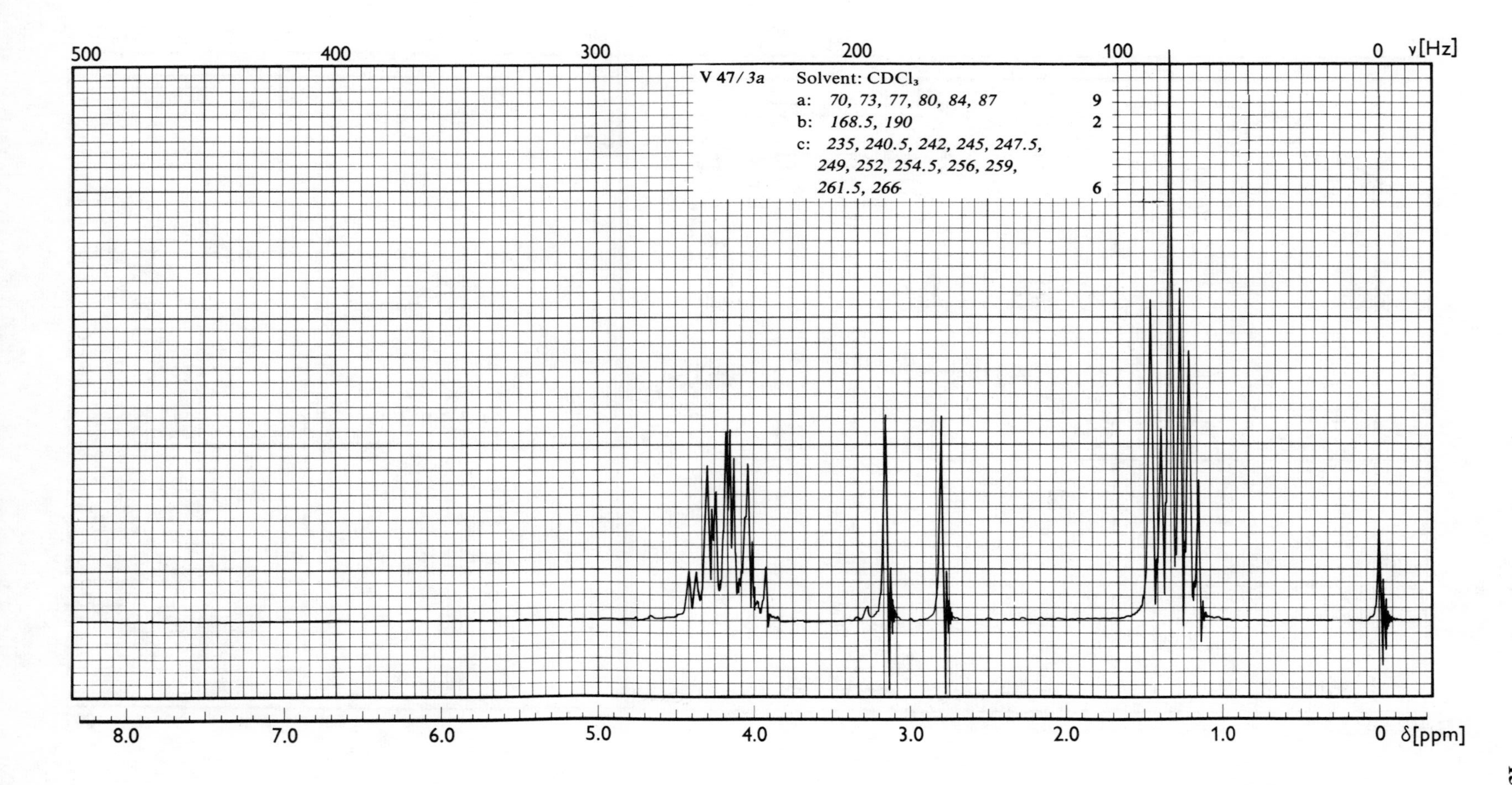
500
400
300
200
100
0
ν[Hz]
V 47/3a
Solvent: $CDCl_3$
a: 70, 73, 77, 80, 84, 87 9
b: 168.5, 190 2
c: 235, 240.5, 242, 245, 247.5, 249, 252, 254.5, 256, 259, 261.5, 266 6
8.0
7.0
6.0
5.0
4.0
3.0
2.0
1.0
0
δ[ppm]

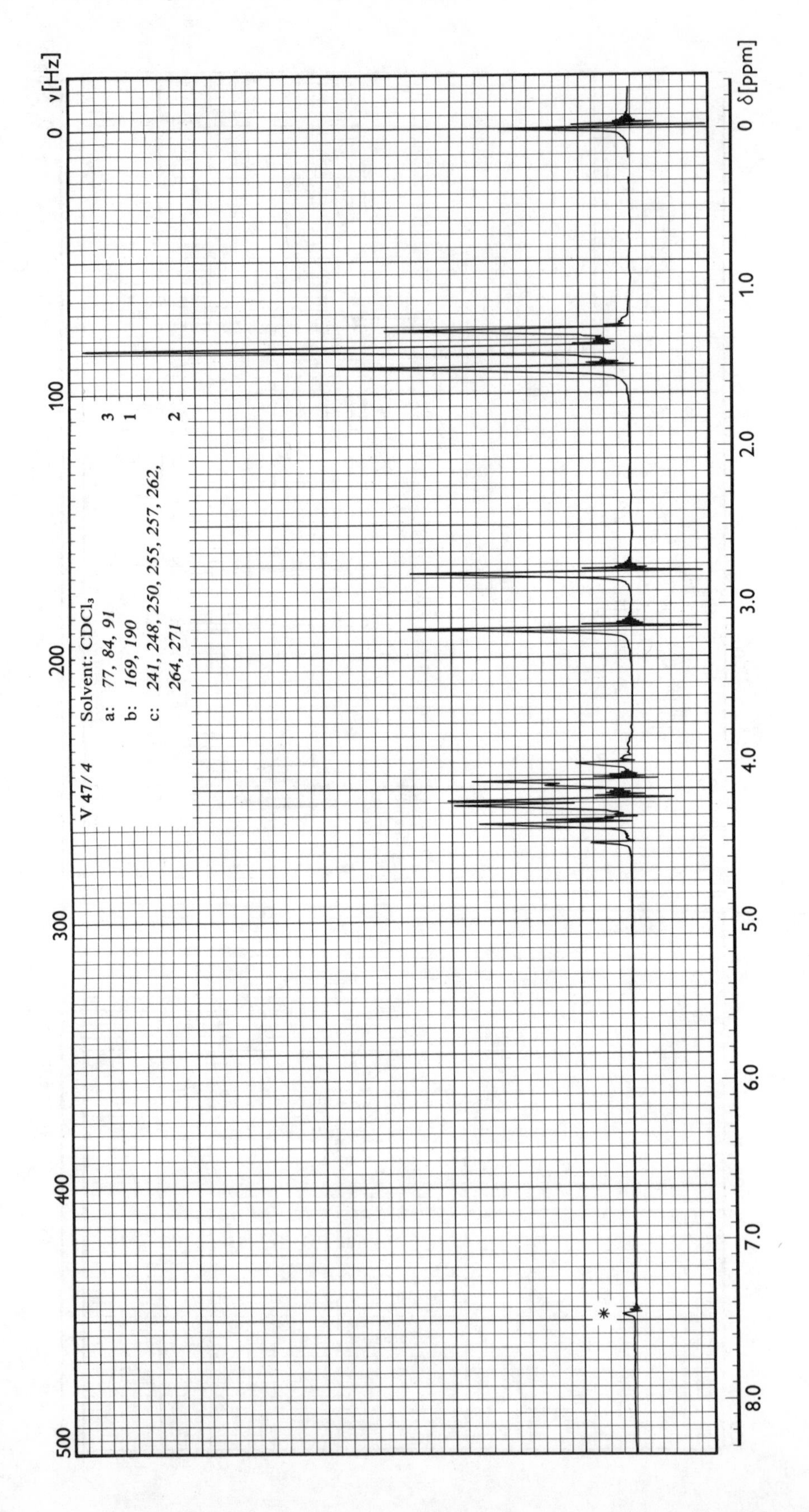

ν[Hz]
0
100
200
300
400
500
δ[ppm]
0
1.0
2.0
3.0
4.0
5.0
6.0
7.0
8.0
V 47/4
Solvent: CDCl₃
a: 77, 84, 91 3
b: 169, 190 1
c: 241, 248, 250, 255, 257, 262, 264, 271 2
*

PROBLEM 48[797,1320]

Decomposition of carbohydrate **I** leads presumably to compound **II**, upon distillation at 150°C. Determine from Spectrum **48**/*1* whether the decomposition yields really the expected compound.

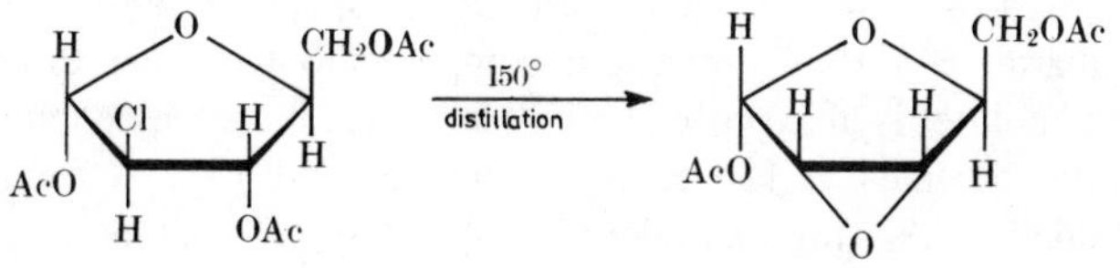

48/I **48/II**

ANSWER

Spectrum **48**/*1* contains four singlest of 3:2:1:1 intensity at 2.03, 4.93, 6.38, and 7.35 ppm. According to the spectrum, the product may not be compound **II**, which contains two acetyl groups, and its $CHCH_2$ group would give rise to an AX_2 spectrum, etc.

The two singlets at large chemical shifts suggest the formation of unsaturated bonds. Since, in addition, the spectrum contains only the signals of the acetoxymethyl group, it is logical to assume that by the elimination of two acetoxy groups and two hydrogens furan derivative **III** has been formed. The assignment is δCH_3 = 2.03, δCH_2 = 4.93, δH-3 = 6.38, and δH-5 = 7.35 ppm.

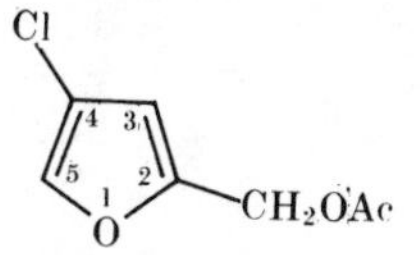

48/III

PROBLEM 49[1450]

The nitration of compound **I** yields several nitro derivatives simultaneously, in various quantities. Determine the structures of four of them on the basis of their Spectra **49**/*1* to *4*.

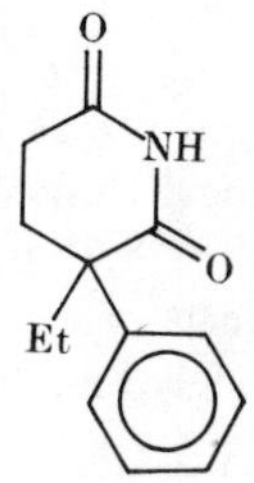

49/I

ANSWER

It is probable that all products are substituted in the aromatic ring, and the saturated ring is unchanged. The structures of the four compounds differ, therefore, presumably in the position and number of nitro groups attached to the benzene ring. This assumption is confirmed by the spectra, as the signals of the protons of the saturated ring of the imide and ethyl groups can be identified in all the four cases.

The methyl triplets of the ethyl groups appear at 0.83, 0.85, 0.80, and 0.75 ppm, respectively, in Spectra **49**/*1* to *4*, while the methylene quartets can be found at 2.00, 2.10, 2.05, and 2.45 ppm. J = 7 Hz for each multiplet. In Spectrum **49**/*4*, the triplet is shifted upfield and the quartet is shifted downfield compared to the other three cases, and the identification of the quartet is also problematic. However, it can be concluded from the line intensities that only the lines at 136.5, 143.5, 150.5, and 157.5 Hz may comprise this quartet. The multiplets of the ring methylene groups at approximately 2.45, 2.55, and 2.45 ppm, respectively, merge into relatively not too broad absorptions in Spectra **49**/*1* to *3*, whereas in Spectrum **49**/*4* the lines of the multiplet are observable between 115 and 195 Hz. As can also be expected, the broad singlet of the imide proton appears in the offset, at almost exactly the same position (around 11.05 ppm) in the four spectra.

The *AA'BB'* multiplet of the ring protons corresponding to four protons in Spectrum **49**/*1* proves the presence of a *p*-disubstituted benzene ring, thus Structure **II**. The chemical shifts of the ring protons are, in the *AB* approximation, 7.68 and 8.25 ppm (the greater value is assigned, of course, to H-3,5 (*ortho* to the nitro group). $J_{AB} \equiv J^o$ = 8.5 Hz.

The doublet-triplet pair of 2:1 intensity of the ring protons in Spectrum **49**/*2* (AX_2 spectrum with roof-structure) indicates a symmetrical trisubstitution of the ring. The splitting of 2 Hz of the lines, characteristic of *meta* coupling, decides in favor of the 3,5-dinitro Structure **III** (2,6-dinitro substitution would otherwise also be possible, but in this case $J_{AX} \equiv J^o$ would be about 9 Hz). This conclusion is also compatible with the fact that all three-ring protons are very strongly deshielded, due to the adjacent nitro groups (in the case of 2,6-disubstitution, deshielding of H-4 would be much smaller).

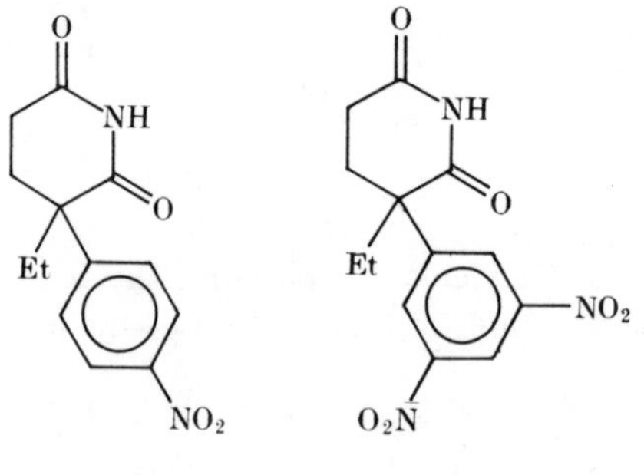

49/II **49/III**

Structure **III** is the more probable sterically as well; the 1,2,3-trisubstituted compound would be much too crowded. The chemical shifts can be read directly from the midpoints of the doublet and the triplet: $\delta A \equiv \delta$H-4 = 8.80 ppm and $\delta X \equiv \delta$H-2,6 = 8.55 ppm. Due to the combined $-I$ and anisotropic effects of the two nitro groups, H-4 situated between two nitro groups is deshielded more strongly than the other two-ring protons, adjacent only to one nitro substituent.

In Spectra **49**/*3* and **49**/*4*, the aromatic protons give *AMX* multiplets, of 3H relative intensities, consisting of two doublets and of one double doublet. These multiplets correspond in fact to *ABX* systems, but one of the couplings is negligibly small, thus the *AMX* approximation is justified. Both spectra correspond therefore to structures containing asymmetrically trisubstituted benzene rings. There are, in principle, four such structures (**IV** to **VII**). The sterically improbable Structure **VII** is also incompatible with the structure of the multiplets. The spectrum of this compound would contain three double doublets, and one of the coupling constants would be approximately 8 to 9 Hz, corresponding to *ortho* coupling (compare Figure 141).

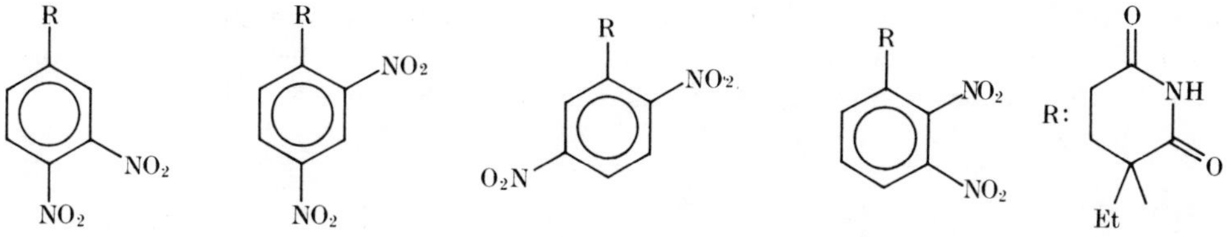

49/IV **49/V** **49/VI** **49/VII**

Accordingly, the two spectra correspond to two of Structures **IV** to **VI**. In order to choose among them it is now necessary to predict the chemical shifts of the ring protons by Equation 277. The substituent constants of Table 39 yield the shifts listed in Table 103. (The effect of the saturated ring was taken the same as that of a methyl group.) The estimated shifts indicate that Structure **VI** is incompatible with either spectrum, because it would give, instead of an *AMX* multiplet, a singlet-like *ABC* spectrum.

Table 103
THE CALCULATED CHEMICAL SHIFTS (PPM) OF AROMATIC HYDROGENS IN COMPOUND 49/IV to VI

Structure Assignment	IV	V	VI
δH-2	8.32	—	—
δH-3	—	9.07	8.37
δH-4	—	—	8.32
δH-5	8.37	8.52	—
δH-6	7.77	7.62	8.32

Assuming that $\delta A > \delta M > \delta X$, from Spectrum **49**/*3* we obtain $\delta A = 8.26$ (the stronger *A* line and the *M* line at higher frequency overlap), $\delta M = 8.17$, and $\delta X = 7.91$ ppm. $J_{AM} < 1$, $J_{AX} = 9$ Hz, and $J_{MX} = 2$ Hz.

From Spectrum **49**/*4* $\delta A = 8.67$ ppm (a doublet, split according to *meta* coupling, in overlap with the weaker *M* doublet of similar splitting; the midpoint of the resulting single maximum was taken as δA), $\delta M = 8.58$ ppm ($J_{MX}/2$ determined from the *X* part was added to the midpoint of the stronger doublet of the *M* part, which is not overlapping with the *A* part), and $\delta X = 8.08$ ppm. The coupling constants are $J_{AM} = 2$, $J_{AX} < 1$, and $J_{MX} = 9$ Hz.

The experimental chemical shifts obtained from Spectrum **49**/*3* agree very well with the values calculated for Structure **IV**, and the mutual position of the multiplets is also as expected. The difference between the experimental and calculated shifts for ring protons H-2,5,6 are 0.15, 0.11, and 0.14 ppm, respectively.

Spectrum **49**/*4* must therefore correspond to Structure **V**. The predicted and observed sequence of the multiplets is the same (the doublet of H-6 with larger splitting owing to the *ortho* coupling is upfield, whereas the less-split doublet of H-3 is downfield to the double doublet of H-5). The differences between experimental and predicted shifts are larger in this case (0.40, 0.06, and 0.46 ppm), but still acceptable taking into account the difference in solvent. The constants of Table 39 pertain to $CDCl_3$ solutions, and the above spectra were measured in DMSO-d_6.

Structure **V** accounts for the irregular methyl and methylene shifts in Spectrum **49**/*4* in comparison with the other spectra. The reason is the anisotropic effect of the nitro group which is close to the ethyl group and the saturated ring in Structure **V**.

PROBLEM 50

In a reaction of PCl_3 and ethanol two substances are formed simultaneously. Determine their structures on the basis of Spectra **50**/*1* and **50**/*2*!

ANSWER

The simpler Spectrum **50**/*1* contains only a triplet and a quintet at 1.28 and 3.90 ppm of 3:2 intensity (J = 7.5 Hz). Consequently, the compound must be triethylphosphite $(EtO)_3P$. The quintet splitting of the methylene signal can be explained by a spin-spin coupling between the phosphorus nucleus of spin quantum number 1/2 and the methylene protons (compare Problem **47**) if it is assumed that the coupling constant $^3J(P,H)$ is practically equal to that of the methylene and methyl protons. (The quintet is the M part of an AM_2X_3 system in which $J_{AM} = J_{MX}$ and $J_{AX} \approx 0$.)

The methylene signal in Spectrum **50**/*2* shows further splitting, and a line pair appears at 63 and 856 Hz both of 1/12 intensity with respect to the ethyl triplet. The two very distant lines ($\Delta\nu$ = 793 Hz) indicate the presence of a PH group, and the large splitting is due to the $^1J(P,H)$ interaction. Thus, $^1J(P,H)$ = 793 Hz, which is near to the expected range (see Table 5). The relative intensity of the doublet and triplet indicates that there are six methyl protons in the molecule. Accordingly, this product is presumably valence isomer $(EtO)_2HP \rightarrow O$ of diethylphosphite $(EtO)_2P\text{–}OH$.

The further splitting of the methylene signal as compared to that observed in Spectrum **50**/*1* is due to the fact that in this compound $^3J(P,H)$ and $^3J(H,H)$ are different (7 and 9 Hz, respectively), giving rise to two quartets at a distance of 9 Hz. These quartets comprise lines **1**, **2**, **4**, and **6** and **3**, **5**, **7**, and **8**, respectively. The chemical shifts are δCH_3 = 1.40, δCH_2 = 4.20, and δPH = 7.66 ppm.

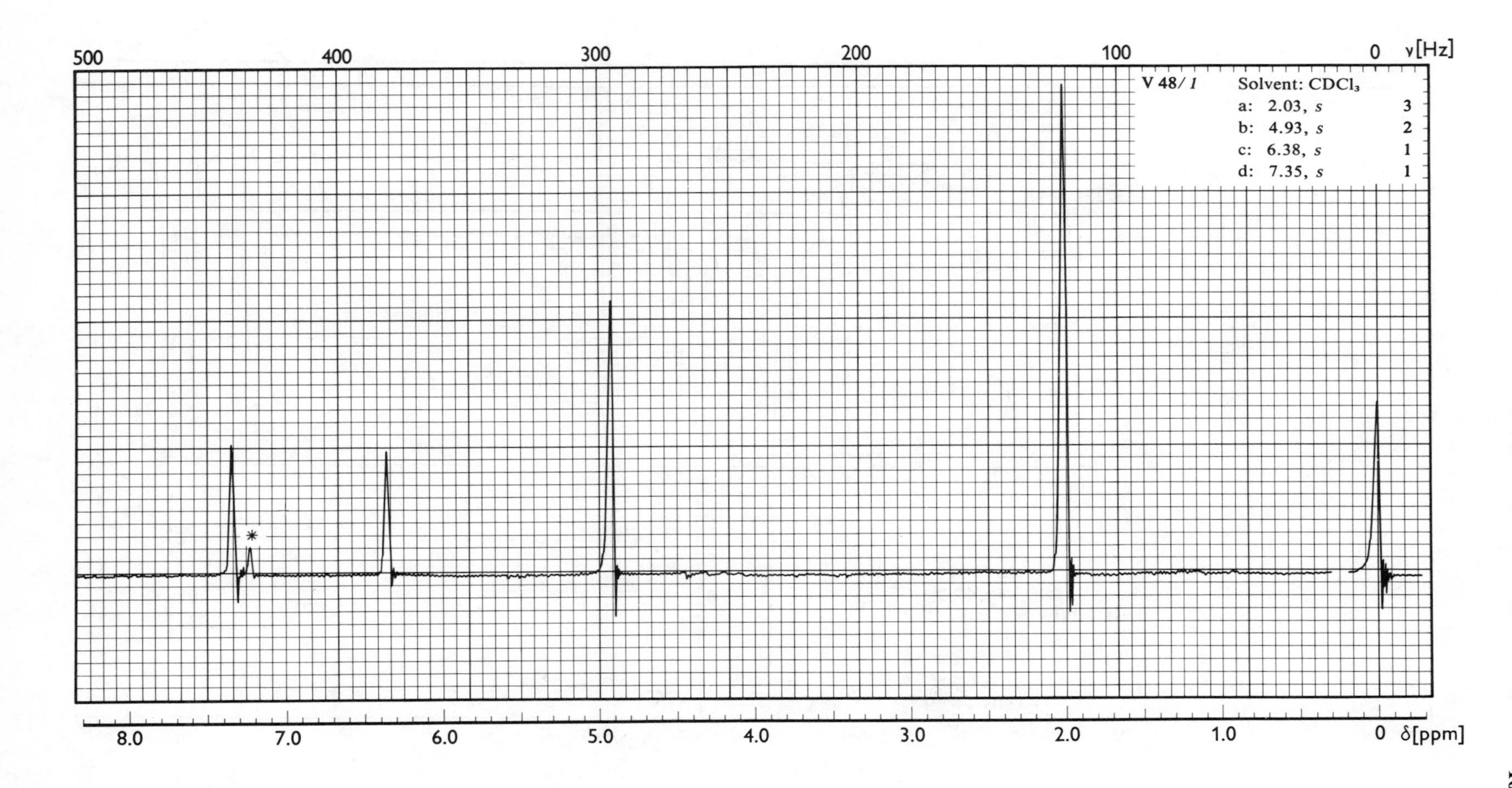
500
400
300
200
100
0
ν[Hz]
V 48/*1*
Solvent: $CDCl_3$
a: 2.03, *s* 3
b: 4.93, *s* 2
c: 6.38, *s* 1
d: 7.35, *s* 1
*
8.0
7.0
6.0
5.0
4.0
3.0
2.0
1.0
0
δ[ppm]

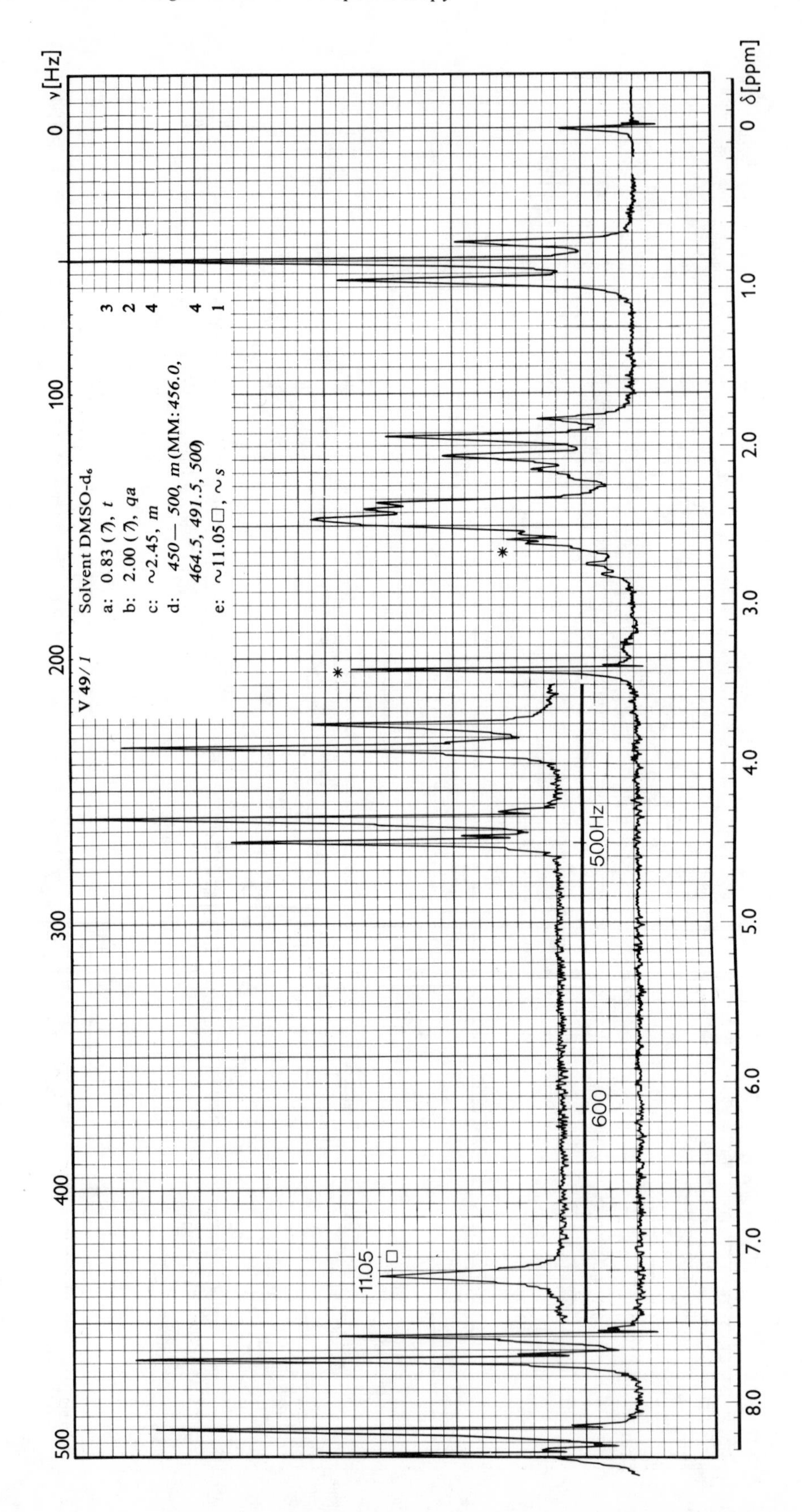
ν[Hz]
0
100
200
300
400
500
V 49/1
Solvent DMSO-d6
a: 0.83 (7), t 3
b: 2.00 (7), qa 2
c: ~2.45, m 4
d: 450 — 500, m (MM: 456.0, 464.5, 491.5, 500) 4
e: ~11.05□, ~s 1
*
*
11.05 □
500Hz
600
δ[ppm]
0
1.0
2.0
3.0
4.0
5.0
6.0
7.0
8.0

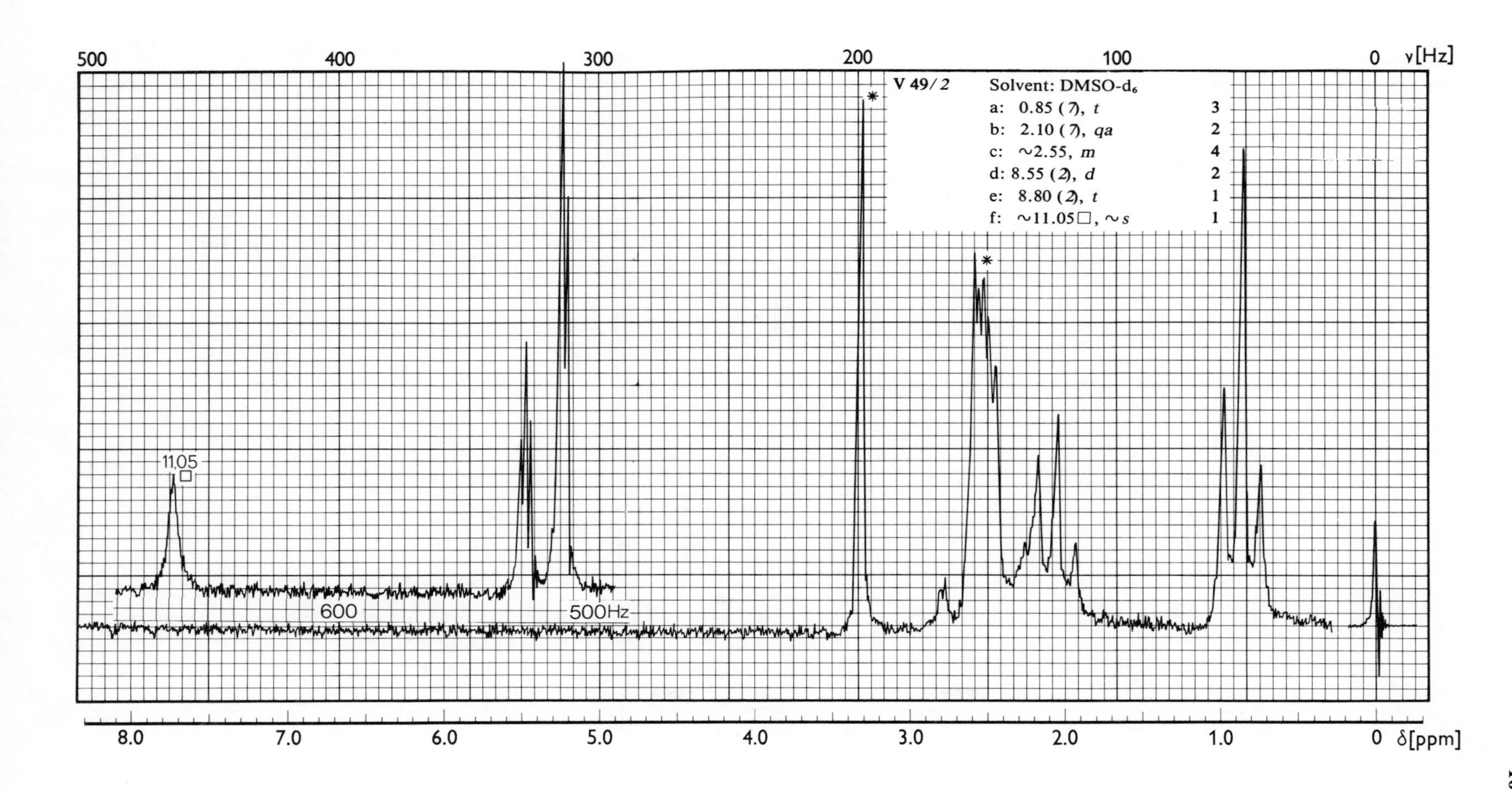

V 49/2
Solvent: DMSO-d₆
a: 0.85 (7), t 3
b: 2.10 (7), qa 2
c: ~2.55, m 4
d: 8.55 (2), d 2
e: 8.80 (2), t 1
f: ~11.05□, ~s 1
ν[Hz]
500
400
300
200
100
0
δ[ppm]
8.0
7.0
6.0
5.0
4.0
3.0
2.0
1.0
0
600
500Hz
11,05

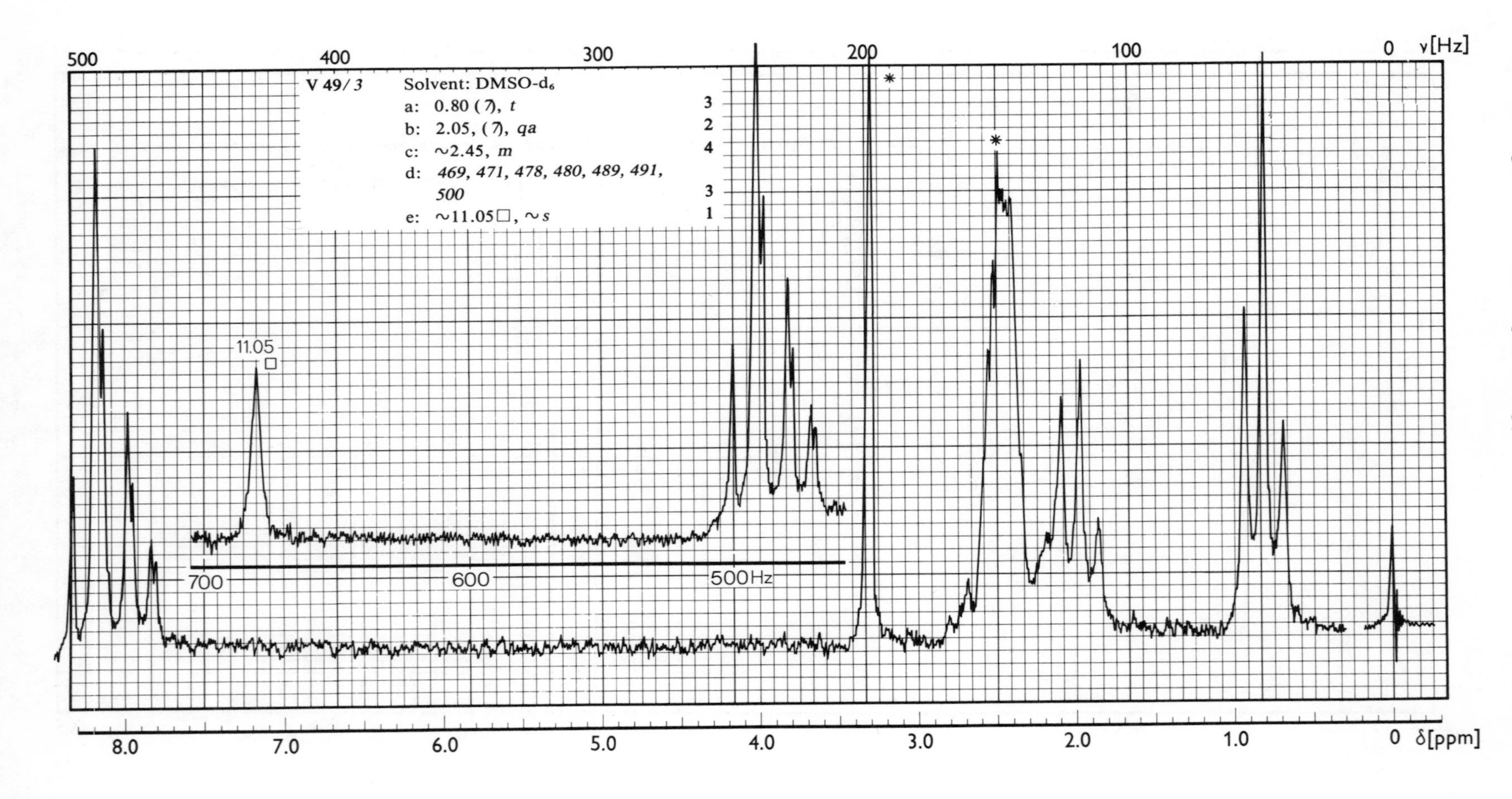
500
400
300
200
100
0
ν[Hz]
V 49/3
Solvent: DMSO-d₆
a: 0.80 (7), t 3
b: 2.05, (7), qa 2
c: ~2.45, m 4
d: 469, 471, 478, 480, 489, 491, 500 3
e: ~11.05□, ~s 1
*
*
11.05 □
700
600
500Hz
8.0
7.0
6.0
5.0
4.0
3.0
2.0
1.0
0
δ[ppm]

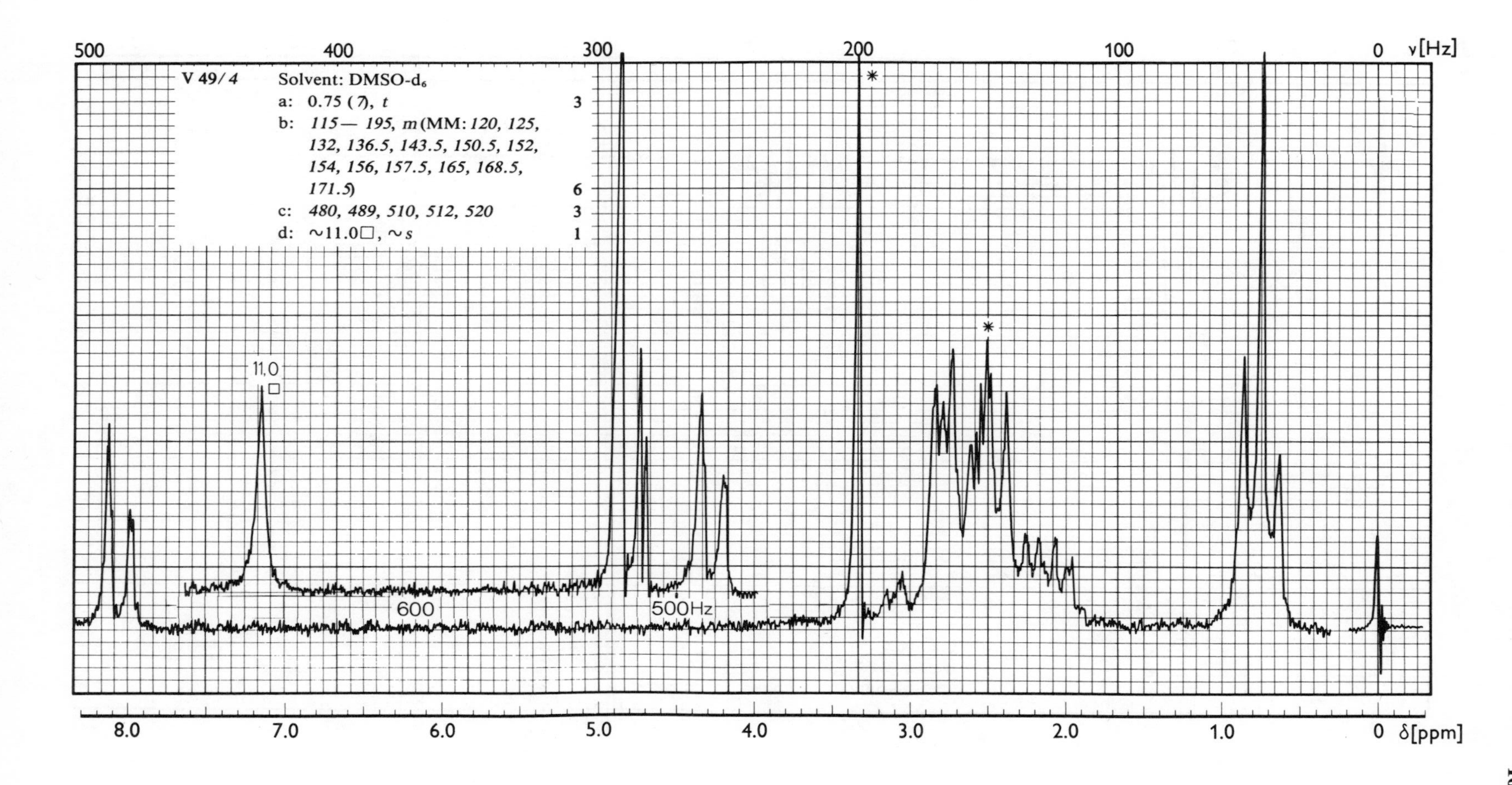
V 49/4
Solvent: DMSO-d6
a: 0.75 (7), t 3
b: 115— 195, m (MM: 120, 125, 132, 136.5, 143.5, 150.5, 152, 154, 156, 157.5, 165, 168.5, 171.5) 6
c: 480, 489, 510, 512, 520 3
d: ~11.0□, ~s 1
500
400
300
200
100
0
ν[Hz]
11,0
□
*
*
600
500Hz
8.0
7.0
6.0
5.0
4.0
3.0
2.0
1.0
0
δ[ppm]

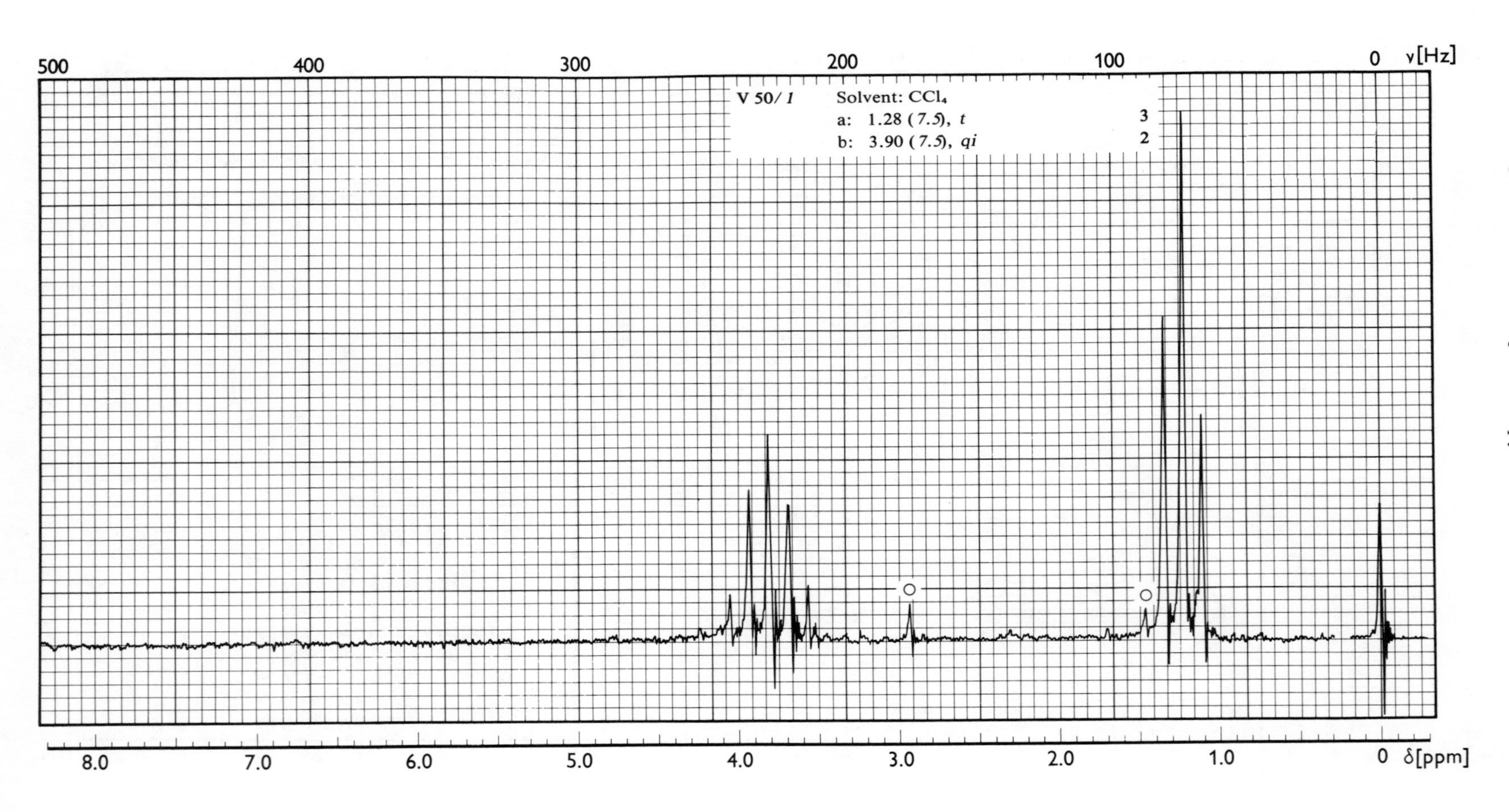
500
400
300
200
100
0 ν[Hz]
V 50/*1*
Solvent: CCl_4
a: 1.28 (7.5), *t* 3
b: 3.90 (7.5), *qi* 2
8.0
7.0
6.0
5.0
4.0
3.0
2.0
1.0
0 δ[ppm]

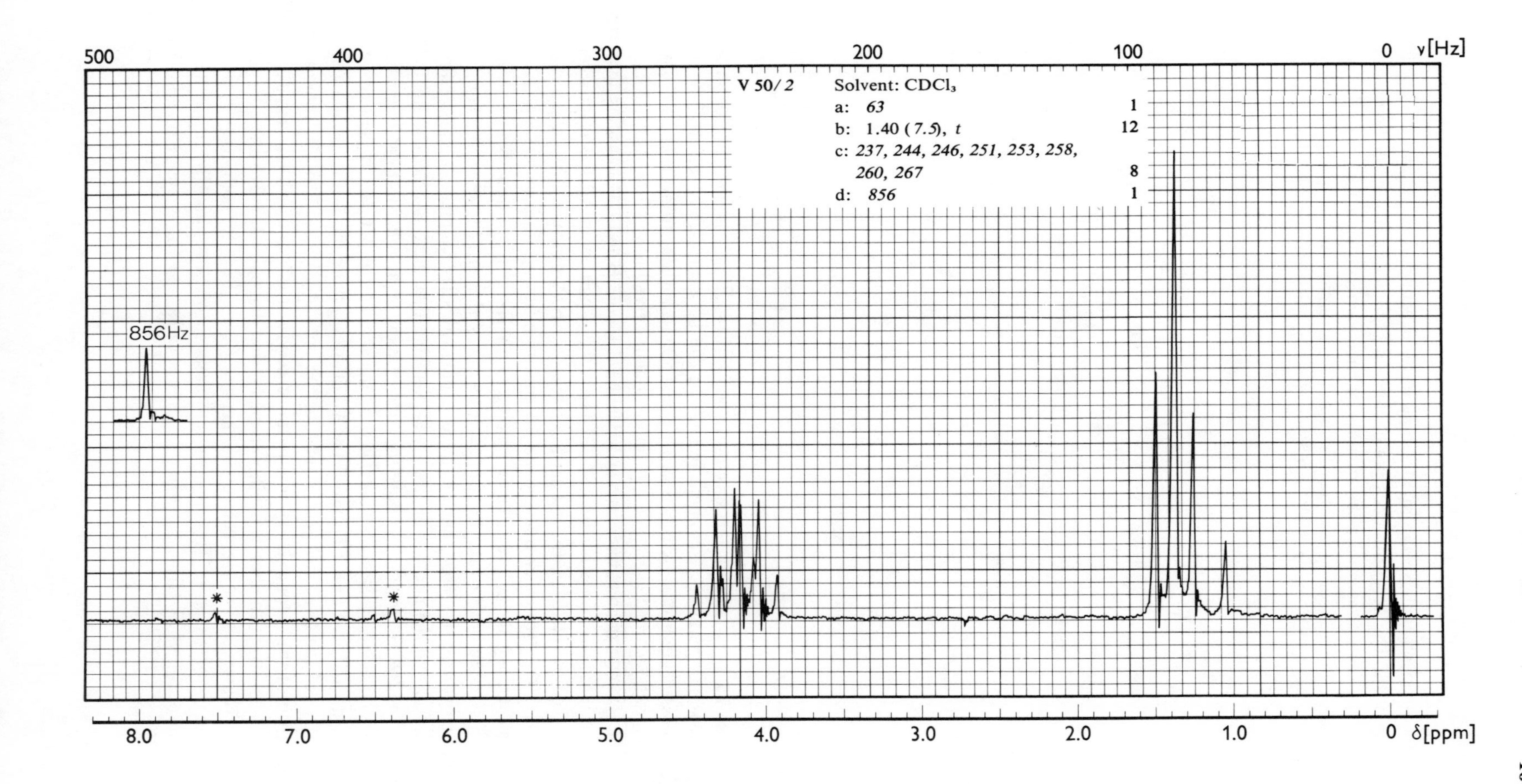

V 50/2
Solvent: $CDCl_3$
a: 63 1
b: 1.40 (7.5), t 12
c: 237, 244, 246, 251, 253, 258, 260, 267 8
d: 856 1
ν[Hz]
500
400
300
200
100
0
856Hz
δ[ppm]
8.0
7.0
6.0
5.0
4.0
3.0
2.0
1.0
0

PROBLEM 51

The compound with Spectrum **51**/*1* may have Structures **I** or **II** both having tautomeric forms as well. The spectrum does not change on D_2O addition. Decide between the alternatives. At room temperature the rotation of methylene group adjacent to the carbonyl is hindered. Also considering this fact, interpret the structure of the spectra recorded at room temperature in $CDCl_3$ and at 90°C in $CDBr_3$ (**51**/*1a*). Calculate $\Delta G‡$ of the hindered rotation for T_c = 80°C.

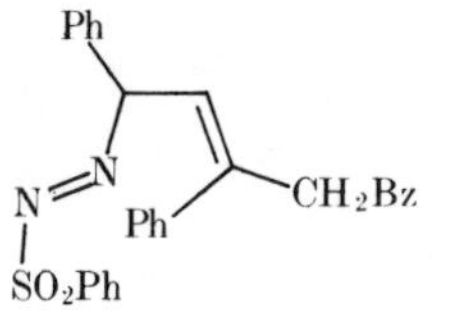

51/Ia 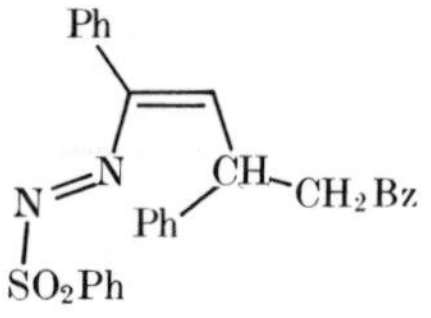 **51/Ib** 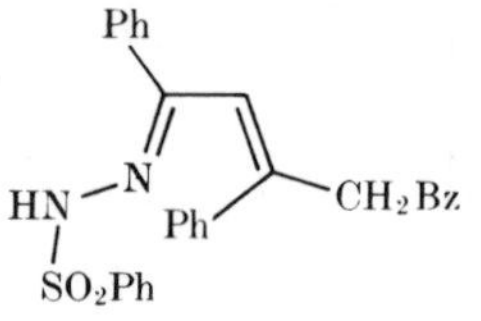 **51/Ic**

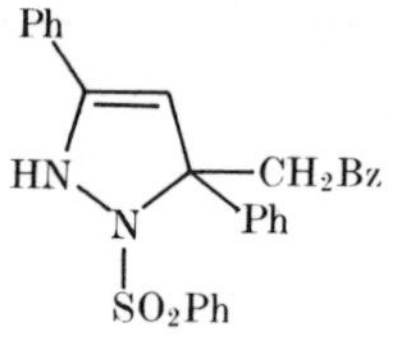

51/IIa 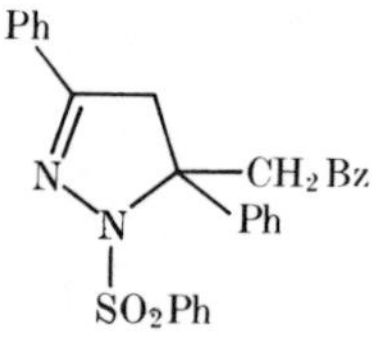 **51/IIb**

ANSWER

Due to the invariance of the spectrum upon D_2O addition, the presence of an NH group is impossible, therefore, tautomers **Ic** and **IIa** can be immediately rejected. In addition to a complicated group of signals between 400 and 500 Hz, arising from the aromatic hydrogens, the spectrum recorded at room temperature contains five lines at 209, 227 (226), 238, 245, and 256 Hz and a doublet with lines at 303 and 321 Hz of 3H and 1H intensities, respectively.

The doublet with lines at 303 and 321 Hz is the *A* part of an *AB* spectrum where J_{AB} = 18 Hz. The lines at 238, 245, and particularly at 227 Hz are strong. The latter has a shoulder at 226 Hz. The two other lines of the *AB* spectrum having lines at 303 and 321 Hz must be found among these lines. Their spacing is 18 Hz and the downfield *B* line must be stronger. On the basis of the line spacing, the pairs of 256-238, 245-227, and 227-209 Hz would be equally possible, but the relative intensities are compatible only with the latter two pairs. From the number of lines it is plausible that there is a second *AB* multiplet in the 200- to 260-Hz region. As only five lines are observable here, it must be assumed that two lines coincide. This is also indicated by the shoulder at 226 Hz. If the lines at 209 and 227 Hz are assigned to the *B* part of the *AB* quartet, giving the *A* lines at 303 and 321 Hz, the *A* lines of the other *AB* quartet must be the ones at 245 and 256 Hz, consequently the *B* line pair of the second *AB* system in this case are at 227 and 238 Hz, and J_{AB} is 11 Hz. There is, however, another possibility, namely, that line pairs 227-245 and 303-321, as well as 209-227 and 238-256 Hz, belong together and J_{AB} = 18 Hz for both *AB* systems. Anyway, the splitting of 18 or 11 Hz suggests the presence of Structure **IIb**, hence the coupling constant of type CH–CH=, occurring in structures **Ia** and **Ib**, is approximately 7 Hz.* In

* See Volume II, p. 52, last paragraph.

compound **IIb** both methylene groups are expected to give *AB* multiplets. The methylene protons in the ring are, of course, chemically nonequivalent (one of them is *cis* to the neighboring phenyl ring and *trans* to the benzoyl group, whereas for its pair the reverse holds). The methylene protons adjacent to the carbonyl group in the side chain also give an *AB* spectrum, since due to the hindered rotation the two hydrogens are chemically non-equivalent also in this case. The lines at 303 and 321 Hz must be assigned to the ring methylene protons, because the doublet is too far from the other signals, and such a large difference in shielding cannot be attributed to hindered rotation.

Since the coupling constants of *geminal* hydrogens are almost always greater than 11 Hz, the corresponding assignment of the lines is most probably wrong. This leads to a structure given schematically in Figure 203. This diagram enables one to derive the following chemical shifts for the methylene protons: δA(ring) = 5.19, δB (ring) = 3.95, $\delta A(CH_2CO)$ = 4.05, and δB (CH_2CO) = 3.67 ppm.

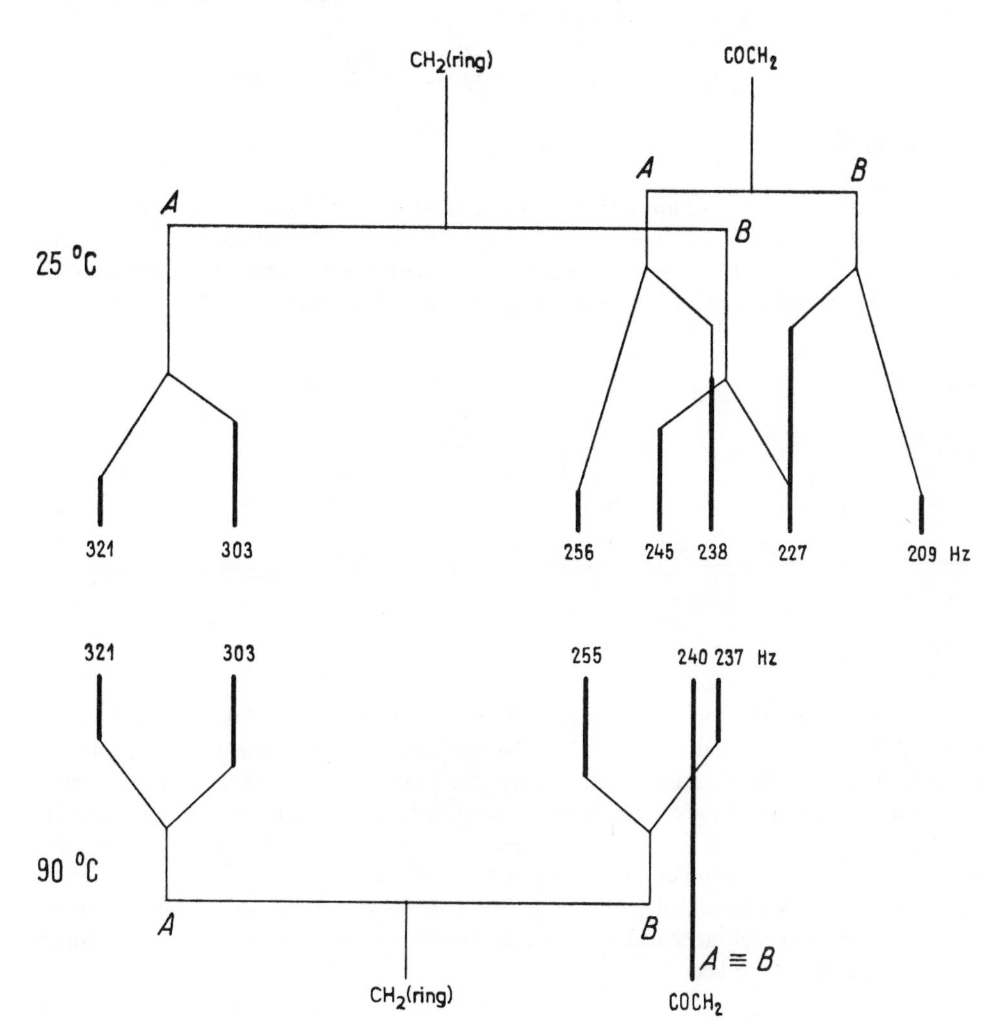

FIGURE 203. The schematic structure of the multiplet of methylene protons of compound 51/**IIb** at 25 and 90°C.

Since J_{AB} = 18 Hz for both groups, the values of $J/\Delta\nu$ are 0.24 and 0.82, and thus both spin systems are of type *AB*. A remaining problem is the steric position of ring protons *A* and *B*. Since in the case of a freely rotating $PhCOCH_2$ group (Spectrum **51**/*1a* taken at 90°C) the shift of nucleus *A* is practically unchanged, while that of proton *B* is significantly different, it can be presumed that the latter is closer to the rotating group, i.e., hydrogen *B* and the $PhCOCH_2$ group are *cis* related to the five-membered ring. In the conformation preferred at room temperature, proton *B* of the $PhCOCH_2$ group is closer to the *vicinal* phenyl ring, and the anisotropy of the latter shields it (proton *B* is ''above'' the phenyl ring in the preferred conformation).

When the rotation becomes free, the shift difference of the methylene hydrogens of the $PhCOCH_2$ group disappears, and the *AB* multiplet collapses into a singlet at 4.00 ppm. The chemical shift of the hydrogen closer to the $PhCOCH_2$ group of ring methylene group slightly increases, whereas that of the other one remains essentially unchanged: $\delta A = 5.20$ and $\delta B = 4.10$ ppm.

Since an *AB* multiplet collapses, for calculation of $\Delta G^{\ddagger}$ Equation 290 must be used. The coalescence temperature is 80°C, and $\Delta\nu = \sqrt{10\cdot 46} = \sqrt{(237 - 227)(255 - 209)} =$ 22 Hz, further $J_{AB} = 18$ Hz. From these data we get:

$$\Delta G^{\ddagger} = 19.13\, T_c\, [9.97 + \log T_c/\sqrt{\Delta\nu AB^2 + 6J^2_{AB}}]$$
$$= 19.13\cdot 353\, [9.97 + \log (353/\sqrt{22^2 + 6\cdot 18^2})] = 74.32 \text{ kJ/mol.}$$

PROBLEM 52

The compound with Spectrum **52/***1* was synthesized from **I** with acetic anhydride. It follows from the reaction conditions that the anhydro ring remains intact, and the IR spectrum indicates the presence of an ester and a conjugated keto group. Derive the structure of the product. No change is observable in the spectrum on D_2O addition.

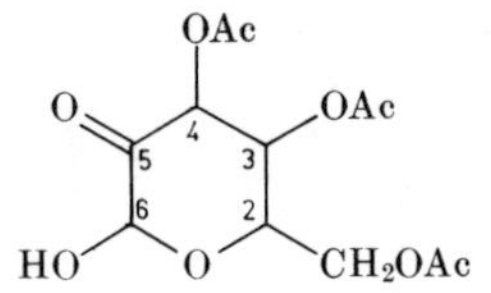

52/I

ANSWER

Spectrum **52/***1* contains five singlets at 2.16, 2.31, 4.97, 6.52, and 7.98 ppm of 3:3:2:1:1 intensity. None of the signals can be assigned to hydroxy protons. Consequently, in addition to the hydrogens of the 6-acetoxymethyl group, the molecule may contain only one further acetyl group and two hydrogens attached to unsaturated carbon atoms. It is certain that there may be no proton on C-2, otherwise the methylene signal would be split. For the same reason, these are no two *vicinal* hydrogens in the molecule.

It is clear from the above that acetic anhydride acted as dehydrant, and two double bonds were formed by elimination of one molecule of water and acetic acid. Consequently, Structures **II** to **IV** may have been formed.

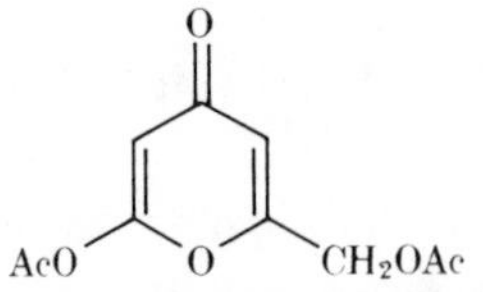

52/II 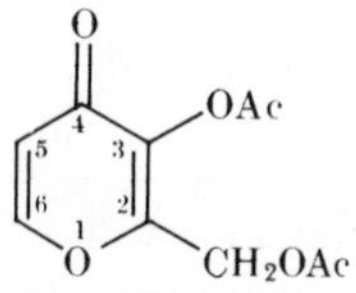 **52/III** 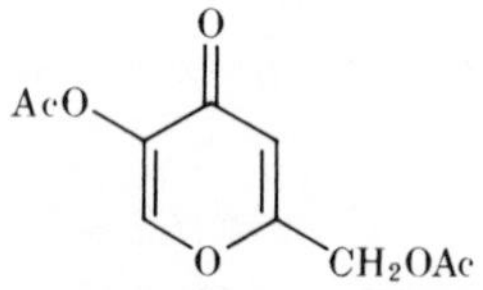 **52/IV**

The only possible location of the keto-carbonyl group is position 3, otherwise it would be impossible for the molecule to have two C=C bonds. Structure **III** can be rejected immediately, because it contains adjacent ring protons, which would have an *AB* multiplet instead of two singlets.

The decision between Structures **II** and **IV** can be based on the chemical shifts of the ring protons. In Structure **IV** one of them is adjacent to the ring oxygen. This is expected to deshield substantially, like with the α-hydrogens of furan derivatives (see Table 40). The shift of 7.98 ppm confirms therefore Structure **IV**. In the spectrum of compound **II** the signals of the H-3,5 atoms would be close to each other. Moreover, the ^4J coupling of H-3 and H-5 would presumably cause a well-observable splitting (2 to 3 Hz), whereas the coupling between H-3 and H-6 in compound **IV** is insignificant (similar to the aromatic hydrogens in *para* position). The singlets of the ring protons are therefore a further evidence for Structure **IV**.

Since the 5-acetoxy group is attached to an unsaturated carbon, δCH_3 is evidently larger than the corresponding shift of the acetoxymethyl group and thus the assignment, in the sequence of increasing shift, is as follows: $\delta CH_3(2)$, $\delta CH_3(5)$, δCH_2, δH-3, and δH-6.

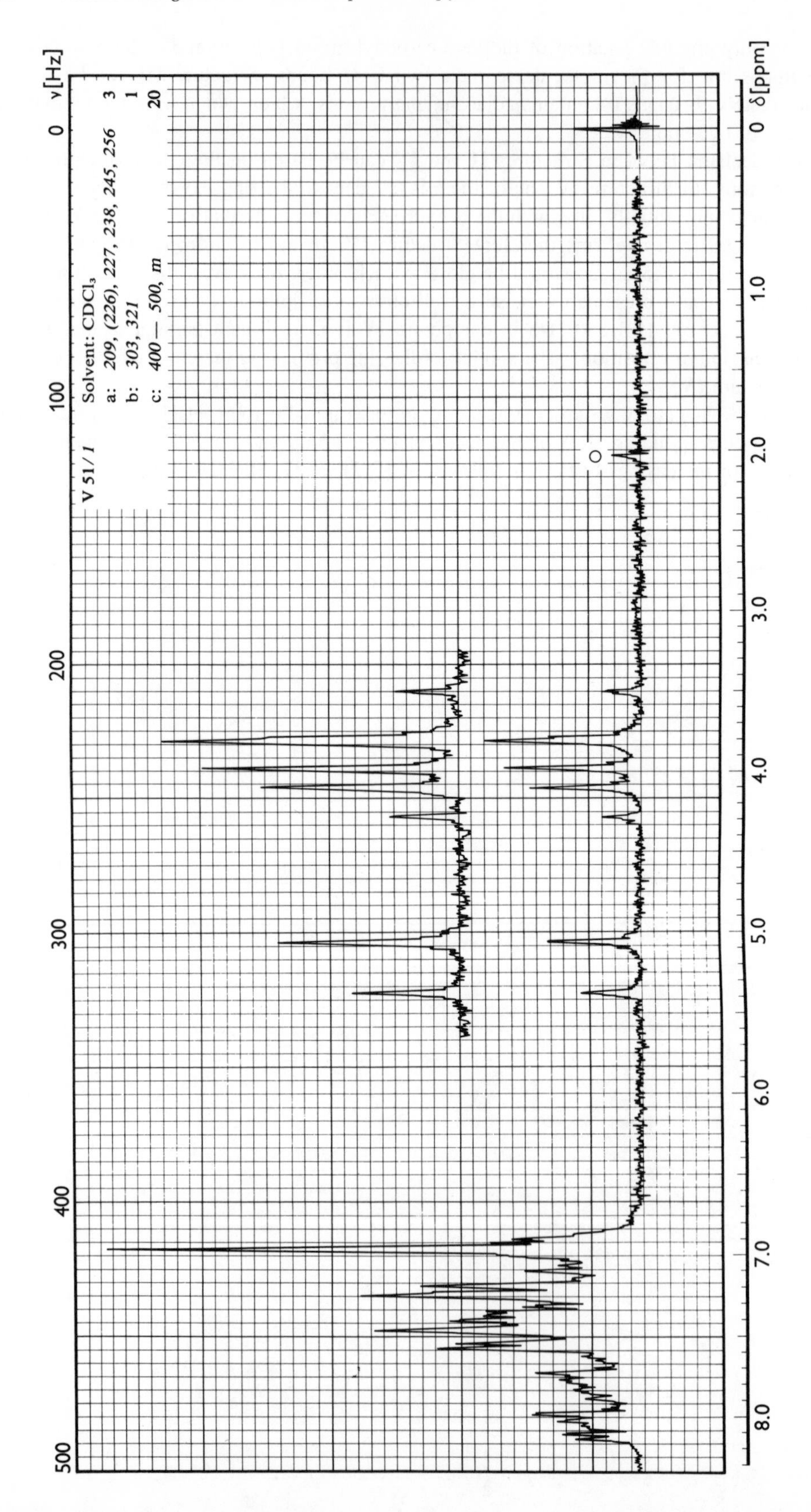
V 51/1
Solvent: CDCl3
a: 209, (226), 227, 238, 245, 256 3
b: 303, 321 1
c: 400 — 500, m 20
ν[Hz]
0
100
200
300
400
500
δ[ppm]
0
1.0
2.0
3.0
4.0
5.0
6.0
7.0
8.0

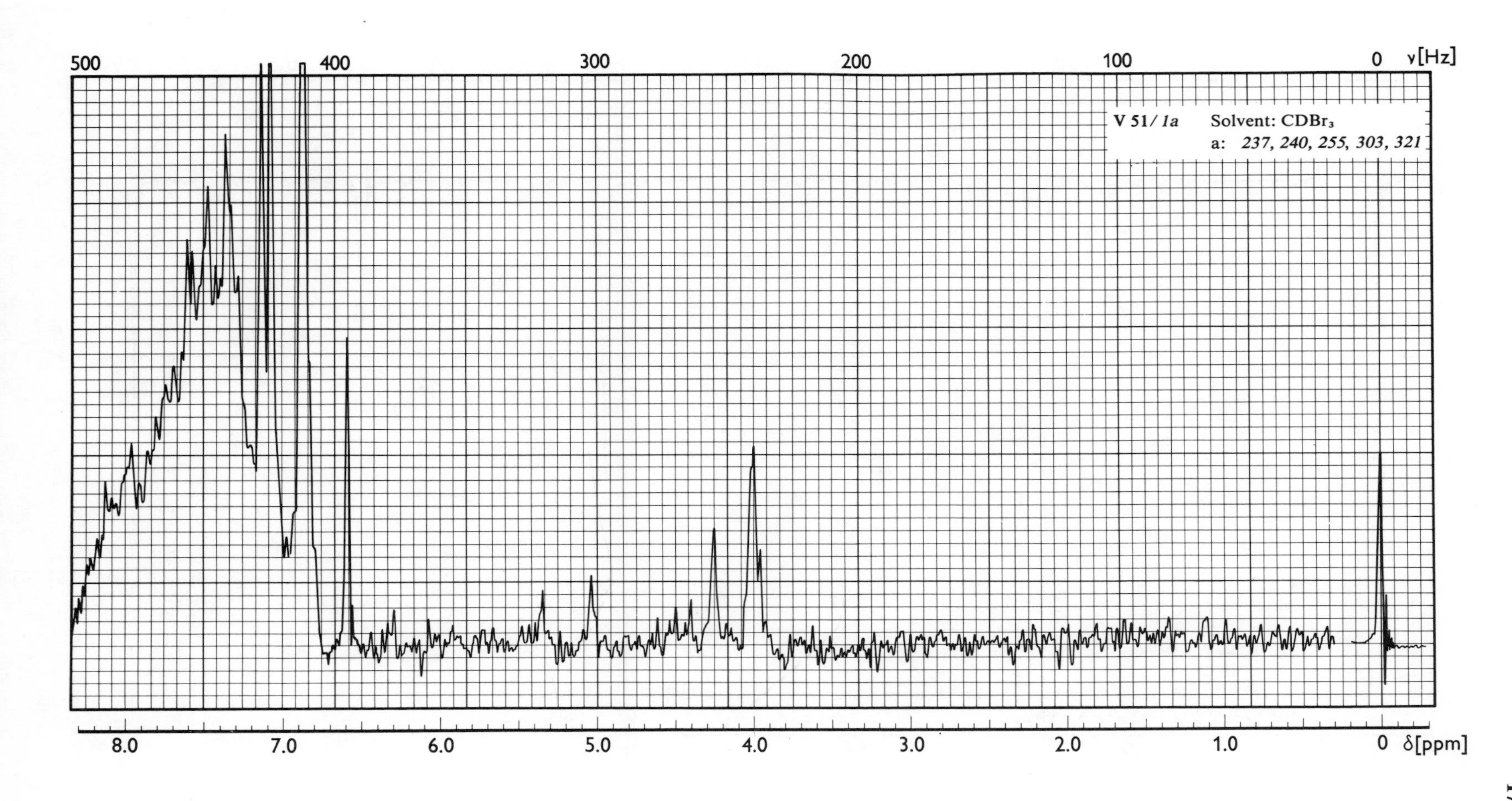

V 51/ *1a*
Solvent: $CDBr_3$
a: *237, 240, 255, 303, 321*
500
400
300
200
100
0 ν[Hz]
8.0
7.0
6.0
5.0
4.0
3.0
2.0
1.0
0 δ[ppm]

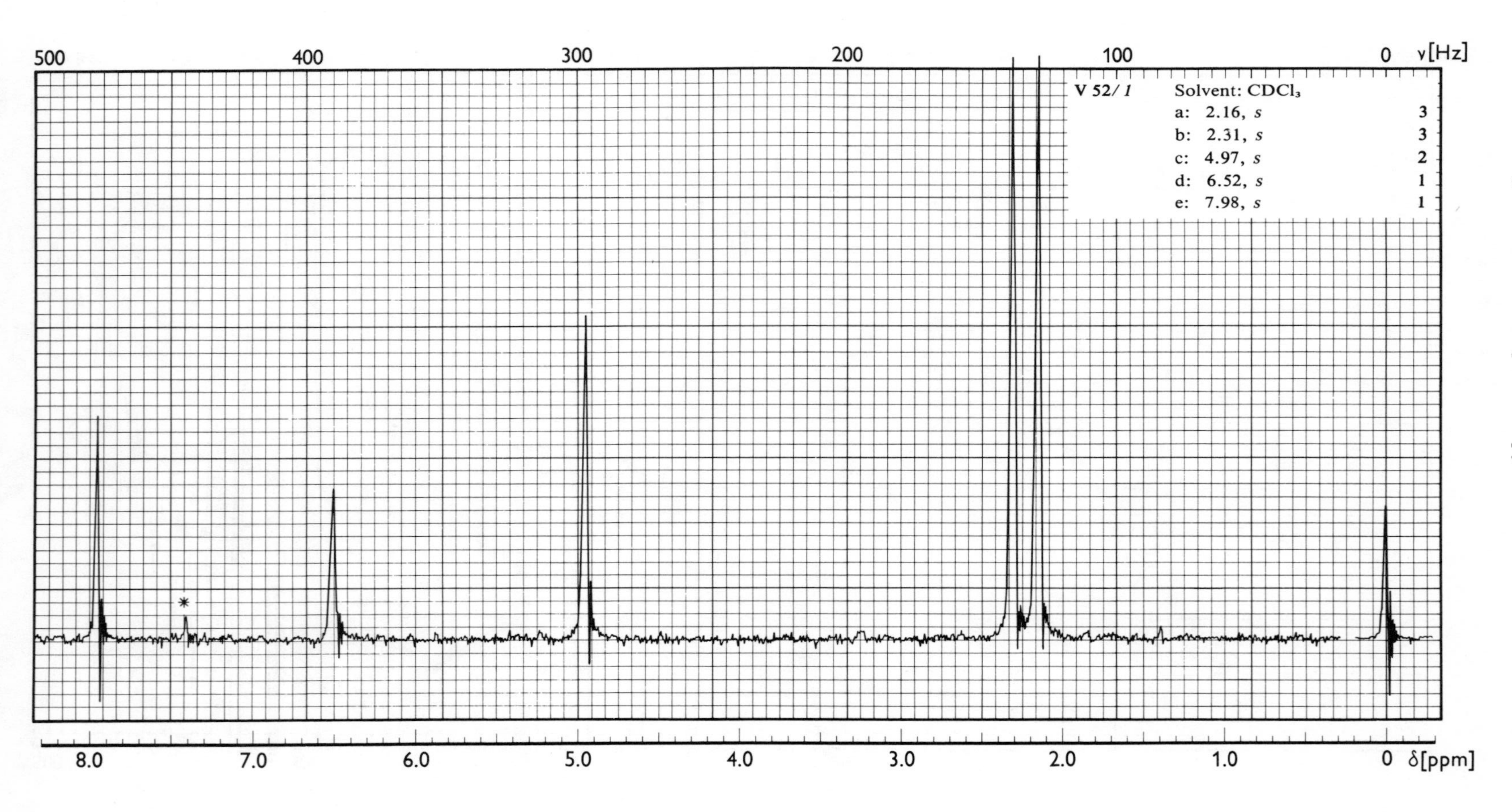

500
400
300
200
100
0 ν[Hz]
V 52/ 1
Solvent: $CDCl_3$
a: 2.16, s 3
b: 2.31, s 3
c: 4.97, s 2
d: 6.52, s 1
e: 7.98, s 1
*
8.0
7.0
6.0
5.0
4.0
3.0
2.0
1.0
0 δ[ppm]

PROBLEM **53**

Interpret Spectrum **53**/*1* of compound **I**. Calculate the chemical shifts and coupling constants.

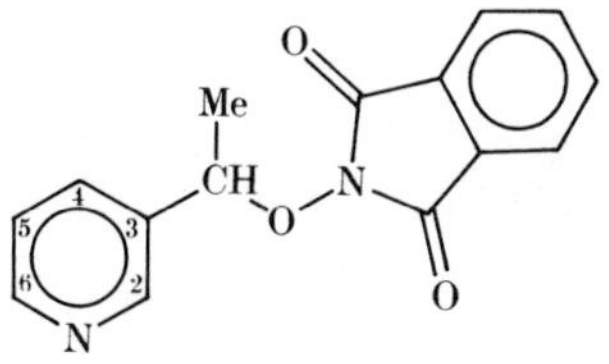

53/I

ANSWER

The doublet at 1.71 ppm arises from the methyl group, which yields, together with the methine hydrogen, an AX_3 multiplet with a coupling constant of $J = 6.5$ Hz. The corresponding methine quartet can be found at 5.50 ppm. The signals due to the CD_2HOD and CD_3OH impurities of the solvent are a quintet at 3.33 ppm and a singlet at 4.75 ppm. The four hydrogens of the phthalimide group are accidentally isochronous, and their singlet is at 7.77 ppm. The remaining signals can be assigned to the hydrogens of the pyridine ring. It will be apparent from the assignment given below that the coupling of the four ring hydrogens is, to a good approximation, of only first order, i.e., the protons form an *AMPX* spin system. The chemical shifts of the two α-hydrogens are, of course, larger owing to the adjacent nitrogen atom. H-2 may interact, to noticeable extent, only with H-4 and H-6, whereas the coupling between H-6 and H-5 is much stronger. Accordingly, the doublet at 8.67 ppm can be assigned to H-2, and the double doublet at 8.52 ppm can be assigned to H-6. From the line positions of this signal, $J_{5,6} = 5$ Hz can be obtained. Since the other coupling constant extracted from this signal is 1.5 Hz and the spacing of the doublet assigned to H-2 is 2.0 Hz, the splitting may not be due to the same coupling (i.e., the 2,6-interaction). Thus, $J_{2,6} < 1$, $J_{2,4} = 2.0$ Hz, and $J_{4,6} = 1.5$ Hz, since $J_{2,5}$ is certainly smaller than 1 Hz. The as yet unknown constant, $J_{4,5}$ is presumably the greatest (*ortho* coupling, which must be larger than $J_{5,6}$ of similar nature, since the latter is decreased by the $-I$ effect of the adjacent nitrogen atom),* and it can be determined from either of the unassigned group of signals. One of the two multiplets belonging to H-4 and H-3 consists of two triplets at a distance of 8 Hz, at approximately 8.07 ppm, whereas the other is a double doublet at 7.44 ppm, also split by 8 Hz. Accordingly, $J_{4,5} = 8$ Hz, the double triplet at 8.07 ppm can be assigned to H-4, and the double doublet at 7.44 ppm can be assigned to H-5. The schematic structure of these signals is shown in Figure 204. The H-4 signal collapses into two triplets, because the two pairs of middle lines are very close to each other. The assignment obtained from the coupling constants is in good agreement with the conclusions drawn from the chemical shifts, because substituent 3 causes a paramagnetic shift in the signal of the adjacent hydrogens, and thus it can be anticipated that δH-2 > δH-6 and δH-4 > δH-5.

* See Volume II, p. 83.

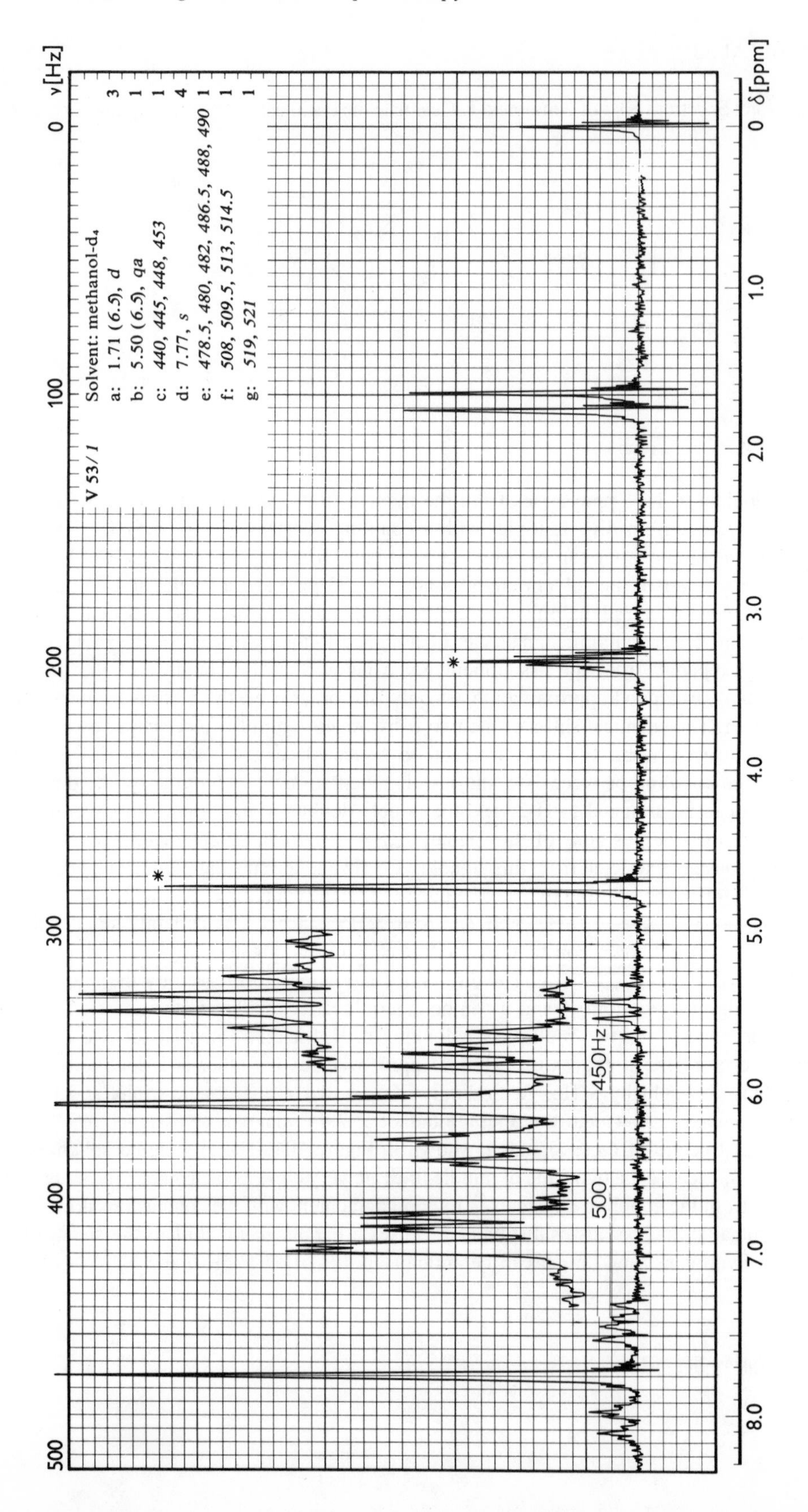

V 53/1
Solvent: methanol-d4
a: 1.71 (6.5), d 3
b: 5.50 (6.5), qa 1
c: 440, 445, 448, 453 1
d: 7.77, s 4
e: 478.5, 480, 482, 486.5, 488, 490 1
f: 508, 509.5, 513, 514.5 1
g: 519, 521 1
ν[Hz]
0
100
200
300
400
500
δ[ppm]
0
1.0
2.0
3.0
4.0
5.0
6.0
7.0
8.0
450Hz
500

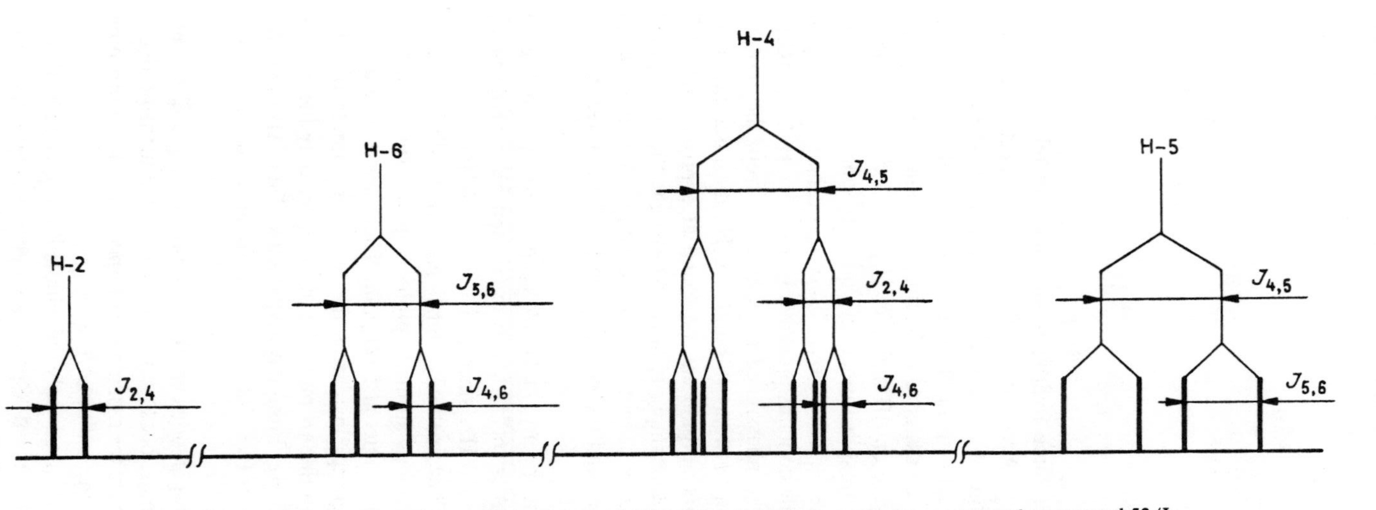

FIGURE 204. The schematic structure of the multiplets of the heteroaromatic ring protons of compound 53/I.

PROBLEM **54**

Determine the structures of compounds with Spectra **54**/*1* to *3* knowing that they are isomers of formula **I**. Interpret the spectrum differences.

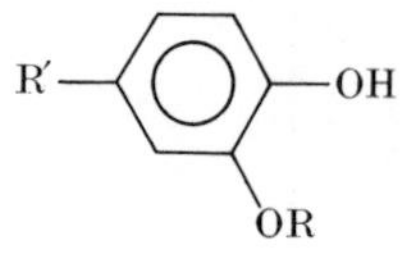

R′—(benzene ring)—OH, with OR on the ring

54/I

ANSWER

All the three spectra contain a singlet of three hydrogens at approximately 3.8 ppm, indicating the presence of one methoxy group. Accordingly, R = Me. In Spectra **54**/*1* and **54**/*2* a line group, corresponding again to three hydrogens, can be found at approximately 1.8 ppm, which is due clearly to a methyl substituent. The remaining signals appear with a $\delta > 5$ ppm, and thus they can presumably be assigned to olefinic or aromatic hydrogens. In Spectrum **54**/*1*, two of these hydrogens give an *AB* multiplet (J_{AB} = 11.5 Hz), close to the *AX* case. The four lines of this *AB* system split further into quartets, and the coupling constants, different for the *A* and *B* lines, are equal to those observed for the methyl signal at approximately 1.8 ppm. Thus, the molecule contains presumably a $-CH{=}CH{-}CH_3$ group. This group must be substituent R′, since the unassigned lines all arise from the protons shown explicitly in formula **I**. In the region between 400 and 410 Hz the three aromatic protons absorb, and the signal of the hydroxy proton can be found at approximately 4.7 ppm, in overlap with the signal of the light isotope impurity of D_2O. Owing to the overlapping signals, the chemical shifts of the ring protons cannot be given exactly. The signal is an *ABC* multiplet, close to the limiting A_3 case, and the chemical shifts of the three protons are between 6.67 and 6.83 ppm.

The multiplets of group R′ are shown schematically, in the AMX_3 approximation, in Figure 205. Assuming that $\delta A > \delta M$, we obtain δA = 6.75, δM = 5.64, and δX = 1.84 ppm, further J_{AM} = 11.5, J_{AX} = 1.5, and J_{MX} = 7.0 Hz, since $J(H^{Me},H^{\alpha}) > J(H^{Me},H^{\beta})$.

Spectrum **54**/*2* is very similar to **54**/*1*. The methoxy signal has practically the same chemical shift (3.75 ppm). The three isochronous aromatic protons yield a singlet at 6.75 ppm. The hydroxy signal appears as a sharp singlet at 5.60 ppm. There are differences in the methyl and olefinic signals of the R′ group, insofar as the former is simple doublet ($\delta CH_3 = \delta X$ = 1.81 ppm, J = 5 Hz) and in the *AB* multiplet of the latter only the *B* lines split into quartets, i.e., $J_{AX} \approx 0$ and J_{BX} = 5 Hz. From the frequencies of the AB lines, δA = 6.23, ppm δB = 5.94 ppm, and J_{AB} = 16 Hz. Note that the difference $\Delta\delta AB$ is much smaller in this case (see below).

Proton *A* must represent the olefinic hydrogen adjacent to the aromatic ring and *B* must represent the one adjacent to the methyl group, because the aromatic substituent produces a strong paramagnetic shift. (This also pertains to the assignment of Spectrum **54**/*1* confirming the conclusions drawn from the structure of the multiplet.)

In the quartet corresponding to the weaker *B* signal, only the middle lines can be identified, but the coupling constants (obtained from the *A* and *X* parts) also permit determination of the frequencies of the outer lines. The two weak lines overlap with the hydroxy signal and the upfield line of the stronger *B* quartet, respectively. The downfield outer line of the latter overlaps, on the other hand, with the stronger line of the *A* part.

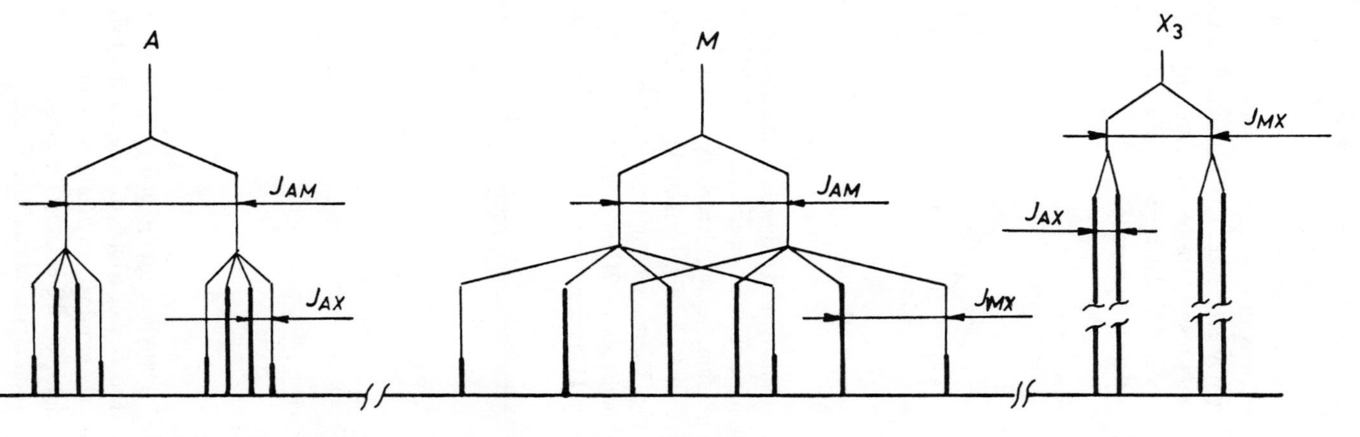

FIGURE 205. The schematic structure of the AMX_3 multiplet of the $-CH=CHCH_3$ group of *cis* isoeugenol (54/II).

Therefore, the molecules corresponding to Spectra **54**/*1* and **54**/*2* are analogous in structure, and they are the *Z* and *E* isoeugenol (**II** and **III**). Because $J^E > J^Z$ for olefins,* Spectrum **54**/*2* belongs to *E* isoeugenol (in this Spectrum $J^E_{AB} = 16$ Hz, whereas in Spectrum **54**/*1*, $J^Z_{AM} = 11.5$ Hz). Note that, in contrast to the rather general empirical rule,** the *transoid* allylic coupling of the *vicinal* olefinic proton with the methyl group is stronger in the *Z* isomer ($J^{transoid}_{AX} = 1.5$ Hz) than the analogous *cisoid* coupling in the *E* isomer ($J^{cisoid}_{AX} \approx 0$). Probably the aromatic ring, close to the methyl group in *Z* isomer, forces the molecule into a conformation favorable for allylic coupling, whereas in *E* isomer the methyl group is free to rotate. It is in good accord with the assignment, however, that $\Delta\delta AB$ is, as mentioned above, smaller in *E* isomer. In Structure **III**, H_β is with respect to the aromatic ring also subjected to the deshielding effect of the coplanar ring located on the same side of the double bond, whereas H_α is exposed to the opposite effect of the methyl group.

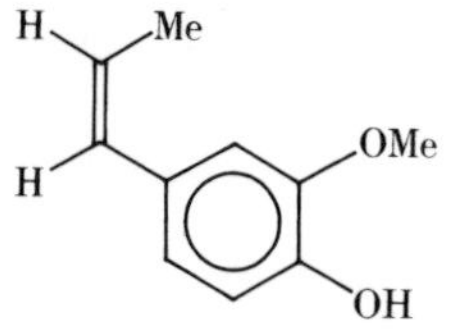

54/II

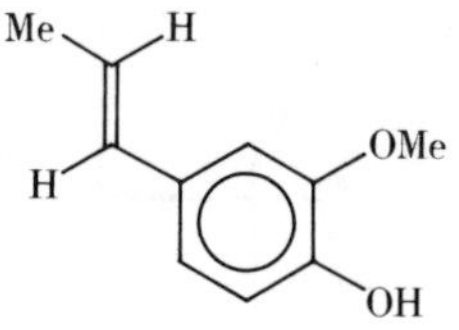

54/III

In Spectrum **54**/*3* of the third substance, the methoxy and hydroxy signals can be observed at 3.81 and 5.68 ppm. The *ABC* spectrum of the aromatic protons appears between 390 and 420 Hz. The signal of the second methyl group is missing, being replaced by a doublet at 3.32 ppm of 2H intensity. Due to an unresolved fine structure, the lines of the doublet are broadened, and the shoulders suggest a triplet splitting. There are complex multiplets in the regions of 290 to 320 and 345 to 385 Hz, with intensities of 2H and 1H, respectively. Consequently, the sample must be structural isomer **IV** (eugenol) containing a terminal double bond, i.e., substituent R′ is an allyl group.

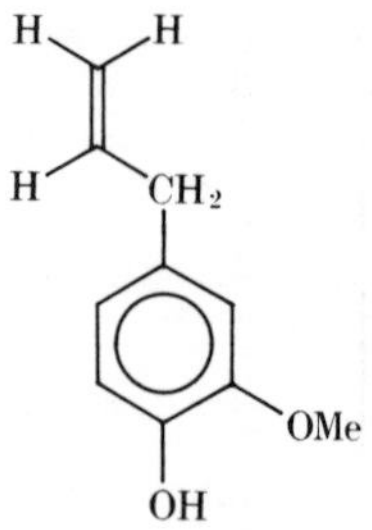

54/IV

The doublet at 3.32 ppm arises from the saturated methylene group, and its splitting of 6.5 Hz is caused by the *vicinal* olefinic hydrogen. The further triplet splitting of the two lines is due to long-range coupling of ~1.5 Hz with the terminal vinylidene hydrogens.

The olefinic protons produce a very complicated multiplet, the complete interpretation of which would require computer methods. (The allyl group represents an $ABCX_2$ spin system close to the limiting AB_2X_2 case.) However, for the structure determination a detailed analysis of the multiplet is unnecessary.

* See Volume II, p. 52 and Table 38.

** See Volume I, p. 63.

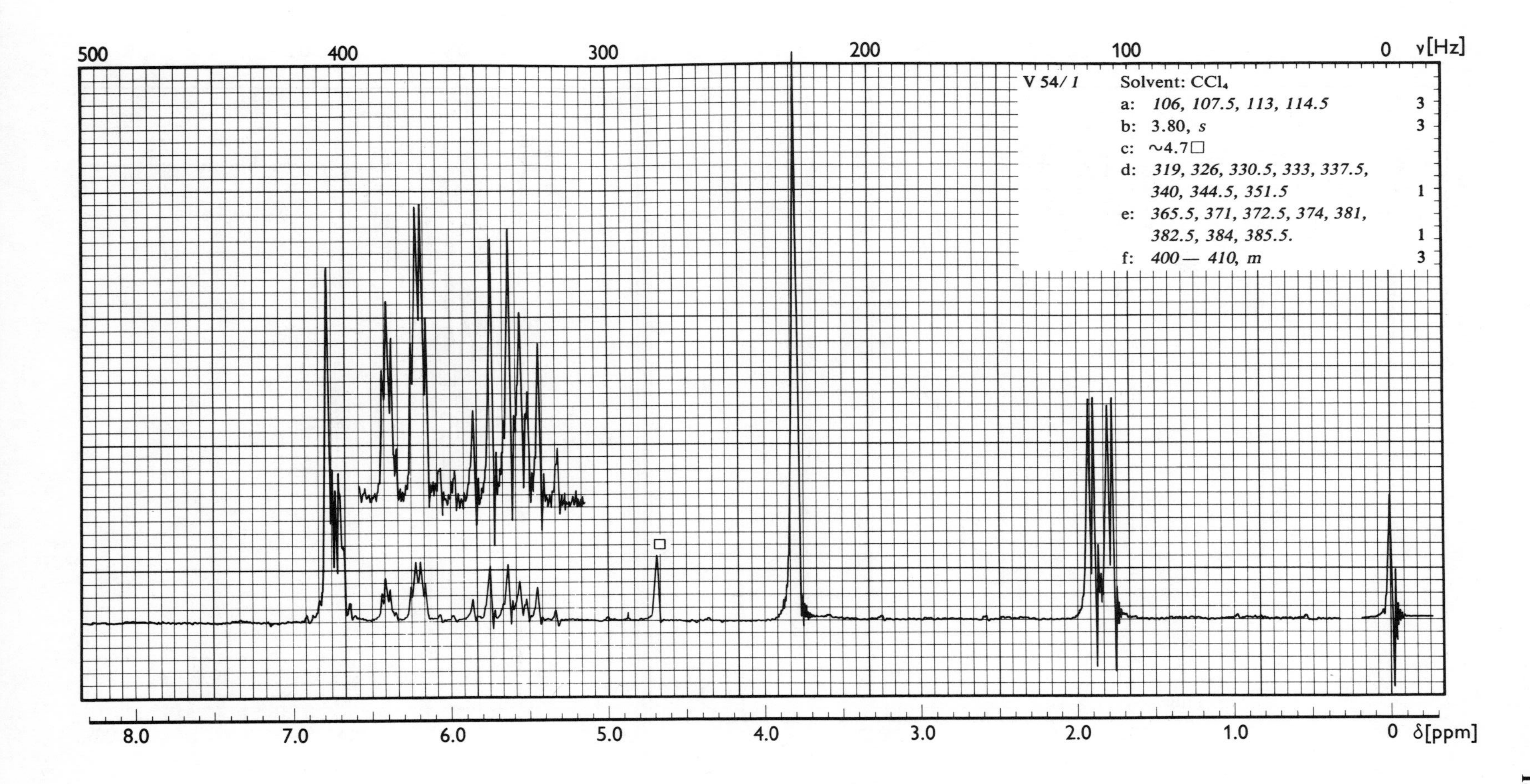
500
400
300
200
100
0 ν[Hz]
V 54/ 1
Solvent: CCl_4
a: 106, 107.5, 113, 114.5 3
b: 3.80, s 3
c: ~4.7□
d: 319, 326, 330.5, 333, 337.5, 340, 344.5, 351.5 1
e: 365.5, 371, 372.5, 374, 381, 382.5, 384, 385.5. 1
f: 400 — 410, m 3
□
8.0
7.0
6.0
5.0
4.0
3.0
2.0
1.0
0 δ[ppm]

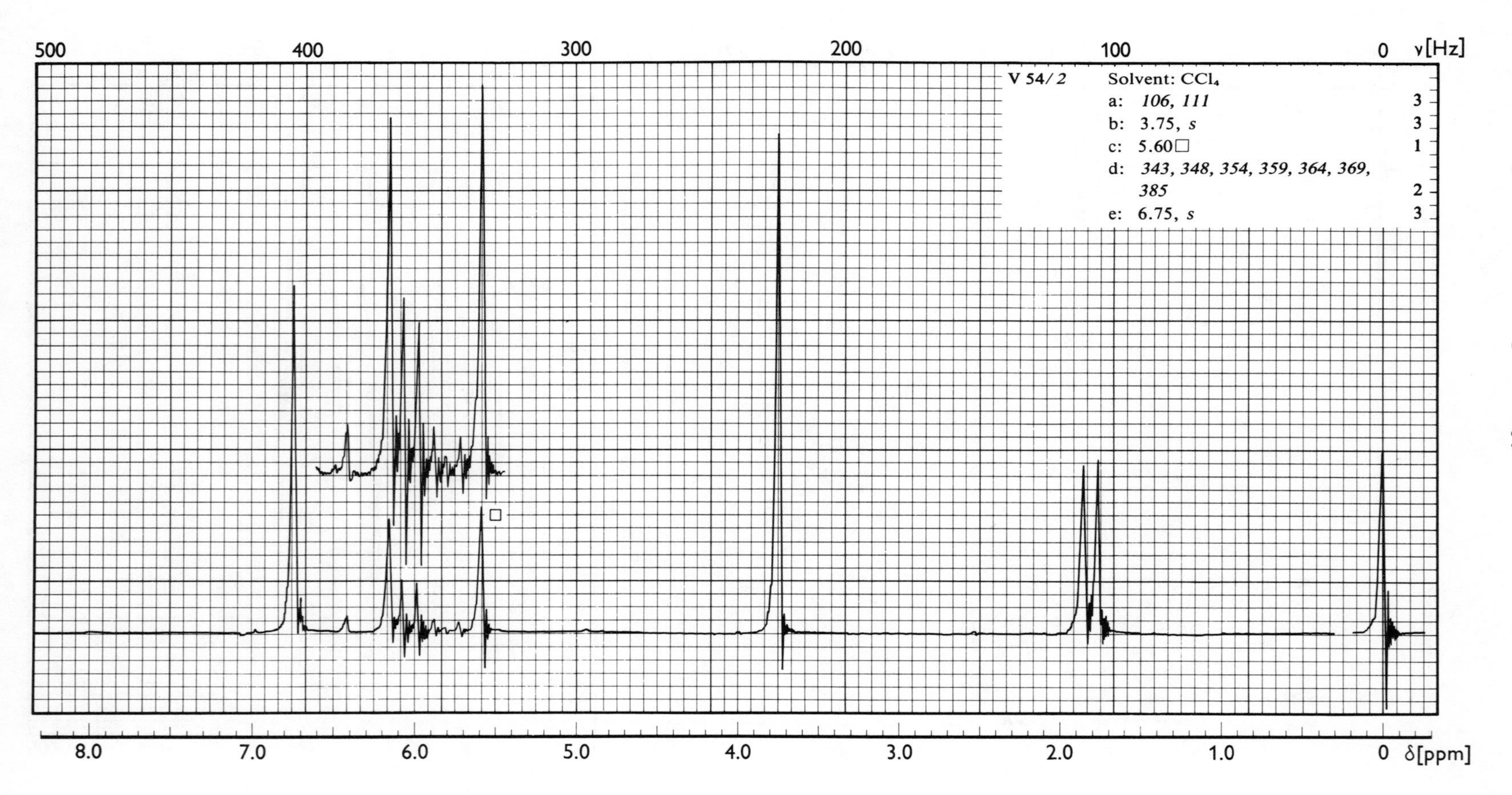
V 54/2
Solvent: CCl₄
a: 106, 111 3
b: 3.75, s 3
c: 5.60□ 1
d: 343, 348, 354, 359, 364, 369, 385 2
e: 6.75, s 3
500 400 300 200 100 0 ν[Hz]
8.0 7.0 6.0 5.0 4.0 3.0 2.0 1.0 0 δ[ppm]

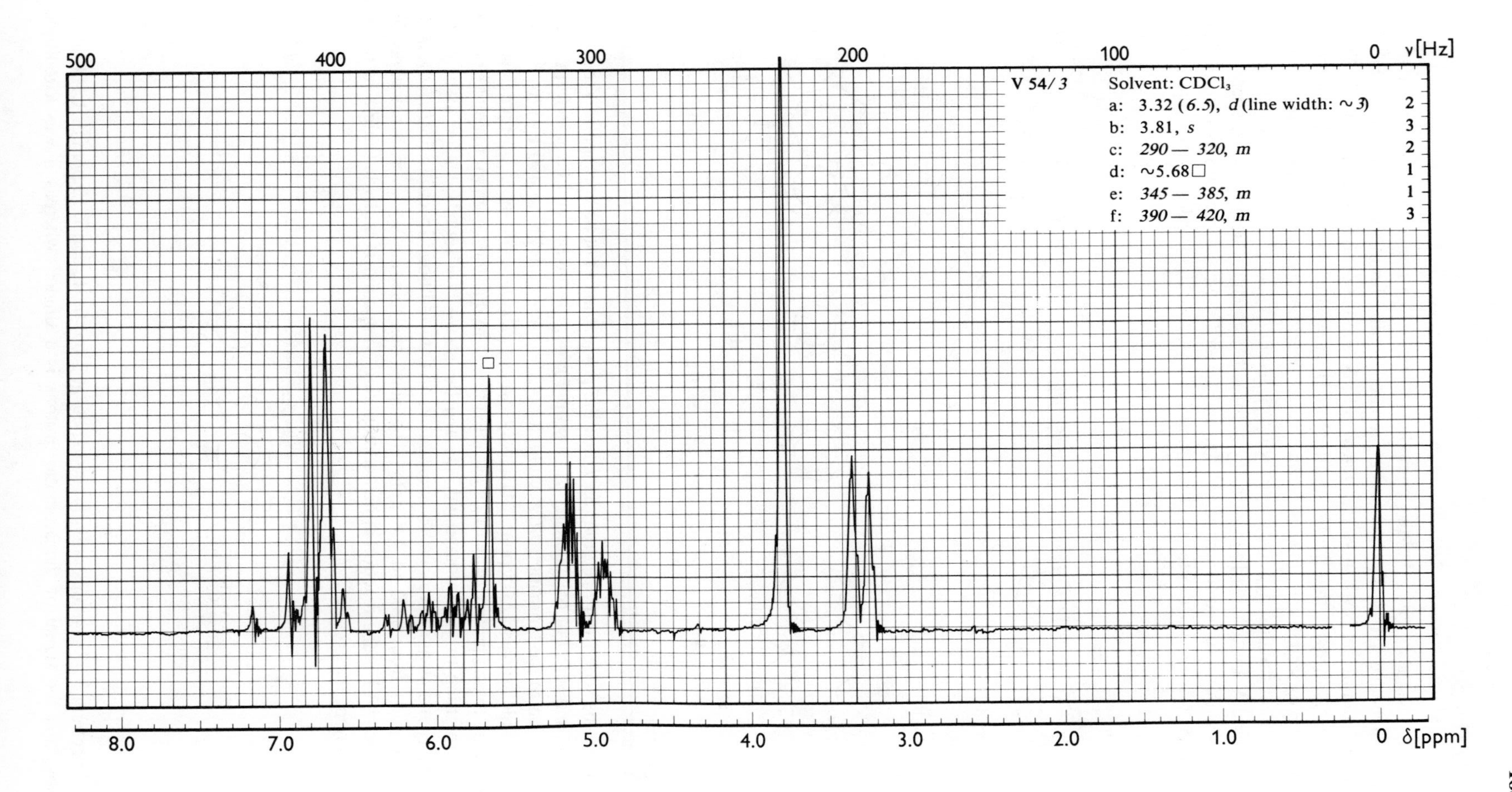
500
400
300
200
100
0
ν[Hz]
V 54/3
Solvent: $CDCl_3$
a: 3.32 (*6.5*), *d* (line width: ~*3*) 2
b: 3.81, *s* 3
c: *290 — 320, m* 2
d: ~5.68□ 1
e: *345 — 385, m* 1
f: *390 — 420, m* 3
□
8.0
7.0
6.0
5.0
4.0
3.0
2.0
1.0
0
δ[ppm]

PROBLEM 55

In the synthesis given below, compound **I** was to be converted into derivative **II**, and then compound **III** or **IV** was to be prepared by the bromination of compound **II**. The original aim of the NMR investigation was to decide between Structures **III** and **IV**. It had turned out, however, that the IR spectrum of **II** shows hydroxy bands, and thus the product is not the expected one, but, presumably, the intermediate **V**. It was therefore necessary also to confirm this hypothesis, proving Structure **V** and **II**, which was obtained from compound **V** by water elimination (heating with elementary iodine). The spectra of compounds with presumed Structures **V**, **II**, and **III** or **IV** were recorded (**55**/*1* to *3*). The first spectrum was recorded after D_2O addition (**55**/*1a*), too, the third is not solvent dependent.

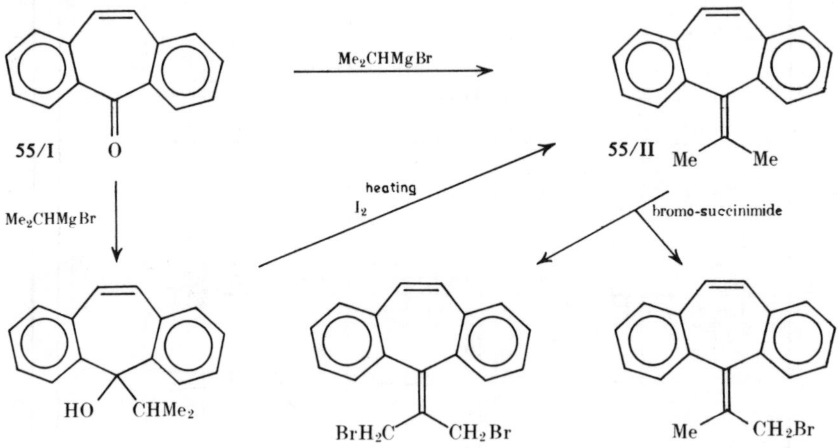

$Me_2CHMgBr$

55/I O

55/II Me Me

heating I_2

bromo-succinimide

HO $CHMe_2$

55/V

BrH_2C CH_2Br

55/III

Me CH_2Br

55/IV

ANSWER

It can be seen from Spectrum **55**/*1* that the product may not be compound **V** because the doublet and septet of 6:1 intensity of isopropyl groups is absent, and the spectrum contains, in addition to the aromatic *ABCX* multiplet (425 to 465 and 470 to 490 Hz) (6 + 2H) and the olefinic signals (7.05 ppm, 2H), only two singlets at 2.06 and 5.03 ppm, each corresponding to one proton. It must be assumed therefore that the Grignard reagent has reduced the oxo-group to yield compound **VI**, and the above two signals can be assigned to the hydroxy and methine protons. In accordance, on D_2O addition (Spectrum **55**/*1a*), the hydroxy signal shifts from 2.06 to 4.8 ppm. The questions remain, however, what happened to compound **VI** in the reaction aimed at preparing compound **II** and what might be the actual structure of the product?

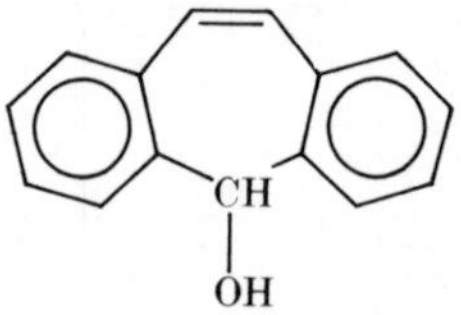

CH

OH

55/VI

Spectrum **55**/2 contains only three singlets at 3.65, 6.91, and 7.15 ppm of 1:1:4 intensity. Note that the signal of the aromatic protons is a singlet, whereas it was a multiplet in

Spectrum **55**/*1*. The two close signals belong to the aromatic and olefinic protons, and this is also confirmed by the intensity ratio of 4:1. This, in turn, involves that the signal at 3.55 ppm arises from two protons, and therefore the product is obviously compound **VII**, the reduced derivative of **VI**. Structure **VII** is compatible with the much smaller shift of the methylene signal relative to the δCH(OH) signal and the smaller upfield shift of the olefinic signal.

The singlet structure of the aromatic signal can be attributed to an accidental isochrony of the ring protons. The RR′CHOH-type compound may disproportionate upon heating in the presence of I_2 crystals as catalyst to yield RR′CH_2 and RR′CO derivatives. Such a process may have taken place to compound **VII**. (Compound **VII** was prepared in an independent way, and its spectrum was identical to **55**/*2*. As a further proof, compound **I** could be separated from the reaction mixture, too.)

Now the structure of the brominated product is to be elucidated. Spectrum **55**/*3* of this substance contains again three singlets at 4.53, 5.90, and 7.28 ppm, of 1:1:4 intensity. Therefore, it is obvious that instead of a bromine substitution on the methylene group, bromine addition took place on the double bond, to yield compound **VIII**. Accordingly, the aromatic and methine signals are shifted downfield in comparison with the spectrum of compound **VII** (owing to the $-I$ effect of the bromine atoms).

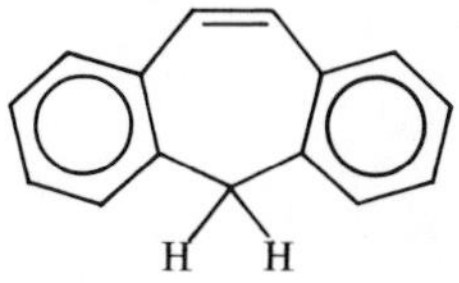

Br Br
CH–CH
H H
H H

55/VII 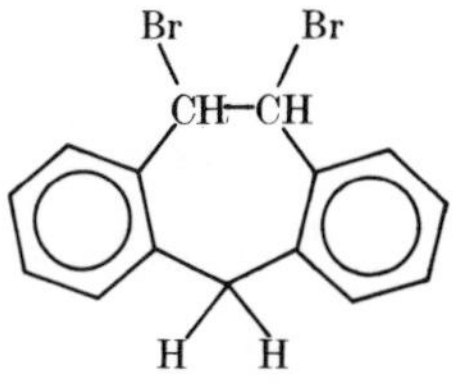 **55/VIII**

As compound **VIII** contains two centers of asymmetry, two diastereomeric structures are possible, the *cis* (**a**) and the *trans* (**b**) isomers. Of these, both for steric reasons (bulky bromine atoms) and from the reaction conditions, the *trans* structure (VIIIb) is more probable (the bromination of olefins with *N*-bromosuccinimide generally yields *trans* dibromo derivatives). The **a** and **b** isomers can be distinguished by analyzing the structure of the methine signal.

Cycloheptatriene is nonplanar, and thus the singlet structure of the methylene signal

Br Br Br Br
a **a′**

Br Br Br Br
b **b′**

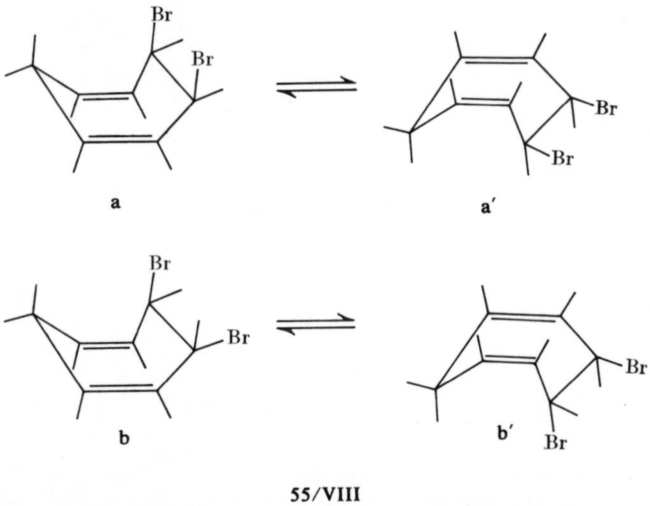

55/VIII

indicates that there is a fast ring inversion at room temperature.* At lower temperatures, however, the methylene hydrogens become anisochronous owing to the slow conformational motions and give an *AB* quartet. This also pertains to the methylene and methine protons of *trans* isomer **VIIIb**. At low temperatures, with slow inversion, they are anisochronous, but became equivalent at room temperature due to the fast ring inversion (**b** $\rightleftharpoons$ *b*′).**

The diastereotopic methylene protons of the *cis* isomer do not become equivalent even in the case of a fast ring inversion. The inverse forms **a** and **a**′ are not identical, since the chemical environments of the methylene hydrogens are different in them (of course, the methine protons are equivalent, independently of temperature). Therefore, if the methylene protons of the *cis* isomer give a singlet signal even at room temperature, it must be due only to accidental isochrony. Since, however, in $CDCl_3$ and DMSO-d_6 solutions the methylene signal of the dibromo compound is invariably a singlet, only the *trans* isomeric structure (**VIIIb**) is possible.

* See Volume II, p. 56.

** For the sake of clarity, the aromatic rings are represented by double bonds in formulas **VIIIa**, **a**′, **b**, and **b**′.

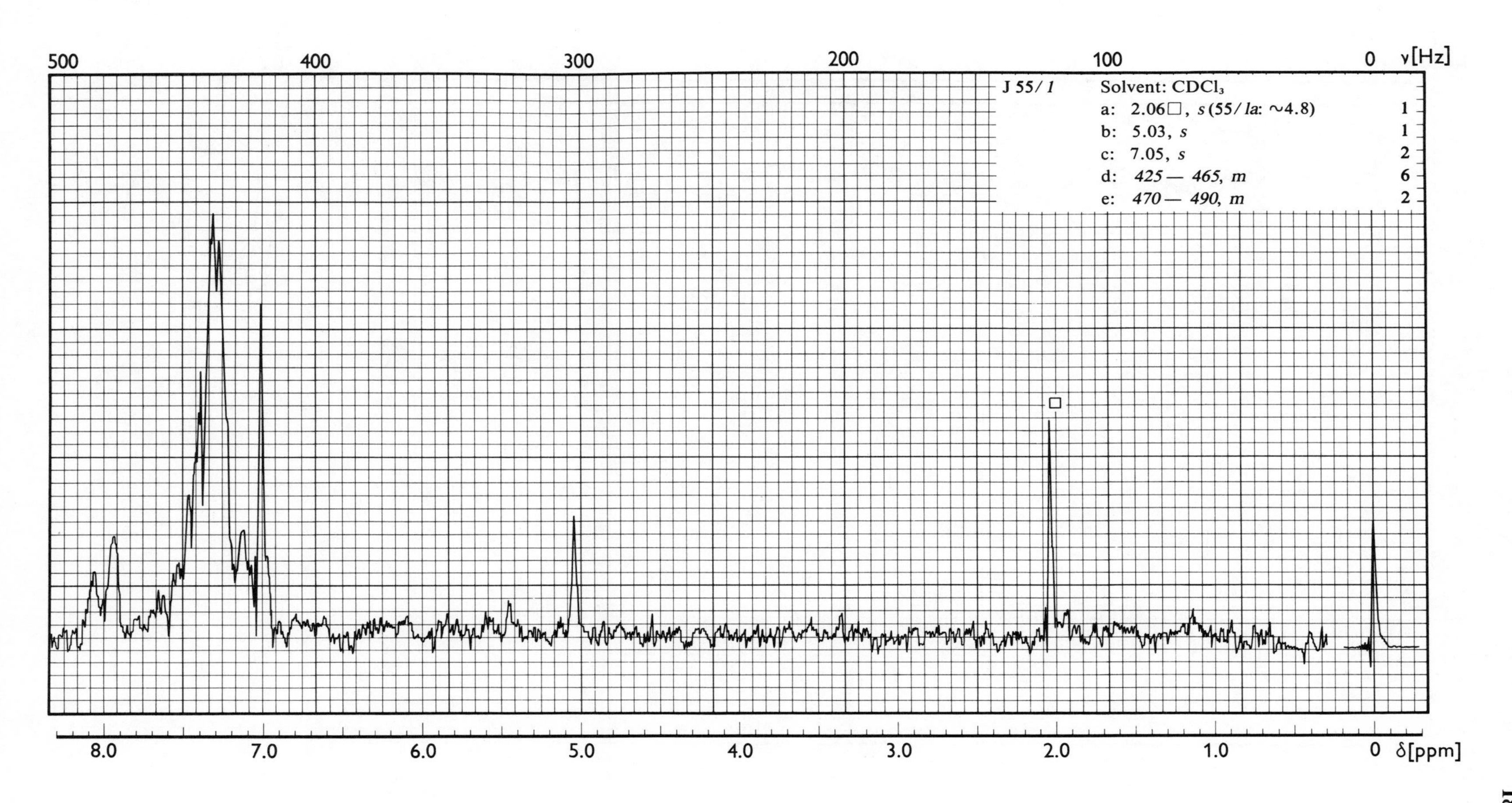
ν[Hz]
500
400
300
200
100
0
J 55/1
Solvent: CDCl₃
a: 2.06□, s (55/1a: ~4.8) 1
b: 5.03, s 1
c: 7.05, s 2
d: 425— 465, m 6
e: 470— 490, m 2
8.0
7.0
6.0
5.0
4.0
3.0
2.0
1.0
0
δ[ppm]

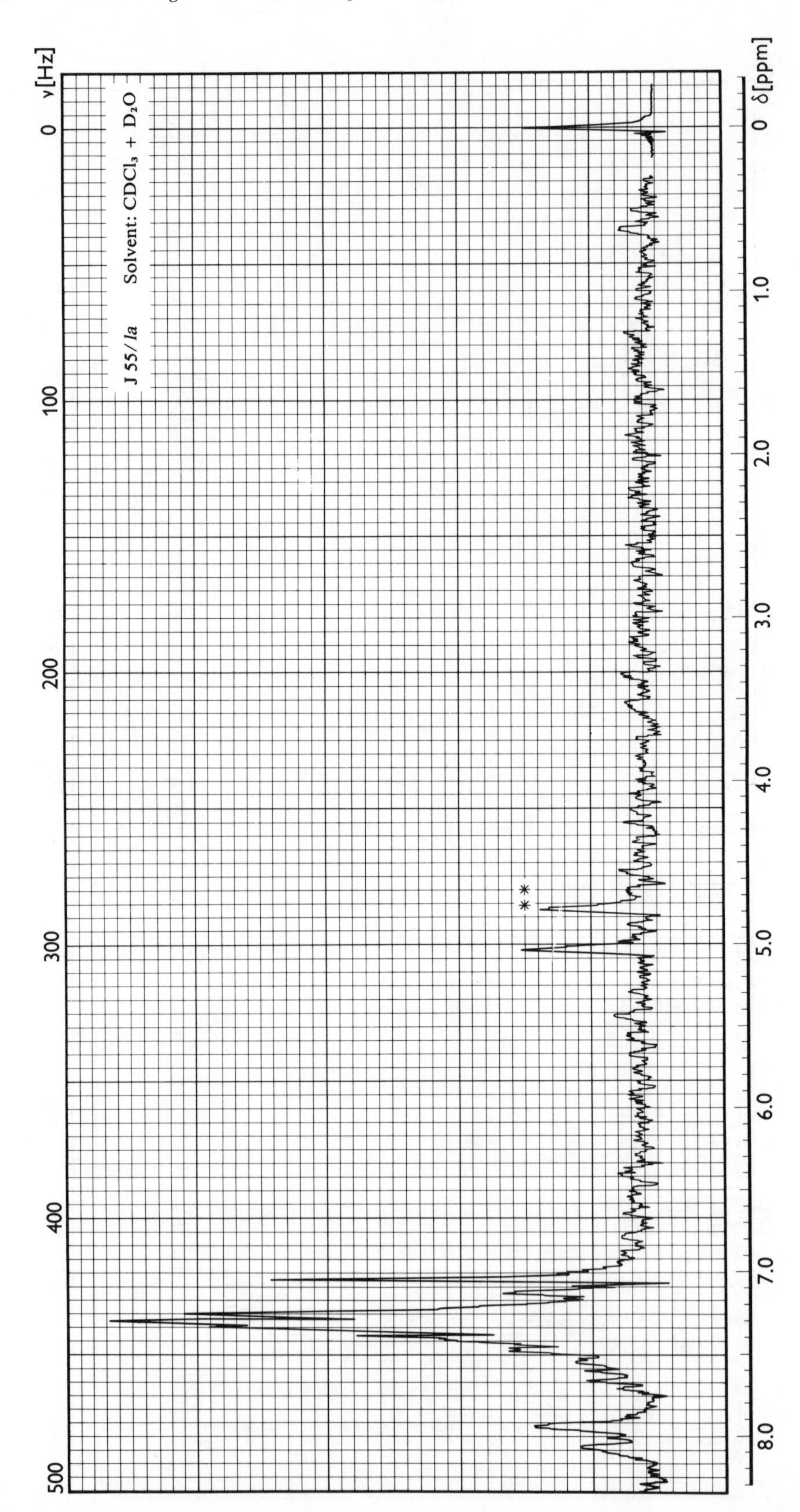
J 55/1a
Solvent: CDCl₃ + D₂O
ν[Hz]
0
100
200
300
400
500
δ[ppm]
0
1.0
2.0
3.0
4.0
5.0
6.0
7.0
8.0
**

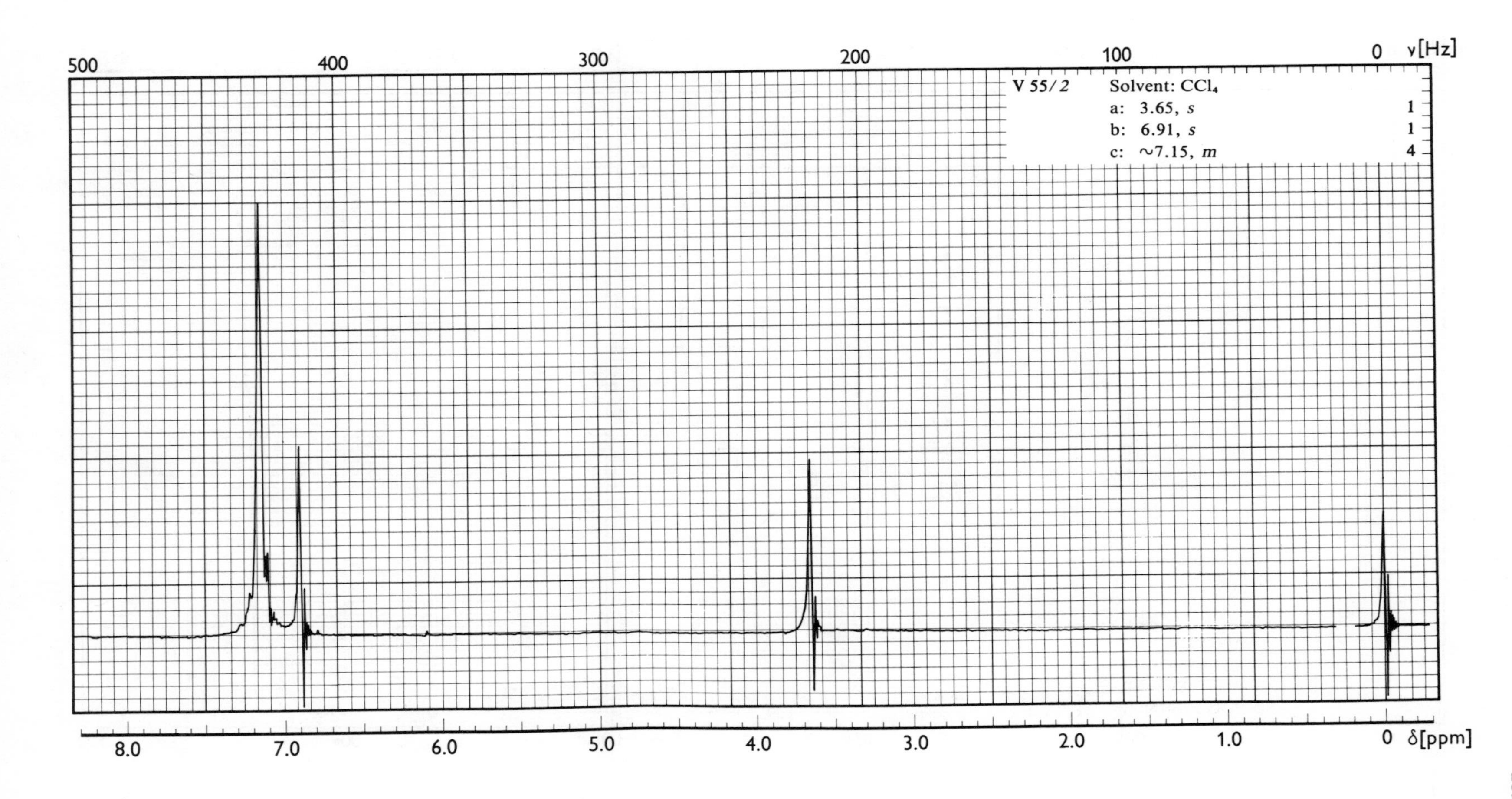

V 55/2
Solvent: CCl_4
a: 3.65, *s* 1
b: 6.91, *s* 1
c: ~7.15, *m* 4
500
400
300
200
100
0
ν[Hz]
8.0
7.0
6.0
5.0
4.0
3.0
2.0
1.0
0
δ[ppm]

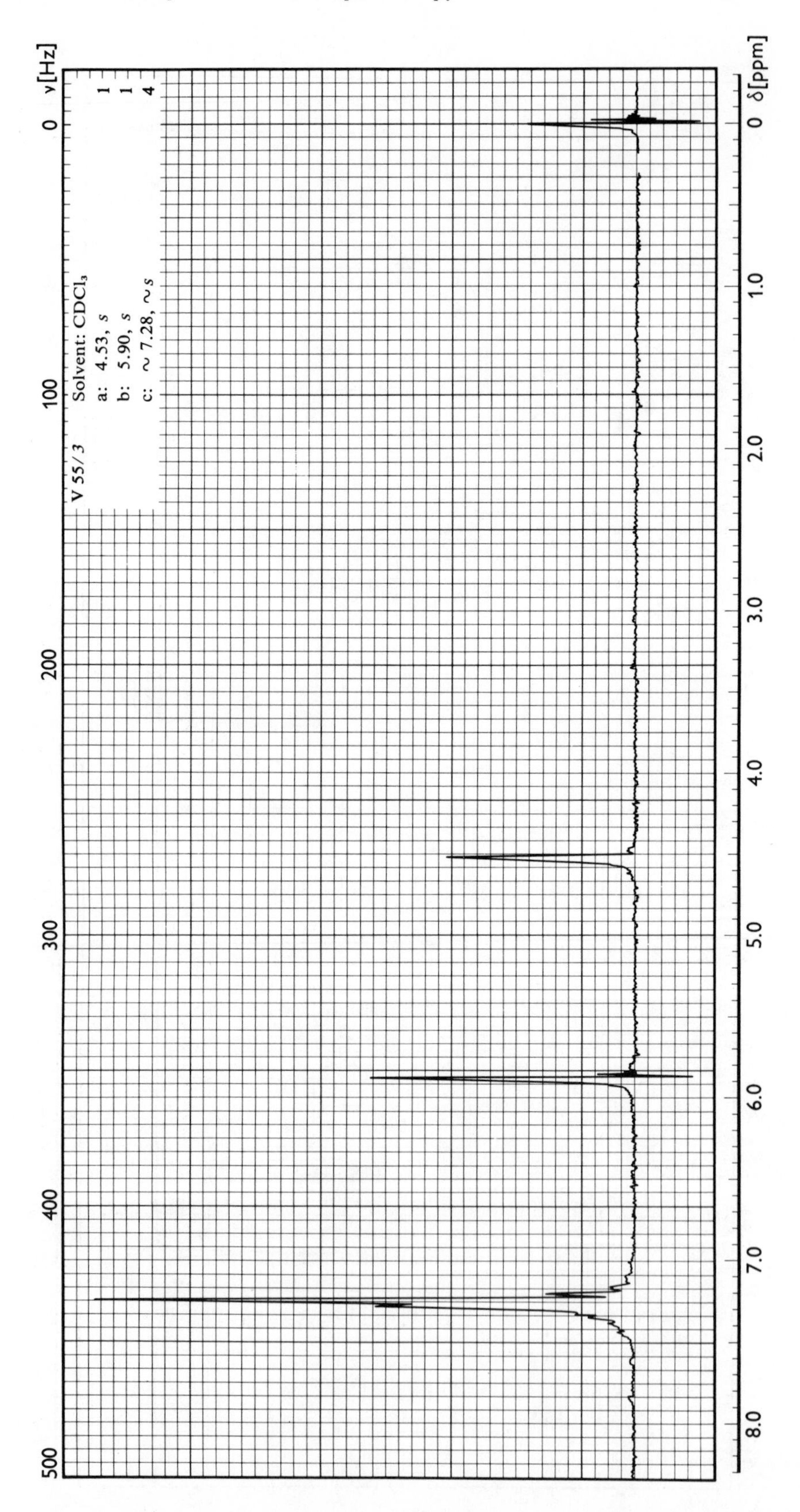

V 55/3
Solvent: $CDCl_3$
a: 4.53, *s* 1
b: 5.90, *s* 1
c: ~ 7.28, ~*s* 4
ν[Hz]
0
100
200
300
400
500
δ[ppm]
0
1.0
2.0
3.0
4.0
5.0
6.0
7.0
8.0

PROBLEM **56**[978]

Interpret Spectrum **56**/*1* of compound **I** and calculate the spectrum parameters, making use of Figure 206a, showing the 435 to 480 Hz part of Spectrum **56**/*1* recorded with a fivefold abscissa expansion.

56/Ia **56/Ib**

ANSWER

The ethyl group produces the usual symmetric triplet and quartet. From the line positions δCH_3 = 1.33 ppm, *t* (J = 7.5 Hz), δCH_2 (partly overlapping with the quintet of the CD_2HOD impurity): 3.53 ppm, *qa* (J = 7.5 Hz), δNH (overlapping with the signal of the CD_3OH impurity): 4.98 ppm, *s*.

The aromatic protons give a symmetric $AA'BB'$ spectrum. (It is not an $AA'XX'$ spectrum, because the A and B parts are not symmetrical separately, only the whole multiplet.) This structure can be explained by assuming fast tautomeric interconversion provided that the mean lifetime of the tautomers is sufficiently short to make the A, A' and B, B' protons chemically equivalent.

The midpoint of the multiplet is 457 Hz. As only 7 of the expected 12 maxima (see Figure 66) appear, it must be assumed that 5 lines overlap with others, or they are too weak to be observable. It is clear from the relative intensities that line pairs **9** to **10**, **5** to **6**, and **3** to **4** are overlapping, and lines **1** and **2** are so weak that they merge into the noise.

The frequencies of the seven observed maxima, measured from the midpoint of the multiplet, are 2.0, 2.8, 5.6, 8.4, 9.5, 11.9, and 15.2 Hz. These data are sufficient to determine the approximate chemical shifts and coupling constants, under the acceptable assumption that $J_{AA'} \equiv J^p \approx 0$. In this case $Q_1 = -Q_2$, $Q_4 = \nu_3 - \nu_9 = 15.2 - 5.6 = 9.6$ Hz,* $\Delta\nu AB^2 + Q^2{}_4 = \nu_3 + \nu_9 = 15.2 + 5.6 = 20.8$ Hz, and $\Delta\nu AB = \sqrt{(20.8^2 - (9.6)^2} = \sqrt{432.64 - 92.16} = \sqrt{340.48} \approx 18.45$ Hz. Accordingly, we have $\nu A = 457 + 18.45/2 = 466$ Hz, $\nu B = 457 - 18.45/2 = 448$ Hz, and $\delta A = 7.77$ and $\delta B = 7.46$ ppm.

The relation $\delta A > \delta B$ is evident since the heteroaromatic ring and the heteroatoms are adjacent to the A protons. Furthermore, $\sqrt{Q_2^2 + Q_3^2} = \nu_5 - \nu_{11} = 11.9 - 2.8 = 9.1$ Hz, $\sqrt{(\Delta\nu AB + Q_2)^2 + Q_3^2} = \nu_5 + \nu_{11} = 11.9 + 2.8 = 14.7$ Hz. Substituting $\Delta\nu AB$ into the latter two equations, we obtain $Q_2 = 5.6$ Hz and $Q_3 = 7.2$ Hz. From Equation 179, $|J_{AB}| = (Q_3 + Q_4)/2 = 16.8/2 = 8.4$ Hz, $|J_{AB'}| = (Q_4 - Q_3)/2 = 1.2$ Hz, and $|J_{BB'}| = (Q_1 - Q_2)/2 \approx Q_2 \approx 5.6$ Hz. Of course, according to our assumption, $J_{AA'} \approx 0$.

Since J_{AB} and $J_{BB'}$ must be in the range characteristic of *ortho* coupling, their approximate values meet the expectations, and so does the value of 1.2 Hz obtained for the *meta* coupling. An iterative computer refinement starting from these values revealed that the approximate spectrum parameters determined above are correct with the experimental error except for $J_{BB'}$ and, of course, $J_{AA'}$ (for which 6.6 and 0.9 Hz were obtained).[978] The theoretically calculated $AA'BB'$ spectrum is shown in Figure 206b.

* Compare Volume I, p. 133.

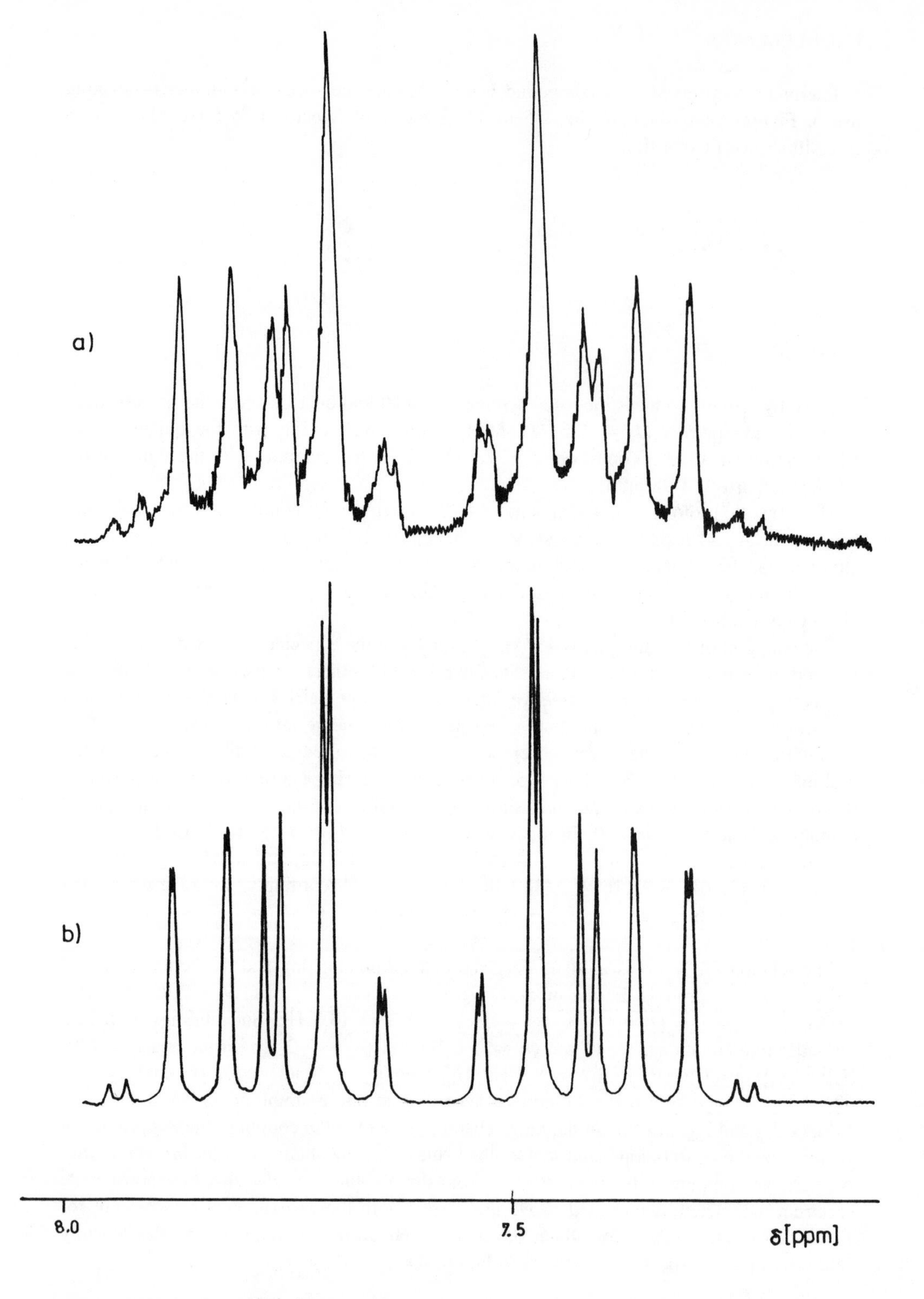

FIGURE 206. The (a) observed and (b) computer simulated *AA'BB'* multiplet of the aromatic protons in Spectrum **56**/*1*.

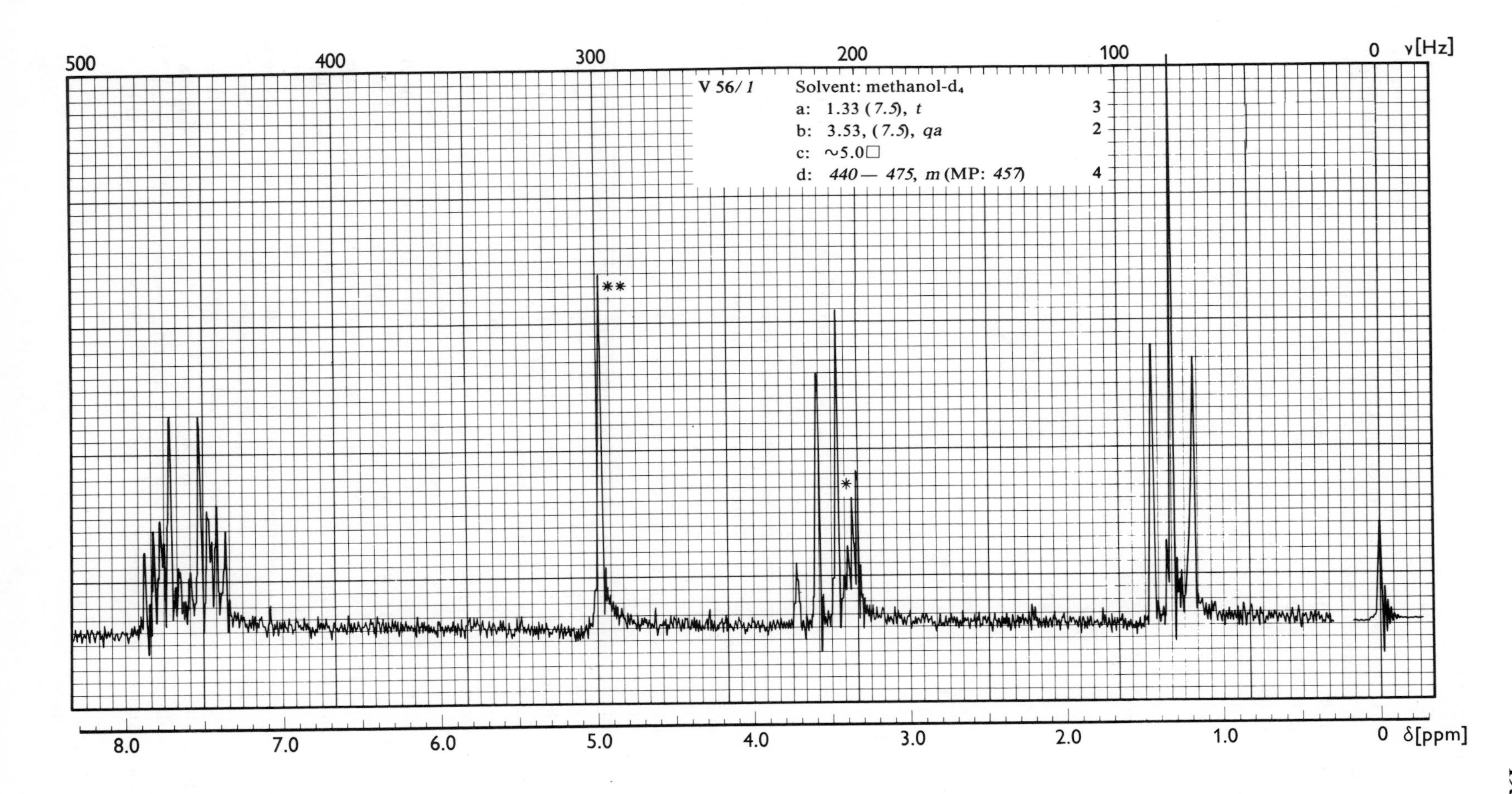
V 56/*1*
Solvent: methanol-d$_4$
a: 1.33 (*7.5*), *t* 3
b: 3.53, (*7.5*), *qa* 2
c: ~5.0□
d: *440— 475, m* (MP: *457*) 4
500
400
300
200
100
0
ν[Hz]
8.0
7.0
6.0
5.0
4.0
3.0
2.0
1.0
0
δ[ppm]
**
*

PROBLEM 57[1342]

Decide from Spectrum **57/*1*** whether the sample is *Z* (**Ia**) or *E* (**Ib**) chalcone derivative. We may use the 60-MHz spectrum of the DMSO-d_6 solution (**57/*1***), the 90-MHz spectrum of the $CDCl_3$ solution (**57/*1a***), and a tenfold abscissa expansion of the latter (**57/*1b***). Why is the 60-MHz spectrum insufficient to elucidate the steric structure? The solubility of the sample in $CDCl_3$ is too low to produce an appropriate 60-MHz spectrum. Why does the same problem not occur at 90 MHz? Sketch the expected structure of the spectrum and decided which part of it enables to elicit the steric structure.

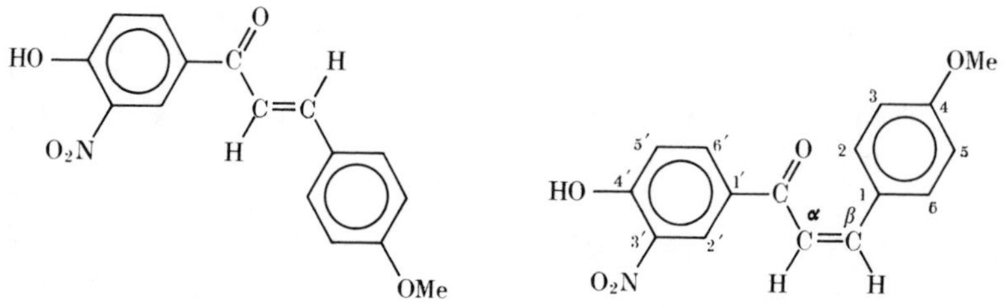

57/Ia **57/Ib**

ANSWER

The olefinic protons are not expected to interact with the other hydrogens of the molecule, and thus they and the hydrogens of the aromatic rings form three independent spin systems. The olefinic hydrogens produce an *AX* or *AB* spectrum, and the answer to the question of *Z-E* isomery may be derived from the splitting of the two doublets, which is 5 to 14 Hz for *Z*, whereas it is 11 to 19 Hz for *E* isomers (see Table 38). Therefore, the key of this problem is to identify the *AB* or *AX* spectrum of the olefinic protons or at least one part *A* or *B* (*X*) of the multiplet. It is known that $\delta A \equiv \delta H_\alpha < \delta B\ (X) \equiv \delta H_\beta$, due to the mesomery of the Ph–CH=CH–CO system.*

The aromatic protons of the 1,2,4-trisubstituted ring have an *ABX* spectrum, since δX of H-2′ is appreciably larger than δA, or δB of the other two ring protons, due to the two adjacent electron-withdrawing substituents. Moreover, if $A \equiv$ H-5′ and $B \equiv$ H-6′, $\delta A < \delta B$, because the former, *para* to the *X* proton (H-2′), is adjacent to a hydroxy group, and the latter to the carbonyl substituent which causes downfield shift.

The expected values of coupling constants** are as follows: $J^o \equiv J_{AB} = 7$ to 10, $J^m \equiv J_{BX} = 2$ to 3, and $J^p \equiv J_{AX} < 1$ Hz. The *B* part of the *ABX* spectrum can thus be distinguished easily from the *AB* spectrum of the olefinic protons, since both maxima of the *B* doublet (and, of course, the signal of the *X* proton as well) are split by 2 to 3 Hz, corresponding to the J^m_{BX} coupling. On the other hand, however, it is easy to mistake the *A* doublet for the *A* part of the *AB* spectrum of olefinic protons, because the small J_{AX} coupling usually causes no significant splitting. Such a false assignment would lead, to the conclusion that the olefinic hydrogens are in *cis* position, because the value of J☆ is quite understandably in the region characteristic of *Z* olefins (see the data above). When, however, the *B* part of the *ABX* spectrum has already been identified, one may determine J_{AB}, and therefore the doublet with a line spacing of J_{AB} can be assigned to nucleus *A* of the *ABX* spin system.

The hydrogens of the *p*-disubstituted aromatic ring have an *AA′BB′* spectrum. The four main maxima of this multiplet look like an *AB* quartet (see Figure 65), and the coupling

* Compare Volume II, p. 52.

** See Volume II, p. 68.

constant $J_{AB} \equiv J^o$ obtained therefrom is also in the range characteristic of *Z* olefins. Nevertheless, the *AA'BB'* multiplet is easy to recognize by the weak lines appearing on both sides of the four main maxima. If $H_A \equiv$ H-3,5 and $H_B \equiv$ H-2,6,$\delta A < \delta B$, because the methoxy substituent causes shielding on the adjacent ring protons, wheras the olefinic double bond deshields the *ortho* hydrogens (see Table 37).

After these preliminaries we may start investigating the spectra, first Spectrum **57**/*1*, recorded at 60 MHz. The quality of the spectrum is rather poor owing to the low resolution and to the line broadening in the strongly polar solvent. The signal with the largest chemical shift is the doublet of H-2′ (the large downfield shift can be attributed to the adjacent nitro and carbonyl groups). The splitting of $J^m \approx 2$ Hz is observable only as a shoulder. The next signal on the downfield side is a double doublet, which corresponds to H-6′ and from the distance of the two doublets J^o (5′,6′) ≈ 9 Hz. Both doublets in the upfield side of the signal groups are split similarly. From the relative intensity of 1:2, the outside doublet belongs to the H-3,5 pair and the other belongs to H-5′. There is only one further strong maximum in the spectrum at 7.82 ppm, arising from the olefinic protons, but it overlaps, as indicated by the intensities, with the multiplet of H-2,6 (the weaker downfield line appears separately at 478 Hz). The olefinic protons are accidentally isochronous (their signal is a singlet in DMSO-d_6), thus the problem of stereoisomers cannot be solved. This is the reason why the 90 MHz spectrum is required.

The signals are expected to be more separated in this spectrum, because the chemical shifts and their differences (in hertz) increase proportionally to the measuring frequency.* Furthermore, $J/\Delta\nu$, and thus the order of coupling decreases, yielding simpler spectra.** Apart from these advantages, the absolute line intensities also increase [by a factor of $(90/60)^{3/2} = 1.84$]*** Therefore, a better spectrum can now be recorded in $CDCl_3$, even at a half concentration. The spectrum of $CDCl_3$ solution is superior to those of DMSO-d_6 solutions, because of the sharper lines. In the Spectrum **57**/*1a*, measured at 90 MHz, all lines are separated, and their assignment is straightforward. Downfield, the first signal, a doublet, can be assigned to H-2′ as discussed above. From the line frequencies of 793.5 and 796.5 Hz, $J_{2',6'} = 2.5$ Hz and δH-2′ = 8.83 ppm (δH-2′ = νH-2′/90 = 794.75/90). The next four lines at 743.5, 746, 752.5, and 755 Hz may belong only to H-6′ because the spacing between two line pairs is 2.5 Hz. The distance of the two doublets yields $J_{5',6'} = 9.0$ Hz. To determine δH-5′ we must find the doublet of H-5′ split by 9.0 Hz and with a more intense downfield line (the inner lines of an *AB* system are stronger). The spectrum contains two such line pairs, and on the basis of the relative intensities, the lines at 651 and 660 Hz can be assigned to H-5′. The other line pair of 2H intensity, at 622 and 631 Hz, must belong to H-3,5, as shown also by the satellite lines (*AA'BB'* multiplet!). From the frequencies of (755.0 + 752.5)/2, (746.0 + 743.5)/2, 660 and 651 Hz, we obtain δH-6′ = 8.32 and δH-5′ = 7.29 ppm.

The doublet, split by 9 Hz, surrounded by satellites, and having stronger upfield maximum, can be found at 684 and 693 Hz. It arises, of course, from H-2,6. Applying the *AB* approximation to the *AA'BB'* multiplet, we obtain $J_{2,3} \equiv J_{5,6} = 9$ Hz and ΔνH-2,3 ≡ ΔνH-5,6 = $\sqrt{(693-622)\cdot(684-631)} = \sqrt{71\cdot 53} = 61$ Hz from which δH-2,6 = 7.64 and δH-3,5 = 6.97 ppm.

Now only four unassigned lines are left at 657, 672, 702, and 717 Hz (the absorption arising from the light isotope content of the solvent is disregarded). The *AB* spectrum close to the limiting *AX* case belongs to the olefinic protons, and the splitting of $J_{\alpha\beta} = 15$ Hz of the doublets indicates unambiguously that the compound is the *E* isomer (**Ib**). From the above frequencies $\delta H_\alpha = 7.40$ and $\delta H_\beta = 7.83$ ppm.

* See Figure 11 and Volume I, p. 26.

** See Volume I, p. 93, first footnote.

***Compare Volume I, p. 23.

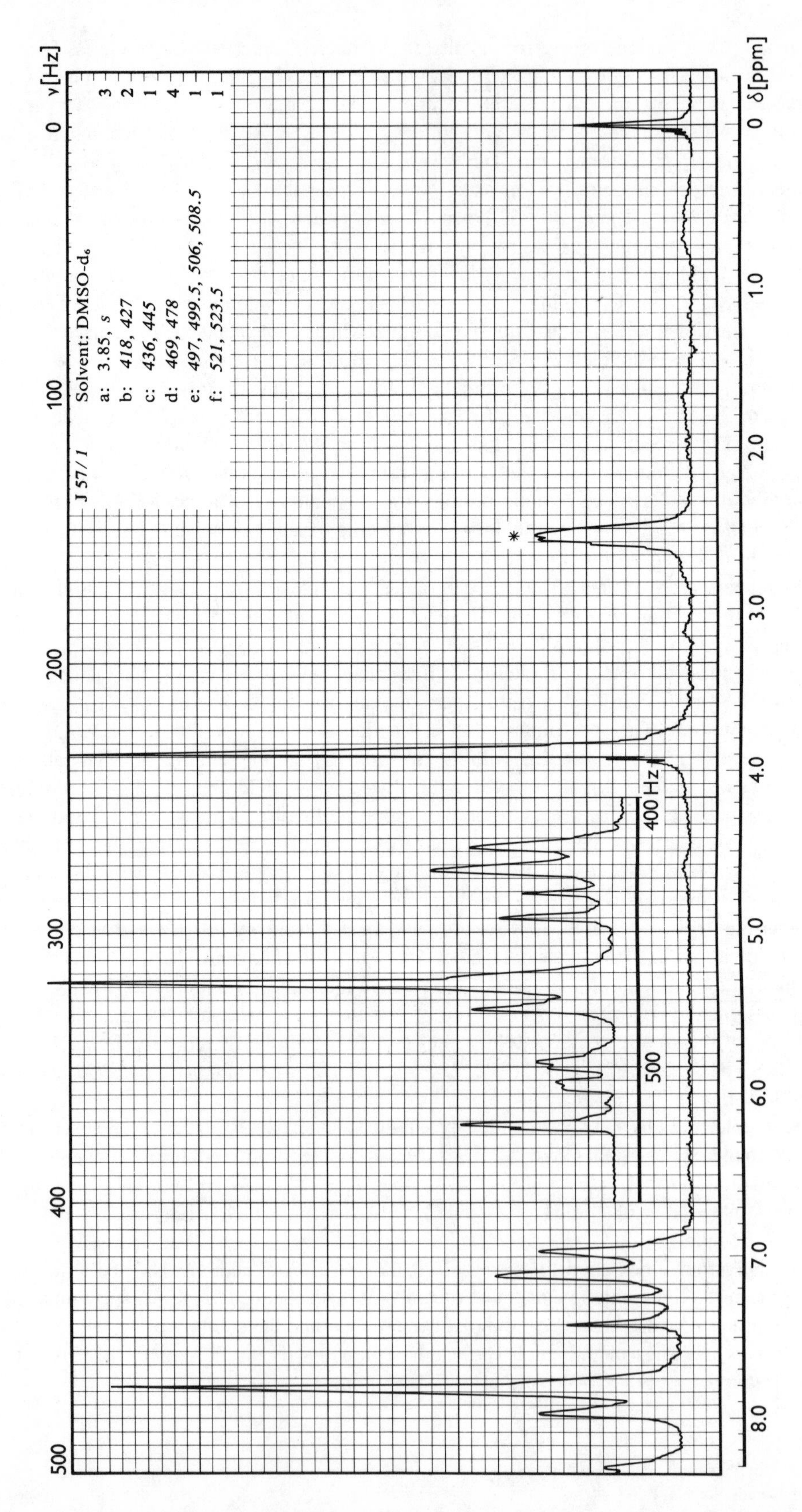
J 57/1
Solvent: DMSO-d6
a: 3.85, s 3
b: 418, 427 2
c: 436, 445 1
d: 469, 478 4
e: 497, 499.5, 506, 508.5 1
f: 521, 523.5 1
ν[Hz]
0
100
200
300
400
500
δ[ppm]
0
1.0
2.0
3.0
4.0
5.0
6.0
7.0
8.0
400 Hz
500
*

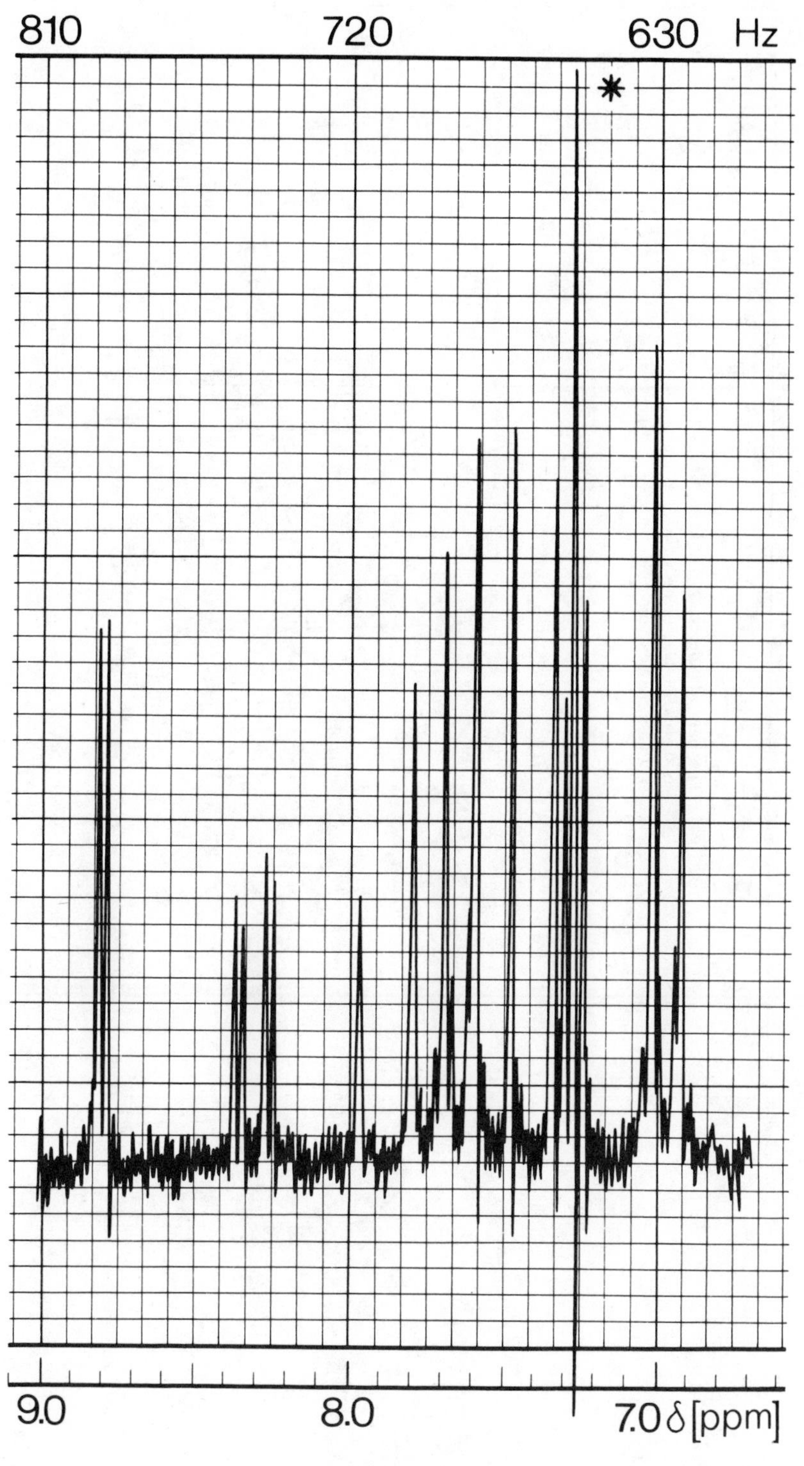

B 57/ *1a*

Solvent: $CDCl_3$

a:	*610 — 640, m* (MM: *622, 631*)	2
b:	*651, 657, 660, 672*	
c:	*680 — 700, m* (MM: *684, 693*)	2
d:	*702.5, 718*	1
e:	743.5, 746, 752.5, *755*	1
f:	*793.5, 796*	1

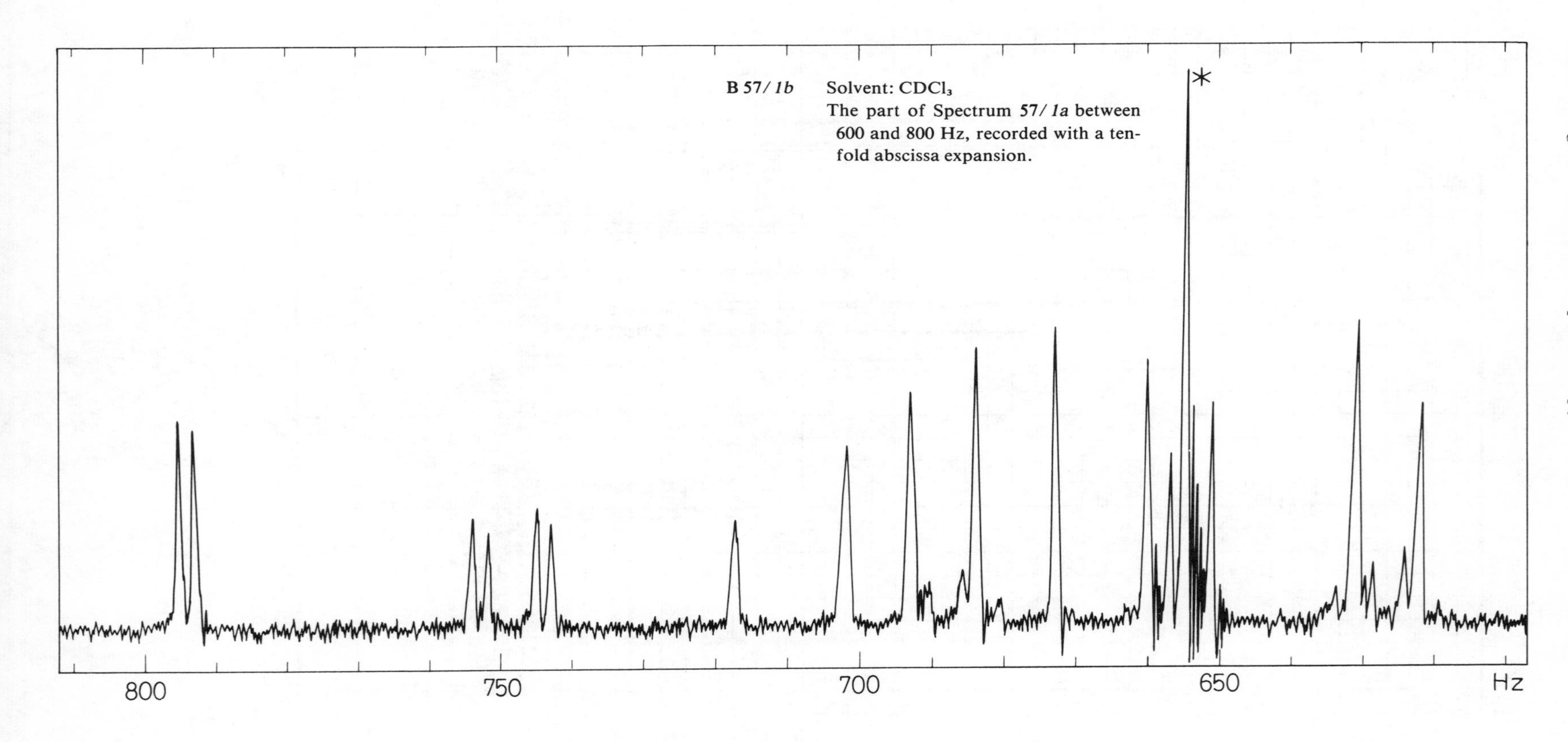

B 57/ *1b* Solvent: $CDCl_3$
The part of Spectrum 57/ *1a* between 600 and 800 Hz, recorded with a tenfold abscissa expansion.

PROBLEM 58[1330]

Interpret the spectra of compound **I** recorded in DMSO-d_6 solution (**58**/*1*) after D_2O addition (**58**/*1a*).

ANSWER

In compound **I** pairs H-1,6, H-2,5 and H-3,4 are chemically equivalent, therefore, their signals coincide pairwise. Accordingly, the structure of Spectrum **58**/*1* can be interpreted considering a CH–CHOH–CH_2 group.

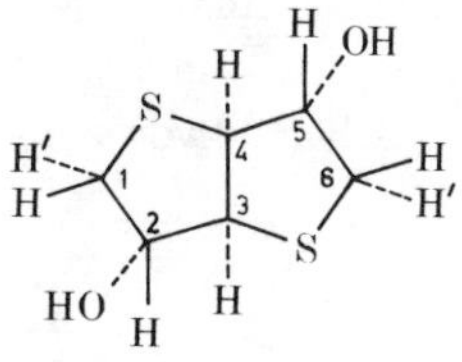

58/I

The chemically nonequivalent methylene hydrogens are in a first-order interaction with H-2 (H-5) giving an *ABX* multiplet, in which $J_{AX} \equiv J^t \neq J_{BX} \equiv J^c$.

The lines appearing at 156.5, 158, 167.5, 169, 195, 198.5, 206, and 209.5 Hz, correspond to the *AB* part of the *ABX* multiplet. The 206-Hz line overlaps with the signal of the water content of the solvent (see Spectrum **58**/*1a*). These frequencies yield the following parameters: $J_{AB} \equiv J_{1,1'} \equiv J_{6,6'} = 11$ Hz, $J_{AX} \equiv J^t_{1,2} \equiv J^t_{5,6} = 3.5$ Hz, $J_{BX} \equiv J^c_{1,2} \equiv J^c_{5,6} = 1.5$ Hz. Furthermore, $\delta A = 3.37$, $\beta B = 2.73$, and $\delta X = 4.38$ ppm.

These parameters can be calculated from the *ABX* spectrum in the following manner:*

1. The two related quartets are selected from the eight lines of the *AB* part (i.e., the two *AB* quartets corresponding to the two possible spin orientations of nucleus *X*). This is easy in the case of Spectrum **58**/*1*, since the even and odd lines have distinctly different intensities. Thus, lines **1**, **3**, **5**,and **7** correspond to one spin orientation of *X* and lines **2**, **4**, **6**, and **8** to the other.
2. The chemical shifts δA and δB corresponding to the two quartets are calculated. The shifts characteristic of the *ABX* system can be obtained as their averages.
3. The coupling constant J_{AB} can be read directly from the spectrum as the spacing of lines **1** and **3**, **2** and **4**, **5** and **7**, **or** **6** and **8** of the *AB* spectrum.
4. The chemical shift δX can also be determined directly as the midpoint of the *X* lines.
5. The coupling constants J_{AX} and J_{BX} will be determined as follows. The half of their sum is equal to the distance between the midpoints of the two *AB* multiplets. Hence, $J_{AX} + J_{BX} = 2(183.75 - 181.25) = 5$ Hz. Their difference is $\sqrt{(2Q_1)^2 - J^2_{AB}} - \sqrt{(2Q)^2 - J^2_{AB}}$, where $2Q_1$ and $2Q_2$ are the spacings of lines **1** and **3** or **2** and **4** in the two quartets, respectively. In our case, $2Q_1 = 40.5$ and $2Q_2 = 38.5$ Hz. This yields $J_{AX} - J_{BX} = \sqrt{40.5^2 - 121} - \sqrt{38.5^2 - 121} = 2$ Hz. Since $J_{AX} + J_{BX} = 5$ Hz, we obtain $J_{AX} \equiv J^t = 3.5$ and $J_{BX} \equiv J^c = 1.5$ Hz.

The chemical shifts and the coupling constants are compatible with the structure. The

* See Section 1.5.5.2.

protons on the same side of the ring as the hydroxy group are deshielded.* The same holds for their coupling constant J^t (see Reference 1330).

There is a first-order coupling between H-2 (H-5) and the hydroxy protons, as proved by the doublet splitting of the hydroxy signal (J = 4 Hz). In accordance, the doublet disappears in Spectrum **58**/*1a* after D_2O addition, and a diffuse absorption appears at higher frequencies (coalesced with the signal of water). Owing to this further coupling, the multiplet of H-2(H-5) occurs as a poorly structured absorption maximum.

The splitting of the X lines of ABX multiplets are $J_{AX} + J_{BX}$ and $2(Q_1 - Q_2)$, respectively, actually 5 Hz and 2 Hz. The width of the complete X multiplet therefore must be 5 + 4 = 9 Hz in the case of a spin - spin interaction with the hydroxy group and 5 Hz without it. The half-band width of the X signal in the two spectra are indeed 9 and 5 Hz, confirming the correctness of our calculations. The detectable CH–OH coupling proves that the exchange process between the hydroxy protons is, as usual,** slow in DMSO-d_6. The singlet of H-3,4 appears at 3.96 ppm. Thus, the anellated hydrogens do not interact with their neighbors as generally.***

* See Volume I, p. 33.

** See Volume II, pp. 100 and 101.

***Compare Volume I, p. 61 and Problem **63**.

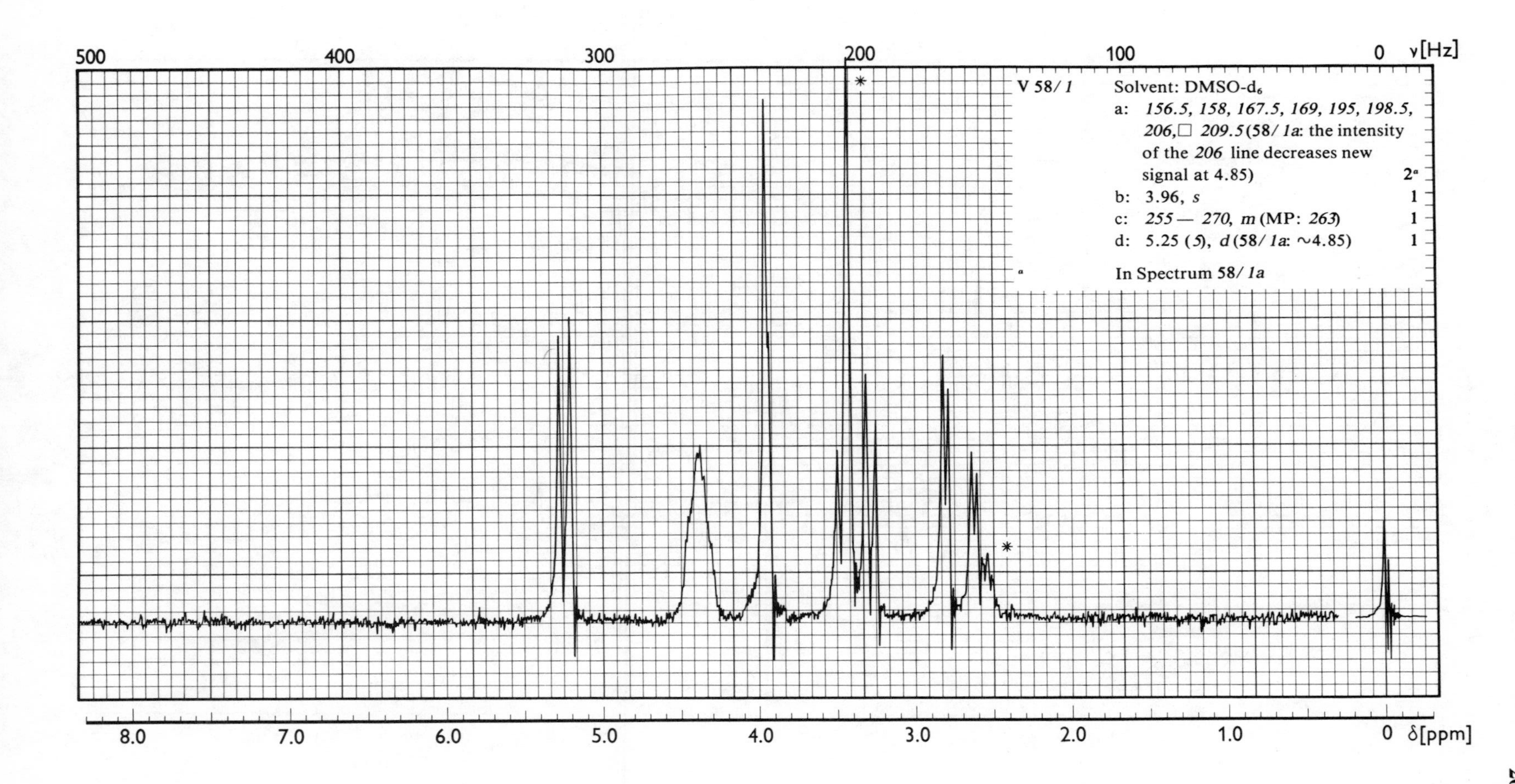
500
400
300
200
100
0 ν[Hz]
V 58/ 1
Solvent: DMSO-d6
a: 156.5, 158, 167.5, 169, 195, 198.5, 206,□ 209.5 (58/ 1a: the intensity of the 206 line decreases new signal at 4.85) 2[a]
b: 3.96, s 1
c: 255 — 270, m (MP: 263) 1
d: 5.25 (5), d (58/ 1a: ~4.85) 1
a
In Spectrum 58/ 1a
*
*
8.0
7.0
6.0
5.0
4.0
3.0
2.0
1.0
0 δ[ppm]

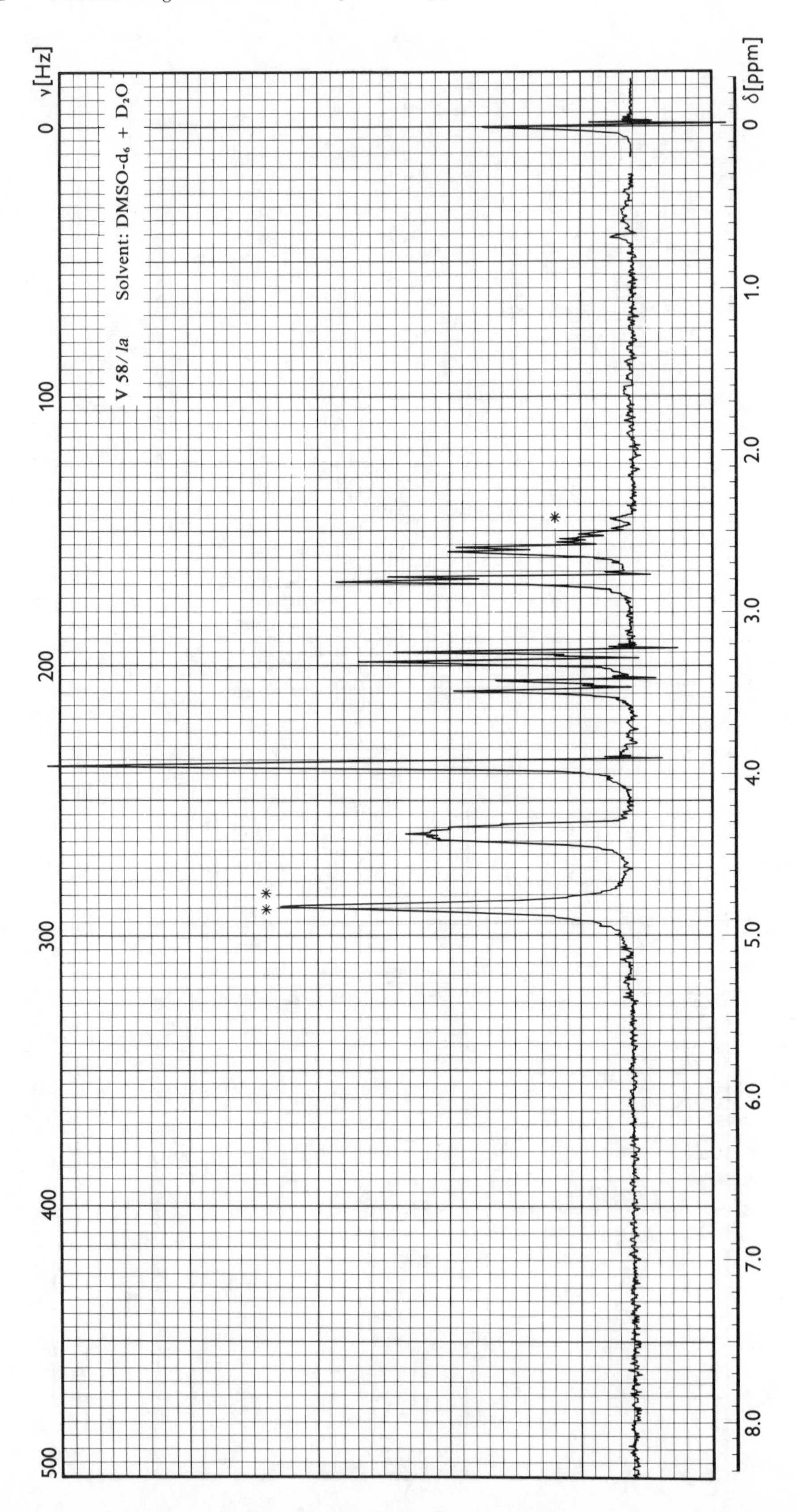
ν[Hz]
0
100
200
300
400
500
V 58/1a
Solvent: DMSO-d_6 + D_2O
δ[ppm]
0
1.0
2.0
3.0
4.0
5.0
6.0
7.0
8.0

PROBLEM 59[802]

In the ''Pummerer rearrangement'' of compound **I**, at higher temperatures in acetic anhydride, the formation of compound **II** can be expected. However, depending on the reaction conditions, five different products could be isolated from the reaction mixtures. On the basis of their IR spectra, the products with Spectra **59**/*1* and **59**/*2* contain an ester group, those with Spectra **59**/*3* and **59**/*4* contain a conjugated keto group, and the product with Spectrum **59**/*5* contains both functional groups. Derive the structures of the products.

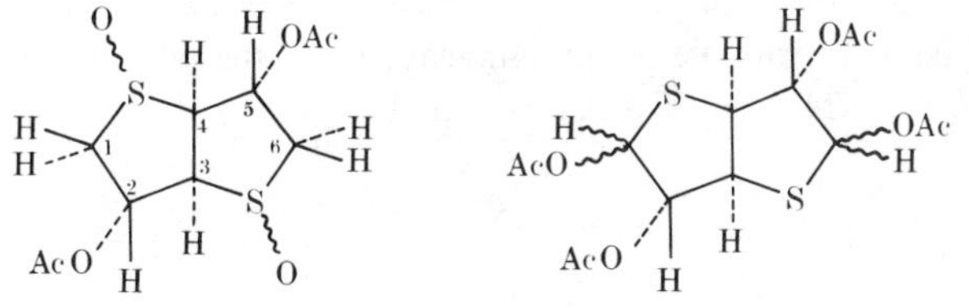

59/I **59/II**

ANSWER

It can be seen immediately from the spectra that none of the reaction products may be **II** or any similar substance because there is no signal in the region (between 4.0 and 6.0 ppm) characteristic of the methine protons of type –CH(OAc). The signals, apart from the methyl signals of the acetyl groups between 2.0 and 3.0 ppm, appear only in the region of hydrogens attached to unsaturated carbons (between 7.0 and 8.0 ppm), except for a singlet at 3.87 ppm in Spectrum **59**/*4*.

According to the IR data, both molecules with Spectra **59**/*1* and **59**/*2* contain acetoxy groups, and the NMR intensities indicate that the number of protons attached to sp^2 carbons is exactly the same as the number of methyl hydrogens. In Spectra **59**/*1* and **59**/*2*, the regions of the aromatic or olefinic proton signals consist of a singlet at 6.95 and 7.34 ppm, respectively, and an *AB* multiplet with lines at 425, 430, 433, and 438 Hz, as well as at 429, 434, 439, and 444 Hz, respectively, of 1:2 intensity ratio. The line frequencies yield δA = 7.24 and 7.35 ppm, respectively, δB = 7.14 and 7.20 ppm, respectively, and in both cases J_{AB} = 5 Hz. The acetoxy signal is at 2.28 and 2.30 ppm, respectively. Accordingly, the molecules contain either one or two acetoxy groups and three or six further hydrogens, respectively. Since the latter hydrogens must be identical to some of the eight hydrogens originally present in the skeleton, the assumption of six such hydrogens would permit the formation of only one double bond. In this case, however, all six skeletal hydrogens could not be attached to unsaturated carbons. We are thus restricted to Structures **III** and **IV**.

These structures are compatible with the spectra. The signal of thiophene protons usually appears between 6.0 and 8.0 ppm (compare Table 41). The coupling of the two adjacent protons produces an *AB* spectrum, and the observed coupling constant of J_{AB} = 5 Hz is also in the range characteristic of thiophenes (see Table 43). The singlet aromatic signal arises from the isolated ring proton, which enables one to distinguish Structures **III** and **IV**. For the latter, ths signal is expected to appear upfield because in Structure **III** the adjacent sulfur atom causes a paramagnetic shift (compare Tables 40 and 41). Therefore, Spectrum **59**/*2* corresponds to Structure **III** and Spectrum **59**/*1* corresponds to isomer **IV**.

Among the spectra of the compounds which contain a conjugated keto group, Spectrum **59**/*3* is very similar to the former ones, the only difference being a downfield shift in the methyl signal (2.50 ppm) and in the aromatic signlet (7.79 ppm). It is evident that the product must be one of isomer **V** or **VI**. This assumption is compatible with the IR spectrum (conjugated keton) and also with the deshielding of the methyl protons. According to the experience, $\delta CH_3(RCOMe) > \delta CH_3(ROOCMe)$ for the same R (see Table 23). The observed strong deshielding of the ring hydrogen adjacent to the acetyl group could also be expected. The problem is again to distinguished the two isomers **V** and **VI**. The decision could based also in this case on the spectrum of the corresponding other conjugated keton — if it were the isomeric pair. The signal of 2H intensity found in Spectrum (**59**/*4*) at 3.87 ppm is, however, incompatible with Structures **V** or **VI**. Fortunately, compound **VI** is known from the literature and its identical melting point and IR spectrum with the sample giving Spectrum **59**/*3* proved their identity. Note that on the basis of Spectrum **59**/*3* alone one could not distinguish Structures **V** from **VI** unambiguously. From the *AB* lines of this spectrum, $\delta A = 7.53$ ppm, $\delta B = 7.24$ ppm, and $J_{AB} = 5.5$ Hz.

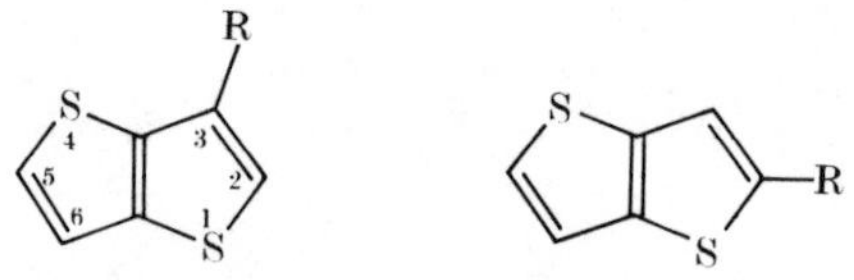

59/III (R = OAc) **59/IV (R = OAc)**

59/V (R = Ac) **59/VI (R = Ac)**

The *AB* quartet ($\delta A = 7.35$ ppm, $\delta B = 7.02$ ppm, $J_{AB} = 5$ Hz) in Spectrum **59**/*4* proves that one of the thiophene rings is still present, and there are no more aromatic protons in the molecule. The molecule contains no acetyl or acetoxy group, and the singlet at 3.87 ppm is due to an isolated methylene group (two isolated methine groups may not be present in the molecule). Since the IR spectrum indicates a conjugated keto group, the product must be **VII**.

The fifth product with Spectrum **59**/*5* contains, according to the IR spectrum, ester and conjugated keto groups. The singlet pair at 2.36 and 2.58, both corresponding to three hydrogens, indicates the presence of an acetyl and an acetoxy group, and due to the two singlets at 6.99 and 7.80 ppm, of 1-1H intensity, the two substituents are not attached to the same ring. The possible structures are therefore the 2-Ac-5-OAc, 3-Ac, 5-OAc, 2-Ac-6-OAc, and the 3-Ac-6-OAc substituted analogues of compounds **III** to **IV**.

It is very probable from the data of the previous spectra and from Spectrum **59**/*5* that the acetoxy group is attached to the carbon adjacent to the sulfur atom (the shitt of one of the aromatic singlets is 6.99 ppm, practically the same as that of the proton adjacent to the acetoxy group in compound **IV**). Moreover, the signal at 7.80 ppm is identical in position to that of the ring proton adjacent to the acetyl group in Spectrum **59**/*3* of compound **VI** (7.79 ppm), which suggests that the product is the 2-acetyl-5-acetoxy-derivatives, **VIII**. (This conclusion was also proved by the mass spectrum.)

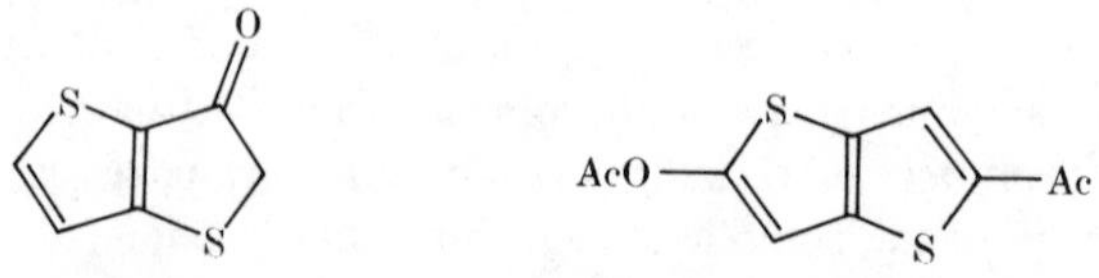

59/VII **59/VIII**

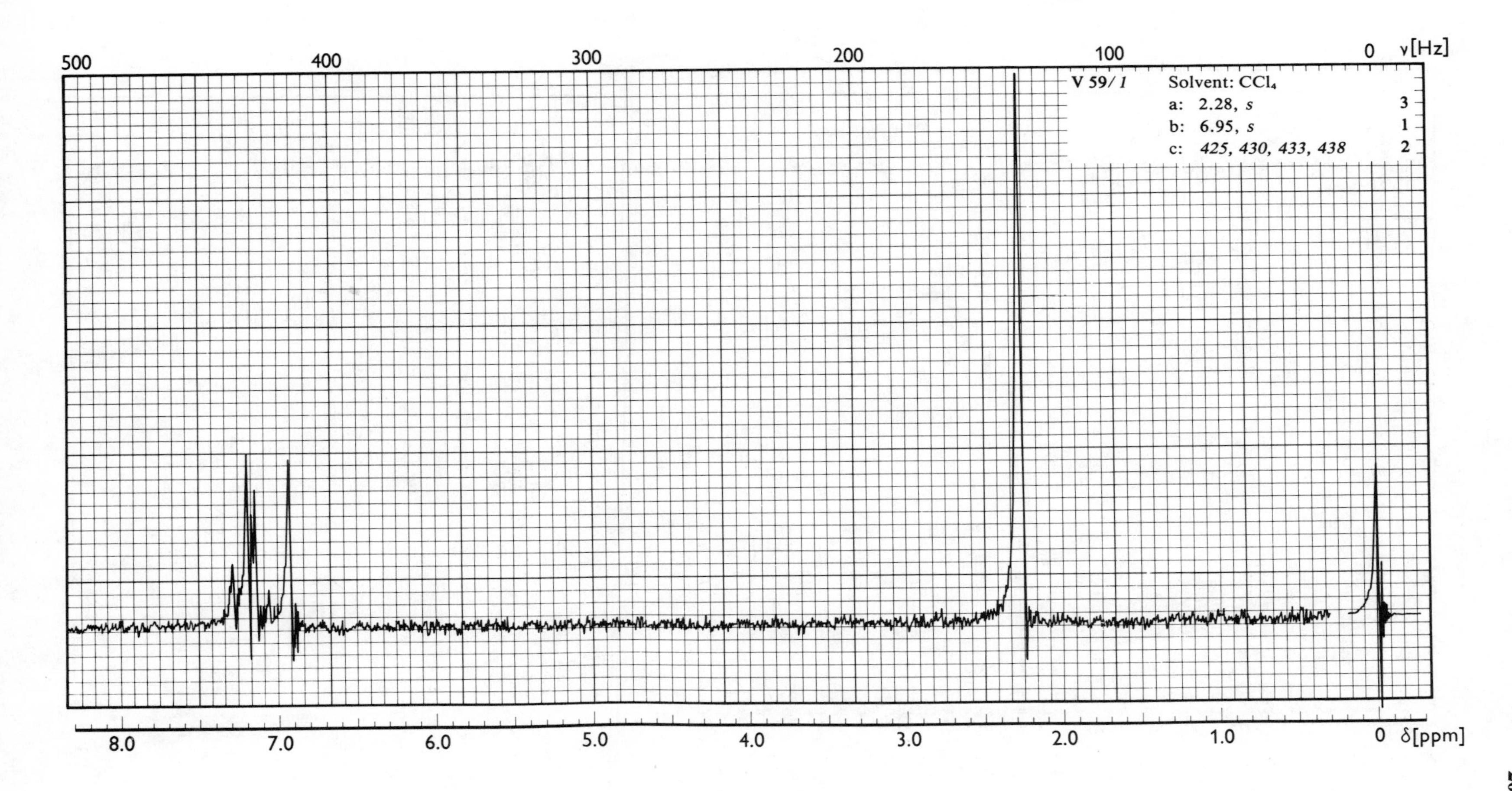
V 59/1
Solvent: CCl_4
a: 2.28, *s* 3
b: 6.95, *s* 1
c: *425, 430, 433, 438* 2
500
400
300
200
100
0
ν[Hz]
8.0
7.0
6.0
5.0
4.0
3.0
2.0
1.0
0
δ[ppm]

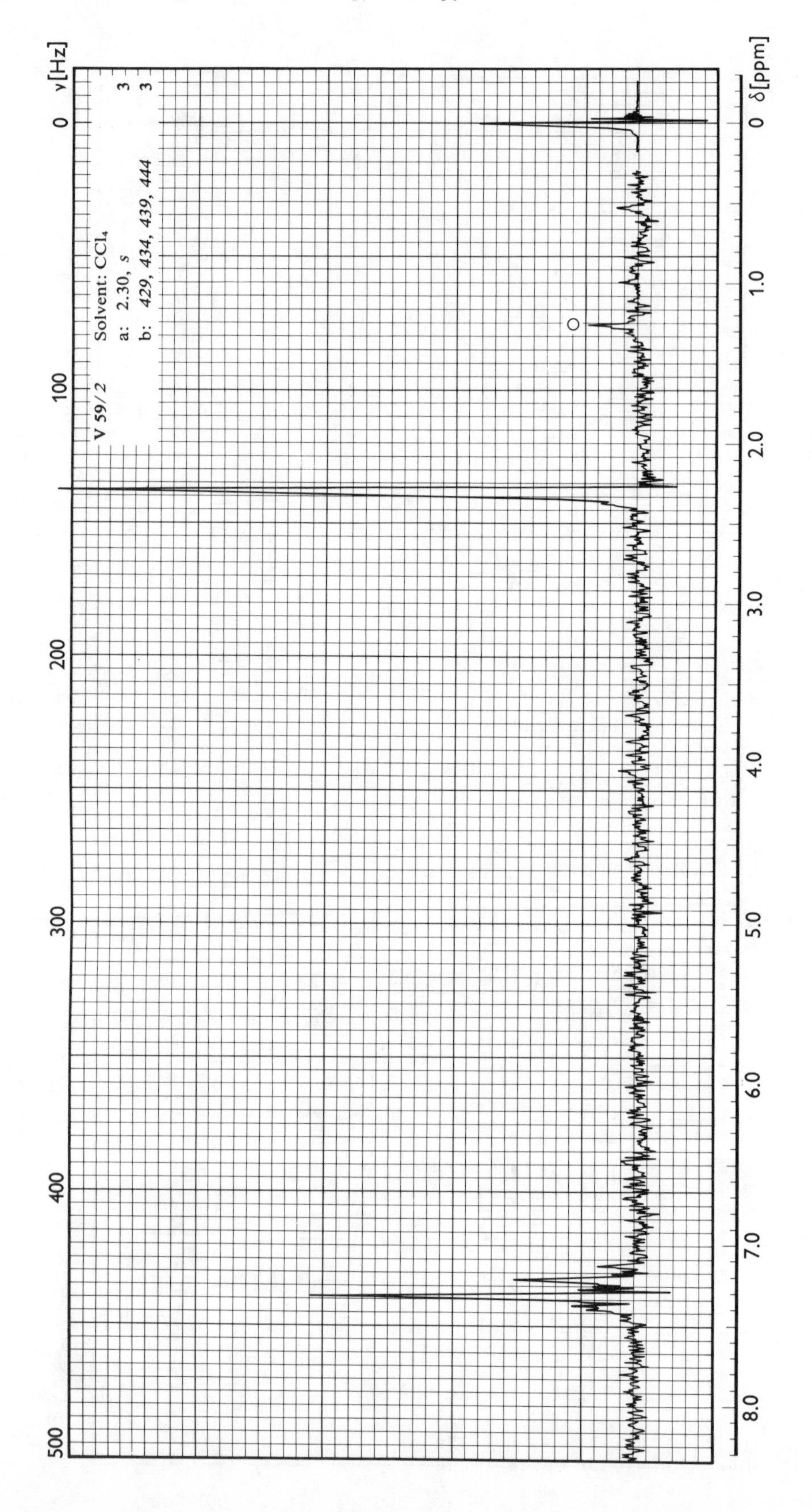
V 59/2
Solvent: CCl4
a: 2.30, s 3
b: 429, 434, 439, 444 3
ν[Hz]
0
100
200
300
400
500
δ[ppm]
0
1.0
2.0
3.0
4.0
5.0
6.0
7.0
8.0

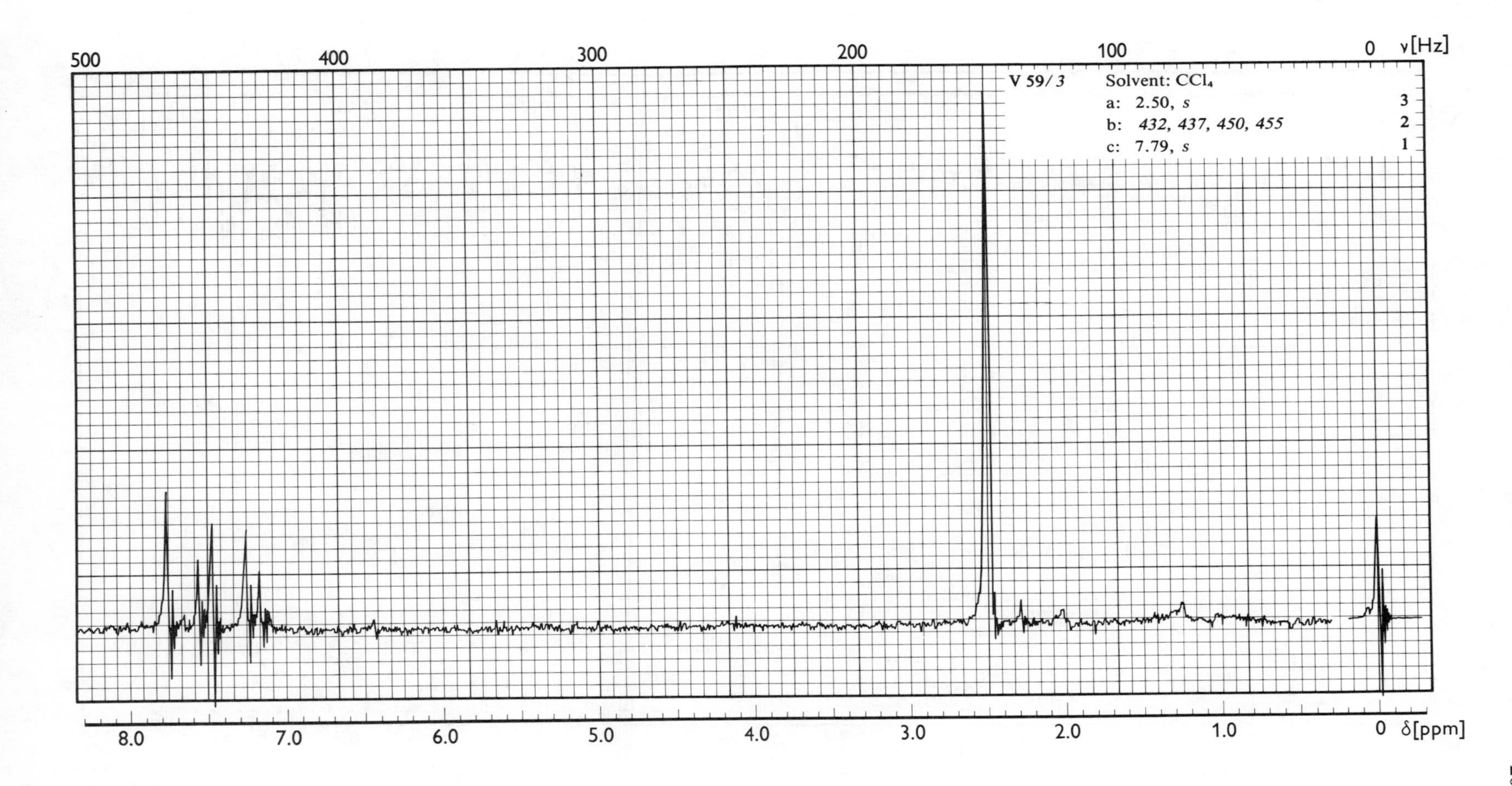

V 59/3
Solvent: CCl_4
a: 2.50, s 3
b: 432, 437, 450, 455 2
c: 7.79, s 1
ν[Hz]
500
400
300
200
100
0
δ[ppm]
8.0
7.0
6.0
5.0
4.0
3.0
2.0
1.0
0

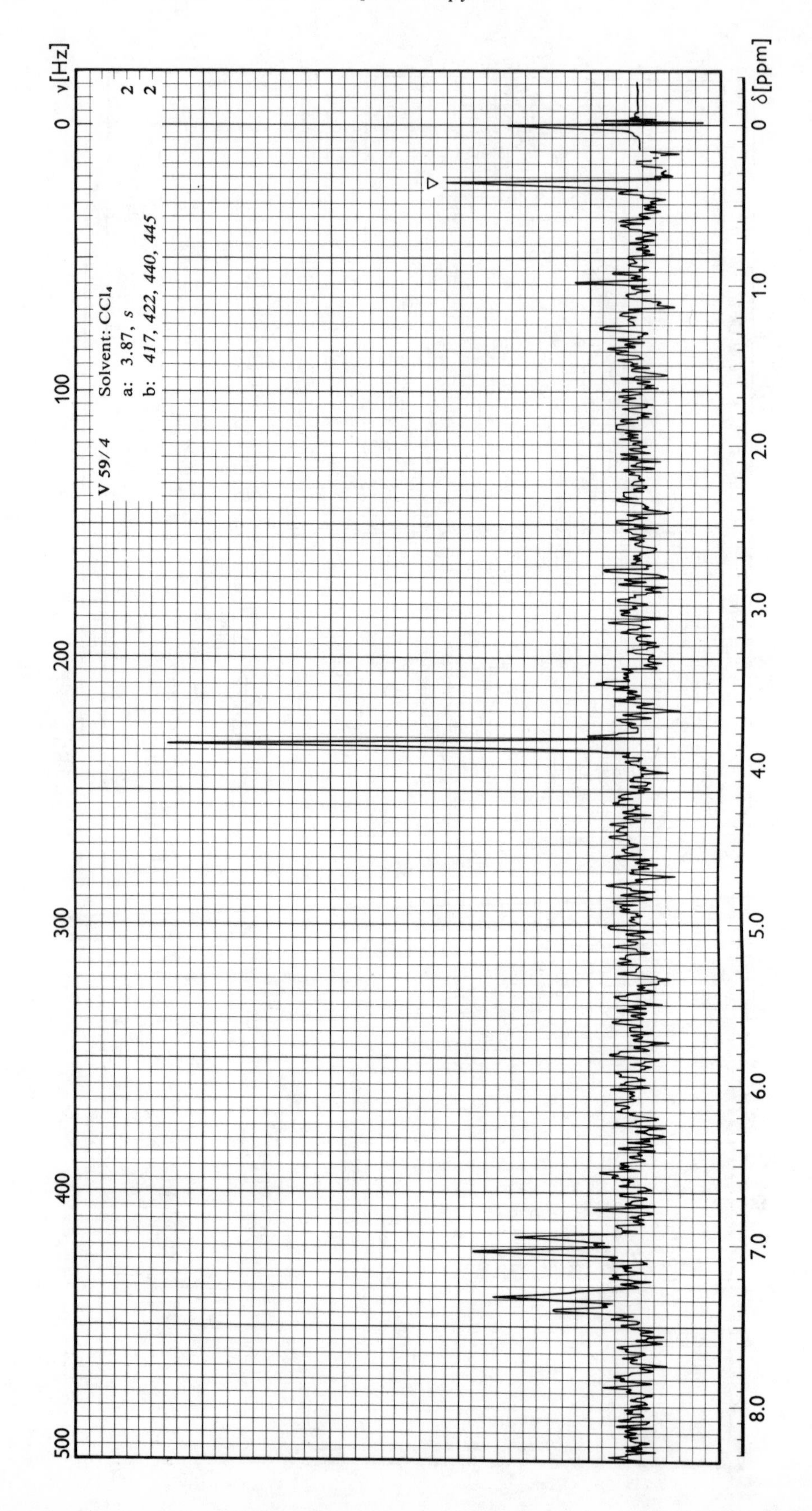

ν[Hz]
0
100
200
300
400
500
V 59/4
Solvent: CCl4
a: 3.87, s 2
b: 417, 422, 440, 445 2
δ[ppm]
0
1.0
2.0
3.0
4.0
5.0
6.0
7.0
8.0

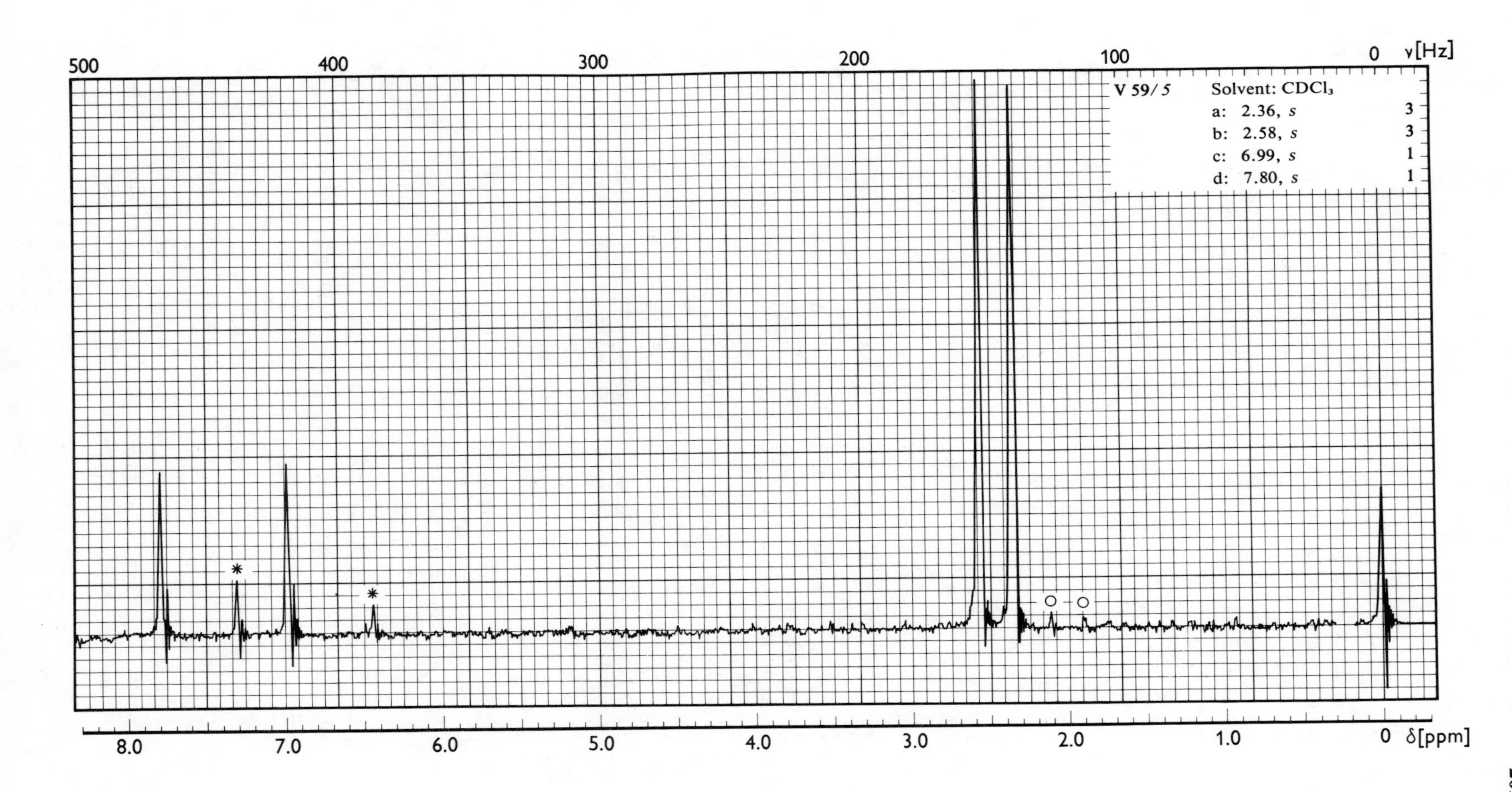

500
400
300
200
100
0
ν[Hz]
V 59/5
Solvent: CDCl₃
a: 2.36, s 3
b: 2.58, s 3
c: 6.99, s 1
d: 7.80, s 1
8.0
7.0
6.0
5.0
4.0
3.0
2.0
1.0
0
δ[ppm]

PROBLEM **60**[132]

3-Amino-5-methylisoxazole (**I**) is prepared by the ring closing reaction of CH_3–CHCl–CHCl–CN and hydroxylamine with elimination of two HCl molecules. The physical properties of the product are, however, strongly varying, indicating the formation of a mixture. Determine from the spectra of six samples of different properties (**60**/*1* to *6*) whether all of them are mixtures or whether compound **I** is also represented in pure form. Derive the structures of the further components (probably by-products) and estimate the ratio of components. Spectrum **60**/*6a* was recorded after acidifying the solution.

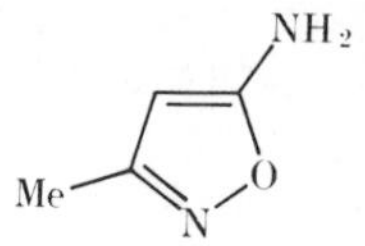

60/I

ANSWER

The spectrum of **I** is expected to consist of only three singlets, in an intensity ratio of 3:1:2, assigned to the methyl, methine, and NH_2 protons. Accordingly, only Spectrum **60**/*1* or **60**/*2* may correspond to Structure **I**, with no regard to the intensities, which cannot be determined owing to a strong broadening of the NH signal and to the overlap of the methyl and solvent signals (CD_2H light isotope content of acetone-d_6). The chemical shifts obtained from the spectra are as follows: δCH_3: 2.25 and 2.04 ppm; δCH: 5.65 and 4.90 ppm; δNH_2: 4.9 and 3.05 ppm. The 3.05-ppm signal overlaps with the maximum of water content in the solvent. The latter signal in case of Spectrum **60**/*1* appears at 3.1 ppm. The assignment of the NH_2 signals can be proved by D_2O addition. In Spectra **60**/*1*, the CH_3 and CH signals split to doublet and quartet, respectively, by ~1 Hz, due to an allylic interaction between them. In Spectrum **60**/*2* these signals are singlets.

It is to be determined which corresponds to compound **I**, and what is the structure of the other sample. Since the spectra were recorded in the same solvent, at the same temperature, and with almost the same concentrations, it is impossible that they belong to tautomeric forms of **I**.

Hydroxylamine, however, may also react with CH_3–CHCl–CHCl–CN, interchanging the roles of its two functional groups, yielding isomer **II**. Structures **I** and **II** are easy to correlate with the spectra, considering that electron delocalization becomes stronger in the ring of **I** involving a more pronounced aromatic character. (The electron shift from the amino group to the ring nitrogen results in more pronounced electron reservoir character of the latter.) Delocalization causes deshielding of the ring hydrogen and of the methyl group. Therefore Spectrum **60**/*1* belongs to isomer **I** and Spectrum **60**/*2* belongs to **II**. The long-range coupling is also consequence of the higher electron density in the ring of **I**.

The sample with Spectrum **60**/*3* is a mixture of isomers **I** and **II** in approximately 2:1 ratio on the basis of the signal intensities. Note that if the ratio were determined from the too closely spaced methyl signals (integral I), a false result (9:2) would be obtained. It can also be seen that a repeated measurement (integral II) leads to a completely different result (5:3). The spectrum also contains two weak signals around 1.35 ppm, indicating the presence of a further component in smaller quantities (the signals around 6.45 and 7.35 ppm arise from the impurity of the solvent). The amount of this component are higher in the other three samples especially in the sample with Spectrum **60**/*6*, which contains, as the other component, **I**.

Thus, for determining the structure of the yet unknown component, Spectrum **60**/*6* is the most appropriate. The two singlets at 2.27 and 5.67 ppm of 3:1 intensity belong clearly to compound **I**. The quintet at 3.33 ppm and the very strong singlet at 4.67 ppm are the signals of the CD_2HOD and CD_3OH impurities of the solvent (methanol). The absorption of the primary amino group of compound **I** overlaps with the latter signal. The unknown compound is therefore responsible for the doublet at 1.30 ppm (J = 6 Hz) and the multiplets in the regions of 140 to 200 and 250 to 290 Hz (the latter multiplet overlaps with the signal of the acidic protons at 4.67 ppm) of 3:2:~1 ratio.

The doublet is, therefore, most certainly the methyl signal of a CH_2–CH–CH_3 group, and the two multiplets belong to the methylene and methine hydrogens, since the chemically nonequivalent methylene protons are also adjacent to one hydrogen (*ABX* spin system). The nonequivalence of the methylene protons suggests the presence of a cyclic structure. Considering the reaction conditions, the only possible structure is **III**. (Its formation can be conceived so that the chlorination of MeCH=CHCN — the first step of the reaction yielding CH_3–CHCl–CHCl–CN— is incomplete, and thus hydroxylamine reacts also with the starting material.)

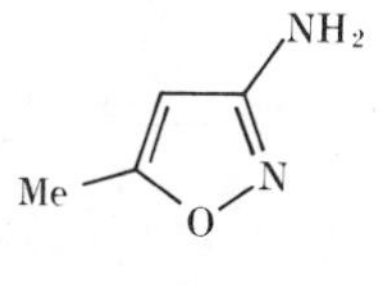

60/II 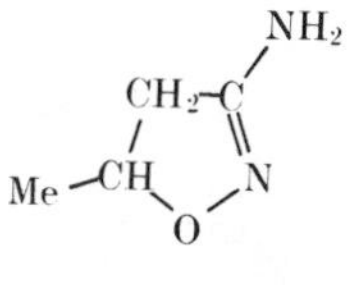 **60/III**

The –CH_2–CH–CH_3 group represents an $ABMX_3$ spin system. The value of J_{AB} (the distance between lines **1** and **3**, **2** and **4**; **5** and **7** or **6** and **8** of the methylene octet) is 16 Hz, and J_{MX} = 6 Hz (see above). The *AB* part of the $ABMX_3$ spectrum can be treated as the *AB* part of a *ABX* system, because $J_{AX} = J_{BX} = 0$, and thus the *X* nuclei affect only the *M* part of the spectrum.

It follows, however, from the frequencies of the eight *AB* lines that $Q_1 = Q_2$* and therefore $J_{AM} = J_{BM}$. Namely, on the basis of the relative intensities, lines **1, 3, 5, 7** and **2, 4, 6, 8** are interrelated, and the spacing of line pairs **1-2, 3-4, 5-6,** and **7-8** is identical (9 Hz). Therefore, the distance between the midpoints of the quartets is also the same, from which $(J_{AM} + J_{BM})/2 = 9$ Hz and $J_{AM} = J_{BM} = 9$ Hz (see Figure 59). From the *AB* part, δA = 3.04 and δB = 2.61 ppm. The *M* part consists of 12 or 11 lines, since J_{AM} and J_{BM} are equal. (The *M* part is a triplet due to the coupling $J_{AM} = J_{BM}$, and the lines split further into quartets by the J_{MX} interaction. The two middle of the resulting 12 lines coincide.) The multiplet overlaps with the signal of the acidic protons, and it can be analyzed only after acidification (see Spectrum **60**/*6a*).

Hence, the mixture having Spectrum **60**/*6* is composed of compounds **I** and **III**. The same two compounds are also present in the sample with Spectrum **60**/*5*. On the other hand, the sample with Spectrum **60**/*4* contains, in addition to the former two components, a substantial amount of isomer **II**.

The relative amount of compound **III** can be determined by comparing the intensity of its methyl doublet to that of the methyl singlet of compound **I** in Spectra **60**/*5* and **60**/*6* and to the total intensity of the methyl singlets of compounds **I** and **II** in Spectrum **60**/*4*. As to the latter spectrum, the ratio **I**:**II** can be determined more accurately from the intensities of the methine signals, because the methyl signals are too close to each other. On the basis of the integrals, the approximate ratios **I**:**II**:**III** of the samples with Spectra **60**/*4*, **60**/*5*, and **60**/*6* are about 4:2:1, 2:0:1, and 2:0:3, respectively.

* Compare Section 1.5.5.2.

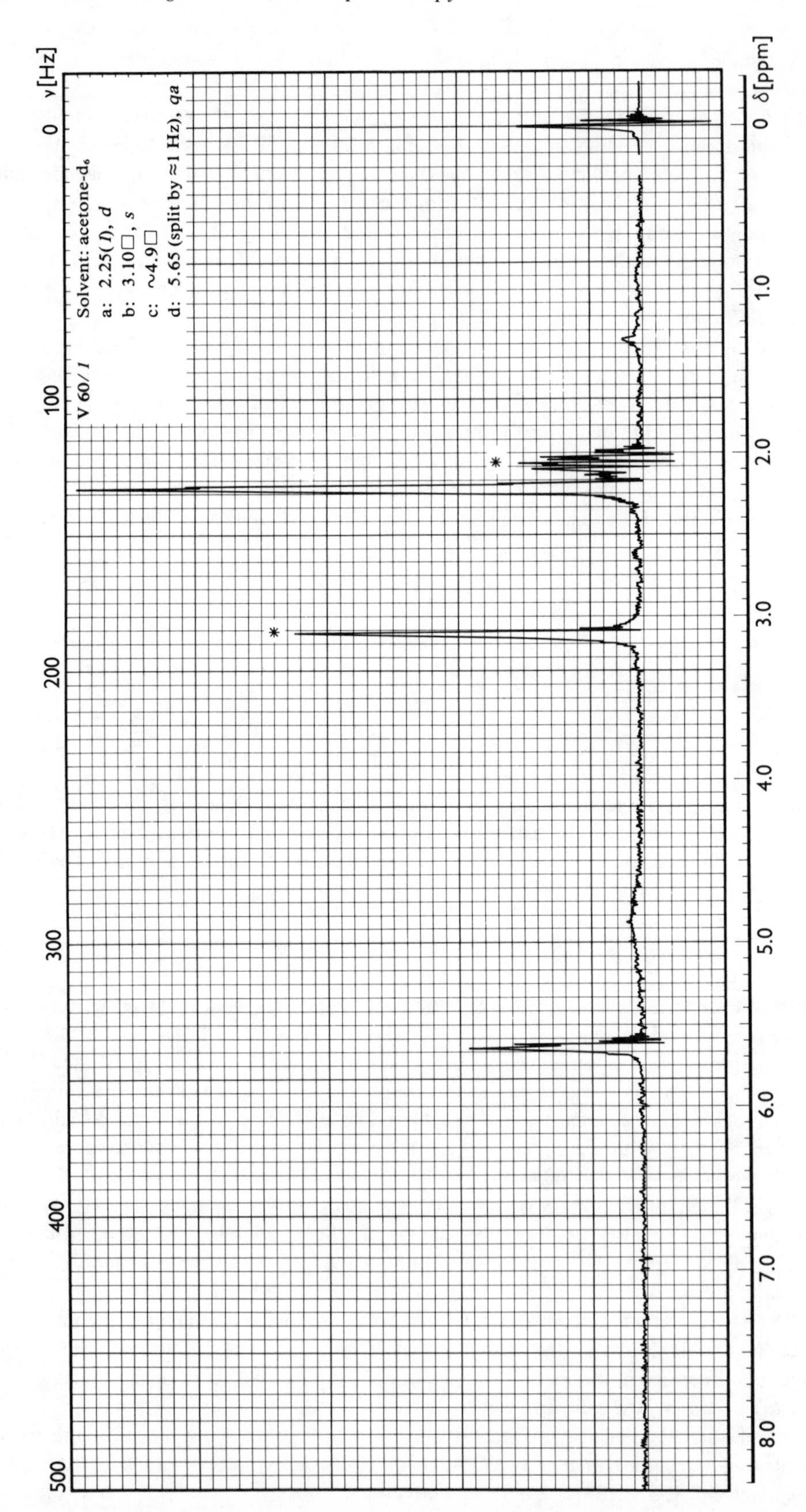
ν[Hz]
0
100
200
300
400
500
V 60/ 1
Solvent: acetone-d₆
a: 2.25(1), d
b: 3.10□, s
c: ~4.9□
d: 5.65 (split by ≈1 Hz), qa
*
*
δ[ppm]
0
1.0
2.0
3.0
4.0
5.0
6.0
7.0
8.0

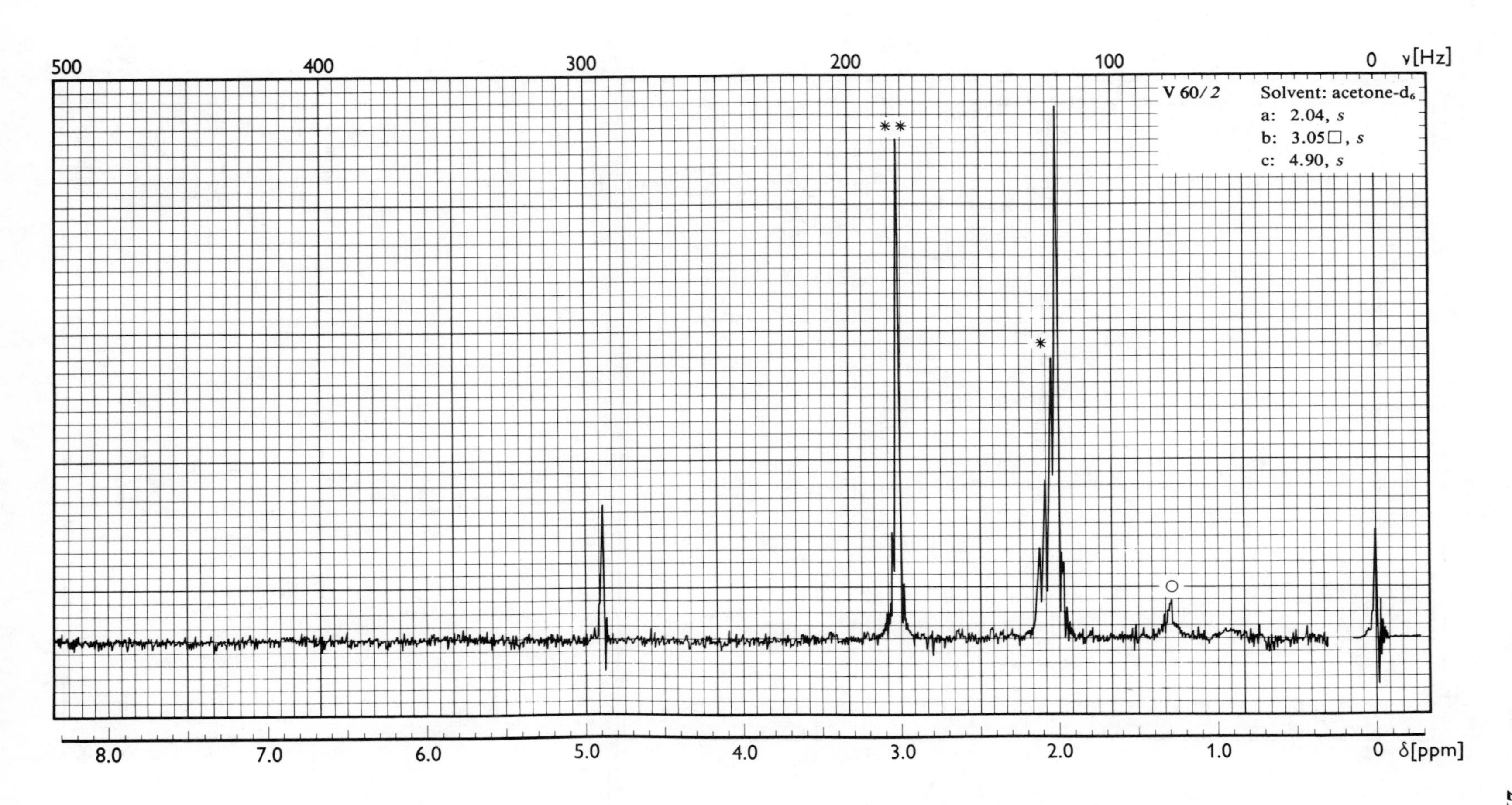
500
400
300
200
100
0 ν[Hz]
V 60/2
Solvent: acetone-d₆
a: 2.04, s
b: 3.05□, s
c: 4.90, s
**
*
○
8.0
7.0
6.0
5.0
4.0
3.0
2.0
1.0
0 δ[ppm]

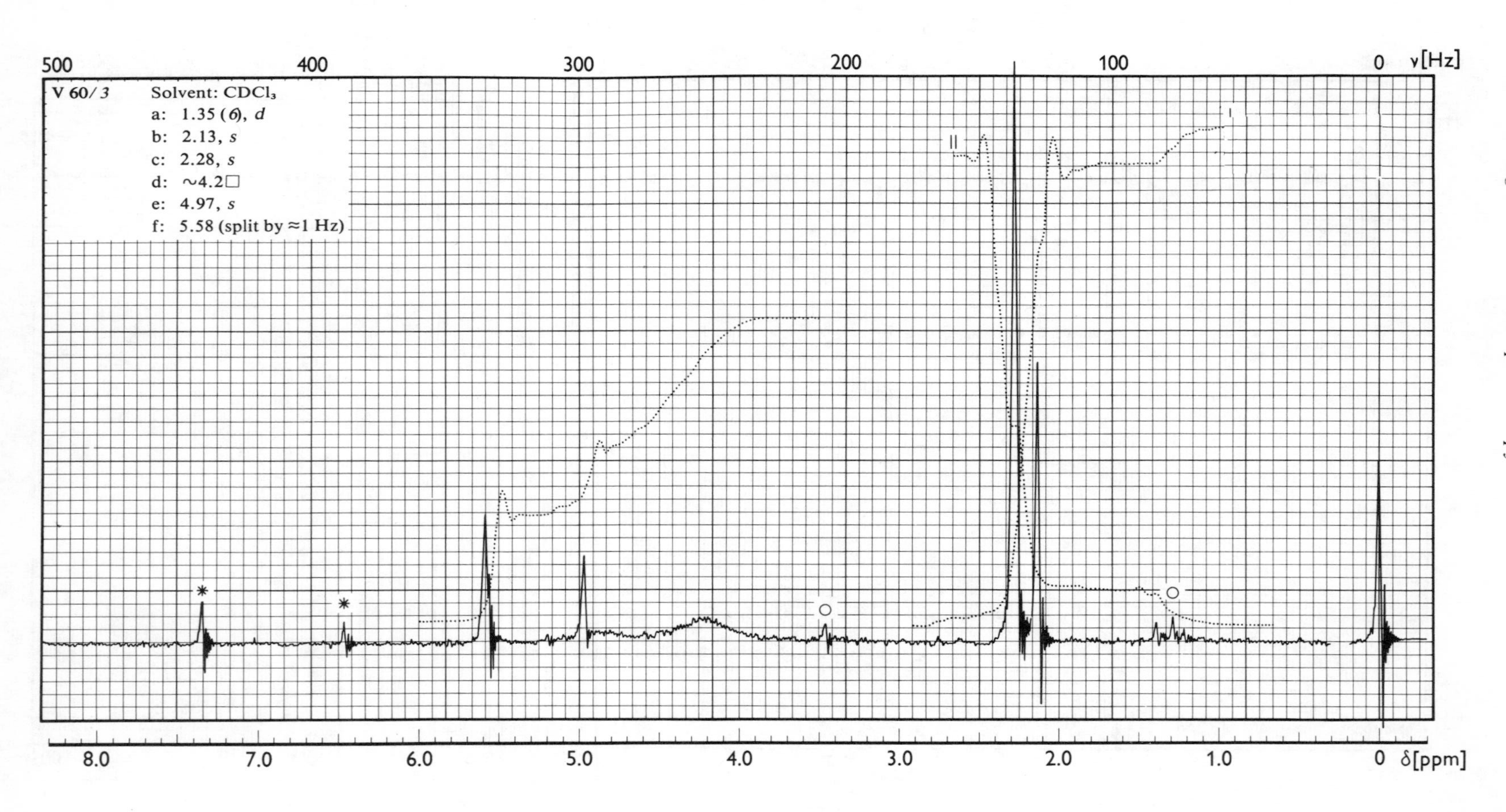

500
400
300
200
100
0
ν[Hz]
V 60/3
Solvent: $CDCl_3$
a: 1.35 (6), *d*
b: 2.13, *s*
c: 2.28, *s*
d: ~4.2□
e: 4.97, *s*
f: 5.58 (split by ≈1 Hz)
8.0
7.0
6.0
5.0
4.0
3.0
2.0
1.0
0
δ[ppm]

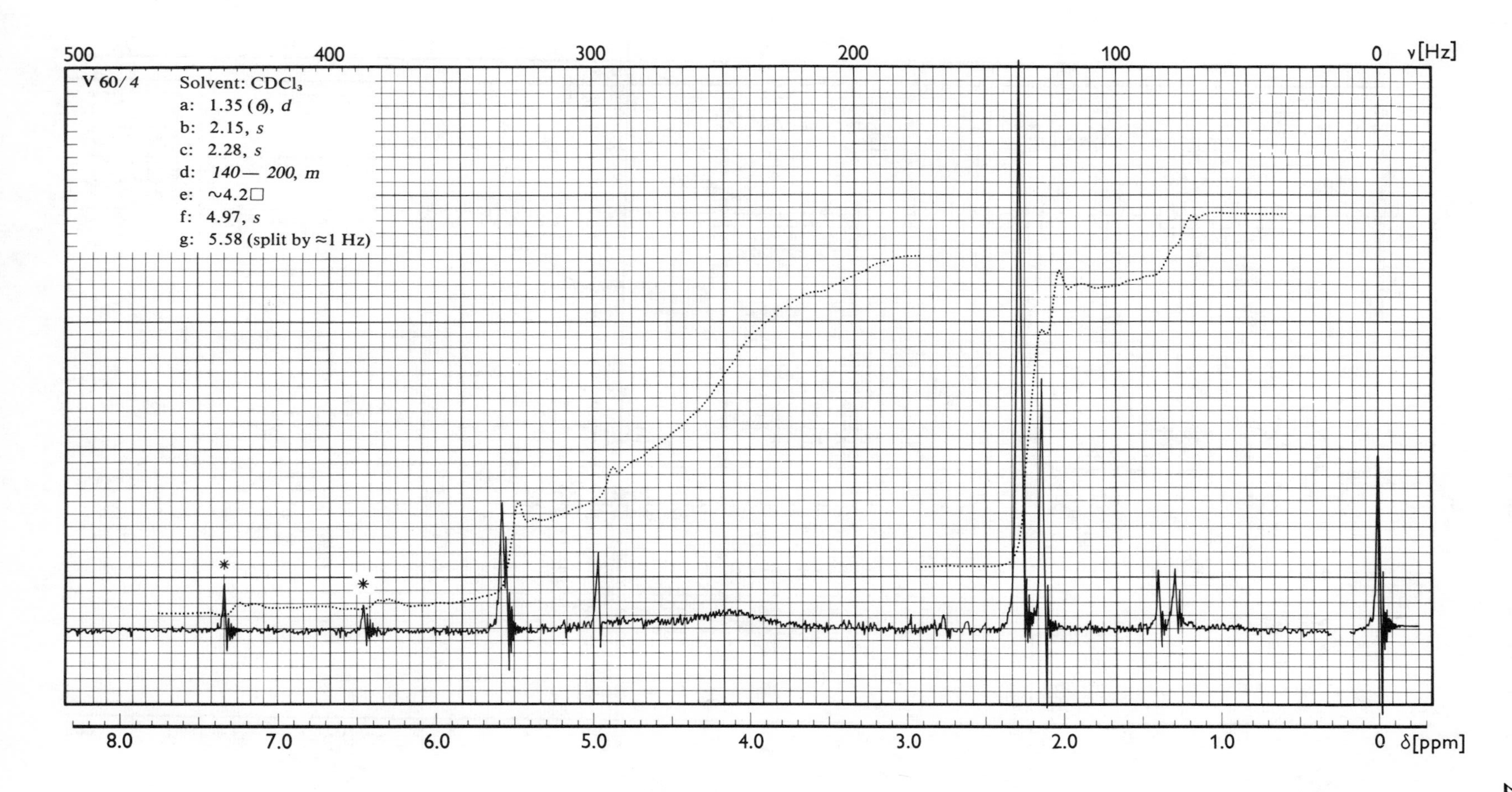

500
400
300
200
100
0 ν[Hz]
V 60/4
Solvent: CDCl₃
a: 1.35 (6), d
b: 2.15, s
c: 2.28, s
d: 140— 200, m
e: ~4.2□
f: 4.97, s
g: 5.58 (split by ≈1 Hz)
*
*
8.0
7.0
6.0
5.0
4.0
3.0
2.0
1.0
0 δ[ppm]

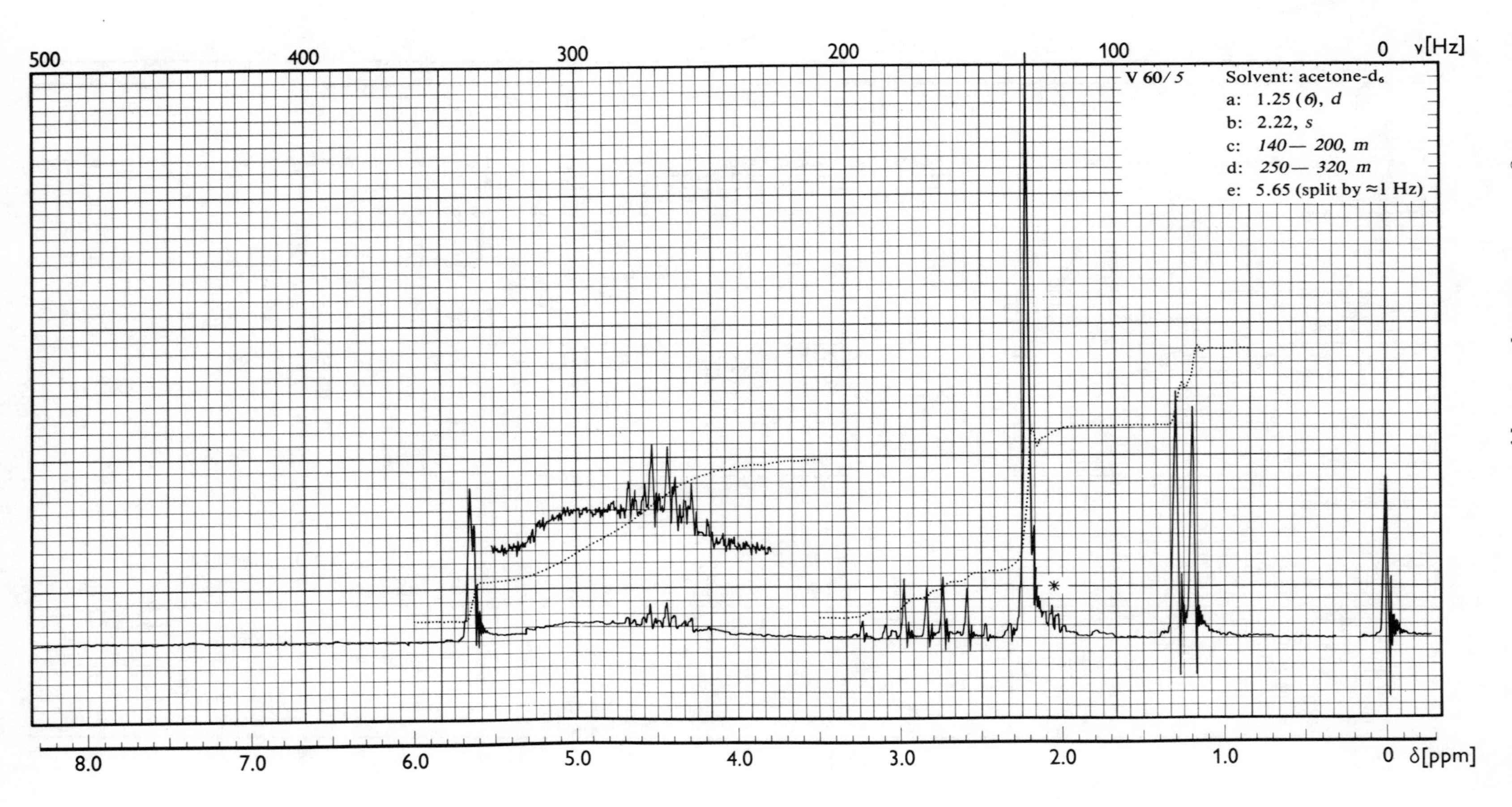
V 60/5
Solvent: acetone-d6
a: 1.25 (6), d
b: 2.22, s
c: 140— 200, m
d: 250— 320, m
e: 5.65 (split by ≈1 Hz)
ν[Hz]
500
400
300
200
100
0
δ[ppm]
8.0
7.0
6.0
5.0
4.0
3.0
2.0
1.0
0
*

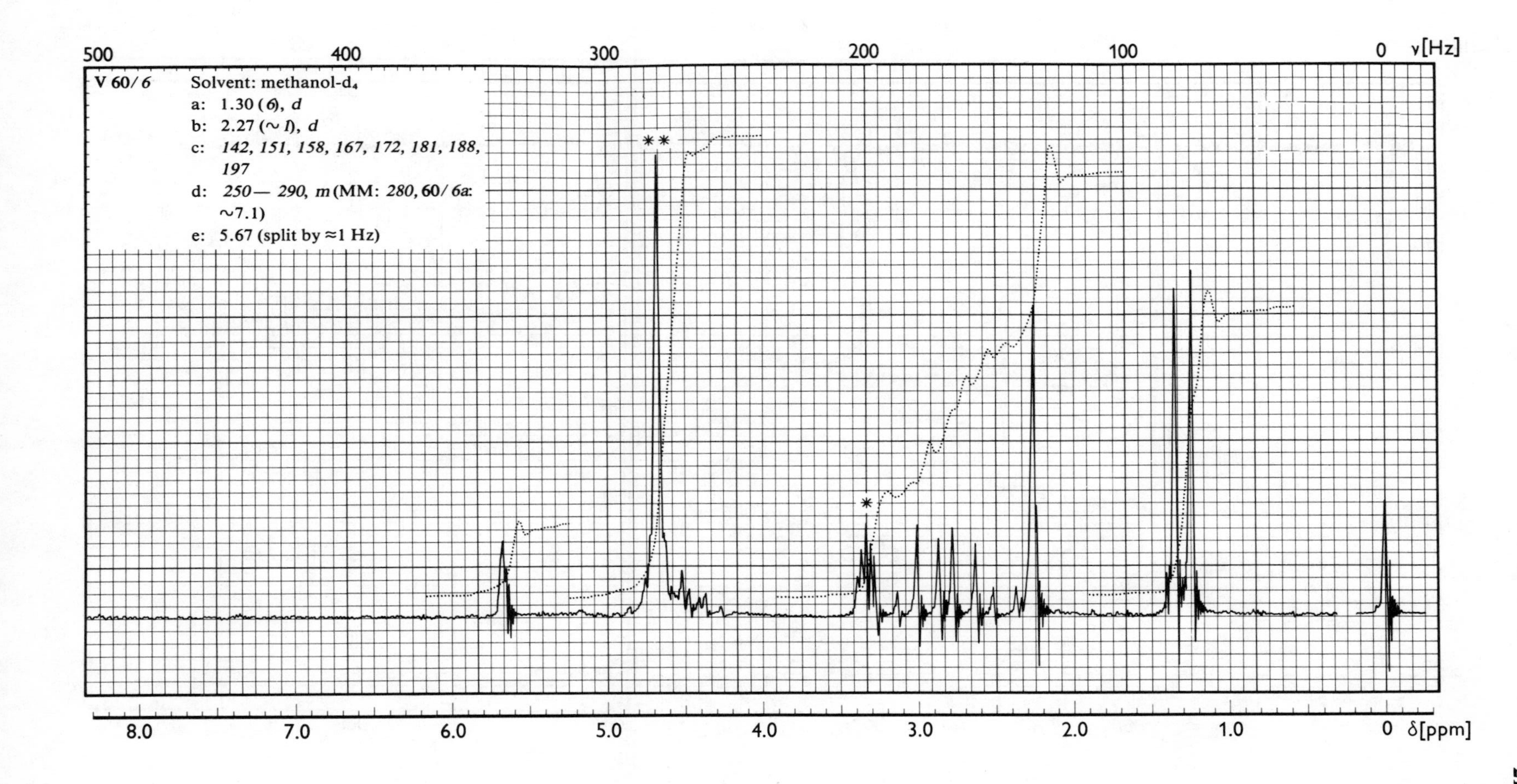
V 60/6
Solvent: methanol-d4
a: 1.30 (6), d
b: 2.27 (~1), d
c: 142, 151, 158, 167, 172, 181, 188, 197
d: 250— 290, m (MM: 280, 60/6a: ~7.1)
e: 5.67 (split by ≈1 Hz)
500
400
300
200
100
0
ν[Hz]
**
*
8.0
7.0
6.0
5.0
4.0
3.0
2.0
1.0
0
δ[ppm]

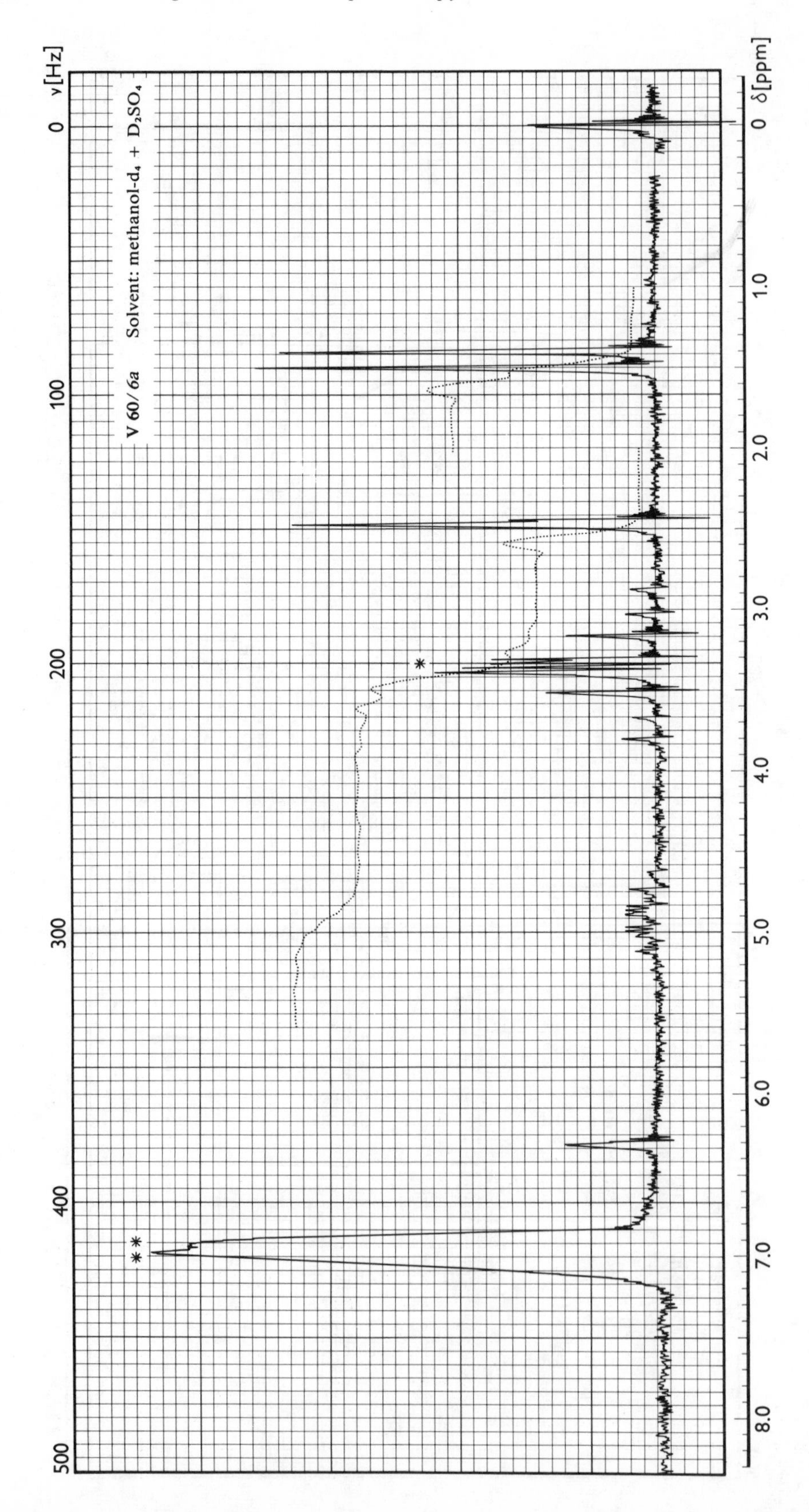
V 60/6a
Solvent: methanol-d4 + D2SO4
ν[Hz]
0
100
200
300
400
500
δ[ppm]
0
1.0
2.0
3.0
4.0
5.0
6.0
7.0
8.0
*
**

PROBLEM 61

Spectrum **61**/*1* is the spectrum of a trisubstituted benzene derivative, with two nitro and a fluorine substituents. What is their relative position?

ANSWER

The spectrum contains three groups of signals: a symmetric triplet at 7.60 ppm, a pair of triplets at 8.60 ppm, and a double doublet at 8.92 ppm of the same intensity. Since there are three signals, the two symmetric of the six possible structures (2,6- and 3,5-dinitro-fluorobenzene) can immediately be rejected.

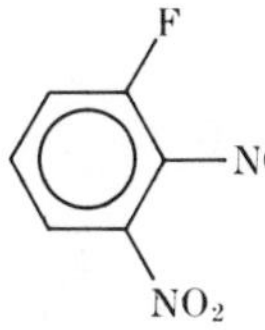

61/I 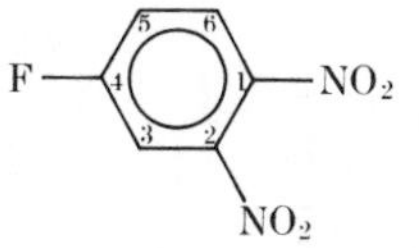 **61/II** 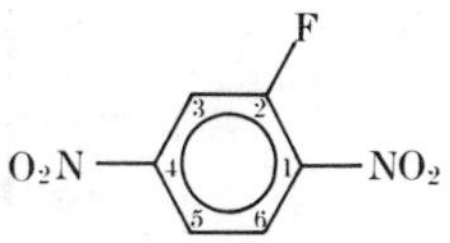 **61/III** 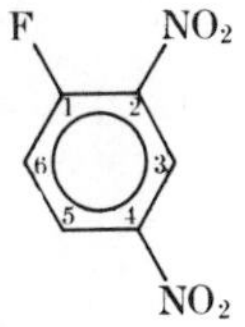 **61/IV**

The further four isomers are **I** to **IV**. The nitro group, due to its $-I$ effect, causes a strong deshielding of the *ortho* ring protons, whereas the analogous effect of the fluorine is overcompensated by its opposite anisotropic effect (compare Problem **26**). Structure **I** can, therefore, also be rejected, because in its spectrum only one signal would show a strong downfield shift. Structure **III** is improbable too, since all the three protons of it would be strongly deshielded, giving signals closer to each other. (This structure can also be discarded on the basis of the splitting pattern; see below.)

Thus, we are restricted to Structures **II** and **IV**, of which the latter is *ab ovo* more probable owing to the very strong deshielding of the downfield multiplet. It is, however, more convincing to construct the schematic spectra and compare them with the experimental results. This is based on the following considerations:

1. In compound **II**, δH-5 < δH-3 < δH-6, whereas in compound **IV**, δH-6 < δH-5 < δH-3.
2. The aromatic protons produce, in both cases, *ABX* spectra in which the coupling with the fluorine atom causes further splitting. Since J^p (H,H) is very small, the complications that make the analysis of an *ABX* spectrum difficult do not occur. As a consequence, the *ABMX* spin system can be treated, to a good approximation, as an *AMPX* case.
3. According to Table 5, J^o (H,H) = 7 to 10, J^o (F,H) = 6 to 10, J^m (H,H) = 1 to 3, J^m (F,H) = 5 to 8, J^p (H,H) = 0 to 1 and J^p (F,H) = 2 to 3 Hz.

In compound **II**, $J^o_{5,6}$ (H,H) ≈ 8.5, $J^m_{3,5}$ (H,H) ≈ 2, and $J^o_{5,6}$ (F,H) ≈ 8 Hz for H-5. Assuming the $J^o_{5,6}$ (H,H) ≈ $J^o_{4,5}$ (F,H), the expected signal is a triplet in which every line splits into a doublet, and the middle doublet may also split further (see Figure 207a).

The coupling constants of H-3 are $J^m_{3,5}$ (H,H) ≈ 2, $J^p_{3,6}$ (H,H) < 1, $J^o_{3,4}$ (F,H) ≈ 8 Hz, resulting in a double doublet, where the lines may split to very close doublets (see Figure 207b). A similar pattern of H-3 signal would yield the splittings in Structure **VI**, too.

The coupling constants of H-6 with the largest shift are $J^o_{5,6}$ (H,H) ≈ 8.5, $J^p_{3,6}$ (H,H) < 1, and $J^m_{4,6}$ ≈ 6.5 Hz. This would produce a double doublet with a larger splitting of the

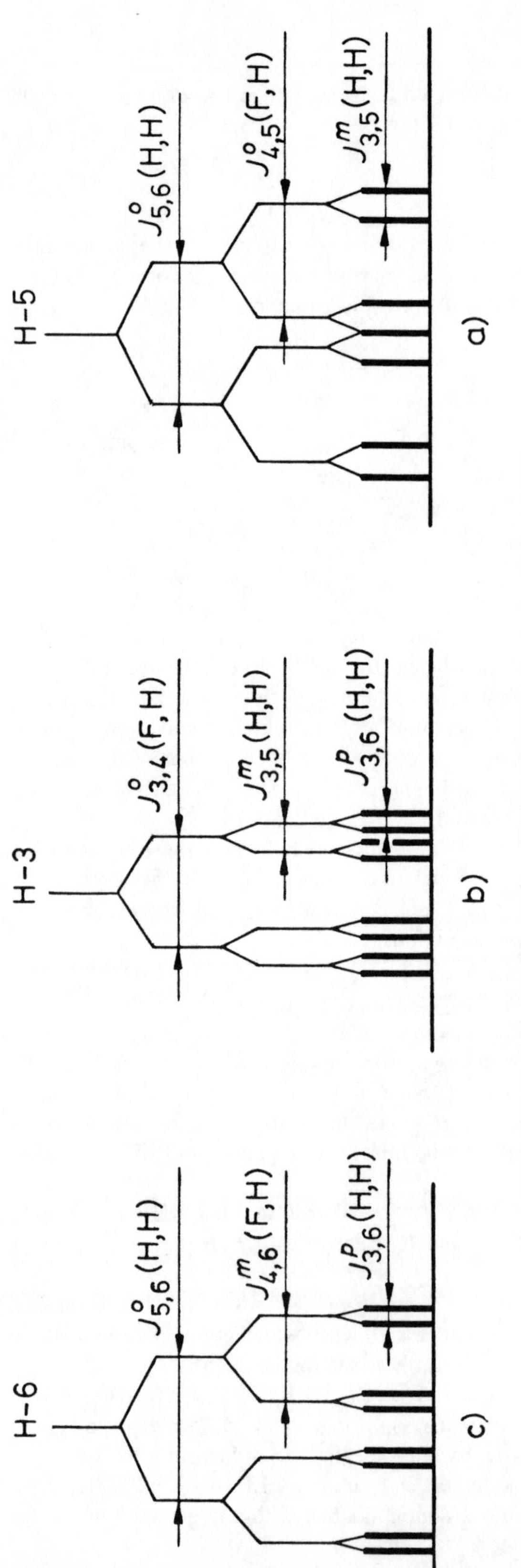

FIGURE 207. The schematic structure of the ^{1}H NMR spectrum of compound **61/II**.

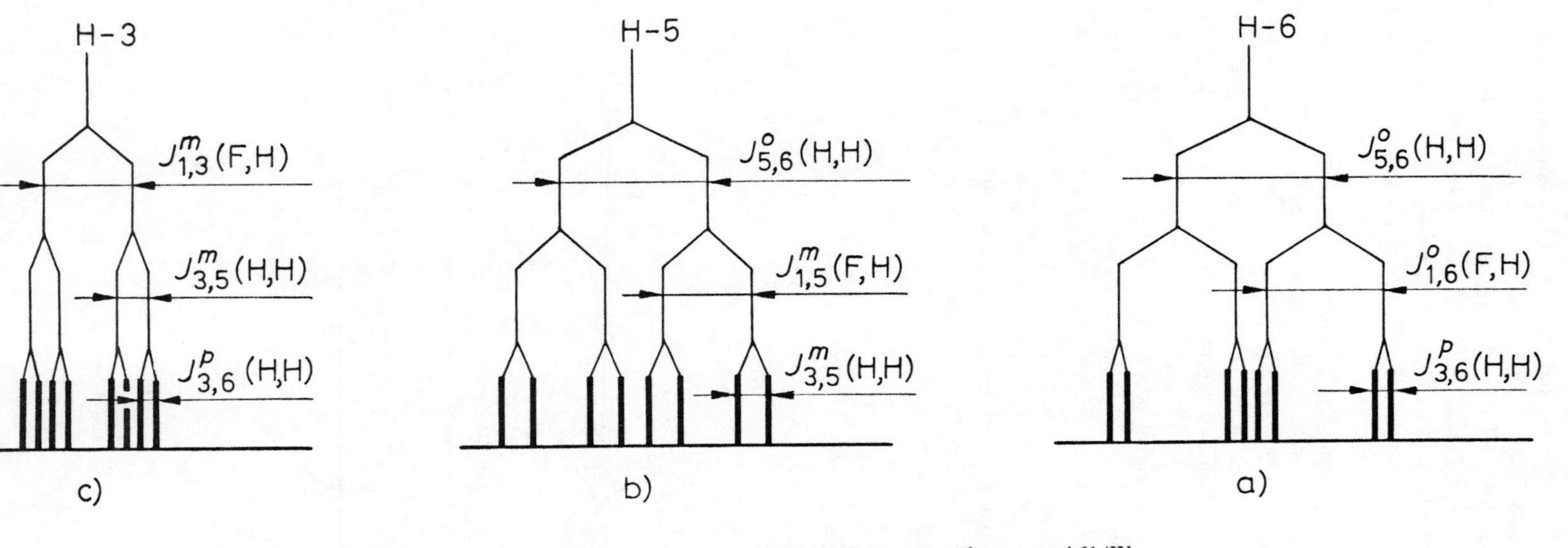

FIGURE 208. The schematic structure of the ^{1}H NMR spectrum of compound **61/IV**.

outer doublets than the distance of the middle lines, and all the four lines may split further into doublets, which is, however, probably insignificant.

A similar analysis of Structure **IV** leads to $J^{o}_{5,6}$ (H,H) ≈ 8.5, $J^{p}_{3,6}$ (H,H) < 1, and $J^{o}_{1,6}$ (F,H) ≈ 8 Hz for the H-6 protons. Provided that $J^{o}_{5,6}$ (H,H) ≈ $J^{o}_{1,6}$ (F,H), the H-6 multiplet is a regular triplet (see Figure 208a). The outer lines of this triplet may exhibit hardly significant doublet splitting, and the middle line is probably split further into a double doublet (if $J_{5,6} \neq J_{1,6}$).

Since $J^{o}_{5,6}$ (H,H) ≈ 8.5, $J^{m}_{3,5}$ (H,H) ≈ 2, and $J^{m}_{1,5}$ (F,H) ≈ 6.5 Hz, the H-5 multiplet consists presumably of four doublets, which might be overlapping (see Figure 208b). The H-3 multiplet, regarding the $J^{p}_{3,6}$ (H,H) < 1, $J^{m}_{3,5}$ (H,H) ≈ 2, and $J^{m}_{1,3}$ (F,H) ≈ 6.5 Hz, consists of two doublets, in which the distance of the two outer lines is smaller than that of the two middle ones. A hardly observable further doublet splitting is also possible for all the four lines (see Figure 208c).

As in the observed spectrum the upfield signal is a triplet and the downfield one is a double doublet in which the outer lines are closer to each other than the inner ones, the structure is **IV**, as already suggested above. From the H-6 multiplet, $J^{o}_{5,6}$ (H,H) ≡ $J^{o}_{1,6}$ (F,H) = 9.5 Hz and $J^{p}_{3,6}$ (H,H) ≈ 0, and from the H-3 signal, $J^{m}_{1,3}$ (F,H) = 6.8 Hz, $J^{m}_{3,5}$ (H,H) = 2.8 Hz, and $J^{p}_{3,6}$ (H,H) ≈ 0. If the H-5 multiplet is constructed with these coupling constants (see Figure 209), the calculated and experimental spectra show the best agreement when $J^{m}_{1,5}$ (F,H) ≈ $J^{m}_{3,5}$ (H,H) ≈ 2.8 Hz. Accordingly, the coupling constants, except for the value of $J^{m}_{1,5}$ (F,H), are in the expected range.

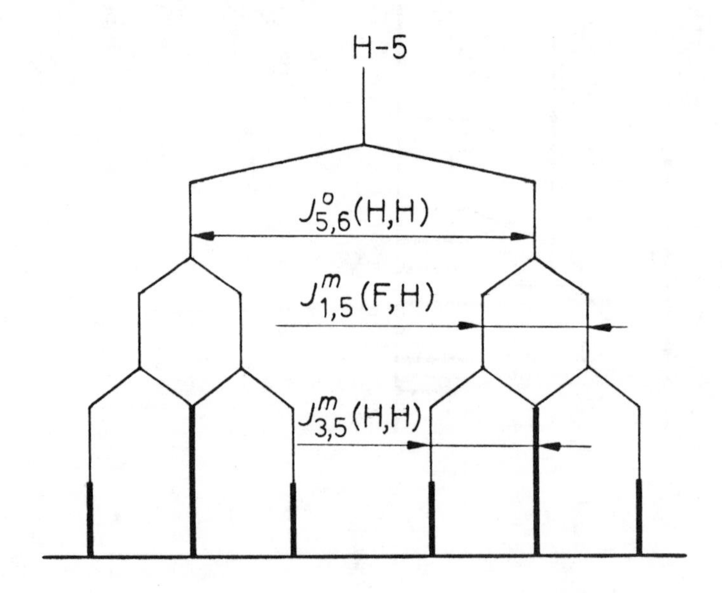

FIGURE 209. The H-5 (*M*) multiplet of compound **61/IV**, constructed from the measured spectrum data.

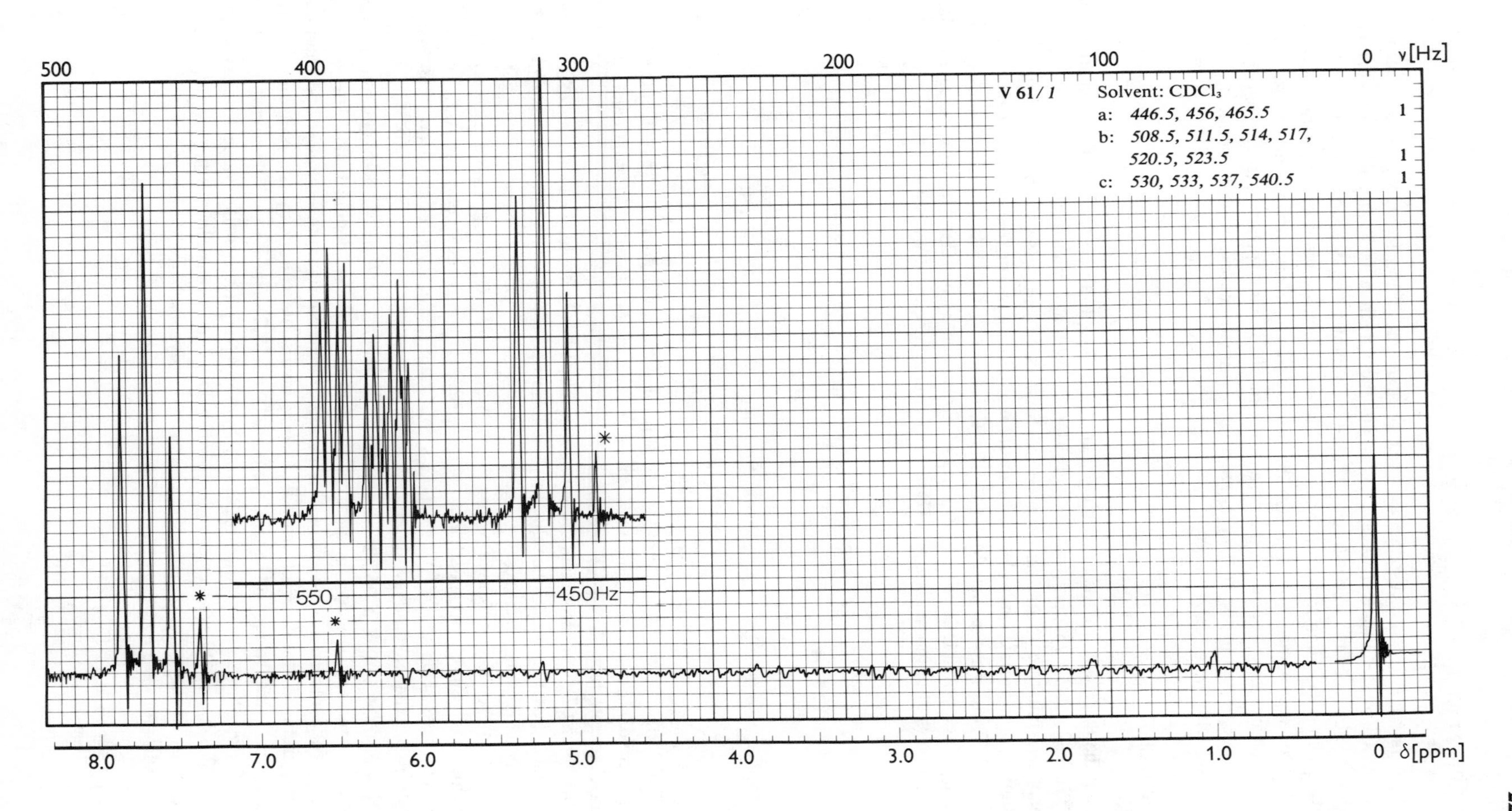

ν[Hz]
500
400
300
200
100
0
V 61/1
Solvent: $CDCl_3$
a: 446.5, 456, 465.5 1
b: 508.5, 511.5, 514, 517, 520.5, 523.5 1
c: 530, 533, 537, 540.5 1
550
450Hz
8.0
7.0
6.0
5.0
4.0
3.0
2.0
1.0
0
δ[ppm]

PROBLEM **62**

Spectra **62**/*1* to *3* correspond to three compounds of the following type of structure: R′OOC–CR=CR–CR=CR–COOR′. What are the substituents R and R′, and what is the difference between the three structures? Are the samples homogeneous? Determine the spin systems corresponding to the observed multiplets. Calculate the spectrum parameters in the first-order approximation, using also the Spectra **62**/*1a*, **62**/*2a*, and **62**/*3a*, recorded with a tenfold abscissa expansion. In the case of the most complex spectrum, long-range coupling must be neglected in the first approximation. Estimate the expectable chemical shifts and compare with the observed ones.

ANSWER

All the three spectra contain a singlet corresponding to three or six hydrogens around at 3.73, 3.82, and 3.78 ppm, respectively, indicating that R′ must be a methyl group. The chemical shifts are in the range expected for methyl esters (see Table 27). There is, in addition, a downfield multiplet, with a total intensity of two or four protons. Consequently, R = H. From the structure MeOOC–CH=CH–CH=CH–COOMe is clear that the samples are the three possible *Z, Z*-, *Z, E*-, and *E, E*- isomers (**I** to **III**). The terminal and middle proton pairs of compounds **I** and **III** are, of course, chemically equivalent.

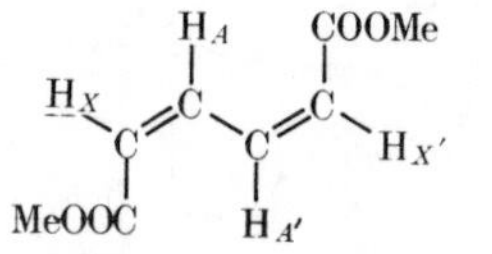

62/I 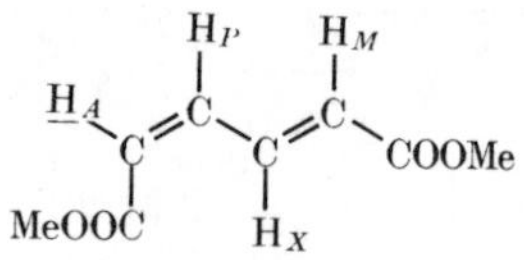 **62/II** 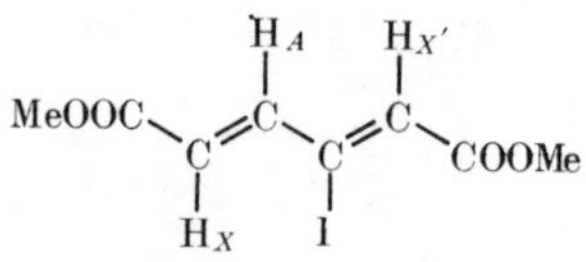 **62/III**

The lines of the olefinic protons are separated into two groups of equal intensities in Spectrum **62**/*1*, which are quite far from each other (at a distance of approximately 2 ppm). Thus, they produce an $AA'XX'$ multiplet.

Spectrum **62**/*3* contains a pair of asymmetric multiplets, completely analogous in structure, but much closer to each other than in Spectrum **62**/*1*. The whole multiplet is symmetric, i.e., the actual spin system is $AA'BB'$.[11]*

The two spectra correspond to the *Z, Z* (**I**) and *E, E* (**II**) isomers, and the asymmetric structure (**II**) must correspond to Spectrum **62**/*2* of asymmetric construction. The latter molecule represents an *AMPX* or *ABMX* spin system; the analysis of the *ABMX* case would require computer methods.

It is possible to pair Structures **I** and **III** with Spectra **62**/*1* and **62**/*3* by determining the coupling constant between the protons attached to the same double bond. This must be smaller for Structure **I**, since in olefins $J^Z < J^E$.** Starting with the simplest Spectrum **62**/*1*, the chemical shifts of the $AA'XX'$ system obtainable simply by the midpoints of the multiplets. Provided that $\delta A > \delta X$, we get $\nu A = 477.25$ Hz, $\delta A = 7.95$ ppm, and $\nu X = 358.75$ Hz; $\delta X = 5.98$ ppm, i.e., $\Delta\nu AX = 118.5$ Hz. The smaller value belongs to the terminal pairs, and the larger chemical shift belongs to the middle proton pairs, of course. According to the experience,*** in conjugated polyene chains the chemical shifts of the interchain protons are always larger than those of the terminal ones.

* See Volume I, p. 133.

** See Volume II, p. 52.

***See Volume II, p. 54.

The shifts can also be estimated by substituting the constants ρ_i of Table 37 into Equation 276. Thus, the calculated shift of the middle protons is 5.25 + 1.24 + 0.46 = 6.95, and of the terminal protons it is 5.25 + 0.78 − 0.05 = 5.98 ppm. Accordingly, label *A* refers to the middle protons, and *X* represents the terminal protons.

The theoretical chemical shift of the terminal protons (δX) agrees with the experimental value (5.98 ppm), but there is a serious discrepancy in δA ($\Delta\delta$ = 1.00 ppm). The reason is that the *A* protons are affected not only by the *Z* ester group (attached to the same double bond), but also by the anisotropic effect of the other ester group, due to its steric proximity.

The coupling constants are calculated from the line spacings read from the expanded Spectrum **62**/*1a* (see Figure 63). Accordingly, Q_1 = 13.3, Q_2 = 10.2, $Q_z = \sqrt{Q_5^2 - Q_1^2} = \pm 13.0$, $Q_4 = 10.5$, $Q_5 = 18.6$, and Q_6 = 16.5 Hz. The coupling constants are, therefore, $J_{AA'} = (Q_1 + Q_2)/2 = 11.75$, $J_{XX'} = (Q_1 - Q_2)/2 = 1.55$, $J_{AX} = (Q_3 + Q_4)/2 = 11.75$, and $J_{AX} = (Q_4 - Q_3)/2 = -1.25$ Hz. The first-order treatment is, therefore, completely justified, because $JAX/\Delta\nu AX = 11.75/118.5 \approx 0.10$.

It is clear from Structure **I** that $J_{AX} \gg J_{AX'}$, and thus $Q_3 = +13.0$ is the correct value. The negative sign of $J_{AX'}$ is reasonable, since it represents a ^4J long-range coupling.* In turn, the value of J_{AX} proves that Spectrum **62**/*1* belongs to the *Z,Z* isomer (see Table 38).

The analogous data can be determined in a completely similar manner for Spectrum **62**/*3* which thus belongs to the *E,E* isomer (**III**). The *trans* position of the hydrogens suggests a greater value for J_{AX}. Since, the ester groups and the hydrogen atoms attached to the other double bond are more distant, the discrepancy between calculated and measured values of δA obtained for the *Z* isomer must decrease. From the spectrum νA = 442.0 Hz, δA = 7.37 ppm, νX = 373.5 Hz, and δX = 6.23 ppm. Consequently, $\Delta\nu AX$ = 68.5 Hz. The theoretical shifts are δA = 5.25 + 1.24+ 1.01 = 7.50 and δX = 5.25 + 0.78 + 0.02 = 6.05 ppm. Thus, the differences are much smaller ($\Delta\delta A$ = −0.13 and $\Delta\delta X$ = +0.18 ppm), indeed.

From Spectrum **62**/*3*, Q_1 = 13.8, Q_2 = 10.8, Q_3 = 17.2, Q_4 = 15.0, Q_5 = 20.2, and Q_6 = 19.7 Hz. The corresponding coupling constants are, in turn, $J_{AA'}$ = 12.3, $J_{XX'}$ = 1.5, J_{AX} = 16.1, and $J_{AX'}$ = −1.1 Hz. The coupling constants of isomers *Z,Z* and *E,E* are, therefore, quite similar, the only appreciable difference being in the value of J_{AX}. In this case, $J_{AX}/\Delta\nu AX = 16.1/68.5 \approx 0.24$ Hz, and therefore, the coupling is not really of the first order, but the first-order treatment is not seriously wrong for the case (e.g., the degenerate lines do not split.)

The higher-order analysis of the *ABMX* spectrum of the *Z,E* isomer is complicated. The actual situation is, however, less problematic because most of the spectrum parameters are but slightly different from those of the two symmetric isomers, and thus we may use them for comparison. First one must construct the schematic structure of the presumable spectrum, and then assign the multiplets to the individual protons.

In the absence of long-range couplings (^4J and ^5J), the signal of the middle hydrogens may split into a double doublet (or, if the coupling constants with the two adjacent protons are nearly the same, into symmetric triplets), and that of the terminal hydrogens split into doublets. The downfield multiplet, however, which must belong to one of the middle protons, consists of eight lines, although four of them are much weaker than the others. Comparing Spectra **62**/*1* and **62**/*2*, it becomes evident that the sample is contaminated with the *Z,Z* isomer. At 7.95 ppm the strongest four lines of the *A* part of its *AA'XX'* spectrum can be observed beside the *X* part of the spectrum of compound **II**, whereas the *X* part is overlapping with the *A* and *M* doublets of the latter isomer.

* See Volume I, p. 55.

Now we must select the lines of the multiplets. Assume that the sequence of chemical shifts is $\delta A < \delta M < \delta P < \delta X$. Then the four high-frequency lines at 490, 501.5, 505.5, and 517 Hz belong to the X proton. From the frequencies, $\delta X = 503.5/60 = 8.39$ ppm, and the coupling constants for proton X are 15.5 and 11.5 Hz. From these values it follows that X must be the middle hydrogen in *trans* position. The signal of nucleus P is a triplet at $\delta P = 6.70$ ppm, and its coupling constants with the two adjacent hydrogens are nearly the same, 11.5 Hz (from the lines at 390.5, 402, and 413.5 Hz). P is therefore the *cis* middle hydrogen.

The lines at 352.5, 359, 364, and 374.5 Hz can be assigned to the A and M terminal protons, actually lines **1** and **3** to nucleus A and lines **2** and **4** to nucleus M. Thus, $\delta A = 5.97$ and $\delta M = 6.11$ ppm. Since the spacing of the former lines is 11.5 Hz and of the latter 15.5 Hz, proton A is in *cis* and proton M is in *trans* position.

It is rather instructive to compare the experimental data to the chemical shift calculated by Equation 276: $\delta A = 5.25 + 0.78 - 0.05 = 5.98$, $\delta M = 5.25 + 0.78 + 0.02 = 6.05$, $\delta P = 5.25 + 1.24 + 0.46 = 6.95$, $\delta X = 5.25 + 1.24 + 1.01 = 7.50$ ppm. Now let us consider the effect of the distant ester groups on the basis of the experimental data of isomers **I** and **III**. Due to its similar relative position, the effect on the X hydrogen is presumably the same as on the middle hydrogens of the Z,Z isomer ($7.50 + 1.00 = 8.50$ ppm). In turn, the effect on proton P is expected to be similar to that on the middle hydrogens of isomer **III** ($6.95 - 0.13 = 6.82$ ppm). There is no analogue in compounds **I** and **III** for the relative positions of hydrogens A, M, and the distant ester groups, and thus we take the average of the observed effects, $(+0.18 + 0.00)/2 = 0.09$ ppm. We obtain $5.98 + 0.09 = 6.07$ and $6.05 + 0.09 = 6.14$ ppm for the A and M protons. There is an excellent agreement between the modified theoretical and the experimental values ($\Delta\delta$ is -0.10, -0.03, $+0.12$, and -0.11 ppm for the A, M, P, and X protons, respectively) and the sequence of shifts — the assignment of the lines to the protons — is the same as that derived from the splittings. Accordingly, $J^E = J_{MX} = 15.5$, $J^Z = J_{AP} = 11.5$, and $J_{PX} = 11.5$ Hz. All the three values are very close to the analogous coupling constants obtained for isomers **I** and **III**.

It is obvious, however, that the long-range couplings J_{AX}, J_{PM}, and J_{AM} also exist in the conjugated system as shown by the shoulders of the A and X lines already in the normal spectrum (**62**/2). The expanded spectrum (**62**/2*a*), in which these splittings are clearly visible, proves the coupling unambiguously. All lines of the X quartet and the P triplet split further into doublets due to the couplings of $J_{AX} = -0.08$ and $J_{PM} = -0.6$ Hz. The lines of the A and M doublets split into triplets, since $J_{AX} \approx J_{PM} \approx J_{AM} \approx 0.6$ Hz.

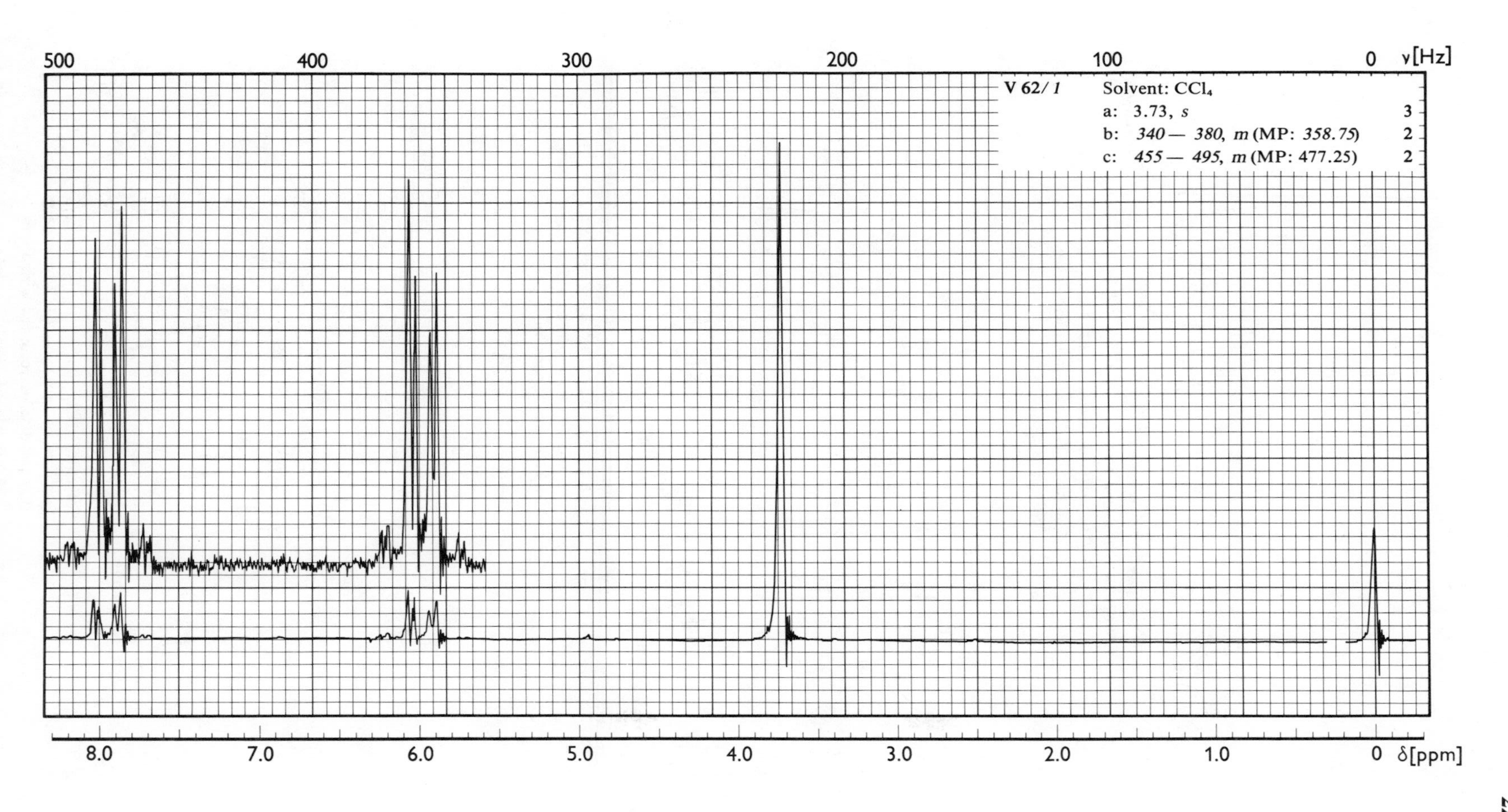

500
400
300
200
100
0
ν[Hz]
V 62/1
Solvent: CCl₄
a: 3.73, s 3
b: 340 — 380, m (MP: 358.75) 2
c: 455 — 495, m (MP: 477.25) 2
8.0
7.0
6.0
5.0
4.0
3.0
2.0
1.0
0
δ[ppm]

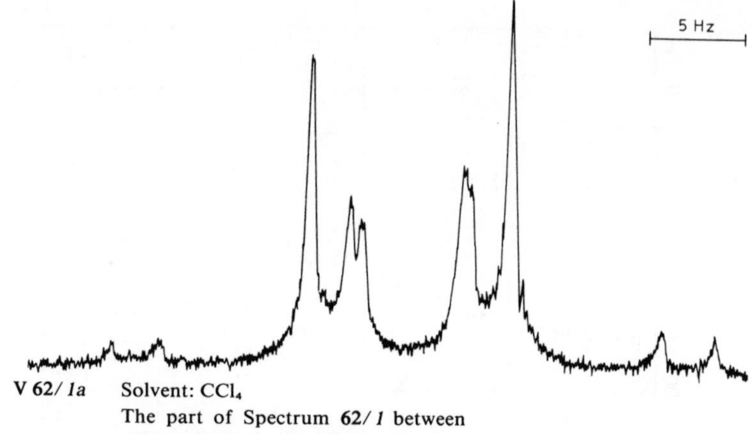

V 62/*1a* Solvent: CCl_4
The part of Spectrum **62/*1*** between 300 and 500 Hz recorded with a tenfold abscissa expansion.

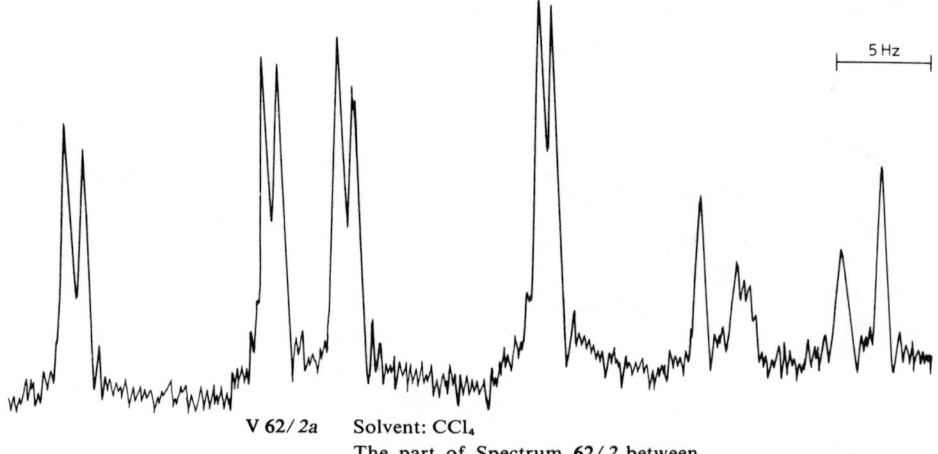

V 62/*2a* Solvent: CCl_4
The part of Spectrum **62/*2*** between 300 and 550 Hz recorded with a tenfold abscissa expansion.

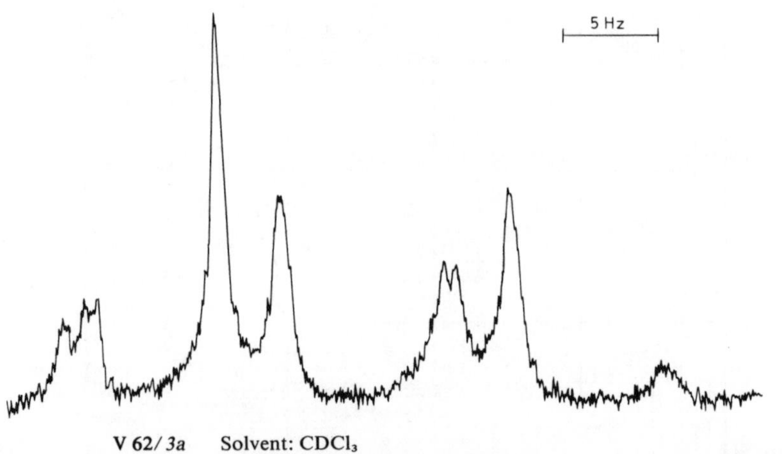

V 62/*3a* Solvent: $CDCl_3$
The part of Spectrum **62/*3*** between 350 and 500 Hz recorded with a tenfold abscissa expansion.

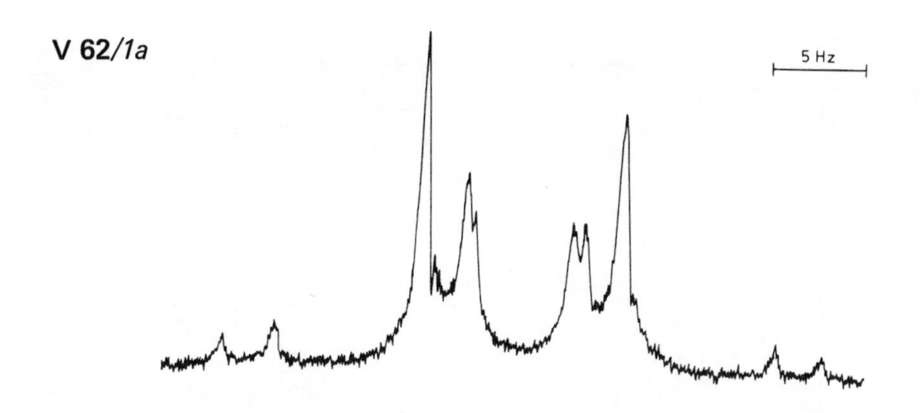
V 62/1a
5 Hz

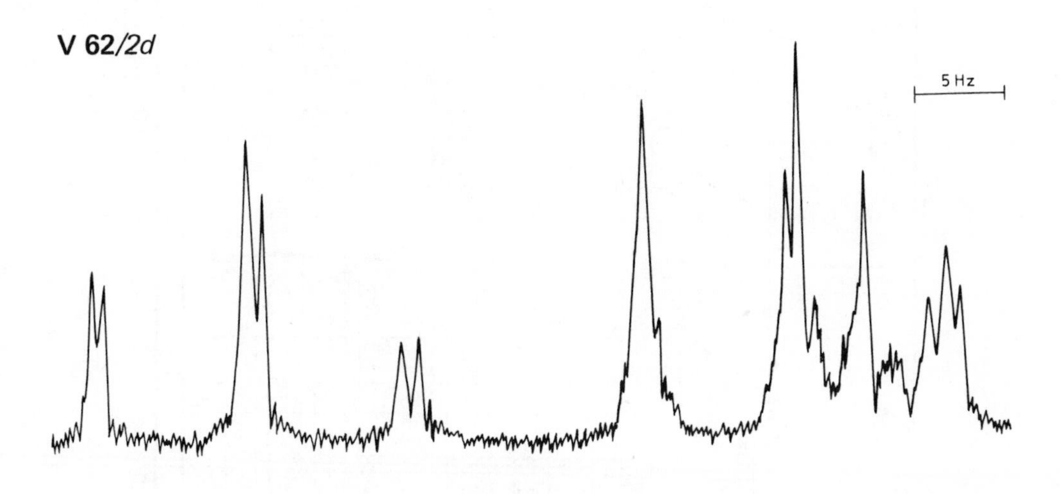
V 62/2d
5 Hz

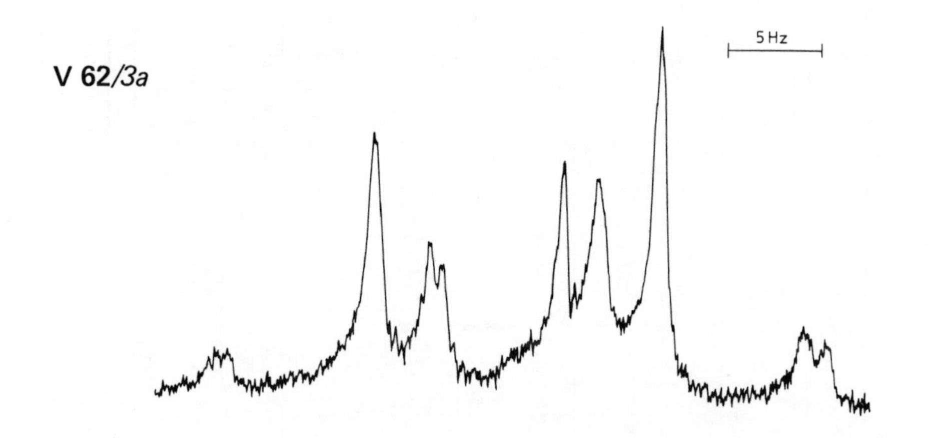
V 62/3a
5 Hz

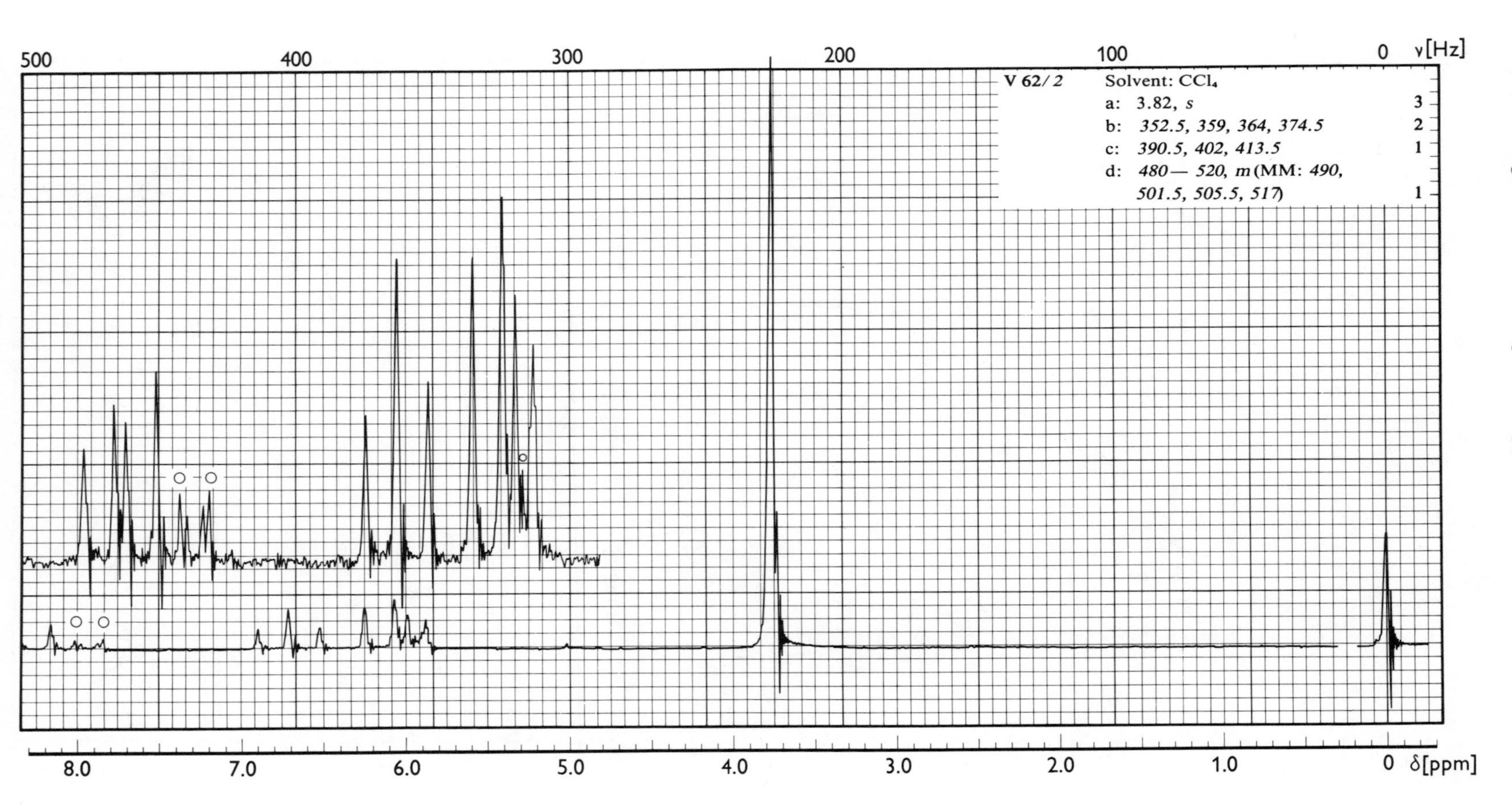
V 62/2
Solvent: CCl_4
a: 3.82, *s* 3
b: *352.5, 359, 364, 374.5* 2
c: *390.5, 402, 413.5* 1
d: *480 — 520, m (MM: 490, 501.5, 505.5, 517)* 1
ν[Hz]
500
400
300
200
100
0
δ[ppm]
8.0
7.0
6.0
5.0
4.0
3.0
2.0
1.0
0

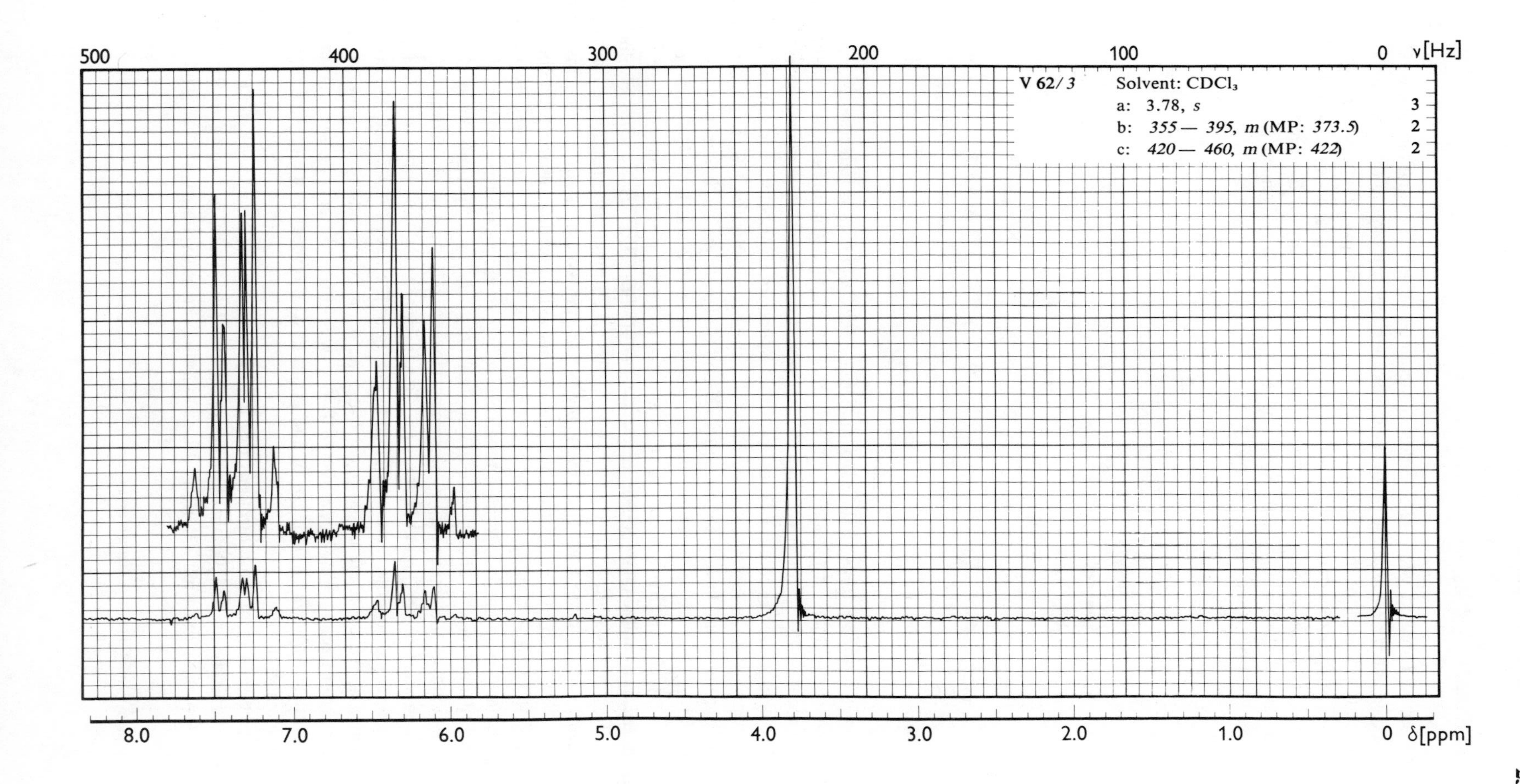

V 62/3
Solvent: $CDCl_3$
a: 3.78, *s* 3
b: *355— 395, m* (MP: *373.5*) 2
c: *420— 460, m* (MP: *422*) 2
500
400
300
200
100
0
ν[Hz]
8.0
7.0
6.0
5.0
4.0
3.0
2.0
1.0
0
δ[ppm]

PROBLEM **63**[800,801,1331]

Compound **I** treated with Na_2S yields compounds **II** and **III** simultaneously. Correlate Spectra **63**/*1* and **63**/*2* with these products and explain the structures of the spectra. Determine the steric arrangement of the thiirane ring and the mesyl groups. Compound **III** was reacted with NaN_3 to prepare a derivative in which one of the mesyl groups is replaced by an azide substituent (**IV**). Using Spectrum **63**/*3*, elucidate the structure of the product. Determine the structure of the product with Spectrum **63**/*4*, which was obtained from the dimesyl derivative (**III**) by a treatment with NaI.

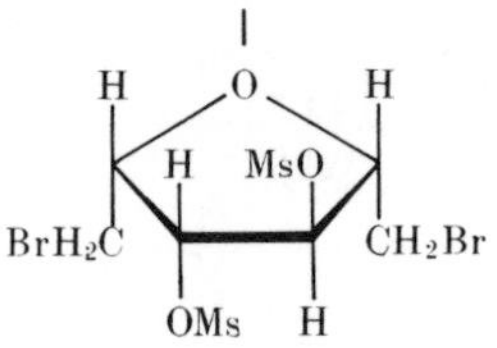

63/I 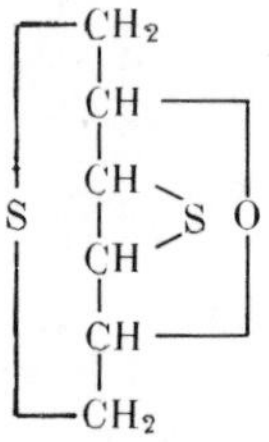 **63/II** 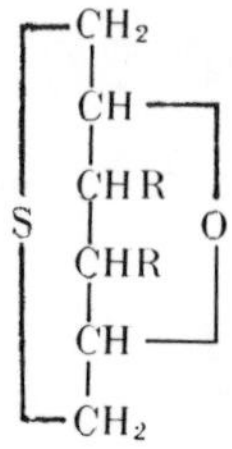 **63/III (R = R′ = OMs)**

63/IV (R = OMs; R′ = N_3)

ANSWER

The pairing of Structures **II** and **III** with Spectra **63**/*1* and **63**/*2* is rather simple. Both the small number of lines in the spectrum, referring to a symmetric molecular structure, and the absence of signals with $\delta > 5$ ppm, which would belong to the protons adjacent to the mesyloxy group (see Problem **64**), indicate that Spectrum **63**/*2* belongs to compound **II**.

Following from the molecular symmetry (C_s), the hydrogen pairs 1-6, 1′-6′, 2-5, and 3-4 are isochronous and have identical shifts. 1′ and 6′ denote the methylene protons (labeled below with *A*) that are in *trans* position with respect to H-2 and H-5. H-1 and H-1′ (H-6 and H-6′) are chemically nonequivalent — as usual in cyclic compounds (compare Problem **58**). In the most stable chair conformation, the *cis* protons (*B* or *M*) are *axial* and the *trans* protons (*A*) are *equatorial*. They give rise to an *AB* spectrum approximating the limiting *AX* case. Due to a coupling with H-2 (H-5), the actual spin system is *ABX*, very close to the limiting *AMX* case, which justifies the use of this approximation in the followings. The anellated H-3 and H-4 atoms do not interact with their neighbors,* and thus they yield an A_2 singlet at 3.47 ppm, indicating a relatively strong shielding of these protons due to the three-membered ring.** The absence of coupling ($J_{2,3} \equiv J_{4,5} = 0$) permits one to derive the steric position of the thiirane ring. Formulas **IIa** and **IIb** are the projections of the two possible isomers of compound **II** from the directions of C-2,3 and of C-4,5, respectively (these atom pairs have the same projection in the figure). Heavy lines denote the bonds that are located above the plane of the paper (i.e., the plane of the odd or even atoms the six-membered ring), approximately parallel to it. The two isomers differ in the steric arrangement of the thiirane ring, which is in *syn* or *anti* position with respect to the oxygen or sulfur of the six-membered ring.

It can be seen from the projections that the dihedral angle of the C_2–H and C_3–H bonds is approximately 30 and 90° in Structures **IIa** and **IIb**. Thus, according to the *Karplus* relation (compare Equation 79), a significant splitting can be expected in the case of Structure

* Compare Volume I, last paragraph on p. 61 .

** Compare Volume II, p. 16.

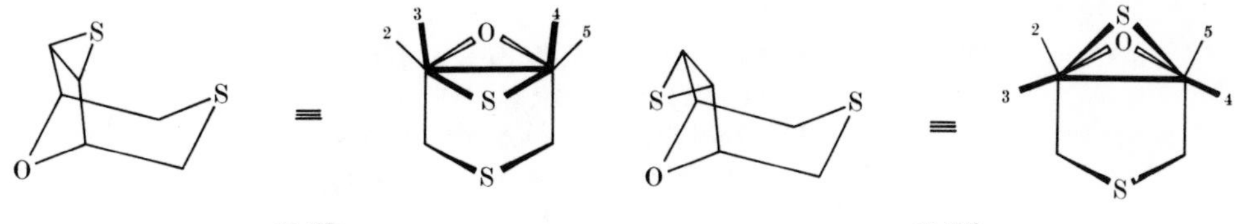

63/IIa

63/IIb

IIa ($J_{2,3} \equiv J_{4,5} > 0$), and only a negligible one can be expected for isomer **IIb** ($J_{2,3} \equiv J_{4,5} \approx 0$). Thus, we can decide in favor of the latter structure.

In the *X* part of the *AMX* multiplet five lines can be recognized, which cannot be attributed merely to an *AMX* splitting (at most four lines may occur). One of the *A* and *M* parts also produces six lines instead of the expected four. It must be assumed, therefore, that the proton pairs 2,5, 1,6, and 1′,6′ are not isogamic, i.e., there are significant long-range couplings. Accordingly, the six protons form an *AA′MM′XX′* spin system. Since, however, the *M* part of spectrum contains four lines just as a simple *AMX* spectrum, the actual spin system is simpler than the most general case (some of the coupling constants are equal or too small to cause observable splitting).

It was assumed implicitly in the above discussion that $\delta M_a > \delta A_e$, in agreement with the literature data, concerning sulfur-containing six-membered saturated rings, where $\delta H_e < \delta H_a$ for the methylene hydrogens adjacent to the sulfur,* in contrast to the general rule pertaining to cyclohexane derivatives.

The H-1′,2,5,6′ atoms can be considered as an *AA′XX′* system in which the *A* and *X* parts are split further by the J_{AM} and J_{MX} interactions, respectively ($J_{MM'} = J_{MA'} = J_{MA'} = 0$).

The *A* and *X* multiplets of *AA′XX′* systems are, as known,** identical. The structure of the multiplet being identical for the *A* and *X* parts can be studied most conveniently in the *A* part. The coupling $J_{AM} \equiv {}^2J_{1,1}{}' \equiv {}^2J_{6,6}{}'$ (14 Hz) is much larger than any other within this system. Thus, the two line groups at a distance of J_{AM} with fine structures due to the interactions of the *AA′XX′* system do not overlap and the complete *A* part of the *AA′XX′* multiplet is repeated at a distance of J_{AM} with a different intensity (roof structure: *AM* → *AB* perturbation). In contrast, since J_{MX} is small and commeasurable with $J_{AX'}$ $J_{AX'}$, $J_{AA'}$, and $J_{XX'}$ the doublet lines of the *XX′* multiplet split by J_{MX} overlap.

The spectrum indicates that the actual *AA′XX′* multiplet is a simplified variant of the general case: instead of ten *A* lines of the latter, two line groups at a distance of J_{AX} appear with triplet-like fine structure. To interpret this structure it must be assumed that the two middle line pairs coalesce and thereby the total intensity of the four lines is higher than that of the originally also degenerate **3** to **4** and **9** to **10** transitions. Hence, the triplet arises from 12 transitions, of which 6 pairs coincide (like in Problem **66**). The two middle ones of the remaining six lines merge into one signal, and the outer two lines are insignificantly weak.

The condition of the above pairwise coincidence is that $Q_1 \approx Q_2$, and thus $Q_5 \approx Q_6$ (see Problem **66**). This holds if $J_{XX'} \approx 0$. In this case the frequencies and relative intensities of the transitions of the *AA′XX′* system are as seen in Table 104. In the table $C^2 = (Q_1 + Q_6)/2Q_5$ and $Q_5 = \sqrt{Q_1^2 + Q_3^2}$ (see Table 19 and Problem **66**). The coincidence of lines **3** and **4** is also bound to the condition $Q_5 \approx Q_1$, which implies that $Q_3 \approx 0$ and $Q^2 \approx 1$. In this case lines **1** and **6** have zero intensities and lines **3** and **4** coincide, giving, indeed, a triplet signal with $Q_4 = 4$ Hz (from the spectrum), and since $Q_3 = 0$, we get $J_{AX} = J_{AX'} = 1.5$ Hz. From $Q_1 = Q_2$ follows that $J_{XX'} = 0$. $J_{AA'}$ cannot be determined from the

* See Volume II, p. 44.

** See Volume I, p. 128.

Table 104
RELATIVE FREQUENCIES AND PROBABILITIES FOR A (OR X) TRANSITIONS OF THE $AA'XX'$ SPIN SYSTEM PROVIDED THAT (a) $Q_1 = Q_2(Q_5 = Q_6)$ AND (b) $Q_1 = Q_5(Q_3 = 0)$

Case	Transition	Line	Frequency	Transition probability
a	1,2	1	$(Q_1 + Q_5)/2$	$1 - C^2$
	3,4	2	$Q_4/2$	1
	5,6	3	$(Q_5 - Q_1)/2$	C^2
	7,8	4	$-(Q_5 - Q_1)/2$	C^2
	9,10	5	$-Q_4/2$	1
	11,12	6	$-(Q_1 + Q_5)/2$	$1 - C^2$
b	3,4	1	$Q_4/2$	1
	5,6,7,8	2	0	2
	9,10	3	$-Q_4/2$	1

spectrum, because under the above conditions the structure of the spectrum is invariant with respect to this constant.

It is now very simple to determine the remaining parameters of the spectrum. From the distance of the two A triplets or M doublets, $J_{AM} = 4$ Hz, and from the midpoints of these multiplets (132, 146, 191, and 205 Hz), $\delta A = 2.33$ and $\delta M = 3.28$ ppm. (A line of one of the M doublets coincides with ths singlet of H-3,4 at 3.47 ppm.) $J_{MX} = 3$ Hz (from the splitting of the M doublets) and δX (H-2,5) $= 4.53$ ppm. The quintet structure of this signal arises from the XX' triplet with a splitting of $J_{MX} = 3$ Hz, according to the scheme shown in Figure 210. The occurrence of $^5J(J_{AX'})$ coupling is due probably to a dihedral angle of almost 180° between the corresponding C–H bonds and to the fact that this interaction may be transmitted by more chains of bonds, containing also heteroatoms.*

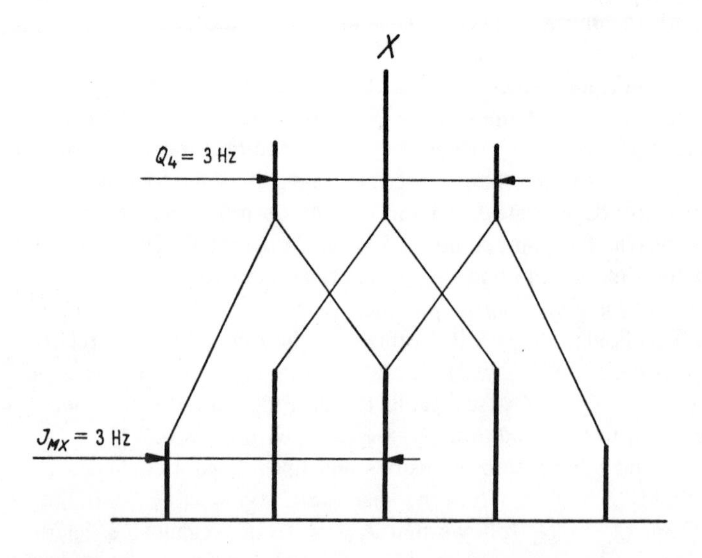

FIGURE 210. The schematic structure of H-2,5 multiplet (X part) in the spectrum of compound **63/II**.

* See Volume I, p. 67.

The assignment of Spectrum **63**/*1* is straightforward, although some regions of the spectrum cannot be analyzed due to the overlapping multiplets and the disturbing solvent absorptions. The methylene groups give rise again to *AB* systems, and all the four lines split into multiplets owing to the interactions with H-2,5 and to the long-range couplings. The H-1,6 and H-1′,6′ pairs are, of course, nonequivalent due to the asymmetric Structure **III**. The signal of the light isotope content of the solvent is close to the *A* lines of the two *AB* spectra (concealing the two middle lines), whereas the downfield lines of the two coincident *B* multiplets (the outer ones) is overlapping with the methyl singlet of the mesyl group. The chemical shifts can, therefore, be estimated only very roughly ($\delta A \approx 2.5$ and 2.35, respectively, and $\delta B \approx 3.15$ ppm).

The sharp maximum at 3.45 ppm arises from the water content of the solvent. δCH_3 (mesyl groups): 3.33 and 3.35 ppm. H-2 and H-5 give rise to two signals around 4.55 and 4.85 ppm. One of them is a triplet-like multiplet, and the other is a much broader signal consisting of two triplet-like signals at a distance of about 7 Hz. The signal of H-3 is a double doublet at 5.33 ppm (J = 7 and 3 Hz), whereas that of H-4 is a doublet at 5.67 ppm (J = 3 Hz). The asymmetry of these signals indicates that the steric positions of the mesyl groups are different (*exo* and *endo*). The configurations of the substituents can also be derived from the line spacings of the H-3 and H-4 signals. The magnitudes of $J_{2,3}$ and $J_{4,5}$, respectively, are different in this case, too, depending on the steric orientation of the mesyl group. If the mesyl group (assume that it is substituent 4) faces the oxygen atom of the six-membered ring (we define this as *exo* position), the dihedral angle between the C_4–H and C_5–H is about 90°. Therefore, according to the Karplus relation, there is no coupling between them (see Formula **IIIa**, obtained by similar projection than Formulas **IIa,b**). If the 3-mesyl group is in *endo* position, the dihedral angle between the two analogous C–H bonds (2 and 3) is 30°, and therefore the anticipated *vicinal* coupling is approximately 7 Hz. Accordingly, the spectrum shows that the mesyl groups have different steric positions in compound **III**, and if substituent 4 is the *exo* and substituent 3 is the *endo* mesyl group, the doublet at 5.77 ppm can be assigned to H-4 and the double doublet at 5.33 ppm to H-3. Moreover, $J_{3,4}$ = 3, $J_{2,3}$ = 7, and $J_{4,5}$ = 0 Hz. Thus, the multiplet around 4.55 ppm arises from H-5, and the broader line system around 4.85 ppm is due to H-2.

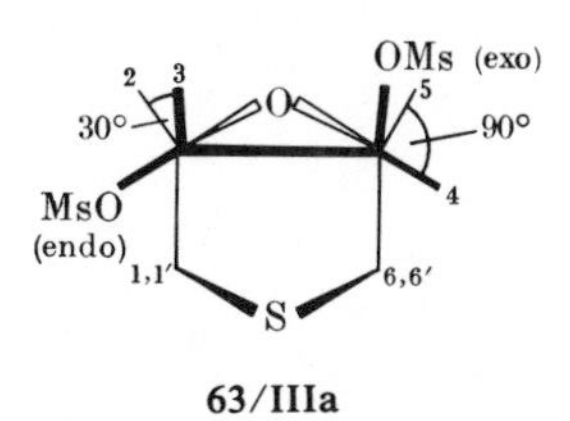

63/IIIa

In Spectrum **63**/*3* of azide derivative **IV**, the methyl signal of the remaining mesyl group can be found at 3.20 ppm, in overlap with the *B* line of the methylene multiplet. The double doublet of the hydrogen adjacent to the mesyl substituent appears at 5.09 ppm. $J_{3,4}$ = 3 and $J_{2,3}$ = 7 Hz. Consequently, the *endo* 3-mesyl group is still present in the molecule, and the azide group is attached in the *exo* position (**IVa**). In agreement with the above, the disappearing methine signal belongs to the hydrogen adjacent to the mesyl group of larger chemical shift. The doublet of H-4 appears at 4.77 ppm, in overlap with the signal of H-2 and H-5 absorbs at 4.40 ppm. The azide group causes, therefore, an upfield shift in the H-2 to 5 signals due partly to the smaller $-I$ effect and partly to the anisotropic shielding. The methylene signals are also well separated in this spectrum and, although the overlapping multiplets do not permit accurate calculations, approximate values can be given for the following parameters: J_{AB} = 14 Hz and δA = 2.37 and δB = 3.23 ppm.

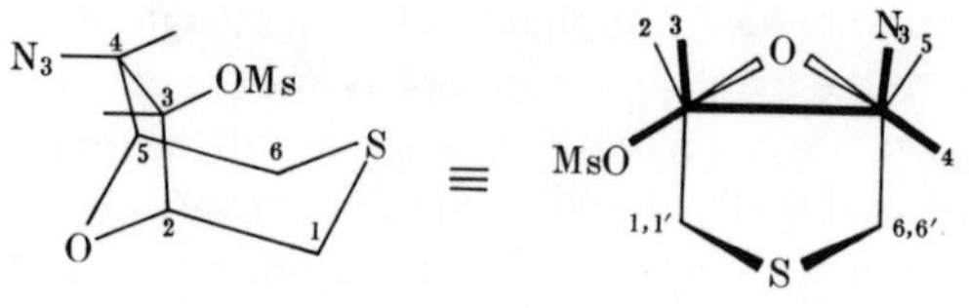

63/IVa

The spectrum (**63**/5) of the unknown product prepared from compound **III** with NaI also contains an *AA'MM'XX'* multiplet. The singlet of H-3,4 is at 6.30 ppm, shifted downfield considerably. The spectrum does not contain further signals (except for those arising from the impurities of the solvent at 5.17, 6.45, and 7.27 ppm). As the chemical shift of the singlet is in the range characteristic of olefinic protons, it can be concluded that an unsaturated derivative (**V**) has been formed by the elimination of the mesyl groups. A strong deshielding due to the $-I$ effect of the double bond is also observable in the signal of H-2,5. The anisotropic effect of the double bond causes an opposite shift in the A multiplet, as shown by the following data calculated from the multiplets: $\delta A = 2.20$, $\delta M = 3.30$, and $\delta X = 4.95$ ppm, $J_{AM} = 13$; $J_{AX} = J_{AX'} = 1.5$, and $J_{MX} = 3$ Hz.

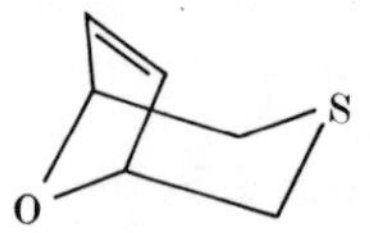

63/V

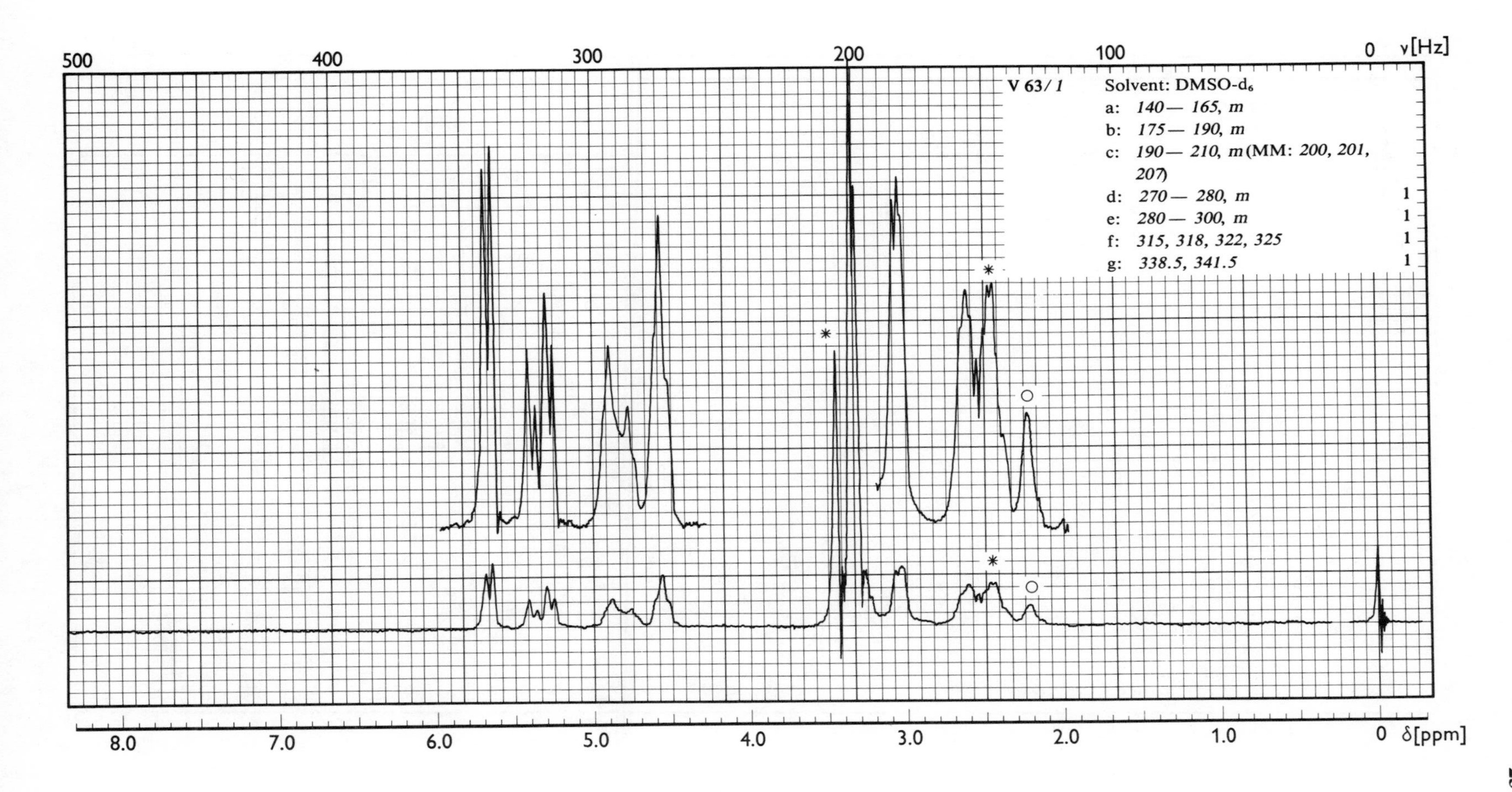
V 63/ 1
Solvent: DMSO-d₆
a: 140— 165, m
b: 175— 190, m
c: 190— 210, m (MM: 200, 201, 207)
d: 270— 280, m 1
e: 280— 300, m 1
f: 315, 318, 322, 325 1
g: 338.5, 341.5 1
ν[Hz]
500
400
300
200
100
0
δ[ppm]
8.0
7.0
6.0
5.0
4.0
3.0
2.0
1.0
0

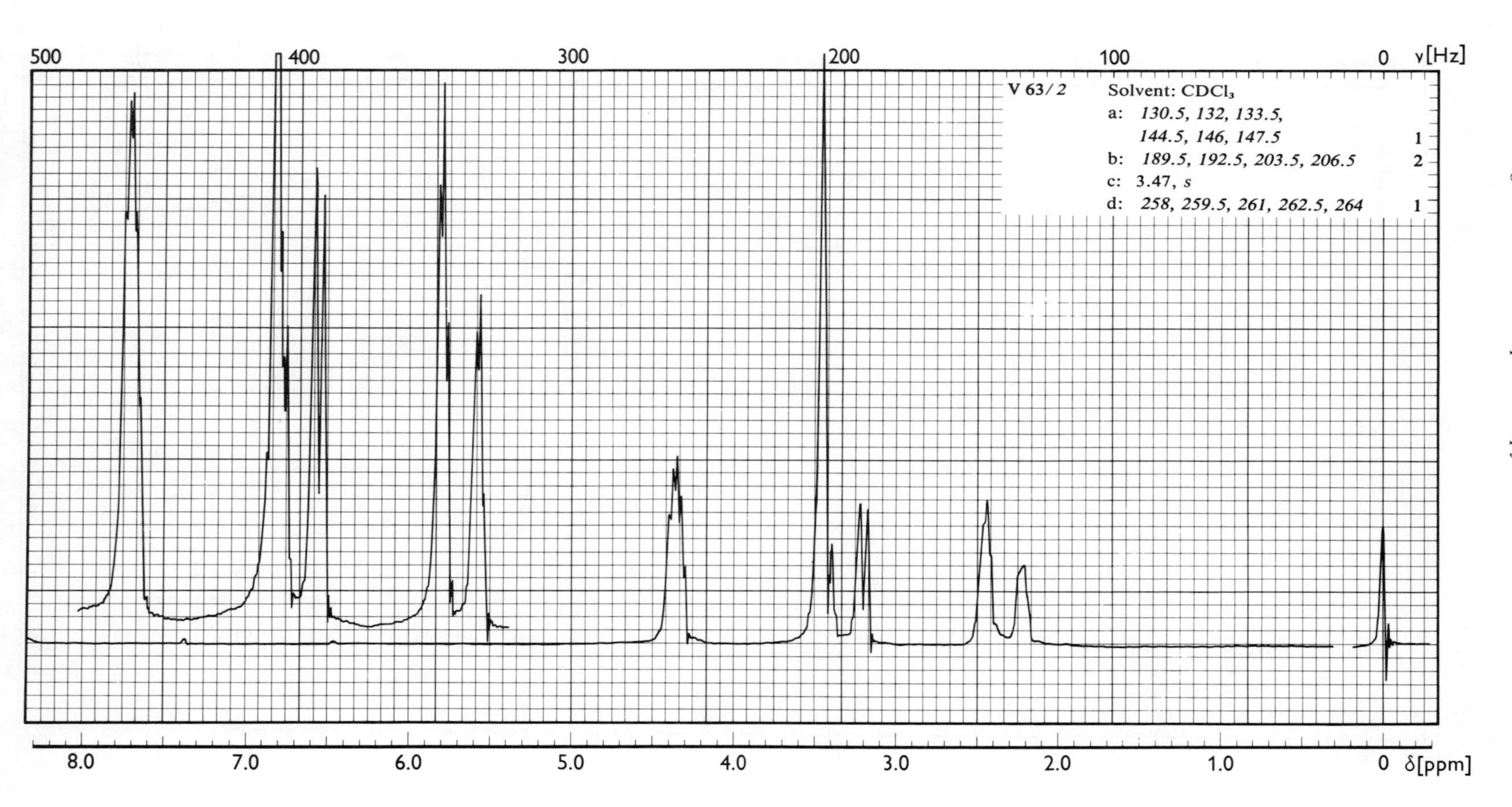
V 63/2
Solvent: CDCl₃
a: 130.5, 132, 133.5, 144.5, 146, 147.5 1
b: 189.5, 192.5, 203.5, 206.5 2
c: 3.47, s
d: 258, 259.5, 261, 262.5, 264 1
ν[Hz]
500
400
300
200
100
0
δ[ppm]
8.0
7.0
6.0
5.0
4.0
3.0
2.0
1.0
0

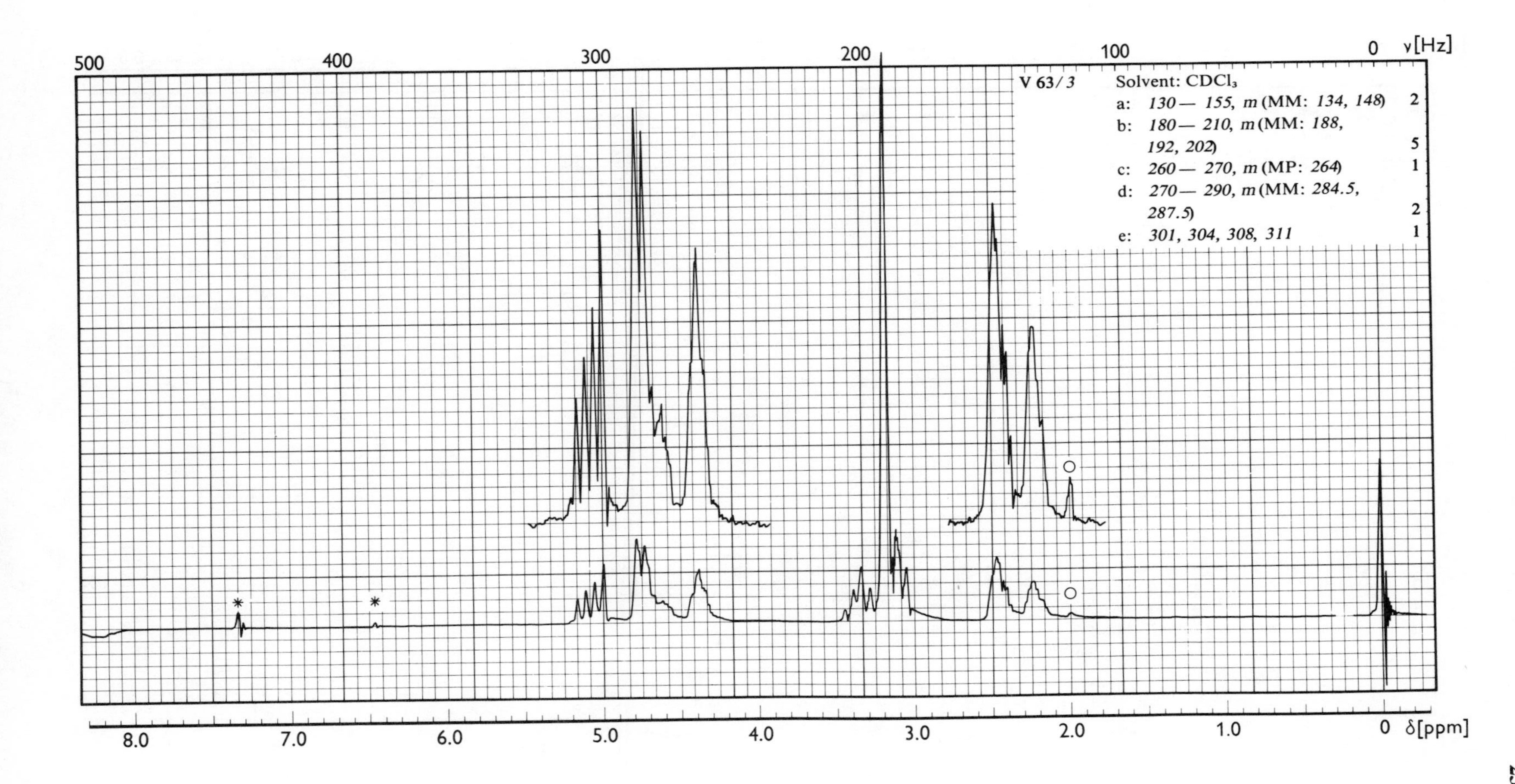
ν[Hz]
500
400
300
200
100
0
V 63/3
Solvent: $CDCl_3$
a: 130 — 155, m (MM: 134, 148) 2
b: 180 — 210, m (MM: 188, 192, 202) 5
c: 260 — 270, m (MP: 264) 1
d: 270 — 290, m (MM: 284.5, 287.5) 2
e: 301, 304, 308, 311 1
δ[ppm]
8.0
7.0
6.0
5.0
4.0
3.0
2.0
1.0
0

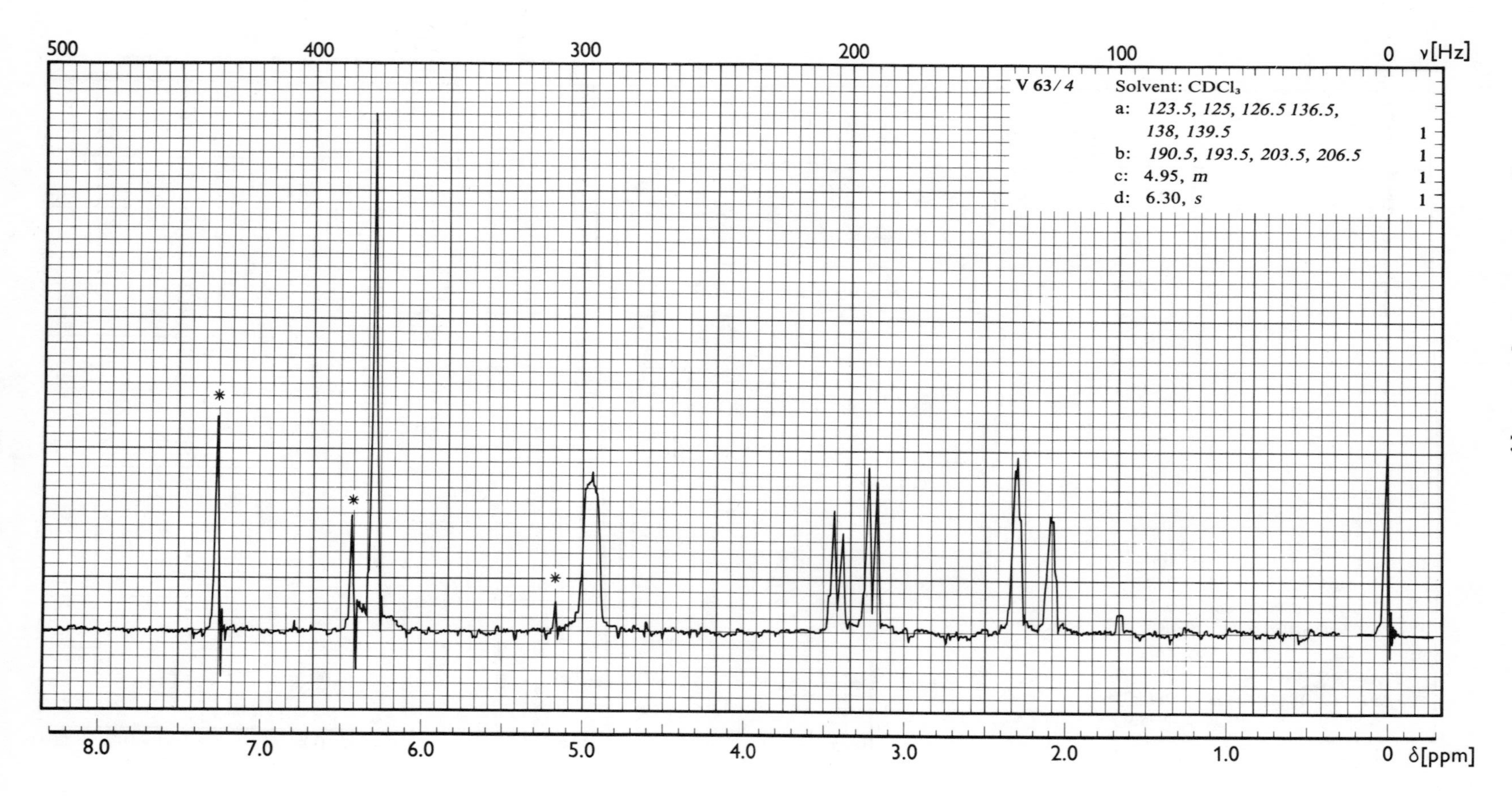
V 63/4
Solvent: CDCl₃
a: 123.5, 125, 126.5 136.5,
138, 139.5
b: 190.5, 193.5, 203.5, 206.5
c: 4.95, m
d: 6.30, s
1
1
1
1
ν[Hz]
500
400
300
200
100
0
δ[ppm]
8.0
7.0
6.0
5.0
4.0
3.0
2.0
1.0
0
*
*
*

PROBLEM 64[1320,1333,1441,1442]

Correlate Spectra **64**/*1* and **64**/*2* with Structures **I** and **II**, using Spectra **64**/*3* and **64**/*4* of the benzoyloxime derivatives **III** and **IV** prepared from isomers **I** and **II** via the corresponding oximes, the above spectra recorded repeatedly after the addition of 1/5 and 1/3 mol of Eu(DPM)$_3$ (**236**, R = Me, Ln = Eu) shift reagent (**64**/*5a,b,c*, **64**/*6a,b*, and **64**/*7a,b*) and the chemical shifts of the signals measured in the latter (Table 105).

64/I (R = O) **64/II (R = O)**

64/III (R = N = OBz) **64/IV (R = N = OBz)**

Table 105
CHEMICAL SHIFTS (PPM) OF THE SINGLETS IN THE SPECTRA 64/*5b,6a,7a* and 64/*5c,6b,7b* RECORDED AFTER THE ADDITION OF 1/5 and 1/3 MOL OF SR Eu(DPM)$_3$, RESPECTIVELY

Spectrum	64/*5b*		64/*5c*		64/*6a*	64/*6b*	64/*7a*	64/*7b*
δ (=CMe)	2.55[a]	2.68[b]	2.76[a]	2.96[a]	2.27	2.46	2.58	2.65
δ OMe	3.49[a]	3.73[b]	3.65[a]	3.92[b]	3.53	3.53	3.95	4.04
δ OMe	3.51[a]	3.81[b]	3.70[a]	4.00[b]	—	3.86	4.12	4.24
δ (=CH)	4.99[a]	5.81[b]	5.22[a]	6.35[b]	5.01	5.55	6.75	7.01

[a] Singlet corresponding to samples having Spectrum **64**/*1*.
[b] Singlet corresponding to samples having Spectrum **64**/*2*.

ANSWER

The assignment of the spectra is simple (see Table 106). It can be seen that the spectra are in accordance with Structures **I** to **IV**. The two methoxy groups give coincident signals in Spectrum **64**/*1*, indicating the accidental isochrony of these substituents. The methyl protons of the N=C–CH$_3$ group are more shielded than the corresponding acetyl hydrogens, owing to the weaker $-I$ effect of the N=C bond in comparison with the carbonyl group. The hydrogens of the condensed aromatic ring give rise to an *ABCD* multiplet in **I** and **II**, which is, overlapping with the *ABB'* part of the *ABB'XX'* multiplet of the benzoyl ring in the case of **III** and **IV**. The *XX'* part is separated, downfield, due to the deshielding of *ortho* hydrogens by the carbonyl group, which is characteristic for benzoyl derivatives (see Table 39 and Figure 140). The isomeric structures cannot be correlated unambiguously with the spectra.

A comparison of the data indicates that in Spectra **64**/*2* and **64**/*4* the three methyl signals and the methine singlet undergo downfield shifts. The shift difference is the largest for the methine signals, particularly for the benzoyloxime derivatives (5.31 − 4.77 = +0.54 and 5.80 − 4.81 = 0.99 ppm, respectively). Consequently, in the compounds corresponding to Spectra **64**/*2* and **64**/*4*, the methine hydrogen must be close to the acetyl or benzoyloxime groups, on the same side of the heteroring. In this position the $-I$ and anisotropic effects of the carbonyl group are probably of the same sign, resulting in a strong deshielding of

Table 106
ASSIGNMENT AND CHEMICAL SHIFTS (PPM) OF THE SIGNALS APPEARING IN SPECTRA 64/*1* to *4*

Spectrum	64/*1*	64/*2*	64/*3*	64/*4*
$\delta CH_3C(sp^2)$, s(3H)	2.33	2.39	2.23	2.31
δOCH_3, 2xs(2 × 3H)	3.38	3.51 3.60	3.37 3.46	3.52 3.72
δ(=CH), s(1H)	4.77	5.31	4.81	5.80
νArH, m	410—460[a]	405—450[a]	405—460[b] 475—495[c]	400—460[b] 475—495[c]

[a] Multiplet of 4H intensity in hertz.
[b] Multiplet of 7H intensity in hertz.
[c] Multiplet of 2H intensity in hertz.

the methine proton. Accordingly, Spectrum **64**/*2* corresponds to Structure **I** and Spectrum **64**/*4* corresponds to Structure **III**.

The larger difference in shielding of the methoxy groups for the isomers **I** and **III** is also compatible with the above conclusion, because their positions with respect to the strongly anisotropic acetyl and benzoyloxime groups are different in these isomers, whereas they are not in the isomeric pairs **II** and **IV**. The above assumptions require further confirmation. The experiments with the shift reagent were performed with this aim. The sense of the anisotropic effect of the carbonyl group is, namely, uncertain due to the free rotation of this substituent. It can be safely assumed, however, that the strongest complexing group of molecules **I** to **IV** is the carbonyl oxygen. Therefore, upon the addition of the shift reagent, the shift of the methine signal must necessarily be larger for the isomers **I** and **III** in which the methine proton and the carbonyl group are closer to each other, at the same side of the dihydrofuran ring. If, therefore, our assumptions concerning the structures are correct, the methine signal must undergo a larger shift in Spectra **64**/*2* and **64**/*4*.

It is preferable to add the shift reagent gradually in little portions, in order that the changes in the spectrum could be monitored easily. It is probable that besides the carbonyl group other electron donors atoms (O, N) may also form coordination bonds with the shift reagent, which might seriously limit the value of the conclusions drawn from the shifts. For this reason, the shift reagent is added only in small quantities, since the probability (statistic weights in the equilibrium) of the above, undesired coordinations increase at higher molar ratios. We did not investigate, therefore, solutions with the usual molar ratio of 1:1, the highest molar ratio of the shift reagent and the sample being 1:3. Due to the peripheral site of the carbonyl oxygen, the axial symmetry (geometric independence) of the complex is very plausible.

It is important to investigate solutions of the same solute and shift reagent concentration. This condition can be ensured by adding the shift reagent to a solution of an approximately 1:1 mixture of isomers. Thus, the shifts belonging to the two isomers can be measured simultaneously from the same spectrum. If the concentrations of the two isomers are not exactly the same, the relative intensities of the displaced signals can be used advantageously to identify the corresponding lines, and the effect of the small difference in concentration can be neglected in the relative shifts.

Spectra **64**/*5b* and **64**/*5c* were taken from $CDCl_3$ solution of a mixture of isomers **I** and **II** containing about 1/5 and 1/3 mol of shift reagent, respectively. Spectrum **64**/*5a* is the spectrum of the pure $CDCl_3$ solution of the mixture. The mixture was richer in isomer **I**, therefore the lines corresponding to Spectrum **64**/*2* are relatively more intense. Note that in Spectra *b* and *c* the own absorption of the shift reagent (denoted by asterisk) has positive δ- value at low concentrations, but it has negative values, in overlap with the sharper TMS signal, at higher concentrations. In Spectra **64**/*5b* and *c* the methoxy signals of isomer **II**, which coincide in Spectrum **64**/*1*, already appear separately, proving their original accidental isochrony. The spectra of the solutions of the benzoyloxime isomers are **64**/*6a,b* and **64**/*7a,b*. The accurate shifts are listed separately in Table 105. The shift differences, obtained from the data of Tables 105 and 106, given in Table 107, definitely confirm the assumption concerning the structures, insofar as the methine signals of compounds **I** and **III** are shifted substantially more than those of isomers **II** and **IV**.

Table 107
SHIFT DIFFERENCES (PPM) OF THE SIGNALS RELATIVE TO THE SHIFTS MEASURED FOR THE $CDCl_3$ SOLUTIONS OF COMPOUND I to IV AFTER ADDITION OF 1/5 AND 1/3 MOL SR $Eu(DPM)_3$, RESPECTIVELY

Presumed structure	I		II		III		IV	
SR to sample molar ratio	1:5	1:3	1:5	1:3	1:5	1:3	1:5	1:3
Δδ(=CH)	0.5	1.0	0.2	0.45	0.95	1.2	0.2	0.75
ΔδOMe	0.2	0.4	0.1	0.3	0.4	0.5	0.15	0.5
Δδ(=CMe)	0.2	0.4	0.1	0.3	0.4	0.5	0.05	0.1
	0.3	0.5	0.2	0.4	0.25	0.35	0.05	0.25

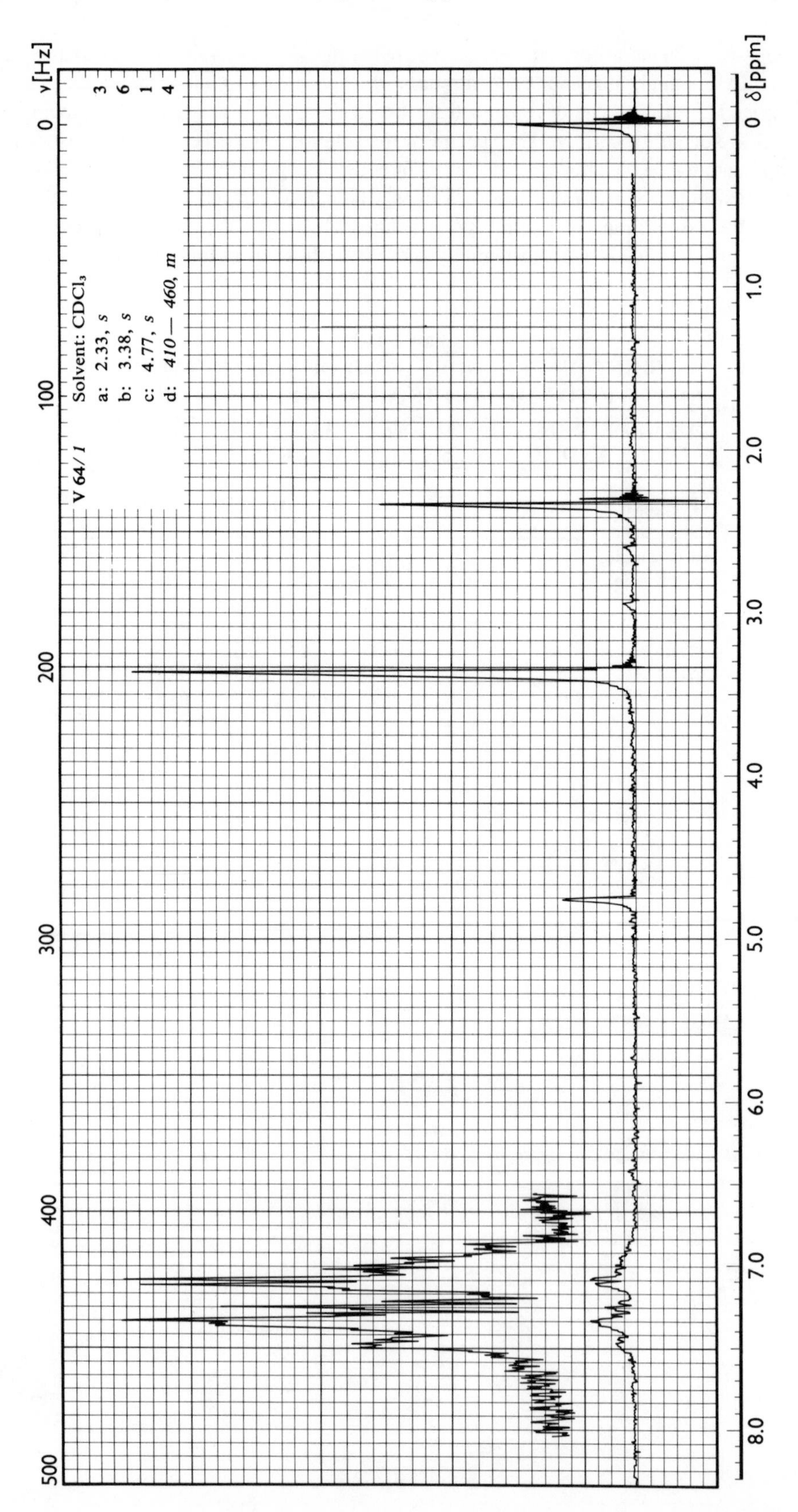
V 64/ 1
Solvent: $CDCl_3$
a: 2.33, s 3
b: 3.38, s 6
c: 4.77, s 1
d: 410— 460, m 4
ν[Hz]
0
100
200
300
400
500
δ[ppm]
0
1.0
2.0
3.0
4.0
5.0
6.0
7.0
8.0

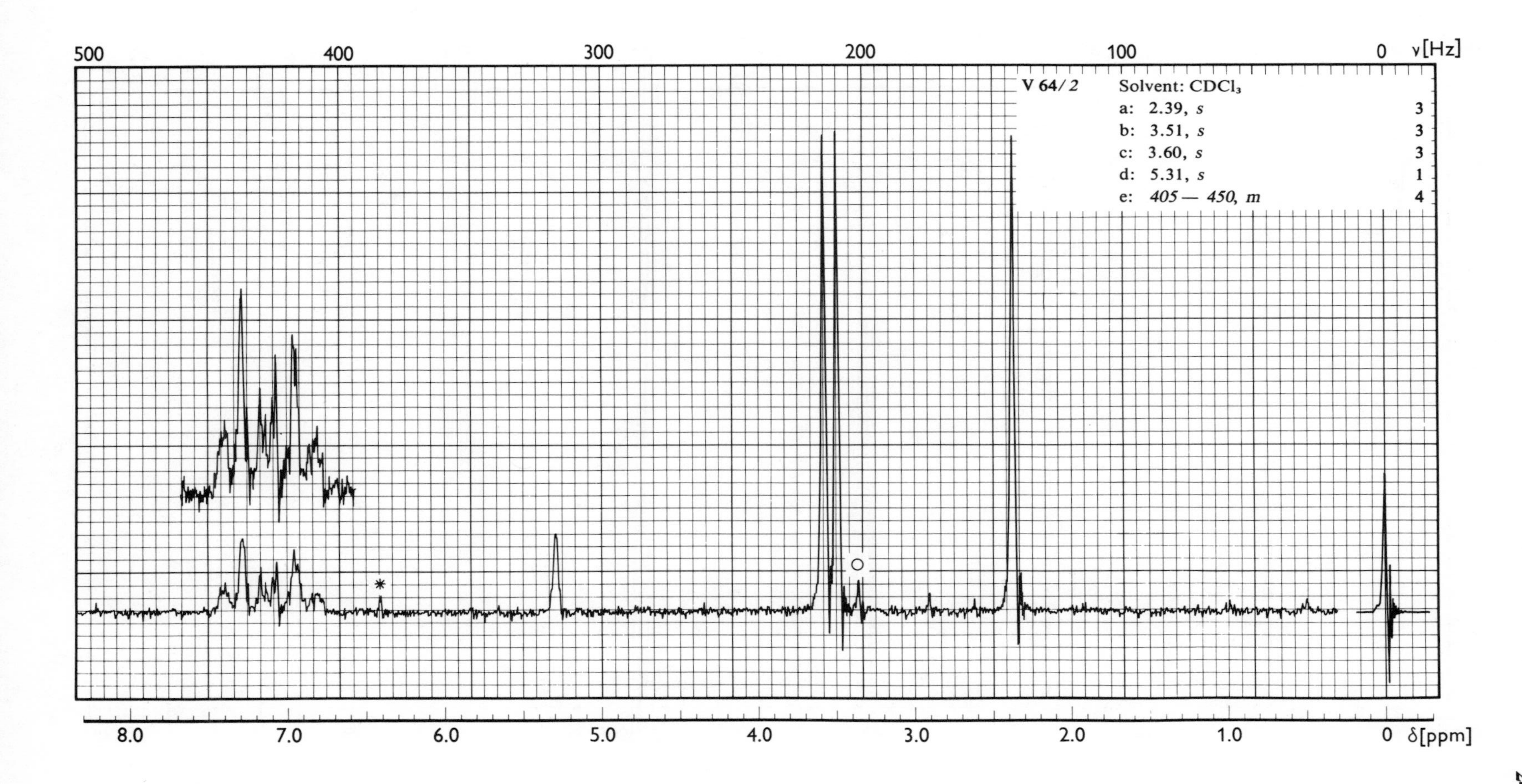
V 64/2
Solvent: $CDCl_3$
a: 2.39, *s* 3
b: 3.51, *s* 3
c: 3.60, *s* 3
d: 5.31, *s* 1
e: *405 — 450, m* 4
ν[Hz]
500
400
300
200
100
0
δ[ppm]
8.0
7.0
6.0
5.0
4.0
3.0
2.0
1.0
0

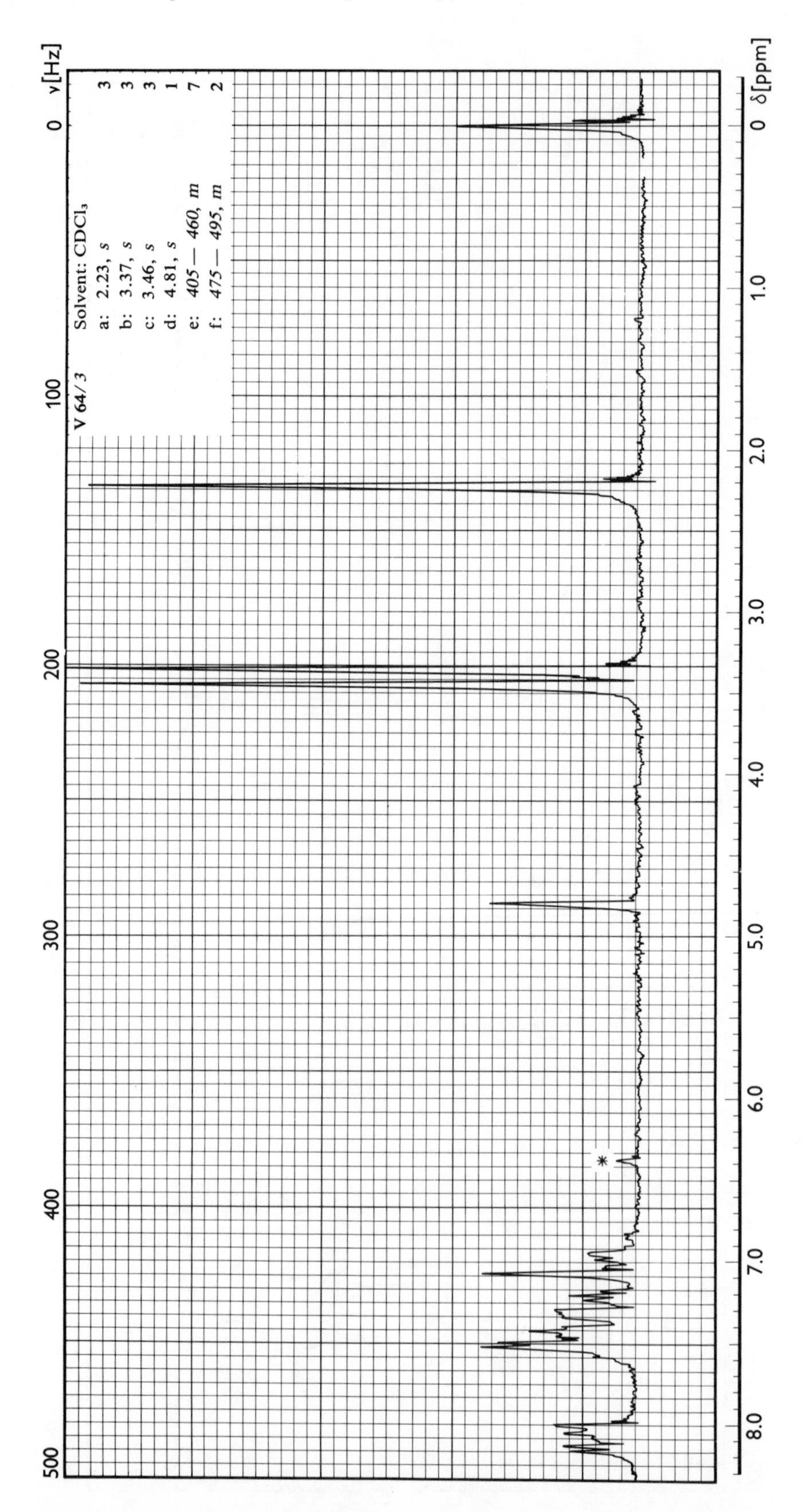
V 64/3
Solvent: CDCl₃
a: 2.23, s 3
b: 3.37, s 3
c: 3.46, s 3
d: 4.81, s 1
e: 405 — 460, m 7
f: 475 — 495, m 2
ν[Hz]
0
100
200
300
400
500
δ[ppm]
0
1.0
2.0
3.0
4.0
5.0
6.0
7.0
8.0
*

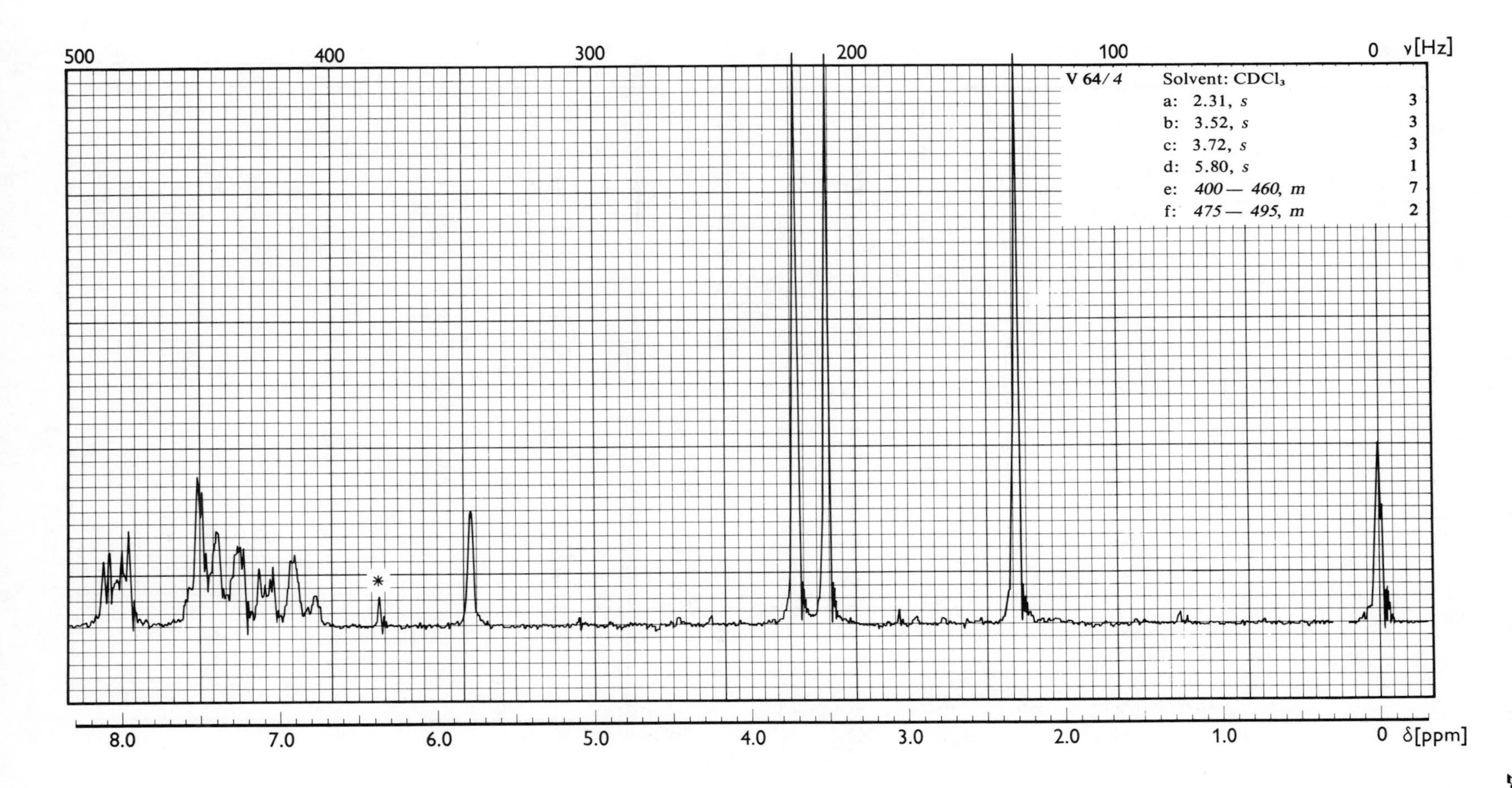

V 64/4
Solvent: $CDCl_3$
a: 2.31, *s* 3
b: 3.52, *s* 3
c: 3.72, *s* 3
d: 5.80, *s* 1
e: *400 — 460, m* 7
f: *475 — 495, m* 2
500 400 300 200 100 0 ν[Hz]
8.0 7.0 6.0 5.0 4.0 3.0 2.0 1.0 0 δ[ppm]
*

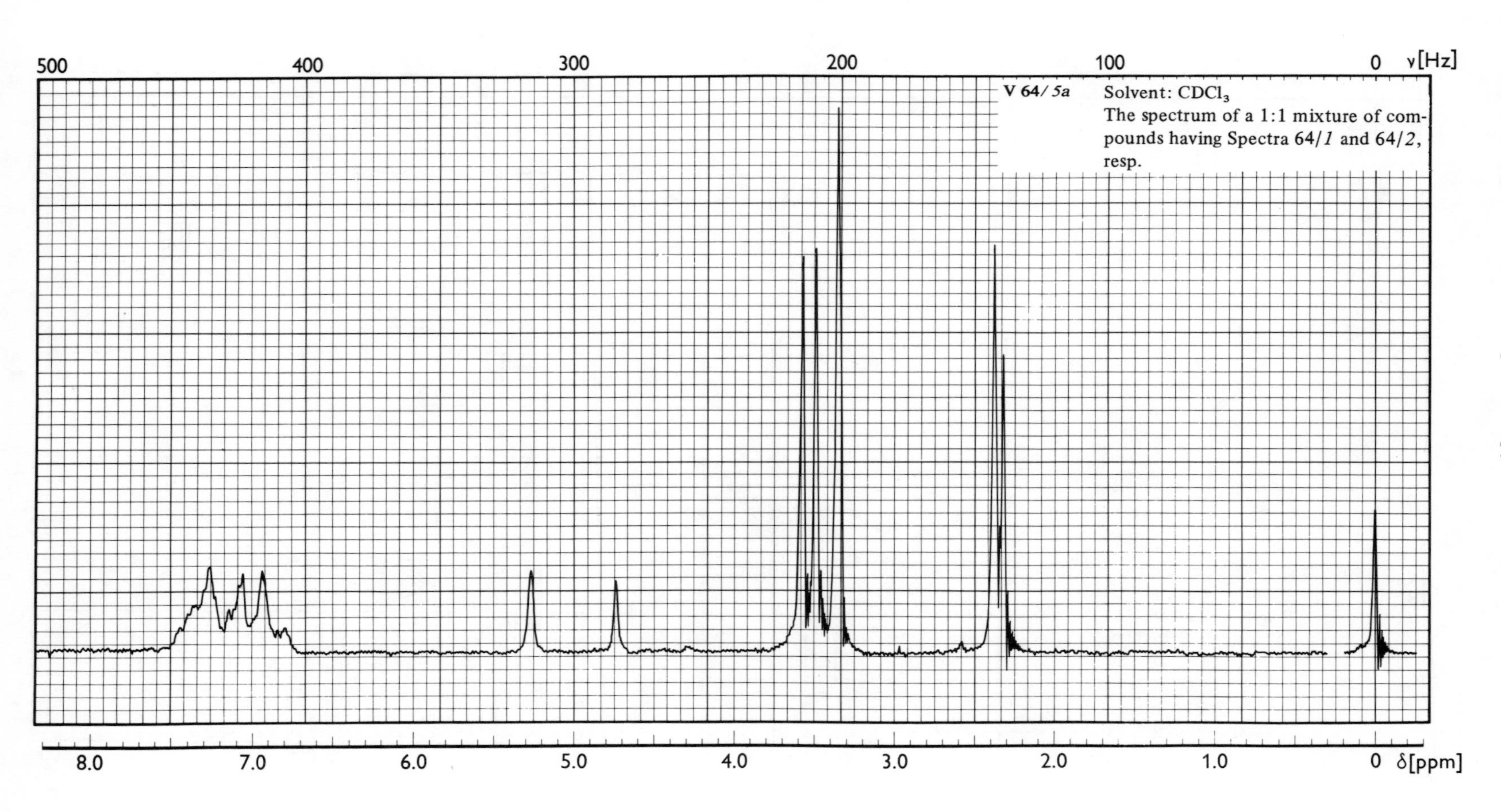
500
400
300
200
100
0 ν[Hz]
V 64/5a
Solvent: CDCl₃
The spectrum of a 1:1 mixture of compounds having Spectra 64/1 and 64/2, resp.
8.0
7.0
6.0
5.0
4.0
3.0
2.0
1.0
0 δ[ppm]

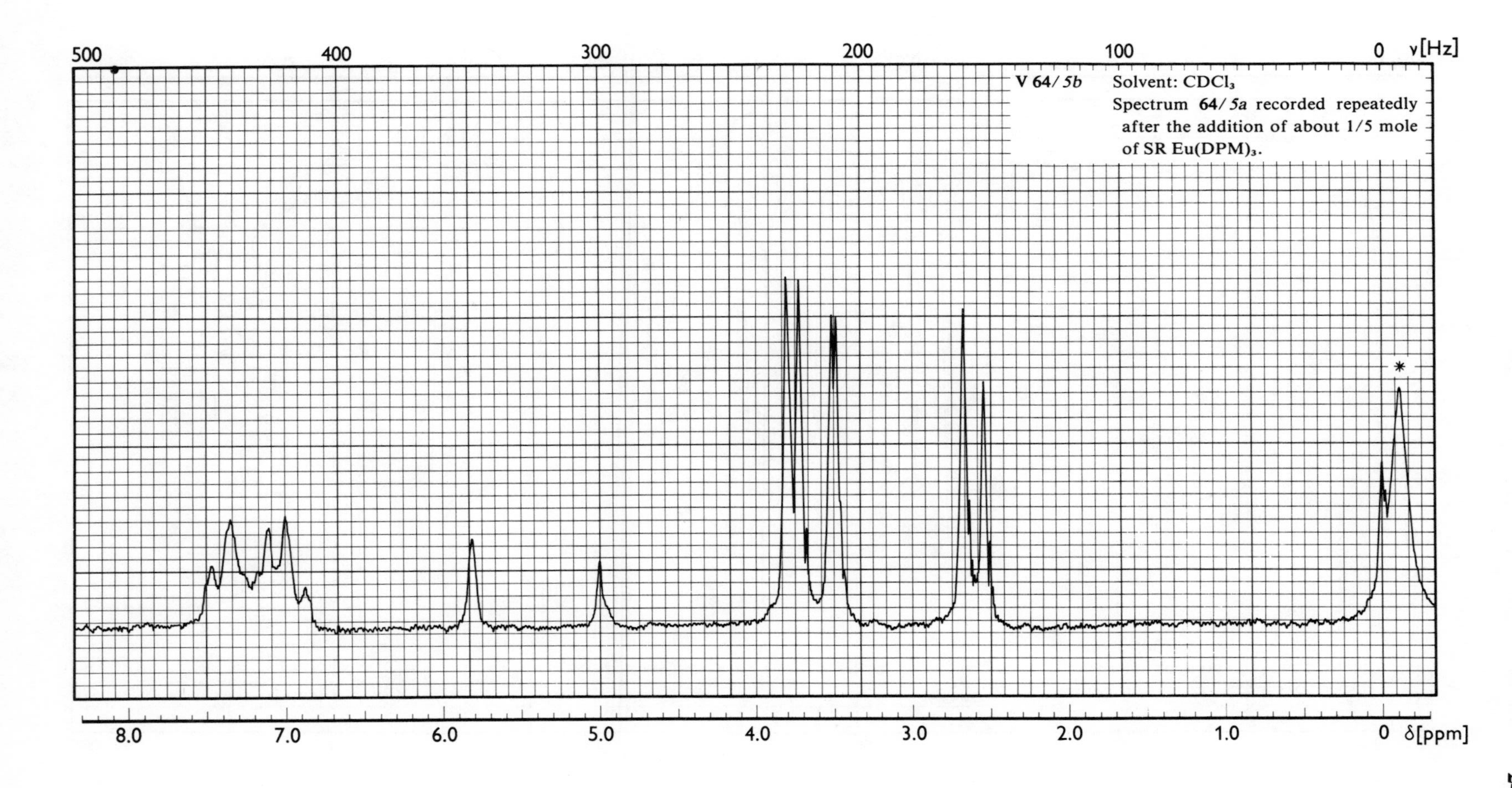

500
400
300
200
100
0
ν[Hz]
V 64/5b
Solvent: CDCl₃
Spectrum 64/5a recorded repeatedly after the addition of about 1/5 mole of SR Eu(DPM)₃.
*
8.0
7.0
6.0
5.0
4.0
3.0
2.0
1.0
0
δ[ppm]

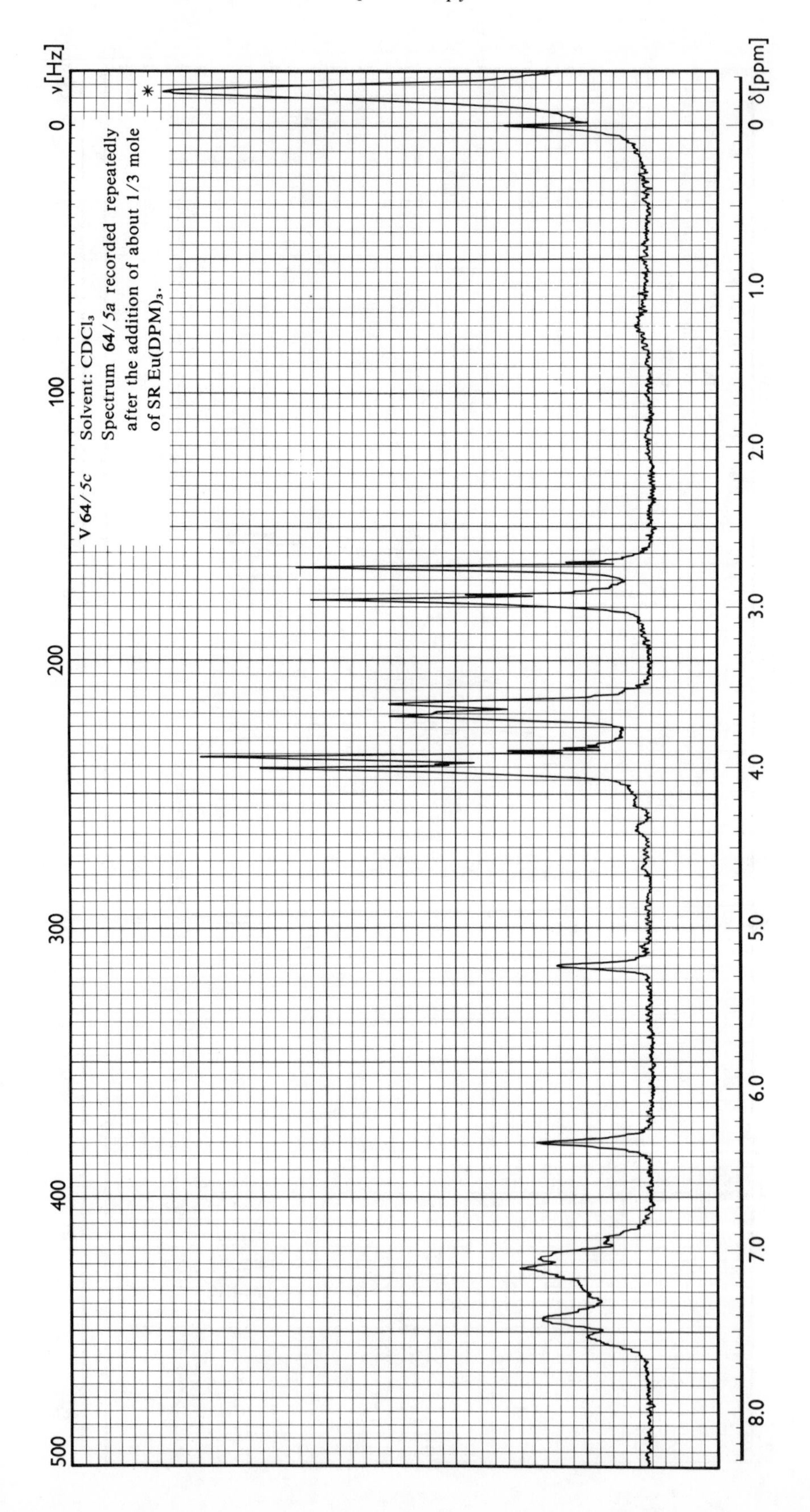
ν[Hz]
0
100
200
300
400
500
V 64/5c
Solvent: CDCl3
Spectrum 64/5a recorded repeatedly after the addition of about 1/3 mole of SR Eu(DPM)3.
*
δ[ppm]
0
1.0
2.0
3.0
4.0
5.0
6.0
7.0
8.0

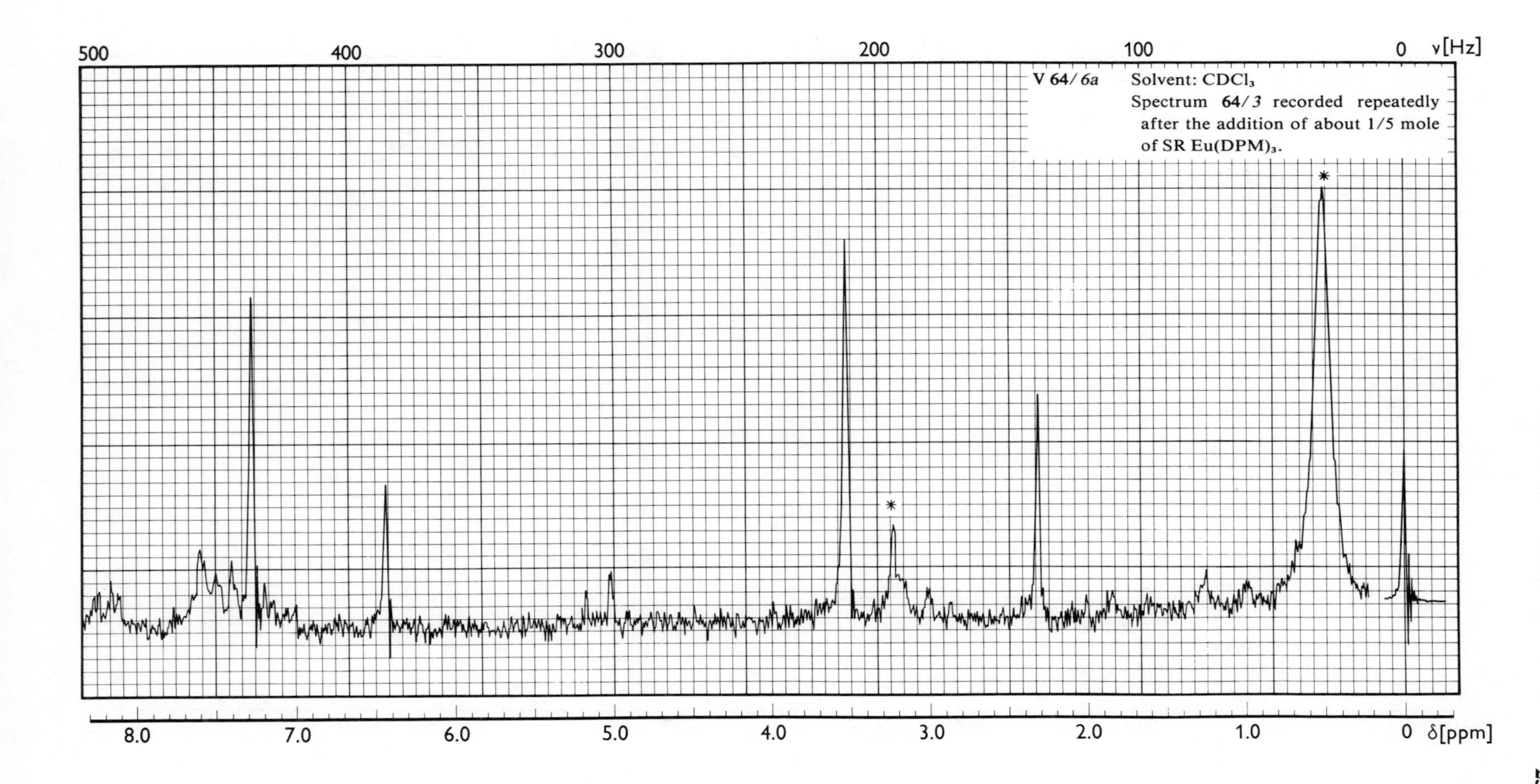
500
400
300
200
100
0
ν[Hz]
V 64/6a
Solvent: $CDCl_3$
Spectrum 64/3 recorded repeatedly after the addition of about 1/5 mole of SR $Eu(DPM)_3$.
*
*
8.0
7.0
6.0
5.0
4.0
3.0
2.0
1.0
0
δ[ppm]

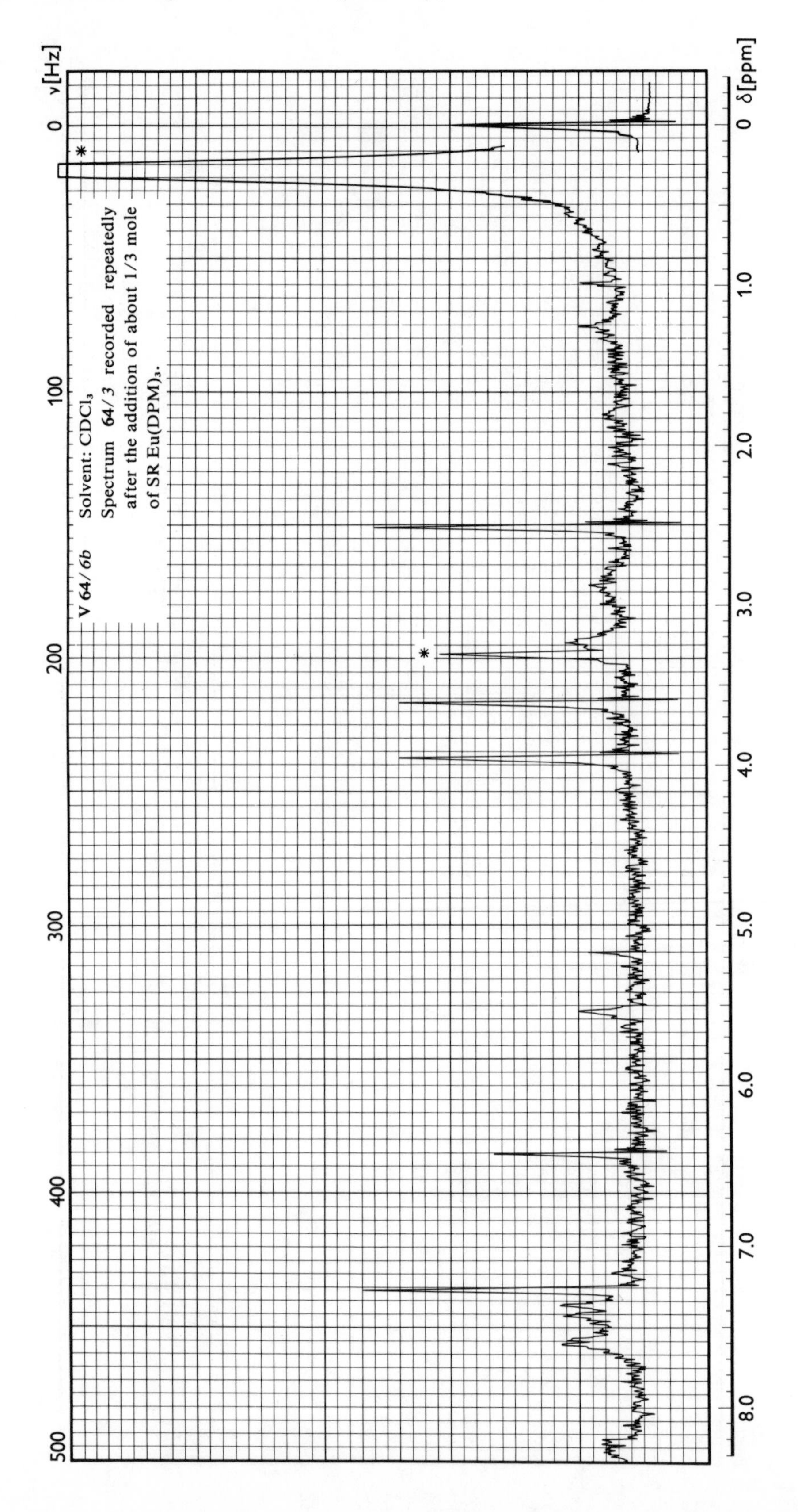
ν[Hz]
0
100
200
300
400
500
V 64/6b
Solvent: CDCl3
Spectrum 64/3 recorded repeatedly after the addition of about 1/3 mole of SR Eu(DPM)3.
*
*
δ[ppm]
0
1.0
2.0
3.0
4.0
5.0
6.0
7.0
8.0

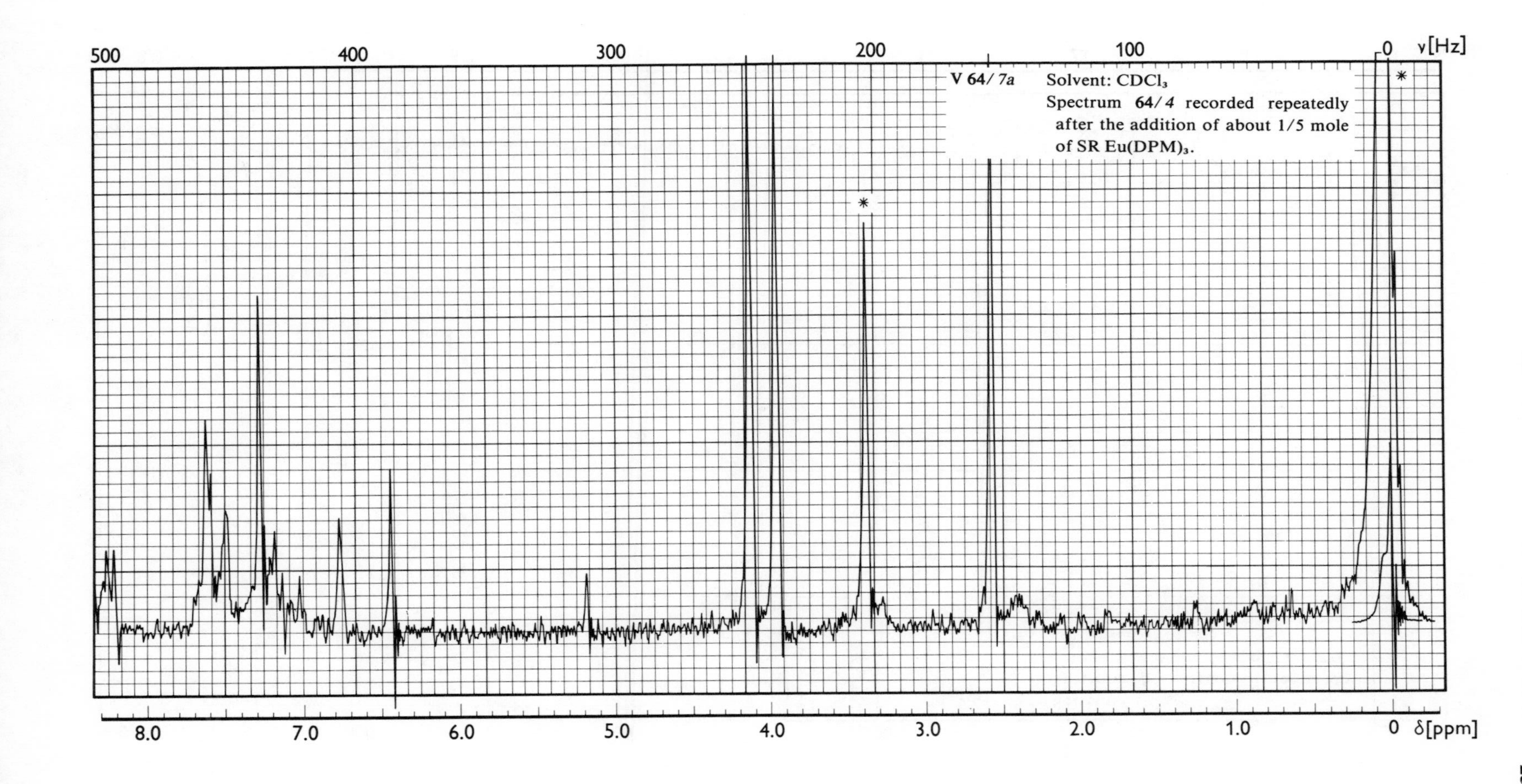
500
400
300
200
100
0
ν[Hz]
V 64/7a
Solvent: CDCl₃
Spectrum 64/4 recorded repeatedly after the addition of about 1/5 mole of SR Eu(DPM)₃.
*
*
8.0
7.0
6.0
5.0
4.0
3.0
2.0
1.0
0
δ[ppm]

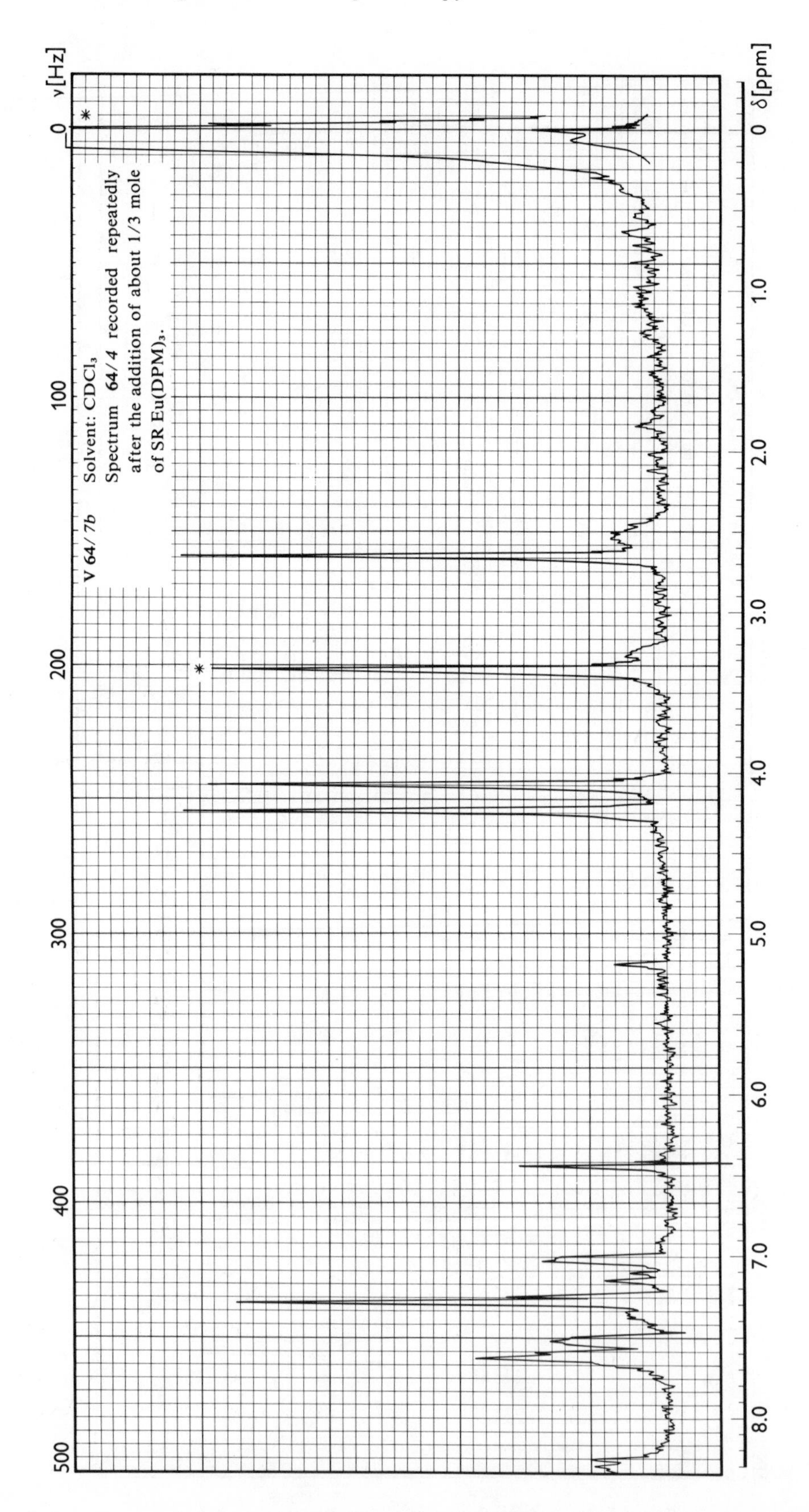
V 64/7b
Solvent: CDCl3
Spectrum 64/4 recorded repeatedly after the addition of about 1/3 mole of SR Eu(DPM)3.
ν[Hz]
0
100
200
300
400
500
δ[ppm]
0
1.0
2.0
3.0
4.0
5.0
6.0
7.0
8.0

PROBLEM 65[799]

The reaction of compound **I** with NaOMe is expected to yield compound **II** (R=H) or **III** (R=H). Is Spectrum **65**/*1* compatible with either of these structures? We may use in the assignment Spectrum **65**/*1a* measured after the addition of about 1/3 mol of $Eu(DPM)_3$ shift reagent and in determining the ring size the results of double resonance measurements (**65**/*1b* to *g*).

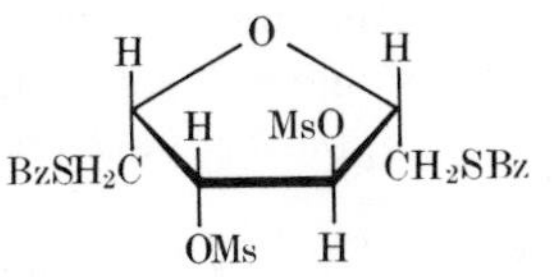

65/I 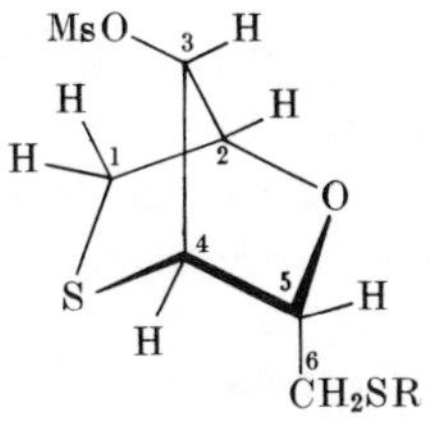 **65/II** 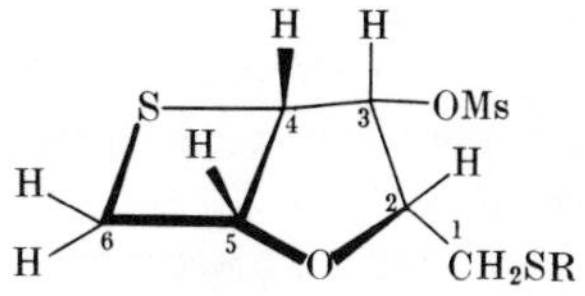 **65/III**

ANSWER

Spectrum **65**/*1*, consisting of broad and overlapping multiplets, indicates that the product may not be compound **II** or **III** with R=H. The multiplets between 435 and 490 Hz arises from aromatic hydrogens, and thus one or both of the *S*-benzoyl groups are still present in the molecule.

The two multiplets of 3:2 intensity in the regions of 435 to 460 and 470 to 490 Hz correspond to the *meta* and *para* protons, giving overlapped signals and to the *ortho* hydrogens, respectively. The paramagnetic shift of the signal of *ortho* protons can be attributed to the adjacent carbonyl group (see Table 39 and Figure 140) and is characteristic of benzoyl derivatives (compare Problem **64**). The presence of the mesyl groups is also certain from the corresponding methyl singlet at 3.15 ppm. It follows from a comparison of the intensity of the above signals (3 + 2 + 3) to the total intensity of the remaining ones of skeletal hydrogens that the molecule contains one benzoyl and one mesyl group, since the above ratio is 1:1 and the number of skeletal hydrogens is eight. Therefore, the sample is compound **II** or **III** with R = COPh, but the former structure is chemically more probable.

These structures are also compatible with the signals of the skeletal protons, as proved by the following assignment. Of the methylene signals, the upfield one arises from the group in the ring, since the other methylene group is adjacent to the *S*-benzoyl group which deshields it. Accordingly, the downfield 1-methylene signal is a doublet split by 6.5 Hz, whereas the other methylene signal is only broadened. This is easy to understand, since the coupling constants in saturated, heterocyclic compounds are usually small, whereas the *vicinal* coupling between hydrogens freely rotating about a C–C bond, is in a magnitude usual for saturated open chain compounds. Accordingly, δCH_2S (ring), *s* (line width approximately 2.5 Hz) $\approx$ 3.0 and δCH_2SR (chain), $d(J = 6.5$ Hz): 3.58 ppm.

The methine protons adjacent to the sulfur atom and mesyloxy group have triplet-like signals at 3.80 and 5.30 ppm. The splitting is about 2.0 Hz in both cases. This assignment follows from the triplet splitting of both signal and from the chemical shifts which is small for the former (adjacent sulfur) and large for the latter signal (adjacent acyloxy). The triplet structure is due obviously to two overlapping doublets, indicating a coupling with two adjacent hydrogens. In contrast, the methine signal of $CH\text{-}CH_2SR$ group would be split to a triplet by 6.5 Hz (as the signal of the neighboring methylene protons) and that of its pair (H-2 or H-5 in Structure **II** or **III**, respectively) would be unresolved (with a line width of ~5 Hz), and they would be closer to each other. Both signals might show a further doublet splitting owing to the $J_{2,3}$ and $J_{4,5}$ couplings, but they probably cause only line broadening. It is therefore obvious that the unresolved absorption between 260 and 280 Hz arises from H-2,5.

The correctness of the assignment is demonstrated well by the spectrum measured after addition of shift reagent (**65**/*1a*) in which the originally overlapping signals of methine hydrogens H-2 and H-5 become separated, and the predicted structures of the two multiplets appear in the spectrum. The upfield signal is a double triplet (J = 2 and 6.5 Hz). The other signal is a broad, unresolved maximum, with a half band width of approximately 7 Hz predicted above (2·2.5 + 2 = 7 Hz).

The ring size can be determined from the results of double resonance measurements. Irradiating the sample with frequencies of the signals of H-2 and H-5 in the case of Structure **II**, first the CH(OMs) signal and then the SCH-triplet should collapse into a doublet (first the $J_{2,3}$ interaction and then, irradiating the frequency of the H-5 signal, the $J_{4,5}$ interaction is removed).

The Spectrum **65**/*1a* of the solution containing shift reagent indicates that in the original spectrum δH-2 < δH-5 (the outer line pair of the double triplet upfield is separated well). The frequency differences of the four methine signals are illustrated schematically in Figure 211.

The decoupling unit of Varian A-60D operates by irradiating the frequency difference of the signals to be decoupled, and the changes in a line pair can be monitored by setting the instrument to one of the signals. If it is set to the downfield line (a "negative" frequency difference is irradiated), the upfield signal will change, and at positive frequency difference, the downfield signal will be simplified.

Thus, if the compound investigated has Structure **II** (R=COPh), both triplets can be expected to simplify upon the irradiation with the smaller frequency differences (the signal spacing is 41.5 and 43.5 Hz). On the other hand, with Structure **III** (R=COPh), the triplets can be expected to collapse into doublets applying larger frequency differences (46.5 and 48.5 Hz). Since saturating the overlapping signal of H-2,5 at a frequency difference of about 40 Hz with respect to either of the triplets, collapse into doublets (see the DR Spectra **65**/*1b* and *c*) and at frequency differences of about 45 or 50 Hz, there is no change (Spectra **65**/*1d* and *e*), the unknown is compound **II** (R=COPh). At a frequency difference of 90 Hz, the interaction $J_{3,4}$ can also be eliminated, and both triplets collapse into doublets (**65**/*1f* and *g*).

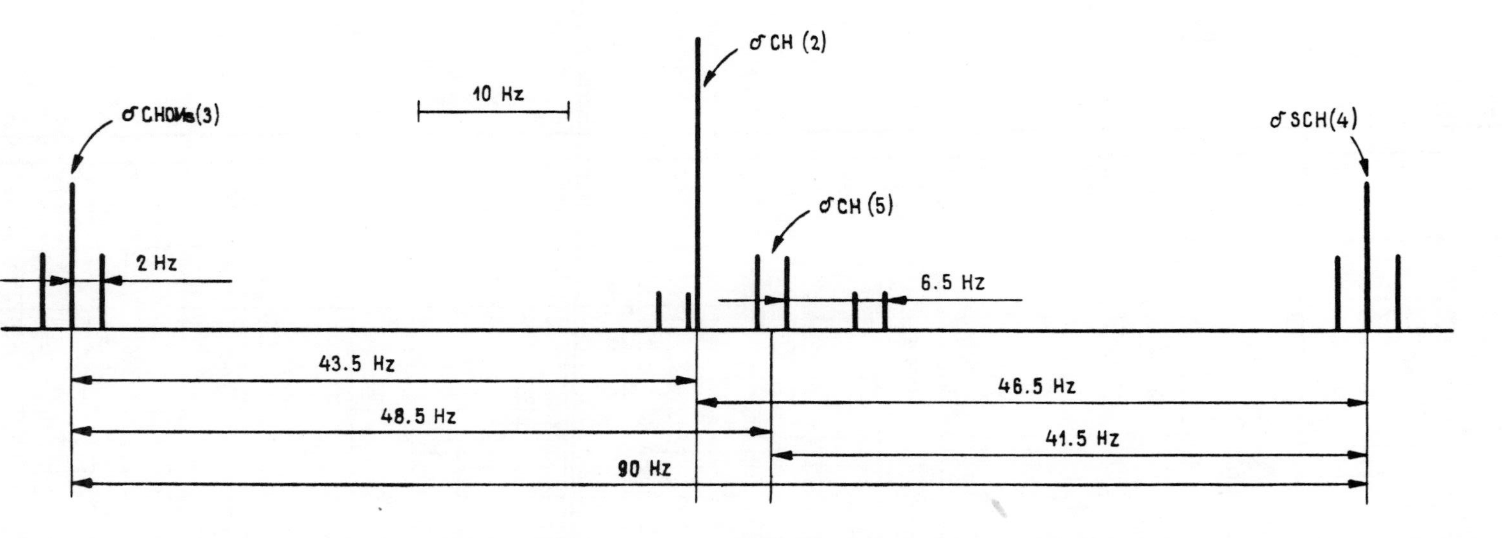

FIGURE 211. The schematic structure of H-2-5 signal system of compound **65/II**; for illustration of the decoupling measurements. Numbering of hydrogens is given corresponding to Structure **65/II**.

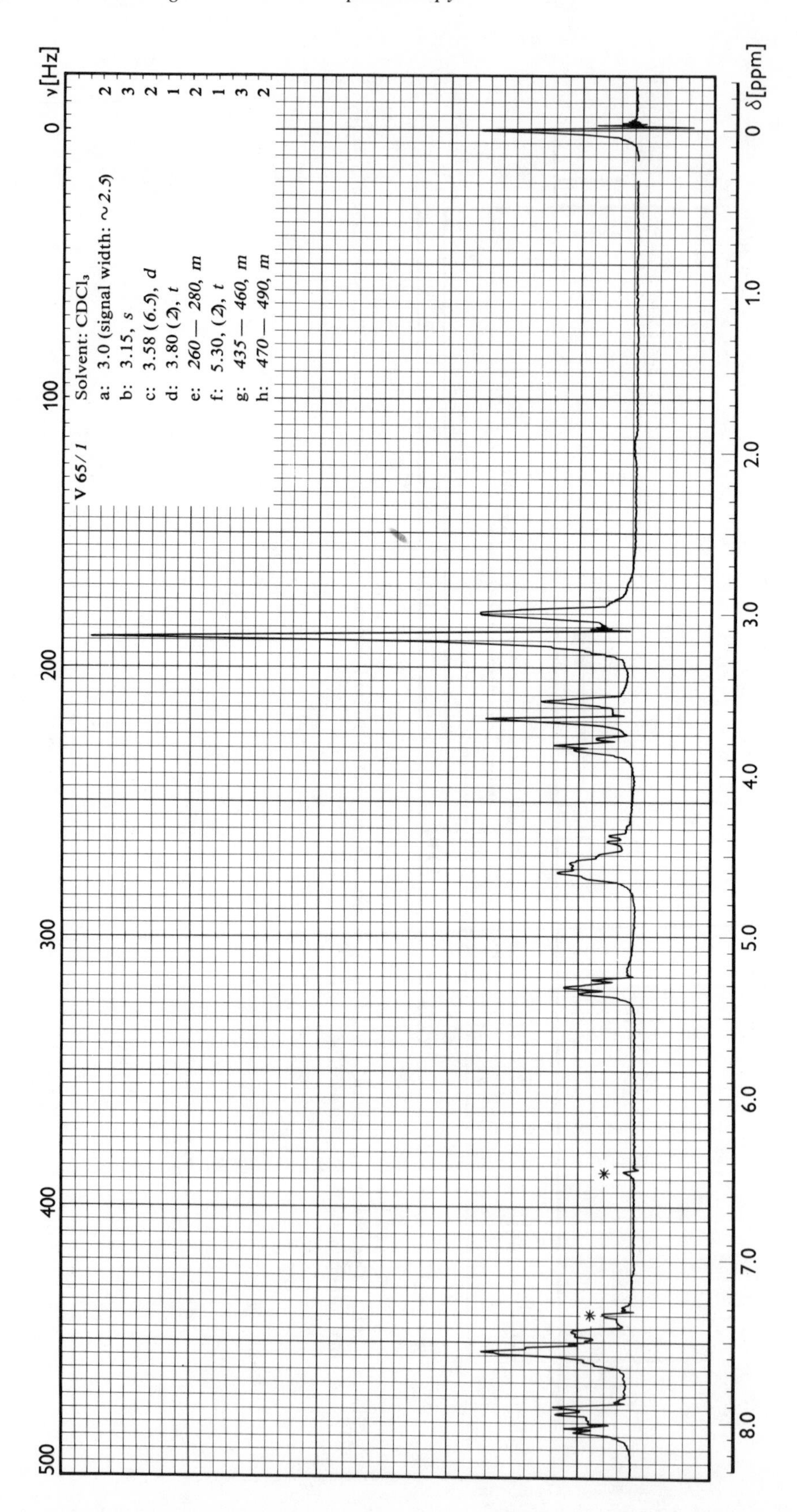

V 65/1
Solvent: $CDCl_3$
a: 3.0 (signal width: ~2.5) 2
b: 3.15, s 3
c: 3.58 (6.5), d 2
d: 3.80 (2), t 1
e: 260 — 280, m 2
f: 5.30, (2), t 1
g: 435 — 460, m 3
h: 470 — 490, m 2
ν[Hz]
0
100
200
300
400
500
δ[ppm]
0
1.0
2.0
3.0
4.0
5.0
6.0
7.0
8.0

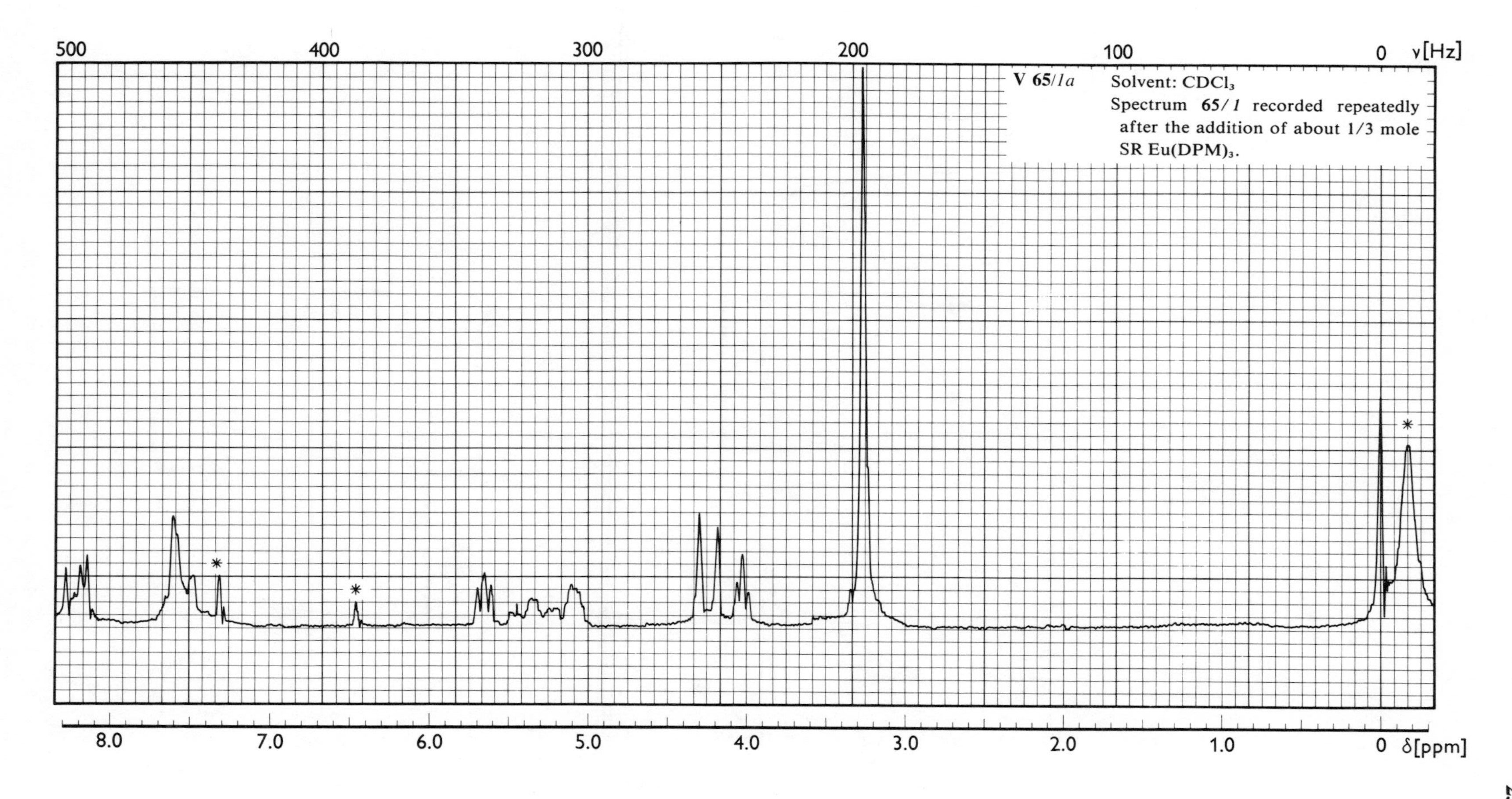

500
400
300
200
100
0
ν[Hz]
V 65/1a
Solvent: $CDCl_3$
Spectrum 65/1 recorded repeatedly after the addition of about 1/3 mole SR $Eu(DPM)_3$.
*
*
*
8.0
7.0
6.0
5.0
4.0
3.0
2.0
1.0
0
δ[ppm]

V 65/*1b-g* Solvent: $CDCl_3$
DR spectra of compound having Spectrum **65/*1*** in the 150-350 Hz region applying ν_2 frequencies of b) −40, c) + 40, d) −51, e) + 51, f) −90 and g) + 90 Hz.

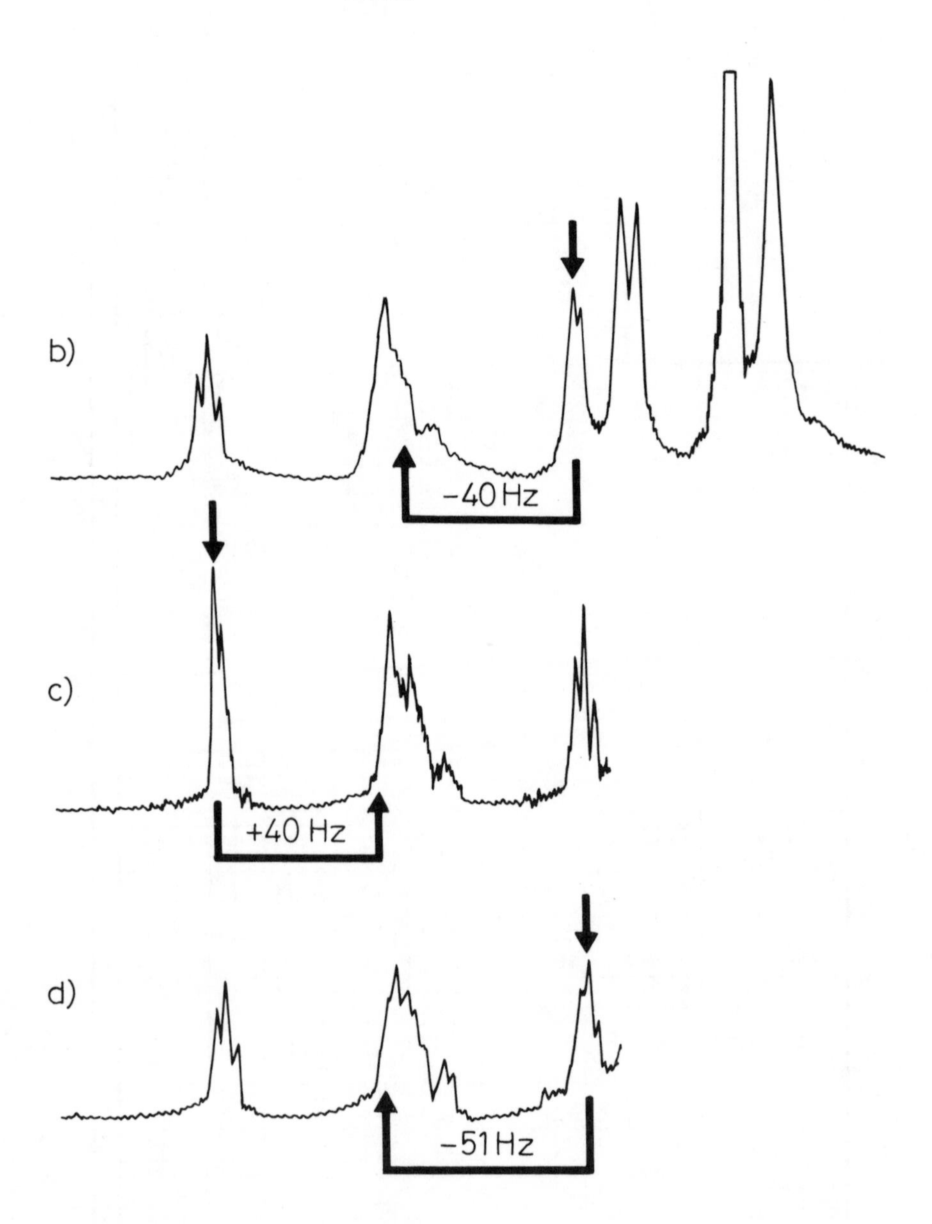

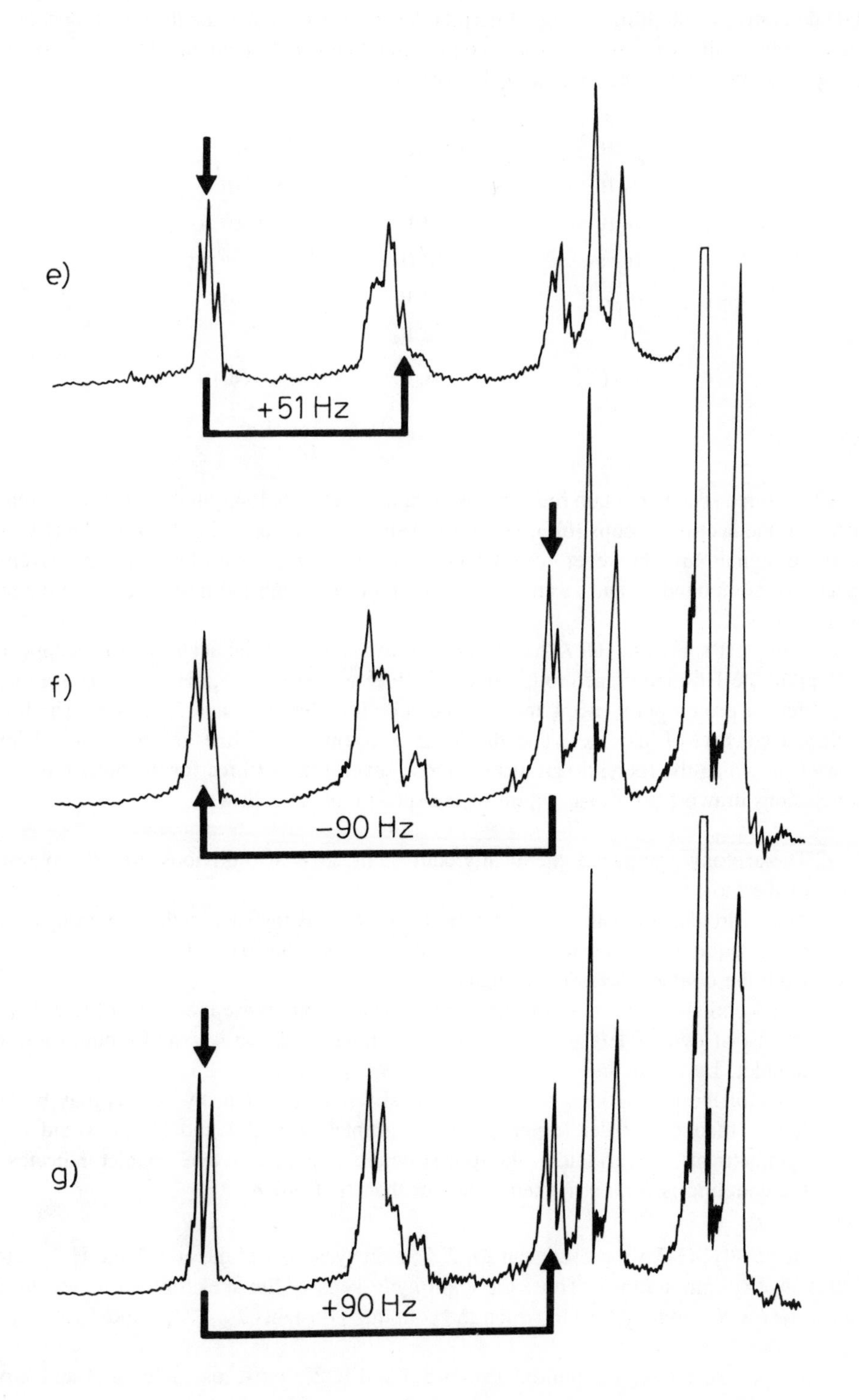
e)
+51 Hz
f)
−90 Hz
g)
+90 Hz

PROBLEM **66**[1320,1329]

Spectra **66**/*1* to *3* are the 90-MHz spectra of D-mannitol (**I**), L-iditol (**II**), and D-glucitol (**III**) derivatives with thiirane rings. Interpret the spectra and calculate the coupling constants and chemical shifts from the spectra of compounds **I** and **II**. Explain the differences between the spectra. Determine the most probable conformations.

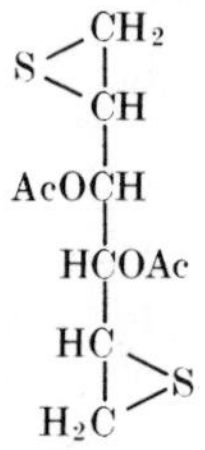

```
  CH2        H2C          H2C
S<|            |>S          |>S
  CH          HC           HC
  |            |            |
AcOCH       AcOCH        AcOCH
  |            |            |
 HCOAc       HCOAc        HCOAc
  |            |            |
 HC           CH           HC
  |>S       S<|             |>S
 H2C          CH2          H2C
```

66/I 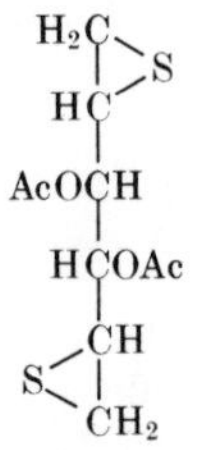 **66/II** 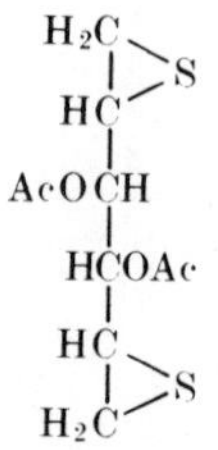 **66/III**

ANSWER

All the three spectra contain four groups of signals. They belong, in the order of decreasing fields, to the acetoxy groups (6H), H-1,6 (4H), H-2,5 (2H), and H-3,4 (2H). The structures of these signals are, however, characteristically different in the three spectra. Even the spectra of compounds **I** and **II** are widely different, although the molecules have the same symmetry (C_2).

The simplest spectrum (**66**/*1*) had **I**. The methyl singlet of the acetoxy groups appears at 2.18 ppm, the 1,6-methylene groups give a doublet at 2.51 (J = 5.5 Hz), and the symmetric multiplets of proton pairs H-3,4 and H-2,5 can be found at 4.75 and 3.01 ppm. The former multiplet consists of six lines, and the latter is composed of line groups arranged into a quartet, in which the four line groups comprise several lines (three for the outer ones). The conclusions drawn from the structure of the spectra are as follows:

1. The acetoxy groups are chemically equivalent, as is also obvious from the symmetry of the molecule.
2. The hydrogens of both methylene groups as well as the two methylene groups (1 and 6) are equivalent, and they interact only with H-2 and H-5, respectively, following from the doublet methylene signals.
3. H-2,5 and H-3,4 are magnetically nonequivalent (otherwise the signal of H-2,5 would consist of two triplets, i.e., of at most six lines, and H-3,4 would produce a double doublet, i.e., four lines).
4. None of the interactions between chemically nonequivalent hydrogens may be much higher than first order (otherwise the multiplet of H-2,5 and H-3,4 would not be symmetrical). Nevertheless, the roof-structure of the methylene doublet indicates that the coupling is, in a strict sense, higher than first order.

Consequently, H-2,5 represents an $AA'XX'$ spin system, and the H-1,2 and H-5,6 atoms represent AX_2 spin systems. Thus, the eight hydrogens of the skeleton can be regarded as an $AA'MM'X_2X'_2$ spin system in which the coupling constants $J_{XX'} \equiv J_{X'X}$ and $J_{AX} \equiv J_{A'X'}$ are zero, further $J_{MX} = J_{MX'}$.

Then, the structure of the multiplet of H-3,4 and H-2,5 must resemble the A and X parts of an $AA'XX'$ spectrum, in which, however, the lines of H-2,5 are split further by the $J_{1,2} \equiv J_{5,6}$ interactions.

The A or X part of an $AA'XX'$ spectrum consists of ten lines,* whereas in the multiplet of H-3,4 only six maxima are separated. It is clear therefore that our system is simpler than the most general $AA'XX'$ case. It is also plausible to assume that $J_{MM'} = 0$, i.e., the coupling between H-2 and H-5 is negligible, causing a rudimentary spectrum to appear. The interaction characterized by $J_{MM'}$ is a typical long-range coupling (^5J), which cannot result in significant splitting for aliphatic, saturated compounds. (Even a ^4J long-range coupling of type $J_{AM'}$ is quite unusual.) It can be shown in full analogy with the case discussed in Problem **63** (compare Table 104) that the 12 allowed transitions are pairwise degenerate, i.e., indeed 6 lines can be expected to appear in the AA' part of the spectrum belonging to H-3,4.

From the sextet, Q_1 can be obtained as the spacing of lines **1** and **3**, Q_4 is the spacing of **2** and **5** and Q_5 is that of lines **1** and **4**. Consequently, $Q_1 = 3.0$, $Q_4 = 7.5$, and $Q_5 = 8.25$ Hz. Since $Q_5 = \sqrt{Q_1^2 + Q_3^2}$, $Q_3 = 7.7$ Hz and the coupling constants are, including the already given values of J_{MX} and $J_{MM'}$:$J_{AA'} = Q_1 = 3$, $J_{AM} = (Q_3 + Q_4)/2 = 7.6$, $J_{AM'} = (Q_4 - Q_3)/2 = -0.1$, $J_{MX} = 5.5$ Hz, and $J_{MM'} = J_{AX} = 0$. Furthermore, $C^2 = 2Q_5/(A_1 + Q_1) \approx 1.4$ (the factor determining the intensity ratios), wherefrom the relative intensities of the lines are 0.6, 2.0, 1.4, 1.4, 2.0, and 0.6.

The multiplets constructed from these parameters are in good agreement with the experimental spectrum. The 6 lines of H-3,4 (see Figure 212a) and the 18 lines arising from H-2,5 (see Figure 212b), obtained from the above multiplet by splitting of the 5.5 Hz lines into triplets (intensity ratio 1:2:1), fit the observed signals well.

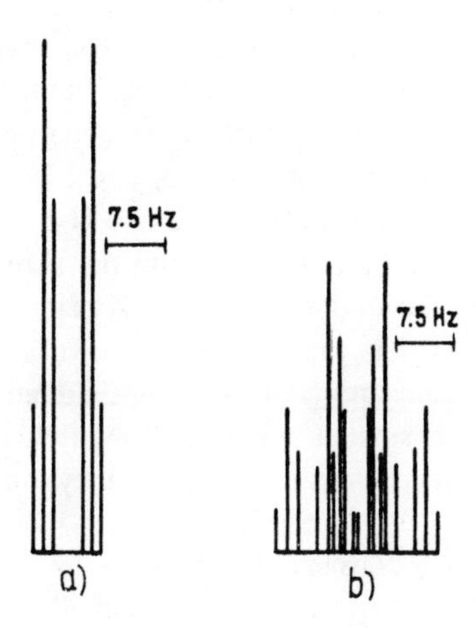

FIGURE 212. The schematic structure of the (a) H-3,4 and (b) H-2,5 multiplets of compound **66/I**.

Spectrum **66**/2 of compound **II** contains more lines, in spite of the unchanged molecular symmetry. The splitting of the signals of 1,6-methylene groups indicates that the methylene protons are nonequivalent. Their signal is a multiplet now, consisting of eight lines, with a midpoint of 148.8 Hz. Consequently, the spin-spin interactions are also influenced by the steric arrangement of the functional groups, i.e., by the configuration and conformation. The singlet of the acetoxymethyl groups appears at 2.17 ppm; the symmetric multiplets of

* See Volume I, p. 130.

H-2,5 and H-3,4, respectively, can be observed at 3.15 and 4.89 ppm. The latter multiplets are very similar to the corresponding signals of **I**. The following conclusions can be drawn:

1. The acetoxy groups are chemically equivalent again.
2. The methylene protons are nonequivalent (i.e., there is a preferred conformation of the molecule, in which the relative positions of the groups are different from those in **I**.
3. H-2,5 and H-3,4 are magnetically nonequivalent.
4. None of the spin-spin interactions between the chemically different hydrogens is substantially higher than first-order (H-1,1′,2 and H-5,6,6′ produce an *AMX* spectrum, see below).

The eight skeletal hydrogens represent, therefore, an $AA'MM'PZP'Z'$ spin system, in which $J_{MM'} = J_{AP} = J_{A'P'} = J_{AZ} = J_{A'Z'} = 0$. Again, the coupling constants can be determined from the multiplets arising from H-3,4, in a way described above. $Q_1 = 4.5$, $Q_4 = 7.5$, and $Q_5 = 9.0$ Hz. This yields $Q_3 = 7.8$, $J_{AA'} = 4.5$, $J_{AM} = 7.65$, and $J_{AM'} = -0.15$ Hz.

The part arising from 1,6-methylene groups consists of four doublets, in which the splitting of all line pairs is 2.0 Hz. This splitting can be attributed to only a coupling between the methylene hydrogens, because in thiirane derivatives the coupling of *vicinal* protons can usually be described only by larger coupling constants.* Of the four doublets, only the distances of the two adjacent line pairs are in the range characteristic of $J^c_{1,2}$ or $J^t_{1,2}$ interactions (e.g., the distance of the first and third doublet is 19 Hz, that of the second and fourth is 18 Hz, and that of the two middle doublets is still 12 Hz, but then the other coupling constant must be 24 Hz). Therefore, these splittings must correspond to J_{MP} and J_{MZ}. Since in thiiranes, $J^c > J^t$,* $J_{MP} = 6.75$, $J_{MZ} = 5.60$, and of course $J_{PZ} = 2.00$ Hz, if *P* labels the methylene protons that have *cis* positions with respect to H-2 and H-5, respectively.

The relative line intensities are in agreement with the above assignment, insofar as the lines of the doublets closer to the midpoint of the *PZ* octet are stronger, because the *P-Z* interaction is somewhat stronger than first order. The downfield doublet is stronger than the next, and the third line pair is again stronger than the upfield outer one. Finally, the difference in intensity within the doublets is smaller in the two downfield doublets, since the mentioned two effects compensate, in part, each other, whereas they are superimposed in the upfield line pairs.

The multiplet of H-2,5, which consists of 6 lines owing to the interaction with H-3 and H-4, is split further by the J_{MP} and J_{MZ} interactions, and all lines are replaced by 4, to yield a total of 24 lines. A part of these lines are, of course, unresolved. The theoretical spectrum constructed with the above parameters (see Figure 213) is in good agreement with the experimental curve. As mentioned above, the fact that compounds **I** and **II** have different spectra suggests the conclusion that the steric structure has a decisive role in the couplings and that there are preferred conformations for these compounds.

It is logical to assume that the bulky acetoxy groups tend to maintain a *trans* periplanar position for steric reasons. This is fully compatible with the small values of $J_{AA'}$ (3.0 and 4.5 Hz in the two compounds), indicating that the dihedral angle of the C–H bonds of H-3,4 is about 60°. The difference between the two compounds may be due to a difference in relative proportion of the preferred conformer for Structures **I** and **II**. The higher coupling constant in Structure **II** suggests that the mean dihedral angle is smaller, which is easy to interpret by a lowering in predominance of the *trans* conformation in the sterically more crowded molecule: the acetoxy groups and the thiirane ring are closer to each other in

* See Volume II, p. 16.

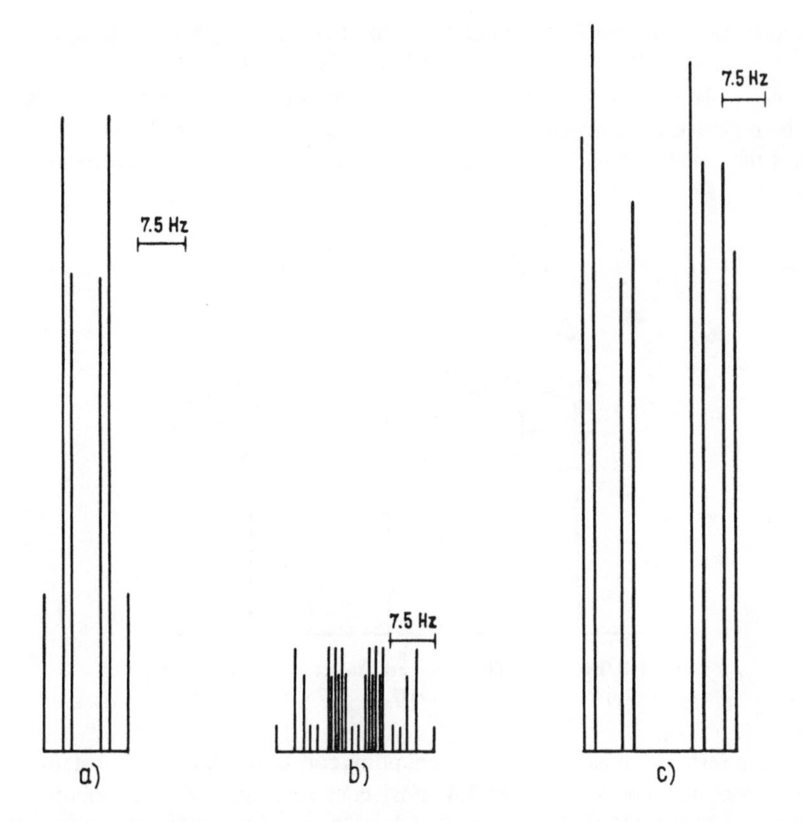

FIGURE 213. The schematic structure of the (a) H-3,4, (b) H-2,5, and (c) H-1,6 multiplets of compound **66/II**.

compound **II**. This is also the reason for the chemical nonequivalence of the methylene hydrogens. The *trans* protons of 1,6-methylene groups are closer to the ester carbonyl group, whereby their chemical shifts decrease: $\delta H_{Z,Z'} = 2.39$ and $\delta H_{P,P} = 2.58$ ppm ($\delta X = \delta X'$ /I/ = 2.51 ppm). A comparison of $\Delta\delta ZP = 0.2$ ppm, i.e., $\Delta\nu ZP = 18$ Hz with $J_{PZ} = 2$ Hz, indicates that their ratio is 0.11, and thus the *PZ* interaction is, to a very good approximation, a first-order coupling.

Since the coupling constants $J_{2,3} = J_{4,5}$ are large, J_{AM} (**I**) = 7.6 and J_{AM} (**II**) = 7.65 Hz; in the most probable conformation the dihedral angel between the C–H bonds of H-2 and H-3 is 180°, similar to that of H-4 and H-5. This fact, together with the already postulated *trans* position of the 3,4-acetoxy groups, leads to Structures **IV** and **V** as the predominant conformations of compounds **I** and **II**.

The molecule of D-glucitol (**III**) has no symmetry elements and thus gives a substantially

66/IV **66/V**

more complex Spectrum **66**/*3*. The signals of the two acetoxy groups appear separately at 2.20 and 2.23 ppm.

The signal of H-3,4 is a symmetric multiplet consisting of eight lines, and thus H-2 to H-5 can be regarded to represent an *ABKL* spin system with $J_{AK} = J_{BL}$ and $J_{AL} = J_{BK} = J_{LK} = 0$, if the symbol of hydrogens H-2 to H-5 is *K*, *A*, *B*, *L* in this sequence (see Figure 214).

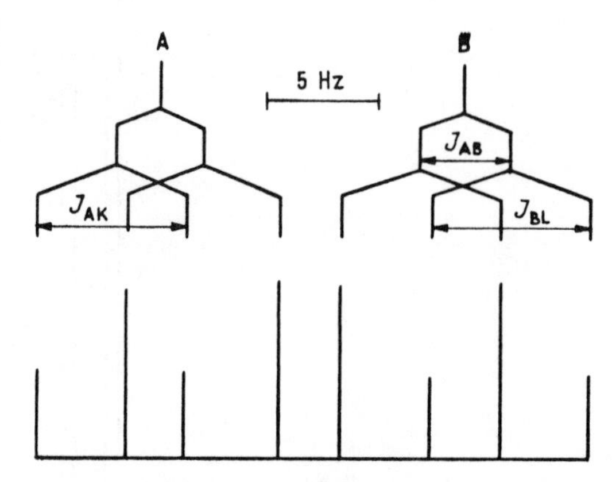

FIGURE 214. The schematic structure of the H-3,4 multiplet of compound **66/III**.

Under these restrictions concerning the coupling constants, the *ABKL* system gives rise to a simple spectrum. The signal of H-3,4 consists of eight lines, due to a doublet splitting of the four lines of the *AB* part by $J_{AK} \approx J_{BL}$ (the *AB* part of the *ABKL* spectrum). We have $J_{AB} = 4.5$ and $J_{AK} = J_{BL} = 7.5$ Hz and $\delta A = 4.98$ and $\delta B = 4.82$ ppm.

As follows from the magnitude of J_{AB}, the most probable conformer is the one containing the acetoxy group in *trans* position. The value of $J_{AK} = J_{BL}$ proves, on the other hand, that H-2,5 and H-3,4 are predominantly also in *trans* position, similar to the other two compounds. The structure of the dominant conformer of **III** is, therefore, **VI**, analogous to the conformations derived from compounds **I** and **II**. The exact analysis of the spectrum of **III** is rather complicated, however, for the conformational analysis this is, as we have seen, unnecessary.

66/VI

CNMR PROBLEMS

Problems **67** to **78** illustrate the use of ^{13}C NMR spectra in structure elucidation. In these problems the list of the chemical shifts of the lines is given with the spectra recorded by BBDR technique. The list also contains the multiplicities observable in the off-resonance or proton-coupled spectra. Integrals are, however, not given since there is no unique relationship between them and the number of equivalent carbons.

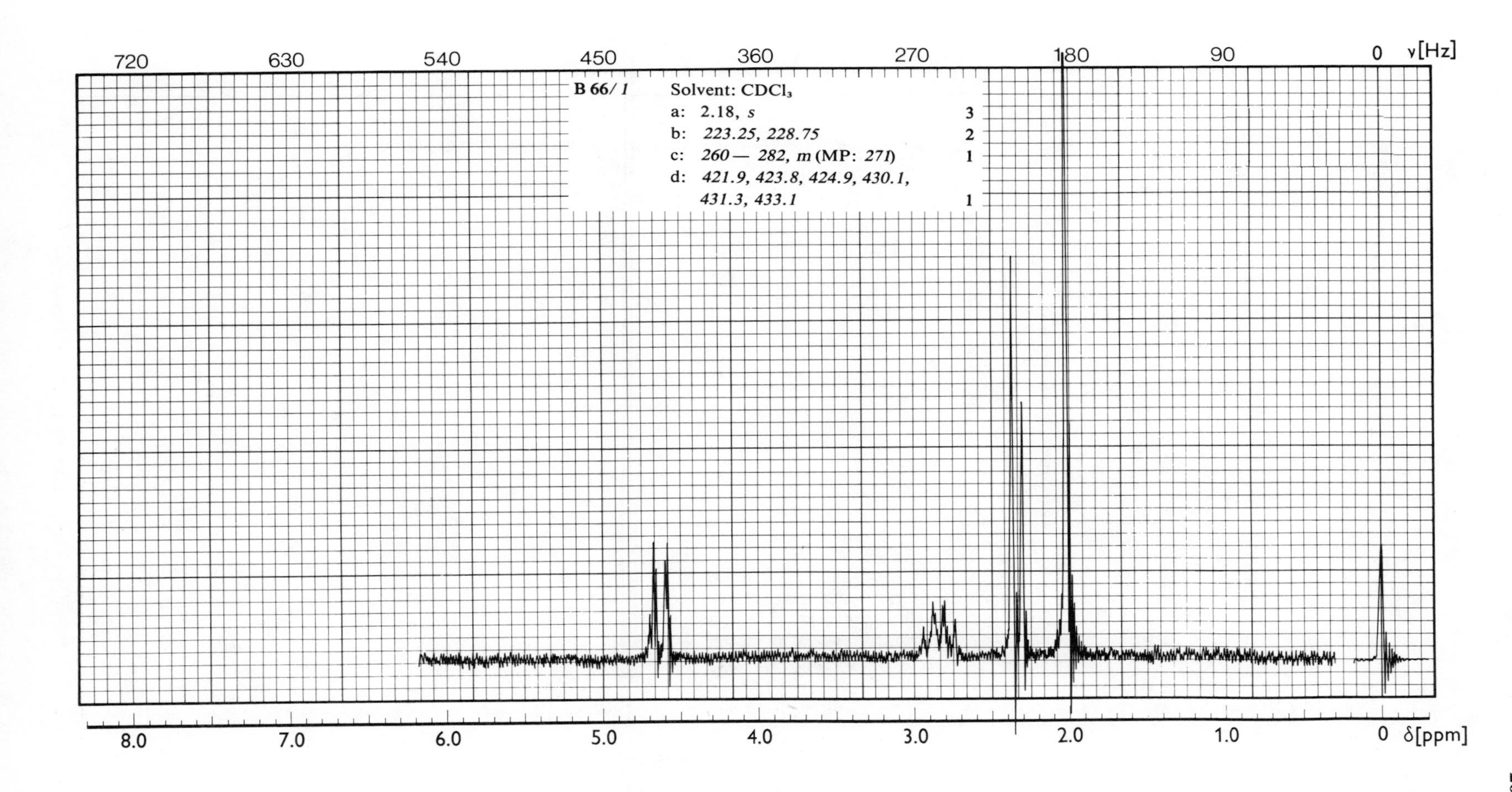

720
630
540
450
360
270
180
90
0
ν[Hz]
B 66/1
Solvent: $CDCl_3$
a: 2.18, *s* 3
b: 223.25, 228.75 2
c: 260 — 282, *m* (MP: 271) 1
d: 421.9, 423.8, 424.9, 430.1, 431.3, 433.1 1
8.0
7.0
6.0
5.0
4.0
3.0
2.0
1.0
0
δ[ppm]

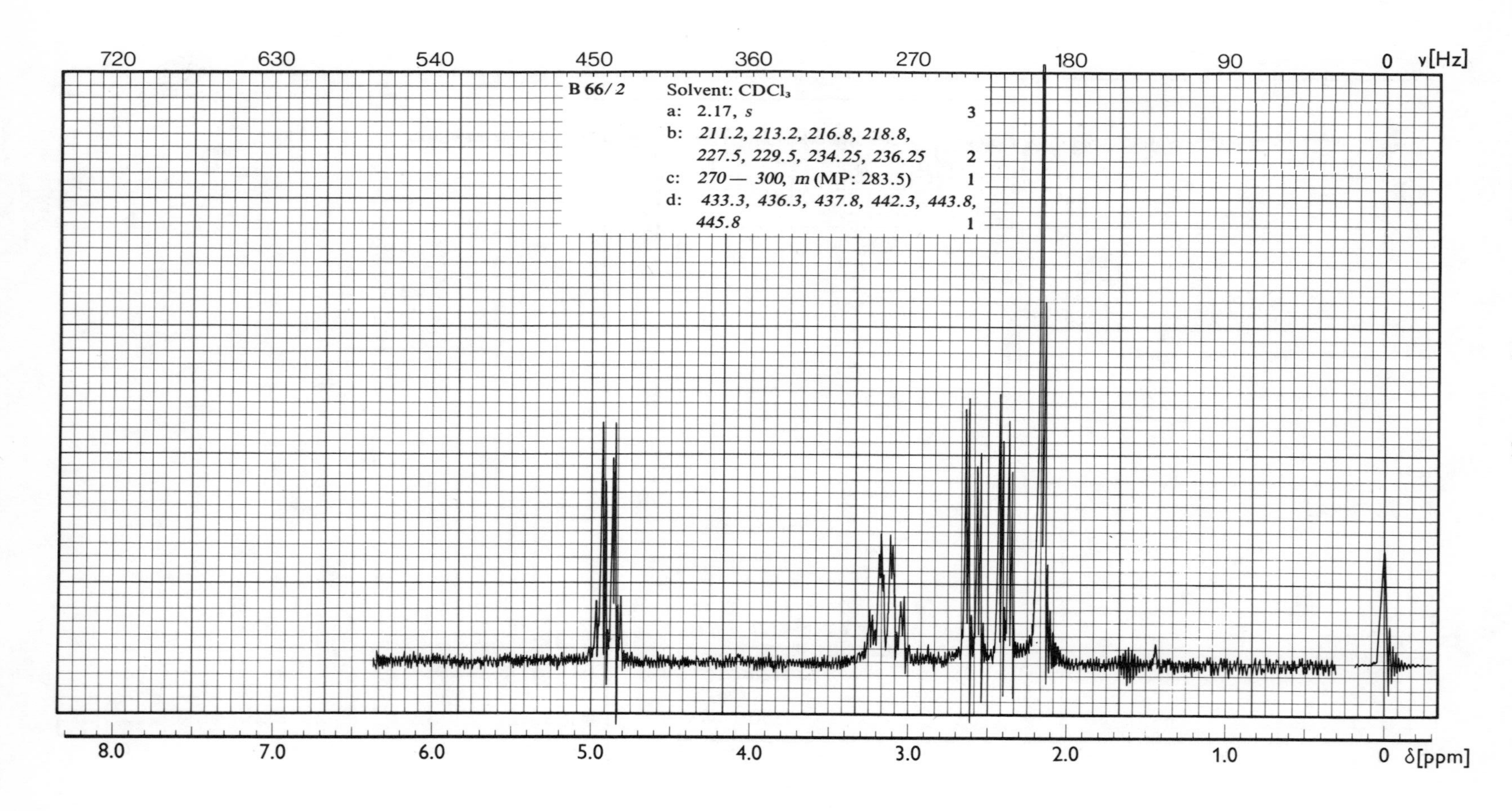
720
630
540
450
360
270
180
90
0 ν[Hz]
B 66/2
Solvent: CDCl₃
a: 2.17, s 3
b: 211.2, 213.2, 216.8, 218.8, 227.5, 229.5, 234.25, 236.25 2
c: 270 — 300, m (MP: 283.5) 1
d: 433.3, 436.3, 437.8, 442.3, 443.8, 445.8 1
8.0
7.0
6.0
5.0
4.0
3.0
2.0
1.0
0 δ[ppm]

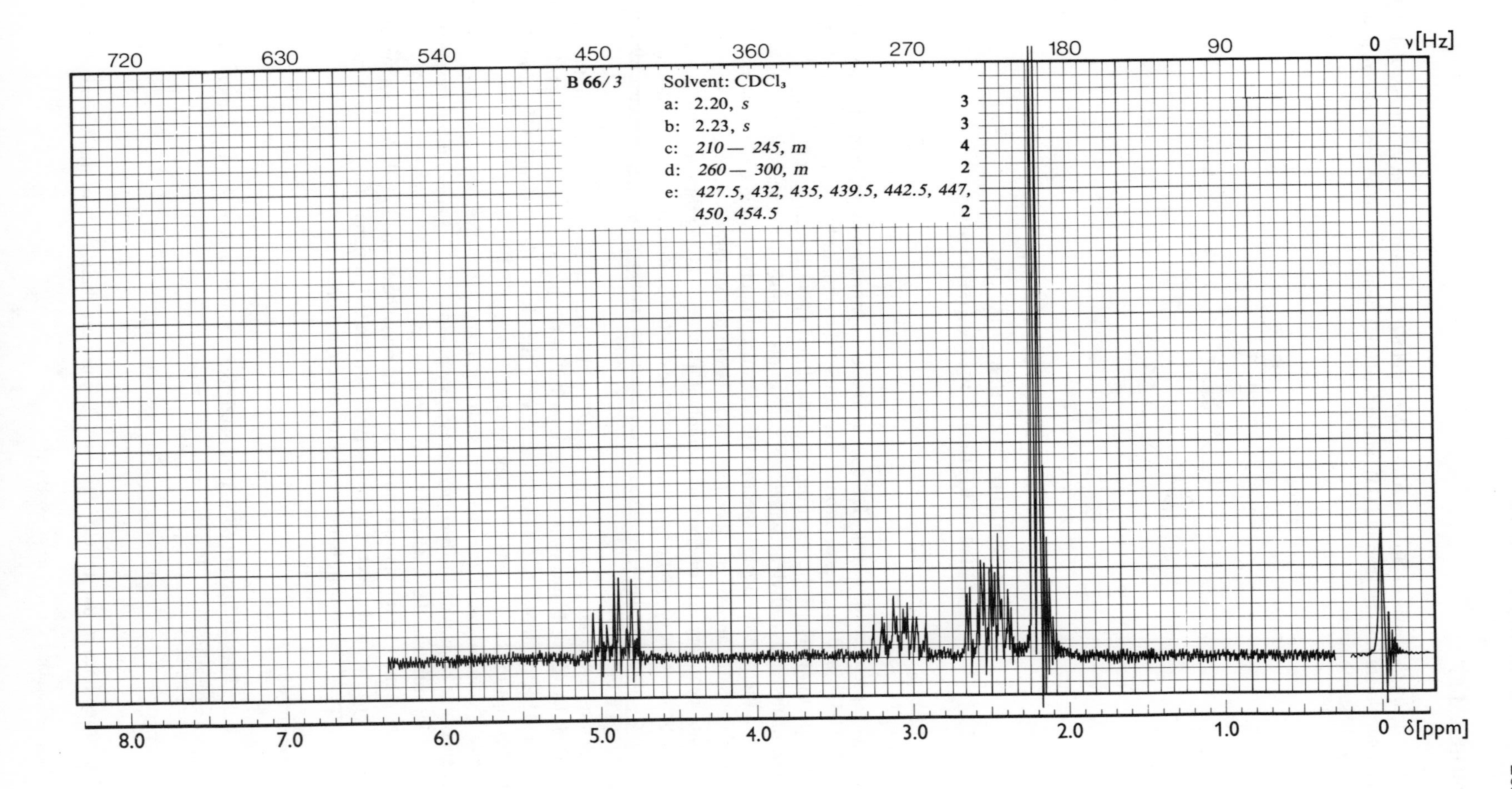

720
630
540
450
360
270
180
90
0 ν[Hz]
B 66/ 3
Solvent: CDCl₃
a: 2.20, s 3
b: 2.23, s 3
c: 210 — 245, m 4
d: 260 — 300, m 2
e: 427.5, 432, 435, 439.5, 442.5, 447, 450, 454.5 2
8.0
7.0
6.0
5.0
4.0
3.0
2.0
1.0
0 δ[ppm]

PROBLEM **67**[1336]

Decide on the basis of Spectra **67**/*1*, **67**/*2*, and **67**/*3*, obtained from 2,5-dimesyl-1,4:3,6-dianhydro-mannitol (**I**), -sorbitol (**II**), and -iditol (**III**), which of the spectra belongs to which of the diastereomers.

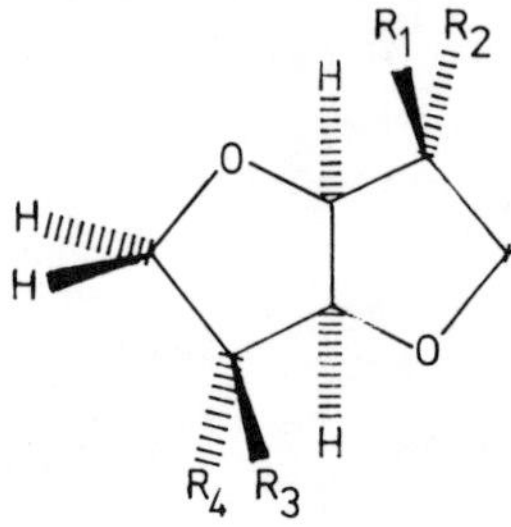

67/I (R_1 = R_3 : OMs; R_2 = R_4 : H)

67/II (R_2 = R_4 : H; R_2 = R_3 : OMs)

67/III (R_1 = R_3 : H; R_2 = R_4 : OMs) (Ms : $-SO_2CH_3$)

ANSWER

In the spectrum of **I** and **III**, due to the C_2 symmetry of these molecules, C-1,6, C-2,5, C-3,4, and the mesyl carbons give coincident signals, hence only four lines appear in the spectrum. Thus, the spectrum of sorbitol derivative **II** may be **67**/*1* only.

The methyl and methylene signals are easy to distinguish on the basis of the multiplicities read from the proton-coupled spectra. Of the equally doublet C-2,5 and C-3,4 signals, the shift of the latter are larger, since they are adjacent to two tertiary carbons, whereas C-2,5 have one secondary and one tertiary neighbor. However, taking into account that with the latter the ether-type oxygen is replaced by a more electronegative ester oxygen, the alternative assignment may be discarded only on the basis of further data (e.g., Spectra **69**/*1* to *3*; compare Problem **69**).

Structures **I** and **III** may be distinguished on the basis of the field effect.* In the case of mannitol derivative **I**, owing to the steric effect of the mesyl groups in *endo* position, the C-2,5 and C-3,4 atoms are deshielded relative to the iditol (*diexo*) isomer **III**. Accordingly, **67**/*3* belongs to isomer **I** and **67**/2 belongs to the isomer **III**.

From a comparison of the chemical shifts of **I** and **III** to the data of sorbitol (**II**), the signals of the rings with *exo* and *endo* mesyloxy groups (C-1,2,3 and C-4,5,6, respectively) may be identified in the spectrum of the latter.

The assignment of the spectra shown on the formulas (Scheme 1) follows from the above. Observe that the field effect is well observable on the C-1,6 signal, too, whereas the methyl group of mesyl substituent is insensitive of configuration.

PROBLEM **68**[1349]

Assign the lines of Spectra **68**/*1*, **68**/*2*, and **68**/*3*. On the basis of this assignment correlate the spectra to Structures **I** to **III**.

ANSWER

The three lines in the upfield region (<70 ppm) are the signals of the *N*-methyl and the

* Compare Section 4.1.1.3, point 4, pp. 154-155.

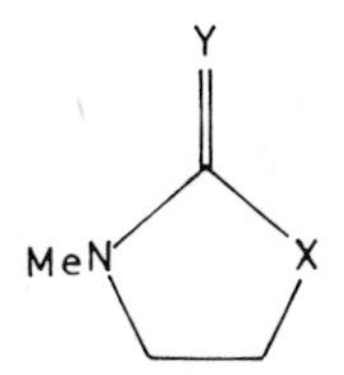

68/I (X : O; Y : NCOPh)
68/II (X : S; Y : NCOPh)
68/III (X : NCOPh; Y : S)

two methylene carbons. The signal of the former is easily selected on the basis of its multiplicity, chemical shift, or intensity. The quartet structure of this signal in the proton-coupled spectrum proves the assignment in itself. Of the two other upfield lines in Spectrum **68**/*1*, the chemical shift of one is too small and the other is too large for an *N*-methyl carbon signal* (similar to the shifts of both lines in **68**/*2* and **68**/*3*). The lower intensity as compared to the methylene signal is due to the free rotation of methyl group, which involves longer T_1 relaxation time and thus partial saturation.**

The two other lines are triplets in all the three proton-coupled spectra corresponding therefore to methylene carbons. The upfield signal in Spectrum **68**/*1* (at 26.6 ppm) corresponds obviously to the carbon of *S*-methylene group (Scheme 2), i.e., this spectrum can be assigned to compound **II**. On the other hand, the strongly deshielded line of Spectrum **68**/*3* corresponds to the *O*-methylene group of **I**. Accordingly, the spectrum of compound **III** is **68**/*2*. The *N*-methylene carbon adjacent to the NMe group is more deshielded (due to the β-effect*** of the methyl group), but the alternative assignment may not be ruled out, either. It can be observed that the shifts of the methylene carbon adjacent to the *N*-methyl group are quite similar in the three spectra (47.0, 51.1, and 48.1 ppm) and that the methylene chemical shifts change in parallel to the electronegativity of the adjacent heteroatom†: 26.6, 45 to 51, and 65.6 ppm, respectively, for the SCH_2, NCH_2, and OCH_2 groups.

The next four lines can be assigned to the benzoyl group. The two strong lines (doublets in the coupled spectrum) are the signals of the *meta* and *ortho* carbons. Although the opposite assignment is also possible, the $\delta C^o > \delta C^m$ relationship is logical due to the electrophylic substituent of the ring. The third, weaker, and (in the coupled spectrum) doublet signal arises from the *para* carbon, both on the basis of its intensity and its downfield position. The most deshielded weak singlet corresponds to the quaternary aromatic carbon.

The two downfield signals belong to the unsaturated carbon between the heteroatoms and to the carbonyl carbon. The former line is weak, as is generally characteristic of quaternary carbons,‡ and its shift, similar to that of the *N*-methyl group, is slightly different in structure **I**, which may also be conceived as urea-type, whereas the same in Structures **II** and **III**, which may be regarded equally as thiourea derivatives. In contrast, the shift of carbonyl carbon is more different for compound **III**: in the case of **I** and **II** the carbonyl group is attached to an imino groups, while in compound **III** it is attached to a diacylamino nitrogen with lower electron density. Accordingly, the carbonyl carbon is more deshielded in Spectrum **68**/*2* of compound **III**. The full assignment is given in Scheme 2.

** Compare Table 50.

** See Volume II, p. 225 and 231.

***See Volume II, p. 153, and Table 52.

† Compare Section 4.1.1.3, point 3, pp. 152-154, and Table 51.

‡ Compare Section 4.1.4.1, pp. 220-221.

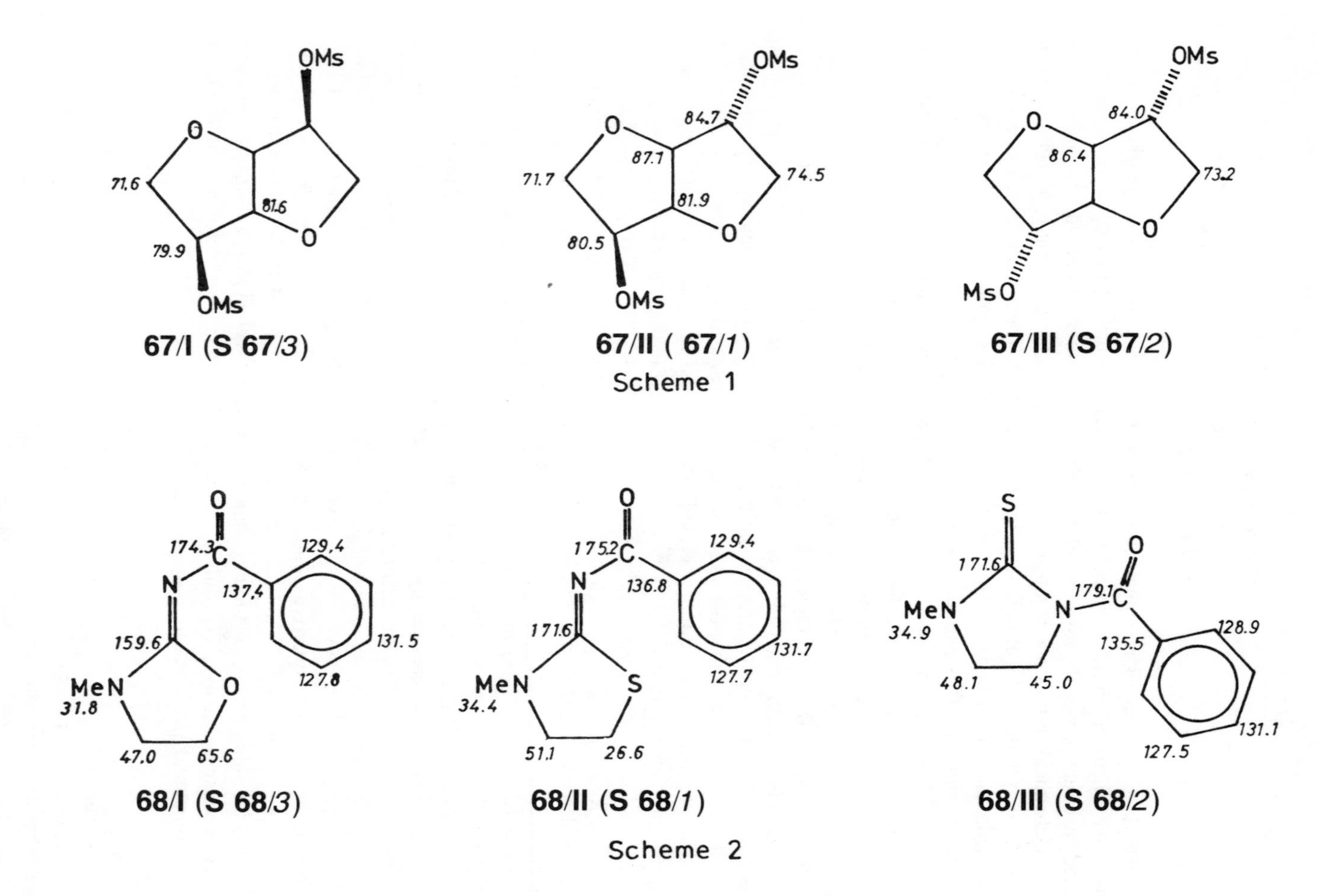

67/I (S 67/3)

67/II (67/1)

67/III (S 67/2)

Scheme 1

68/I (S 68/3)

68/II (S 68/1)

68/III (S 68/2)

Scheme 2

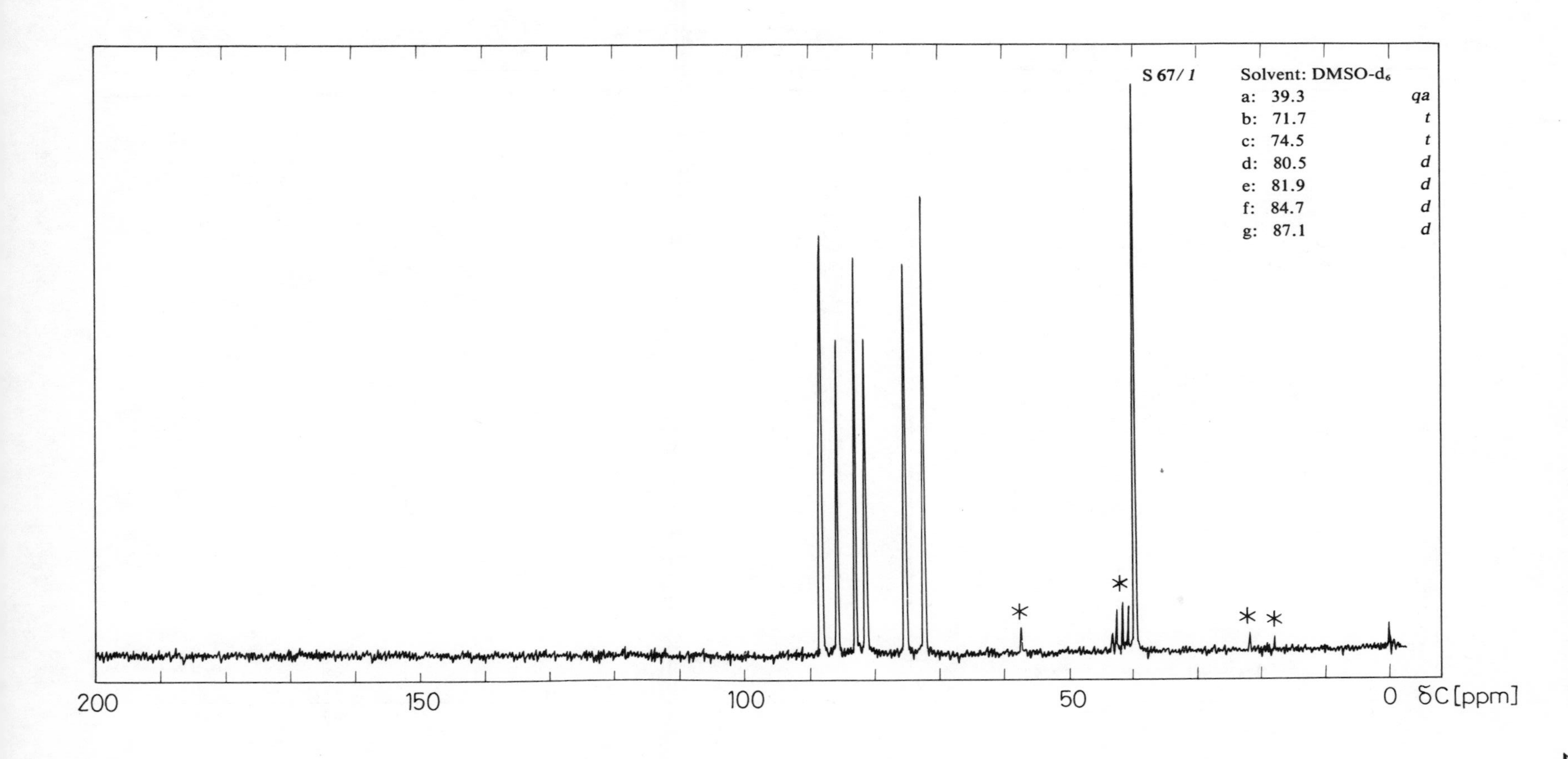
S 67/1
Solvent: DMSO-d$_6$
a: 39.3 qa
b: 71.7 t
c: 74.5 t
d: 80.5 d
e: 81.9 d
f: 84.7 d
g: 87.1 d
200
150
100
50
0
δC [ppm]

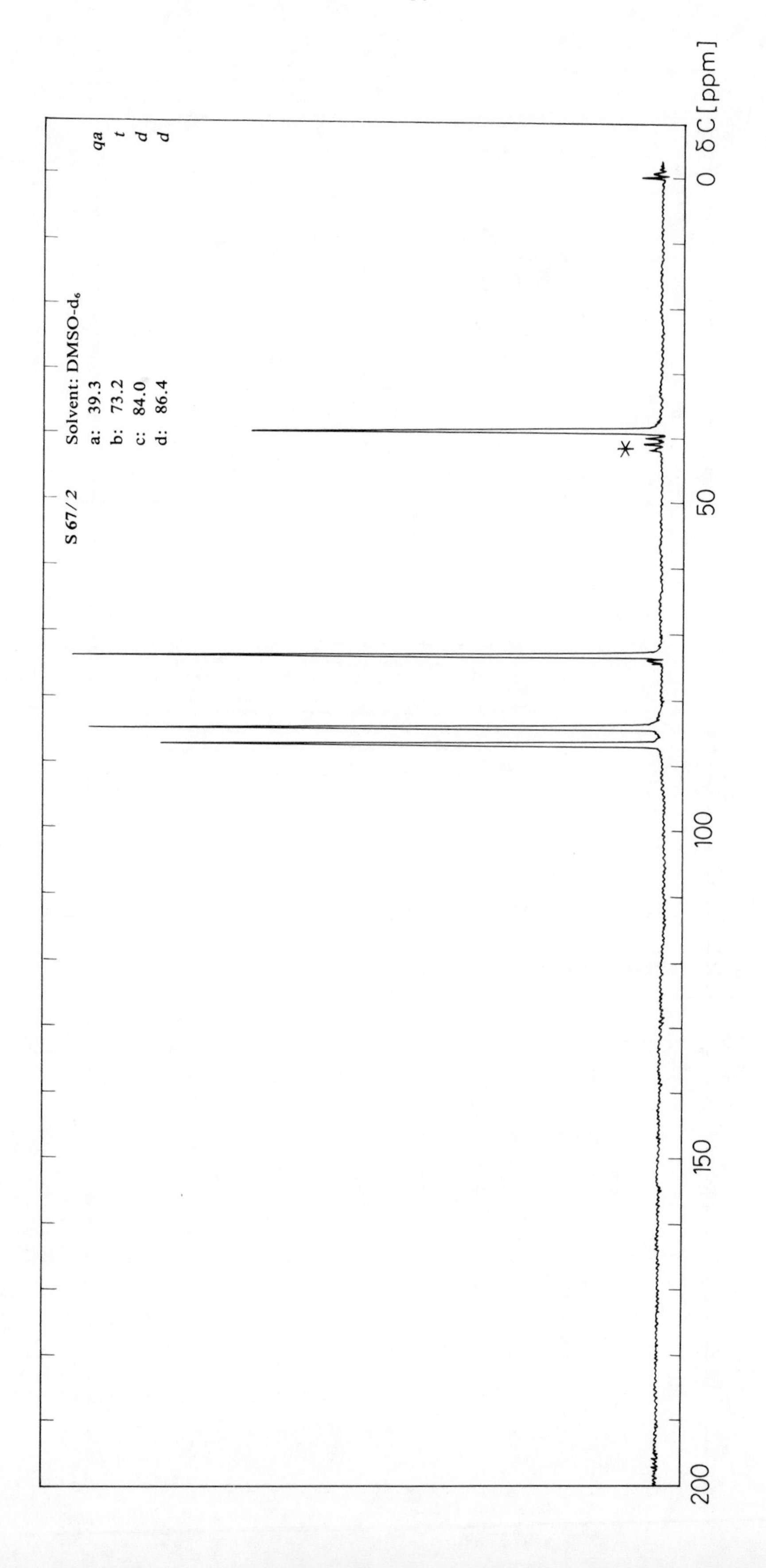
S 67/2
Solvent: DMSO-d6
a: 39.3 qa
b: 73.2 t
c: 84.0 d
d: 86.4 d
*
0 50 100 150 200
δC [ppm]

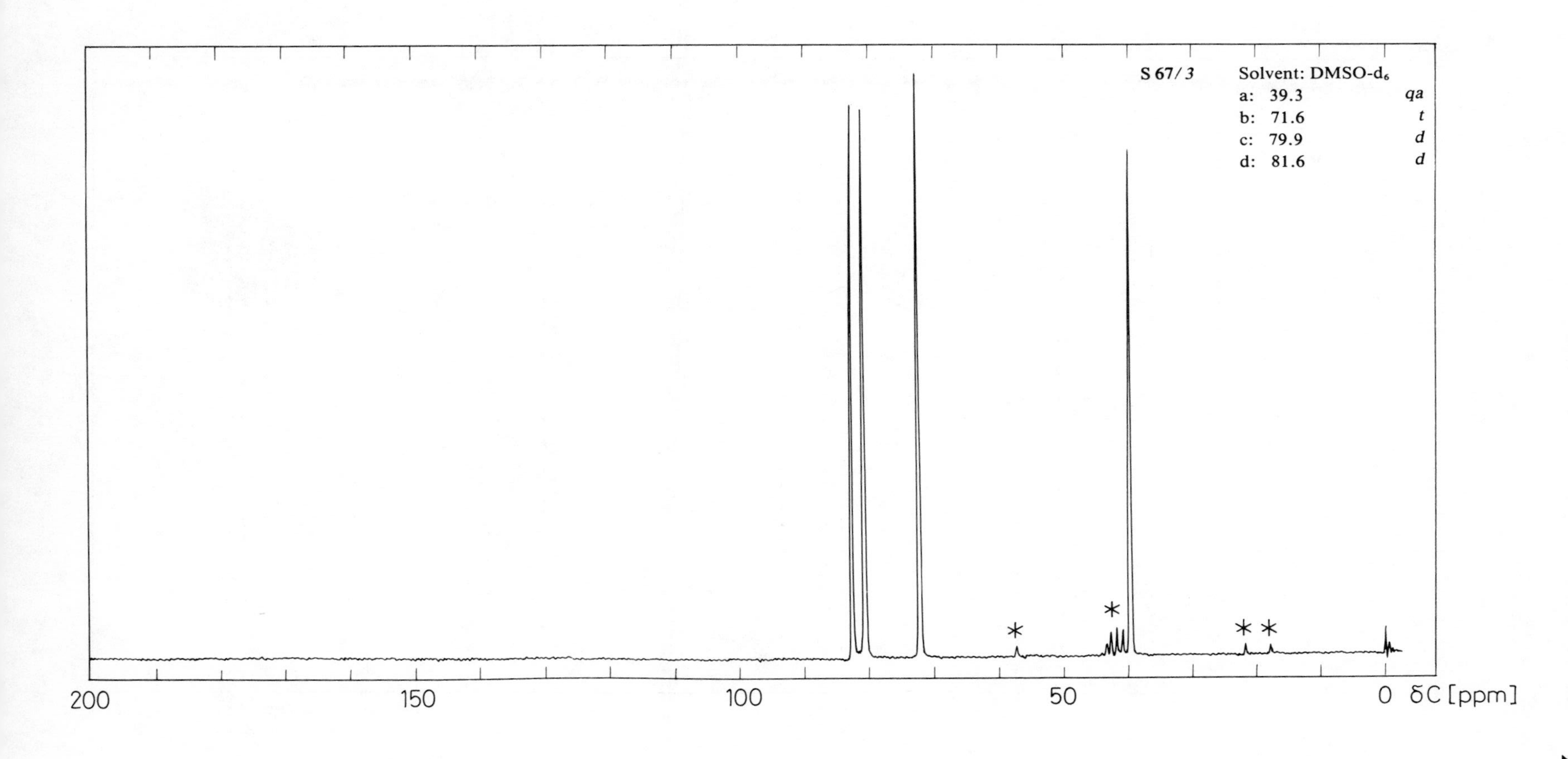
S 67/3
Solvent: DMSO-d₆
a: 39.3 qa
b: 71.6 t
c: 79.9 d
d: 81.6 d
200
150
100
50
0 δC [ppm]

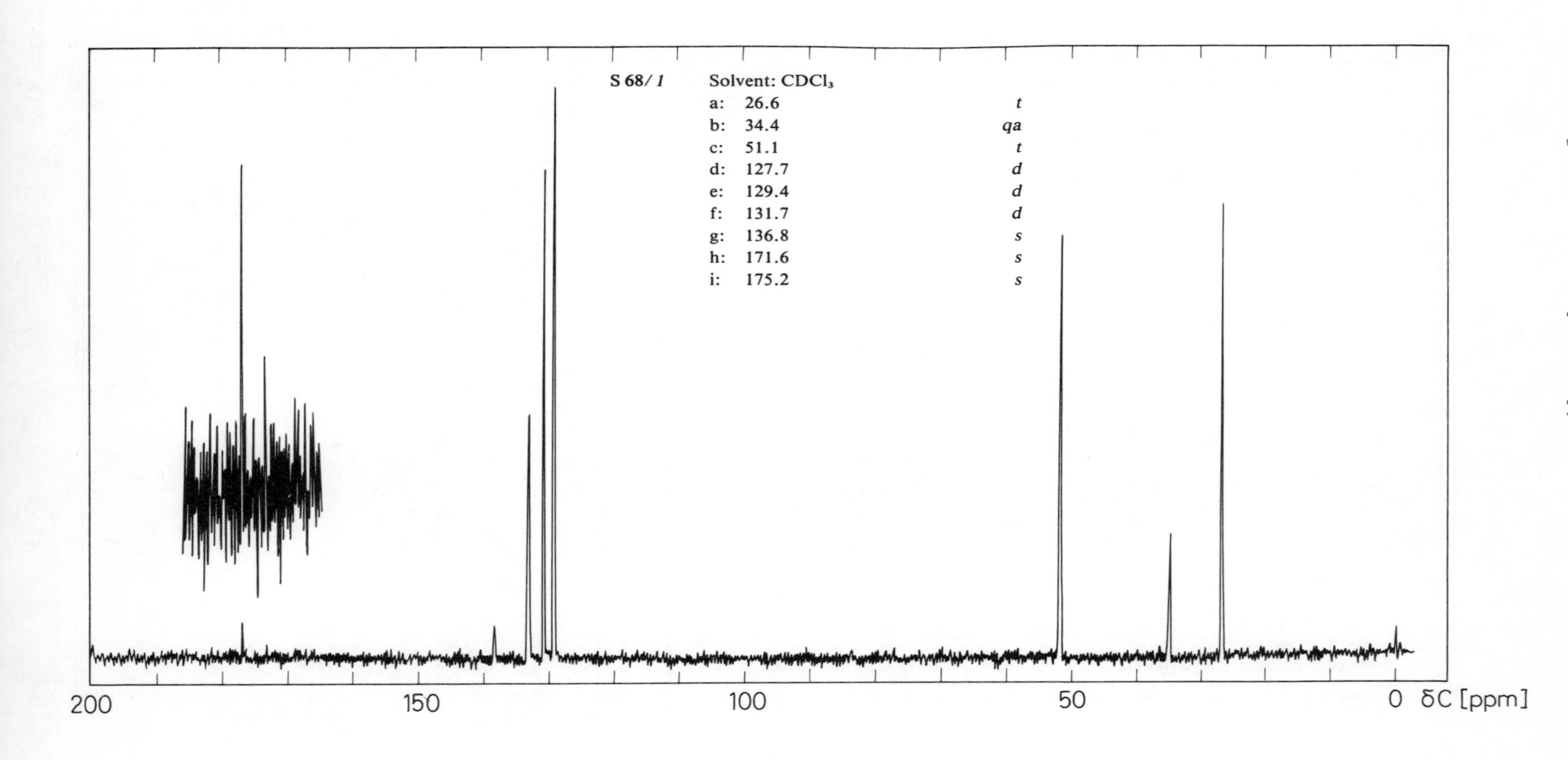
S 68/1
Solvent: $CDCl_3$
a: 26.6 *t*
b: 34.4 *qa*
c: 51.1 *t*
d: 127.7 *d*
e: 129.4 *d*
f: 131.7 *d*
g: 136.8 *s*
h: 171.6 *s*
i: 175.2 *s*
200
150
100
50
0
δC [ppm]

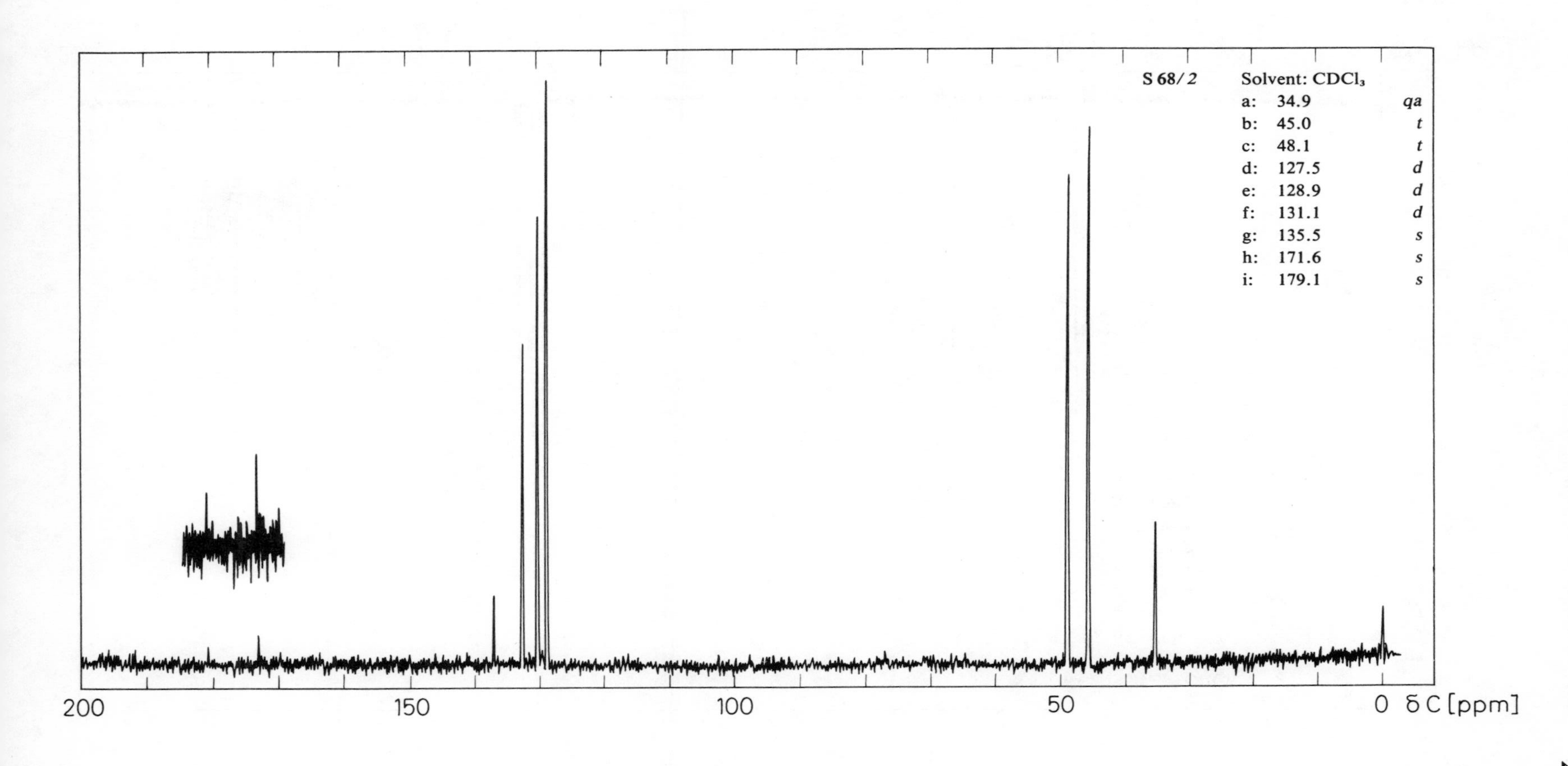
S 68/2
Solvent: $CDCl_3$
a: 34.9 qa
b: 45.0 t
c: 48.1 t
d: 127.5 d
e: 128.9 d
f: 131.1 d
g: 135.5 s
h: 171.6 s
i: 179.1 s
200
150
100
50
0 δC [ppm]

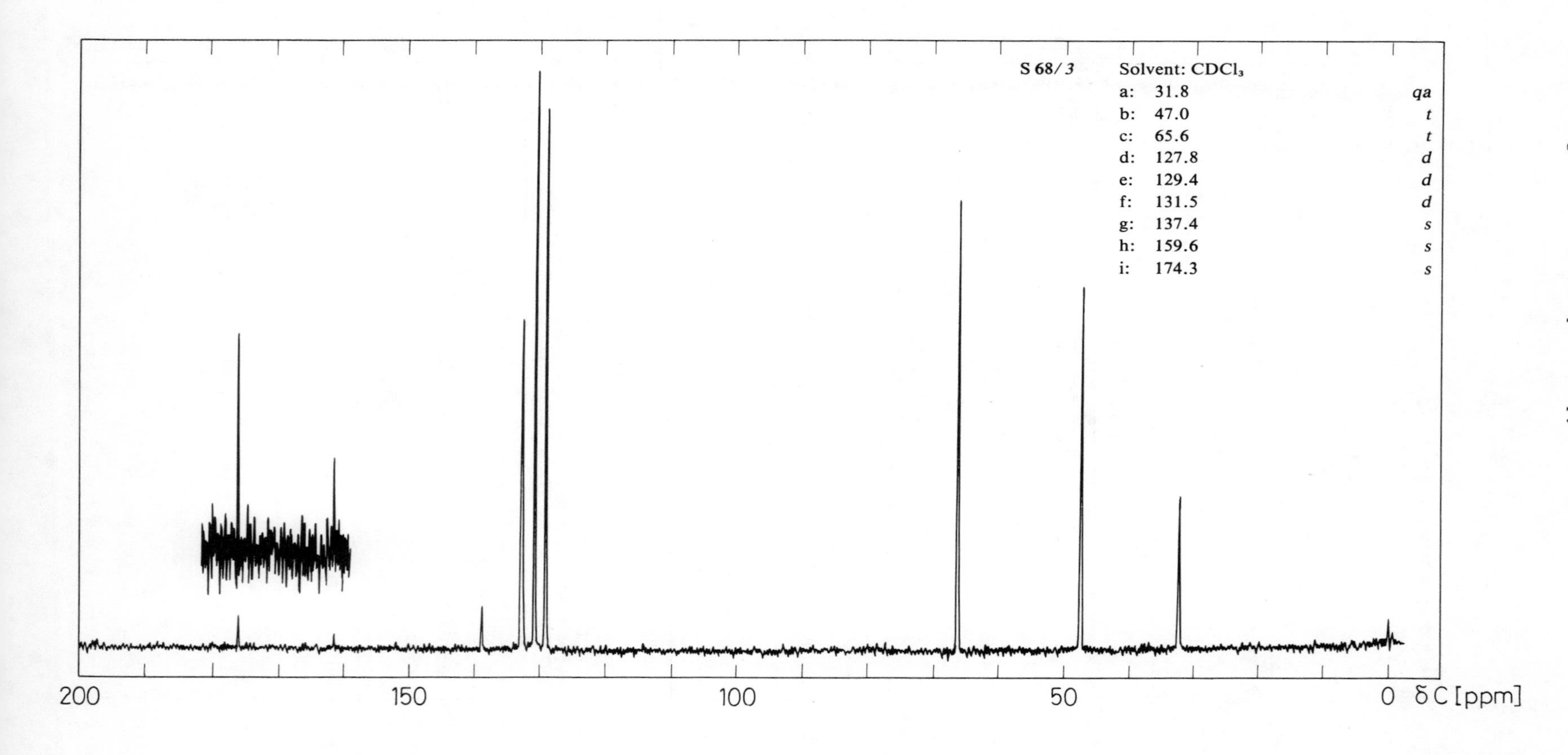
S 68/3
Solvent: $CDCl_3$
a: 31.8 *qa*
b: 47.0 *t*
c: 65.6 *t*
d: 127.8 *d*
e: 129.4 *d*
f: 131.5 *d*
g: 137.4 *s*
h: 159.6 *s*
i: 174.3 *s*
200
150
100
50
0
δC [ppm]

PROBLEM **69**[1336]

Assign Spectra **69**/*1*, **69**/*2*, and **69**/*3* to Structures **I** to **III** and explain the differences in chemical shift. Compare the shifts of C-2,5 and C-3,4 to those obtained in Problem **67** and confirm the assignments ambiguous in that problem.

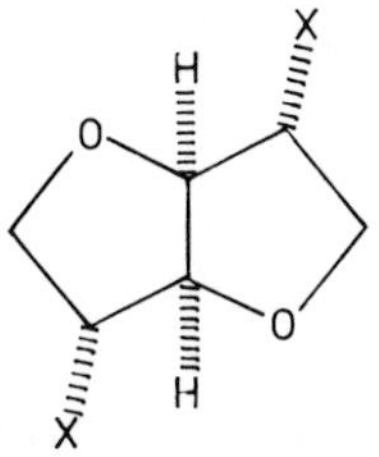

69/I (X : Cl)
69/II (X : Br)
69/III (X : I)

ANSWER

All the three spectra contain three lines, since the signals of C-1,6, C-2,5, and C-3,4 atom pairs coincide, owing to the C_2 molecular symmetry. The most deshielded lines (doublet) hardly differ in the three spectra and correspond to atoms C-3,4 (Scheme 3). The chemical shift of the triplet C-1,6 signal is not too sensitive, either, to the exchange of halogens (77.6, 76.4, and 75.7 ppm).

However, the C-2,5 line undergoes a significant shift upon the effect of halogen substituent, and the chemical shift decreases in the order of Cl, Br, and I.* Accordingly, to Structures **I**, **II**, and **III**, Spectra **69**/*3*, **69**/*2*, and **69**/*1* may be assigned.

On comparing the chemical shifts of C-3,4 to the analogous data of dimesyliditol (Structure **67/III**), it is obvious that of the two methine signals (84.0 and 86.4 ppm) only the one with the larger shift may correspond to the C-3,4 atoms, in accordance with the assignment of signals given in the answer to Problem **67**. The analogous chemical shifts of the carbon atoms of the ring carrying the *exo* substituent of sorbitol derivative **67/II** are even closer to the shift (87.1 ppm) of the halogen derivatives **I**, **II**, and **III**. Comparing the C-2,5 shifts of compound **67/III** (84.0 ppm), it can be seen that already the chlorine atom causes a significant shielding with respect to the mesyl derivatives (23.8 ppm), although here the lower electronegativity of chlorine (compared to the mesyloxy group), has a significant contribution to this effect. The outstandingly strong shielding of iodine-substituted carbons is, however, a characteristic feature, which makes it easy to recognize iodine-containing compounds by ^{13}C NMR.**

PROBLEM **70**[1323]

After the assignment of signals, decide on the basis of the differences which of Spectra **70**/*1* and **70**/*2* belongs Structures **I** and **II**. Using the spectroscopic data of these compounds, choose between tautomers **III** and **IV** on the basis of Spectrum **70**/*3* of the unsubstituted derivative.

* Heavy atom effect, compare Chapter 4.1.1.3, point 3, pp. 152—154.

** Compare Volume II, p.156.

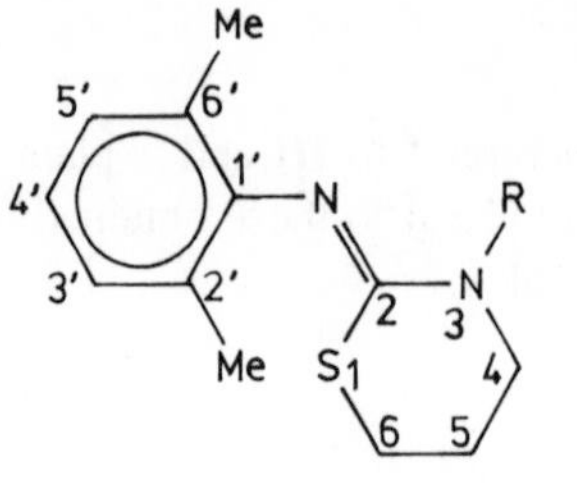

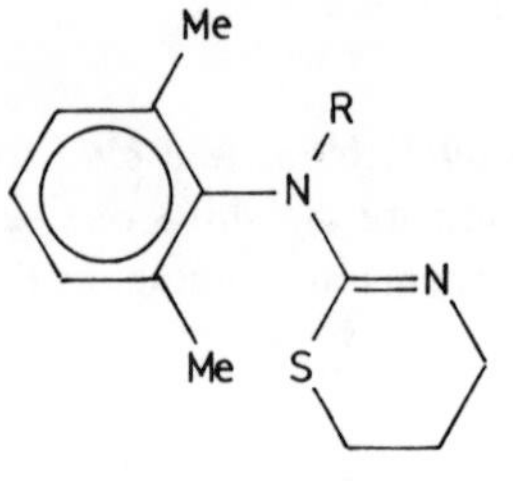

70/I (R : Me) **70/II (R : H)**

70/III (R : Me) **70/IV (R : H)**

ANSWER

Of the two methyl signals selected on the basis of quartet splitting, the most shielded line of the spectra belongs to the carbon attached to the aromatic ring and the other one belongs to the *N*-methyl carbon.

Of the three triplet methylene signals, the upfield one is the line of C-5 atom, not adjacent to heteroatoms. Since the electron affinity of nitrogen is higher than that of sulfur, the downfield triplet is the C-4 signal.* C-2, the most deshielded atom, gives the left-wing signal of the spectra at approximately 150 ppm.

The remaining four lines (two doublets and two singlets in the coupled spectrum) are due to aromatic carbons. The strongest (doublet) signal belongs to C-3′,5′ and the other strong (doublet) signal belongs to the C-4′ atom. Taking into account the effect of nitrogen and methyl substituents (compare Table 74) and the intensities, the two remaining (singlet) lines arise from the C-2′,6′ (upfield line) and C-1′ (downfield line) quaternary carbon atoms.

The three spectra are rather similar, and by comparing the data of the two *N*-methyl substituted derivatives, it can be seen that the most significant difference is observable in the position of the aromatic signals. In Spectrum **70**/*1*, the shifts of C-3′,5′ and C-4′ are similar to one another and to that of benzene, whereas the C-2′,6′ signal is close to the position expected when the effect of methyl substitution is also taken into account (137.4 ppm). The C-2′,6′ and C-4′ lines are shifted considerably upfield in Spectrum **70**/*2* (by 9.2 and 5.8 ppm, respectively), the C-3′,5′ shift hardly changes (decreases by 0.7 ppm), and the C-1′ signal shifts in the opposite direction (increases by 6.0 ppm). These differences are interpretable if Spectrum **70**/*1* is assigned to Structure **II** and Spectrum **70**/*2* is assigned to Structure **I** (Scheme 4). In the latter, the C=N bond is conjugated with the aromatic ring, and thus the electron density around C-2′,6′ and C-4′ increases. The C-1′ atom, is however, obviously deshielded since this carbon is inside of conjugation chain.** Since in Spectrum **70**/*3* of the *N*-unsubstituted compound the shifts of the aromatic signals are very similar to those of Spectrum **70**/*2* assigned to Structure **I**, it is reasonable to assume that under the conditions of measurement ($CDCl_3$ solution) the tautomeric equilibrium is shifted in favor of form **III**.

* Compare Table 50.

** Compare Volume II, p. 176.

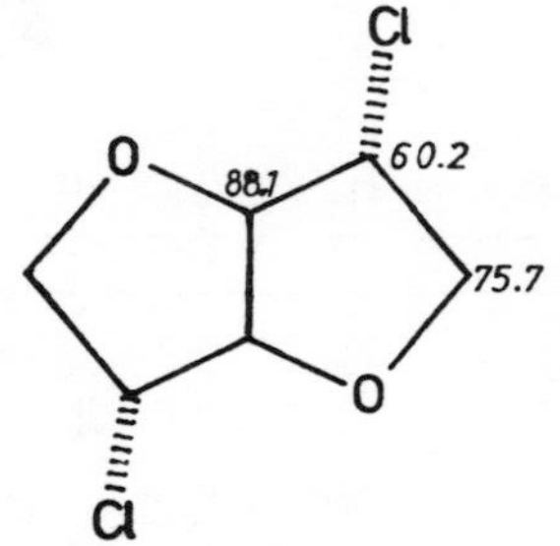

69/I (S 69/*3*)

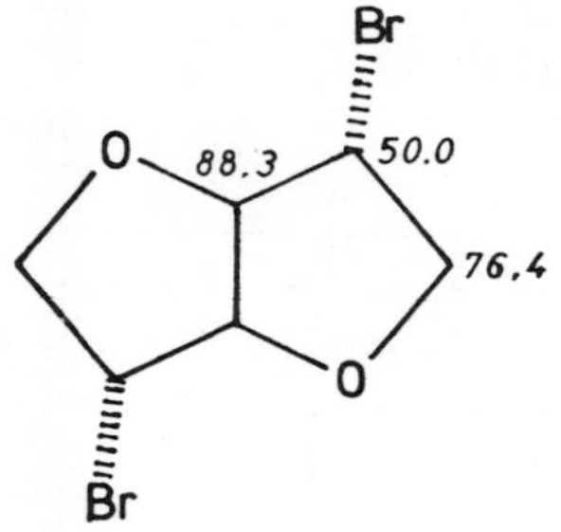

69/II (S 69/*2*)

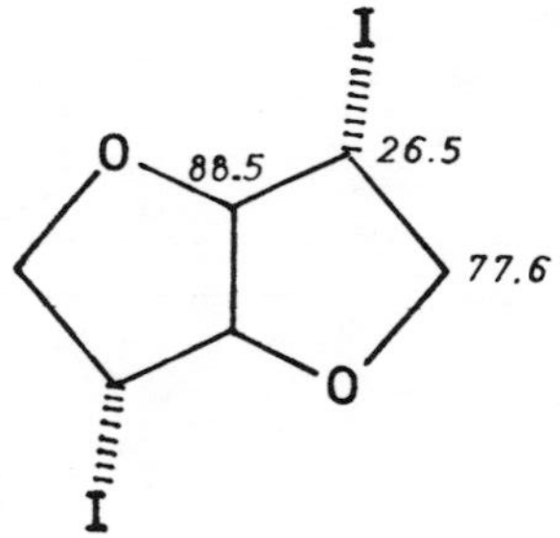

69/III (S 69/*1*)

Scheme 3

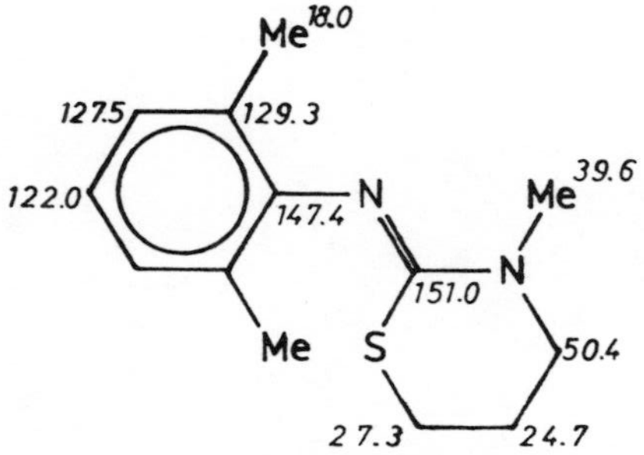

70/I (S 70/*2*)

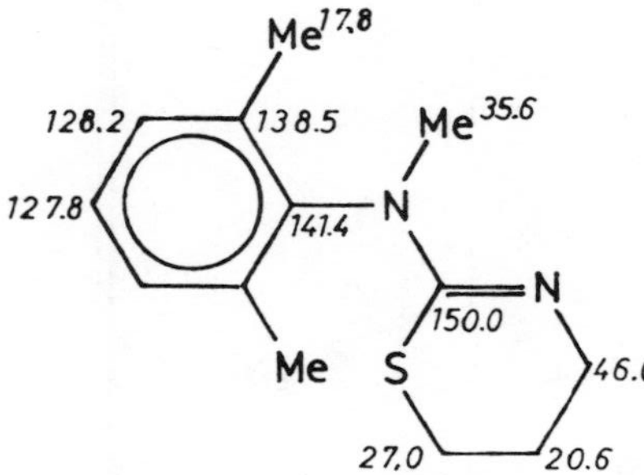

70/II (S 70/*1*)

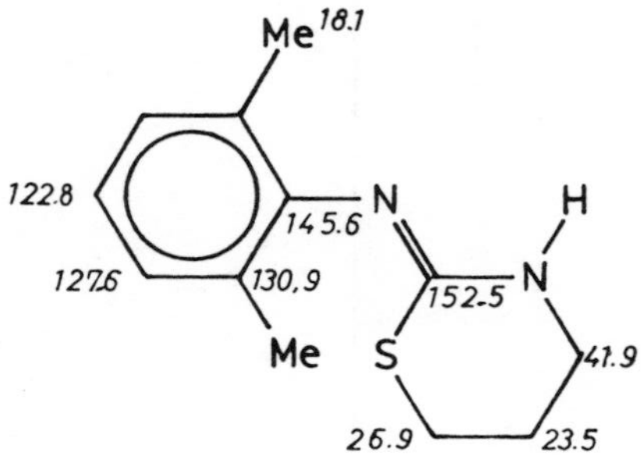

70/III (S 70/*3*)

Scheme 4

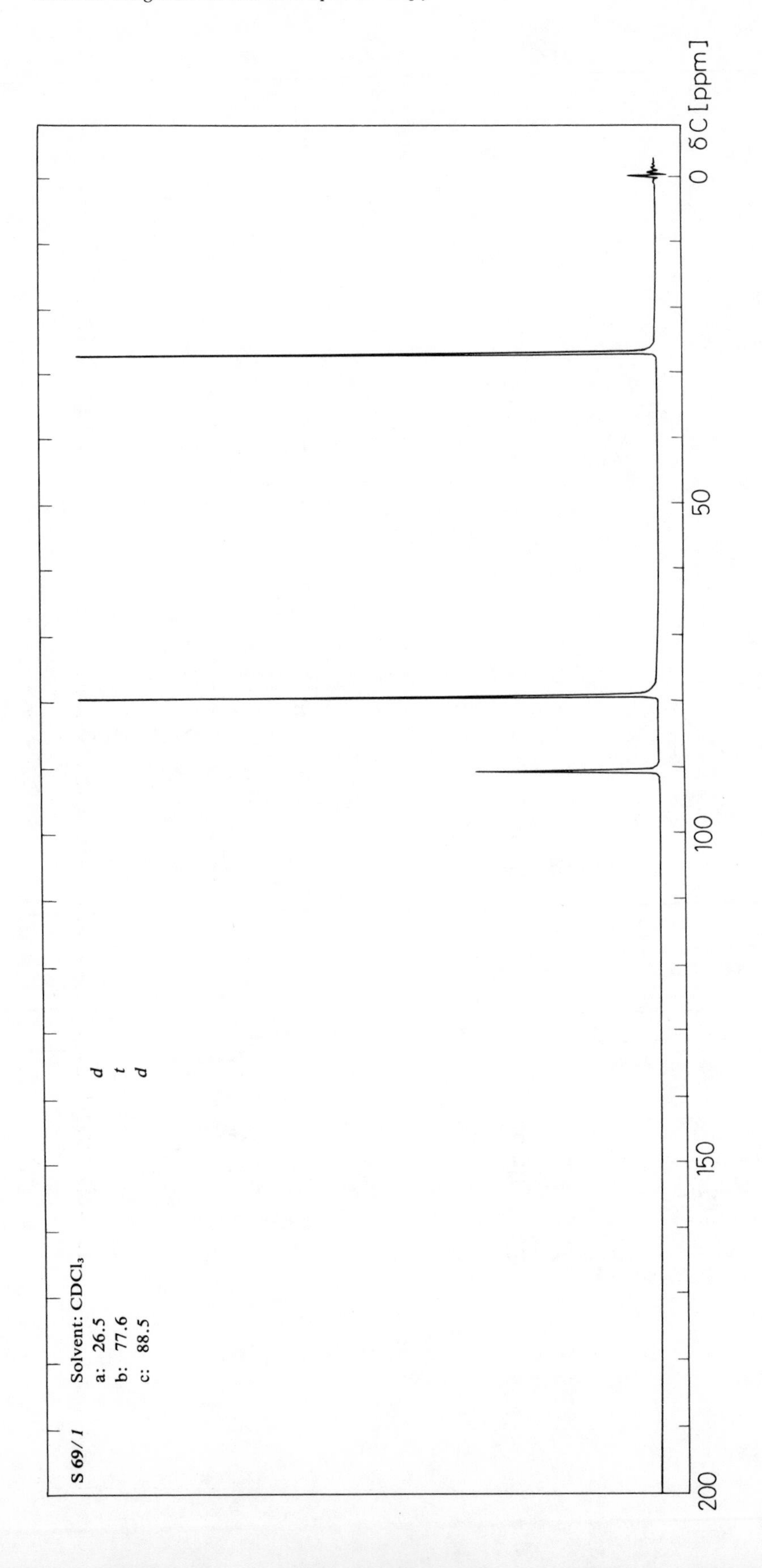
S 69/1
Solvent: $CDCl_3$
a: 26.5 d
b: 77.6 t
c: 88.5 d
0 δC [ppm]
50
100
150
200

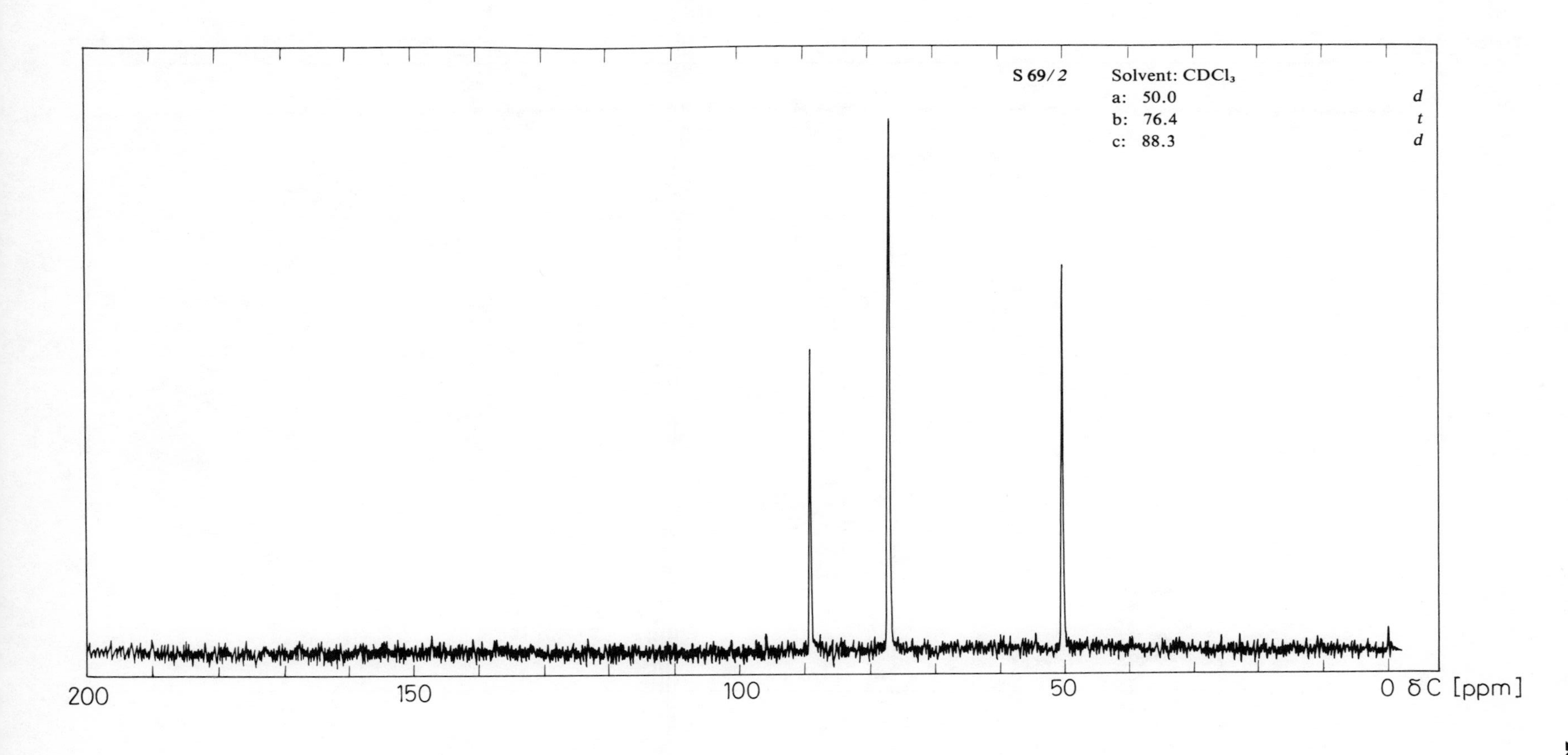
S 69/2
Solvent: $CDCl_3$
a: 50.0 d
b: 76.4 t
c: 88.3 d
200
150
100
50
0 δC [ppm]

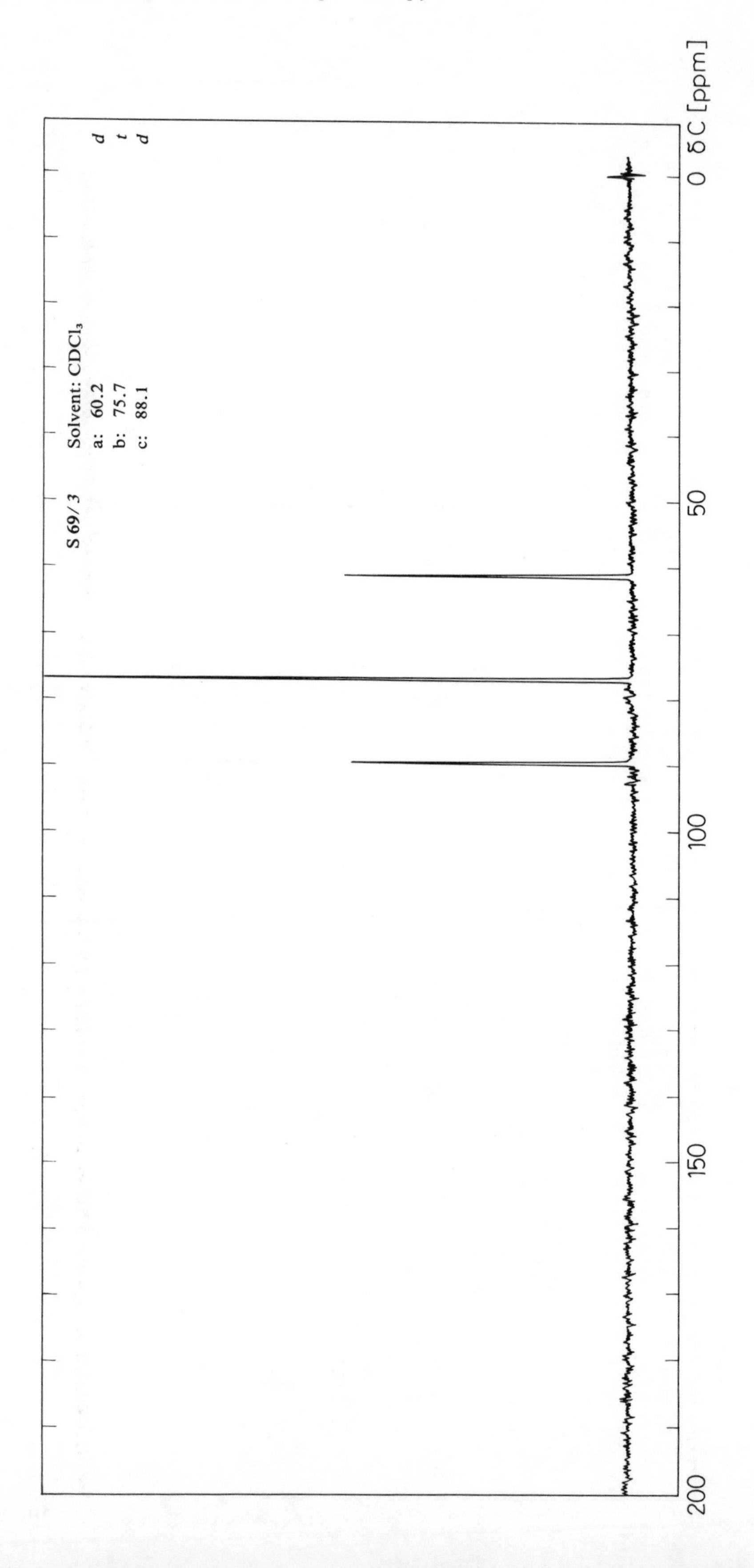
S 69/3
Solvent: CDCl₃
a: 60.2 d
b: 75.7 t
c: 88.1 d
0 δC [ppm]
50
100
150
200

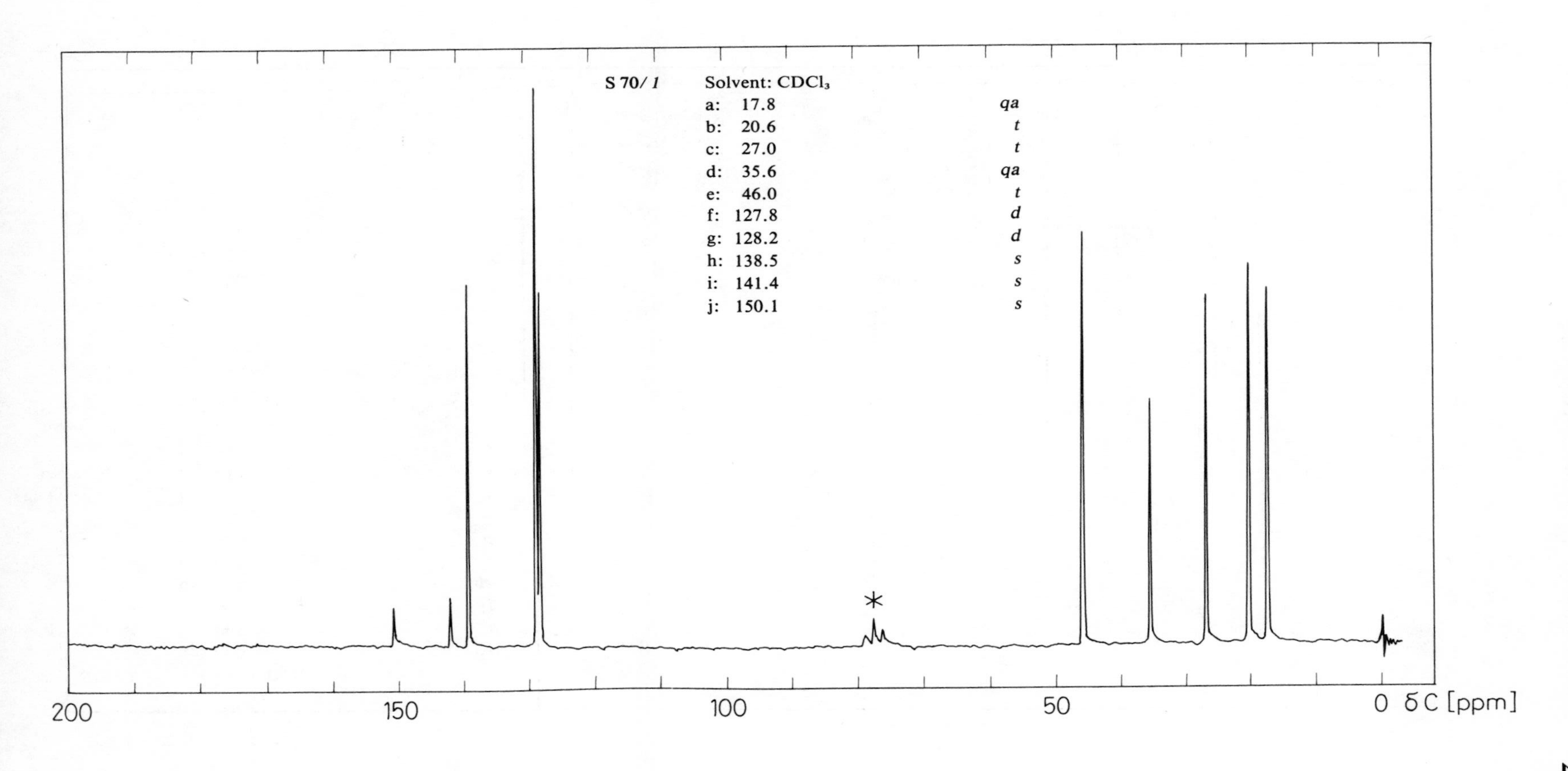

S 70/1
Solvent: CDCl3
a: 17.8 qa
b: 20.6 t
c: 27.0 t
d: 35.6 qa
e: 46.0 t
f: 127.8 d
g: 128.2 d
h: 138.5 s
i: 141.4 s
j: 150.1 s
*
200
150
100
50
0 δC [ppm]

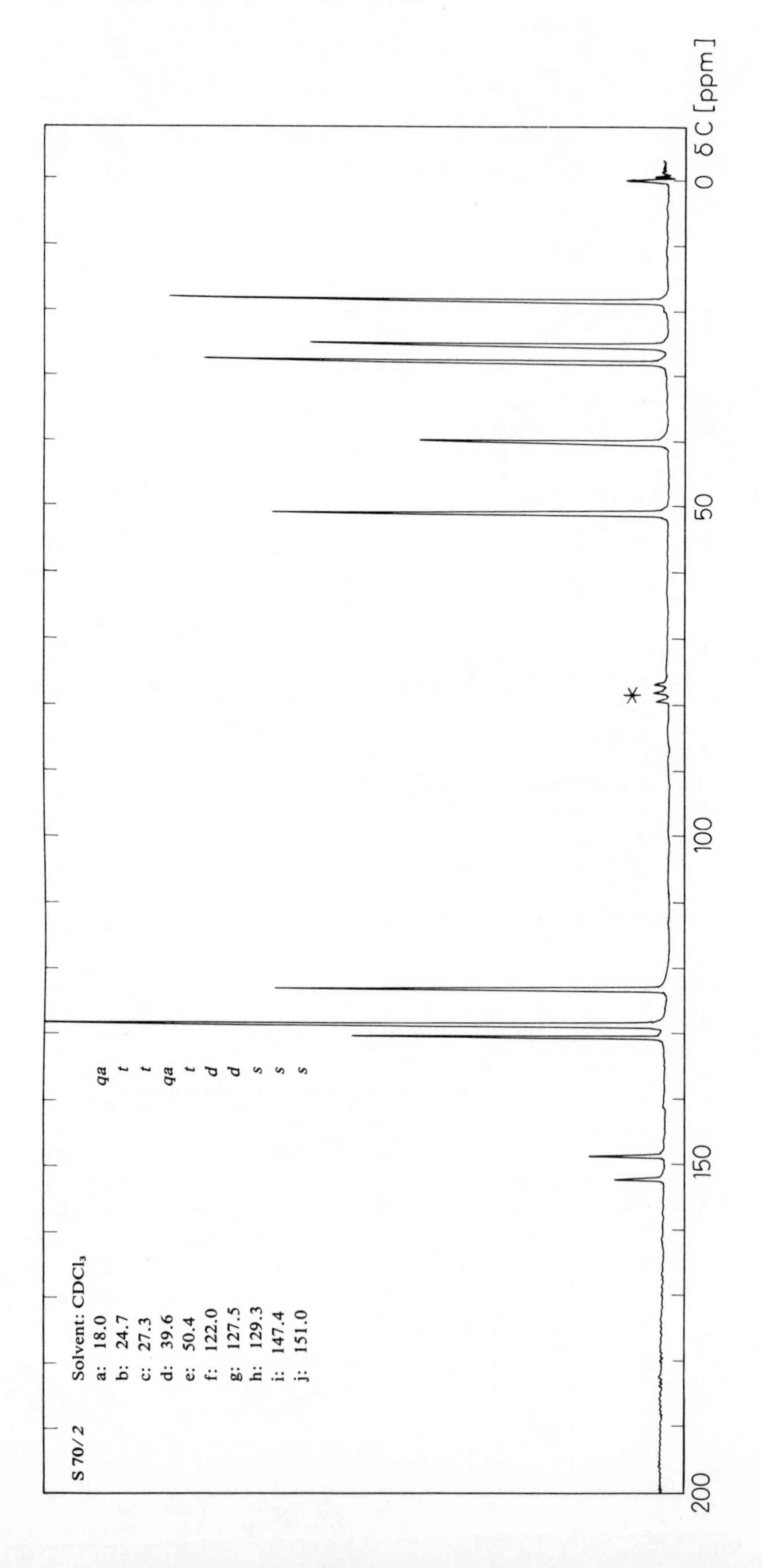
S 70/2
Solvent: CDCl3
a: 18.0 qa
b: 24.7 t
c: 27.3 t
d: 39.6 qa
e: 50.4 t
f: 122.0 d
g: 127.5 d
h: 129.3 s
i: 147.4 s
j: 151.0 s
*
200
150
100
50
0 δC [ppm]

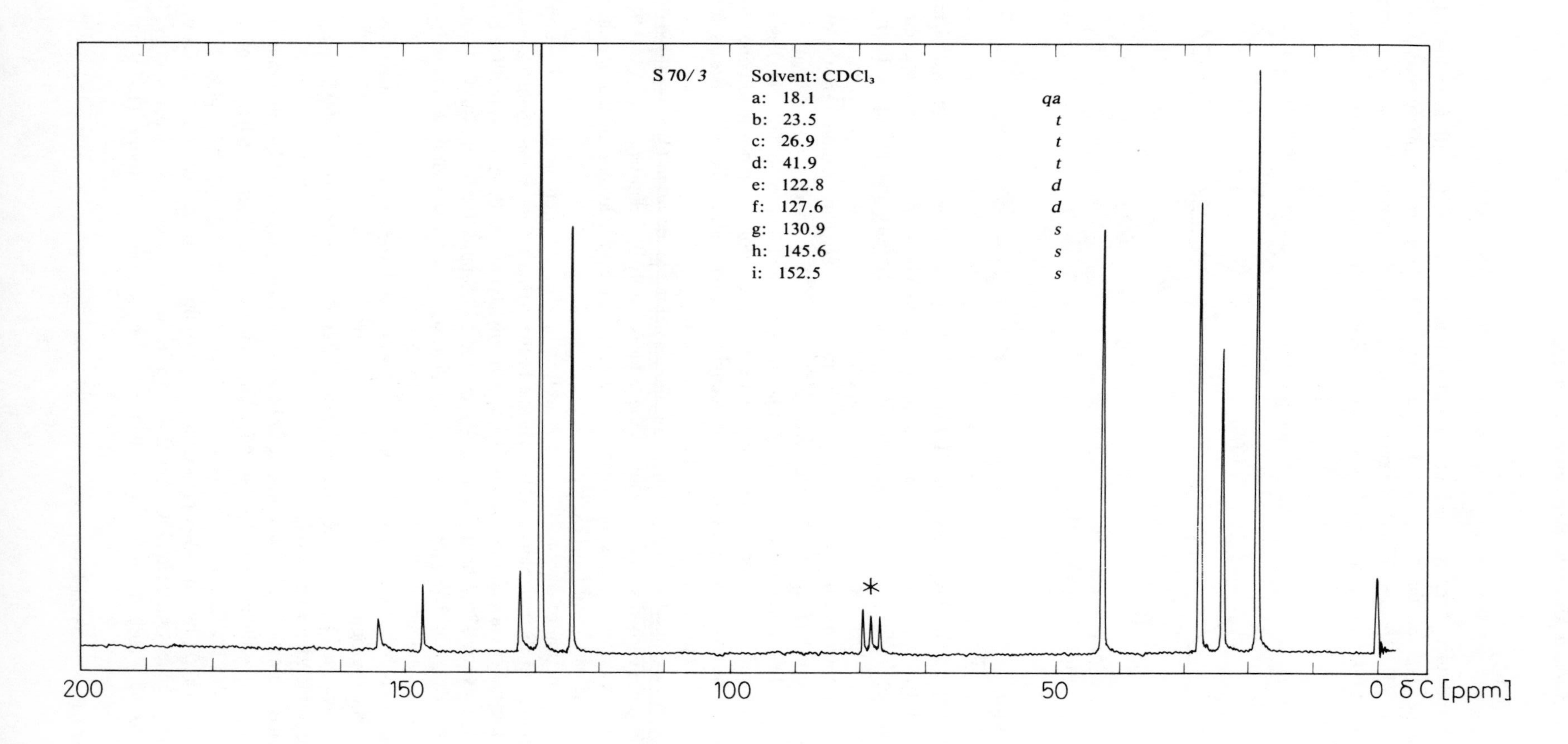
S 70/3
Solvent: $CDCl_3$
a: 18.1 qa
b: 23.5 t
c: 26.9 t
d: 41.9 t
e: 122.8 d
f: 127.6 d
g: 130.9 s
h: 145.6 s
i: 152.5 s
*
200
150
100
50
0 δC [ppm]

PROBLEM **71**[1334]

A by-product of the Mannich reaction of styrene (Ph–CH=CH_2) with formaldehyde and ammonia is, on the basis of IR and ^{1}H NMR spectra, probably compound **I**. **71**/*1* and **71**/*2* are the ^{13}C NMR spectra of the *N*-methyl-*O*-acetyl (**II**) and the aromatized derivative (**III**) of compound **I**, respectively. Assign the spectra and decide whether the spectra support Structures **II** and **III** and thus Structure **I**.

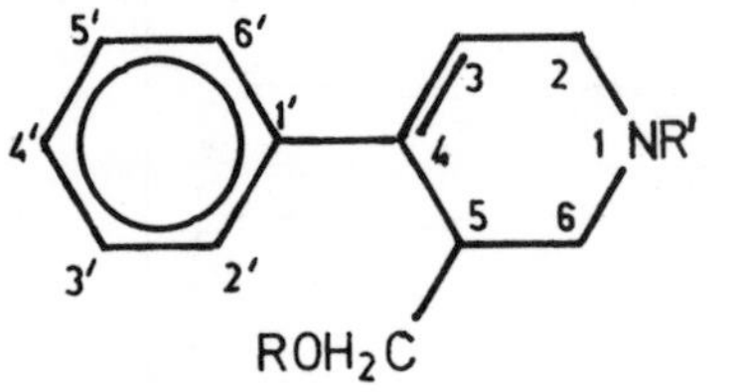

5′ 6′ 3 2 1′ 4′ 4 1 NR′ 5 6 3′ 2′ ROH_2C

71/I (R = R′ : H)
71/II (R : Ac; R′ : Me)

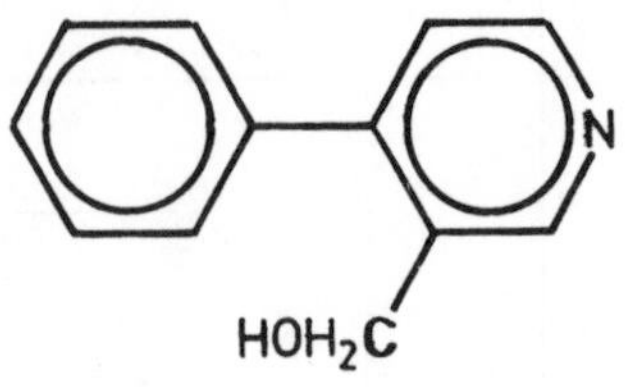

N HOH_2C

71/III

ANSWER

The upfield signal of Spectrum **71**/*1* at 20.5 ppm is due certainly to the acetyl methyl group, which is also proved by the multiplicity. On the basis of quartet splitting, the other methyl signal (*N*-methyl) is the 45.5-ppm line. The 37.2-ppm signal belongs to the methine carbon (doublet in the coupled spectrum), and the small chemical shift may be attributed to the neighborhood of the two secondary carbon atoms and the electron donor *N*-methyl group. Methylene carbons C-2 and C-5 produce the signals at 53.9 and 55.1 ppm. Due to the closeness, an alternative assignment is also possible. However, it is sure that of the signals belonging to the saturated carbon atoms, the downfield one (at 65.0 ppm) can be assigned to the exocyclic methylene group, since the adjacence of tertiary carbon and oxygen equally has deshielding effects.

The two strongest lines of the aromatic signals belong to the *ortho* (125.6 ppm) and *meta* (128.4 ppm) carbon atoms; the shift of the latter is hardly different from the ^{13}C NMR shift of benzene,** since conjugation changes primarily the electron density and thus the chemical shift of the *ortho* and *para* carbons.

The distinction of carbons 1′ and 3 as well as of 4′ and 4 is problematic. (However, the interchange of the two pairs is ruled out due to the higher intensity and doublet splitting of the latter pair.) On the basis of the data of α-methylstyrene,[360] the 127.1-ppm line belongs to C-4 (thus the 125.3-ppm line to C-4′) and the 139.7-ppm line belongs to C-3 (and thus C-1′ produces the 135.1-ppm line). As mentioned, a reverse assignment cannot be ruled out. The assignment is summarized in Scheme 5.

The most-shifted downfield line of the spectrum, at 170.3 ppm, arises, of course, from the carbonyl group.

In Spectrum **71**/*2* only one saturated carbon signal can be observed at 60.1 ppm, which is due to the methylene group.

In the aromatic region nine lines can be expected and observed. The two more intense lines can be assigned to the *ortho* and *meta* phenyl carbons (128.5 and 128.7 ppm). It is also clear that the three most-shifted lines arise from the α- and γ-carbon atoms of pyridine, and the weakest line at 149.3 ppm is the signal of the substituted carbon, C-2. Of the two doublets, the less shielded (150.0 ppm) belongs to the α-carbon and the other belongs (148.0 ppm) to the γ-carbon. Of the two further quaternary signals, the stronger (137.9 ppm) can

** See Table 49 and Volume II, p. 187.

be assigned to the phenyl ring and the weaker one at 134.3 ppm can be assigned to the substituted β-carbon of the pyridine ring. This is also supported by the chemical shift of the former signal (with respect to compound **II**, the heteroaromatic ring replacing the conjugated double bond can only increase the shift of the quaternary phenyl carbon). Of the two remaining signals (124.4 and 128.4 ppm), the stronger and more downfield shifted one arises from the *para* phenyl carbon and the other arises from the unsubstituted β-carbon of pyridine ring. In addition to the intensities (the signals of the phenyl group are all stronger than the analogous signals of the pyridine ring), the assignment is also supported by the chemical shifts: the shift of the *para* phenyl carbon may only increase with respect to Spectrum **71**/*1* (125.3 ppm). Proving the Structures **II** and **III** assumed for the derivatives, the spectra also support the assumption on the unknown by-product, i.e., Structure **I**.

PROBLEM 72[1324]

Spectra **72**/*1* to *4* were recorded from the pairs of isomers **I-II** and **III-IV**. Assign the spectra to the isomer pairs differing in the R substituent and then assign the spectra within the pairs to the 3- and 5-substituted derivatives. Is it possible to conclude on the conformation of the molecules from the spectra?

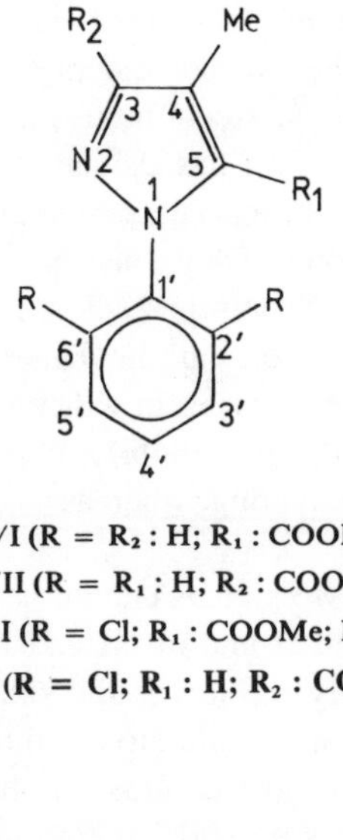

72/I (R = R_2 : H; R_1 : COOMe)
72/II (R = R_1 : H; R_2 : COOMe)
72/III (R = Cl; R_1 : COOMe; R_2 : H)
72/IV (R = Cl; R_1 : H; R_2 : COOMe)

ANSWER

The methyl and methoxy signals are easy to select, since in region characteristic of saturated carbons only these two lines occur. Of course, the line around 10 ppm belongs to the *C*-methyl and the one around 52 ppm belongs to the *O*-methyl group.

The identification of carbonyl signals as the lines of most deshielded carbon at 160 and 163 ppm, respectively, is also straightforward. With the phenyl-substituted derivatives, too, with the 2′,6′-dichlorosubstituted analogues one line of outstanding intensity is expected in the aromatic region, corresponding to the *ortho* (2′,6′) and *meta* (3′,5′) carbons in the former and to the *meta* carbons in the latter. Consequently, Spectra **72**/*2* and **72**/*4* belong to isomeric Structures **I** and **II**. The lines closer to the line of benzene (128.5 ppm) at 128.5 and 129.4 ppm can be assigned to C-3′,5′, and the more substituent sensitive lines at 119.6 and 125.7 ppm can be assigned to C-2′,6′.* Therefore, the strongest lines of Spectra **72**/*1* and **72**/*3* at 128.3 and 128.6 ppm belong to C-3′,5′ atoms adjacent to the chlorine-substituted carbons.

* Compare Volume II, p. 187, also Problem **71**.

The most upfield, weak line of the aromatic region may correspond to C-4 of the pyrazole ring only, since the two other substituted pyrazole carbons are adjacent to nitrogens and thus they are much less shielded (for pyrazole δC-4 = 105.4 and δC-3,5 = 134.6 ppm).* Thus, the shifts of the C-4 signal in the four spectra are, in turn, 123.3, 122.3, 121.4, and 123.8 ppm.

Of the not yet assigned carbons, C-4′ and the also unsubstituted (C-3 or C-5) pyrazole carbon may be expected to give stronger signals. Therefore, in the order of Spectra *1* to *4*, the following lines correspond to these carbons: 130.6 and 142.6, 127.2 and 127.6, 131.2 and 132.2, moreover 128.2 and 141.2 ppm. These chemical shifts give the key for the distinction of the isomers. Pyrazole carbon C-3 adjacent to the sp^2 nitrogen atom is obviously more deshielded in the case of Structures **I** and **III** than C-5 adjacent to an sp^3 nitrogen in Structures **II** and **IV**. Thus, Spectra **72**/*1* and **72**/*4* correspond to the 5-substituted isomers and Spectra **72**/*2* and **72**/*3* correspond to their 3-carbomethoxy pairs. Also taking into account the conclusions drawn from the intensities of phenyl signals on substituent R, Structures **I** to **IV** correspond to Spectra **72**/*4*, **72**/*2*, **72**/*1*, and **72**/*3* (Scheme 6). The distinction of the two signals in question is simple in **72**/*1* and **72**/*4*, the 142.6- and 141.2-ppm lines may arise from C-3 of the pyrazole rings only, and since this signal is weaker in both cases than that of C-4′, it is probable that in Spectra **72**/*2* and **72**/*3* the 127.2- and 131.2-ppm lines correspond to the latter, whereas the weaker lines at 127.6 and 132.2 ppm are the C-3 signals.

The identification of the signals of chlorinated carbons in the Spectra of **III** and **IV** is easy, since the total intensity of the two coincident lines of two identical atoms is higher than for the other substituted carbons. Accordingly, the 134.5-ppm line of Spectrum **72**/*1* and the 134.4-ppm line of Spectrum **72**/*3* has also been assigned. The still missing two lines are due to the carbon atoms of phenyl rings adjacent to nitrogen and to the carbon of pyrazole ring bearing the carbomethoxy group. They must be weak and downfield-shifted signals. Both of them can be recognized in Spectrum **72**/*2* at 139.6 and 142.4 ppm, and the larger shift corresponds probably to the pyrazole carbon. This assignment is supported by Spectrum **72**/*3* arising from the isomer of similar type, in which the two lines are at 136.0 and 142.7 ppm (the latter is very weak, but still discernible): chlorine substitution causes evidently a shift in the signal assigned to the phenyl ring, whereas the line of pyrazole carbon is insensitive in this respect.

In Spectra **72**/*1* and **72**/*4* these two signals cannot be identified. Comparing the shifts of the C-3 lines in the spectra of **I** and **II** and of **III** and **IV**, respectively, it can be seen that the carbomethoxy substituent hardly influences the shift (141.2, 142.4, 142.6, and 142.7 ppm). It is therefore reasonable that the situation is the same with the C-5 signal, i.e., in the spectra of **I** and **III** this signal must be close to the value measured for Structures **II** and **IV** (127.6 and 132.2 ppm). It is possible, therefore, that the very weak lines at 130.4 and 131.1 ppm in Spectra **72**/*4* and **72**/*1*, respectively, are the signals of C-5. The signal of the phenyl carbon substituted with nitrogen may perhaps overlap with the C-3 signal. It is worth observing that in Spectrum **72**/*2* the C-2′,6′ signal is upfield shifted relatively to the case of isomer **I** (Spectrum **72**/*4*). This is due to the stronger conjugation in Structure **II**, for which the coplanar conformation, possible for steric reasons, due to the hindrance of phenyl ring and the carbomethoxy group, and there is no such effect on the *ortho* carbon (an analogous, though weaker change in shift may also be observed on the signals of *para* carbon atoms).**

In the case of dichlorophenyl substitution coplanarity is impossible for either isomers; accordingly, the chemical shifts of the phenyl ring signals are practically the same. For the isomers the shifts of 4-methyl and carbonyl signals also show systematical differences: δC_{Me}

* Compare Table 76.

** Compare Problem **76**.

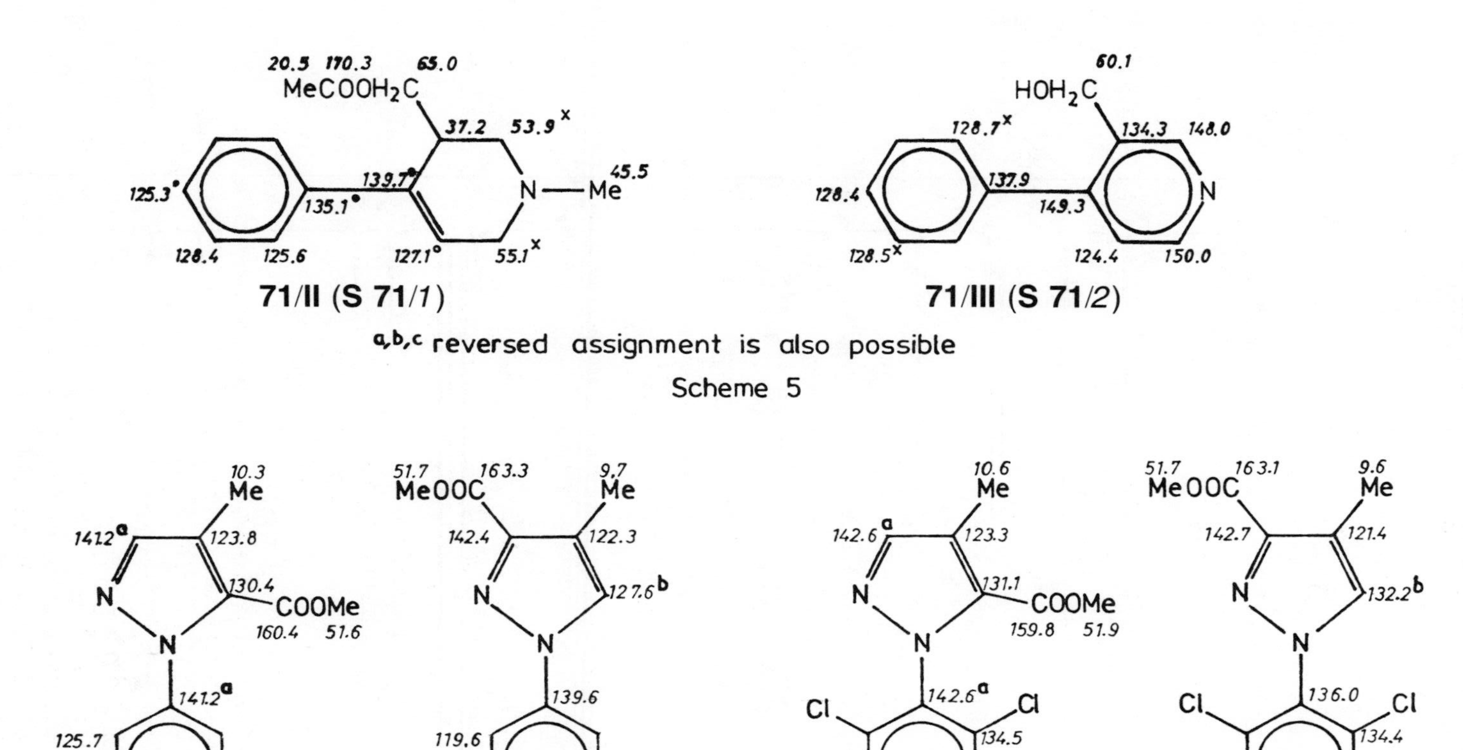

71/II (S 71/1) **71/III (S 71/2)**

[a,b,c] reversed assignment is also possible

Scheme 5

72/I (S 72/4) **72/II (S 72/2)** **72/III (S 72/1)** **72/IV (S 72/3)**

[a] overlapped signals [b] reversed assignment is also possible

Scheme 6

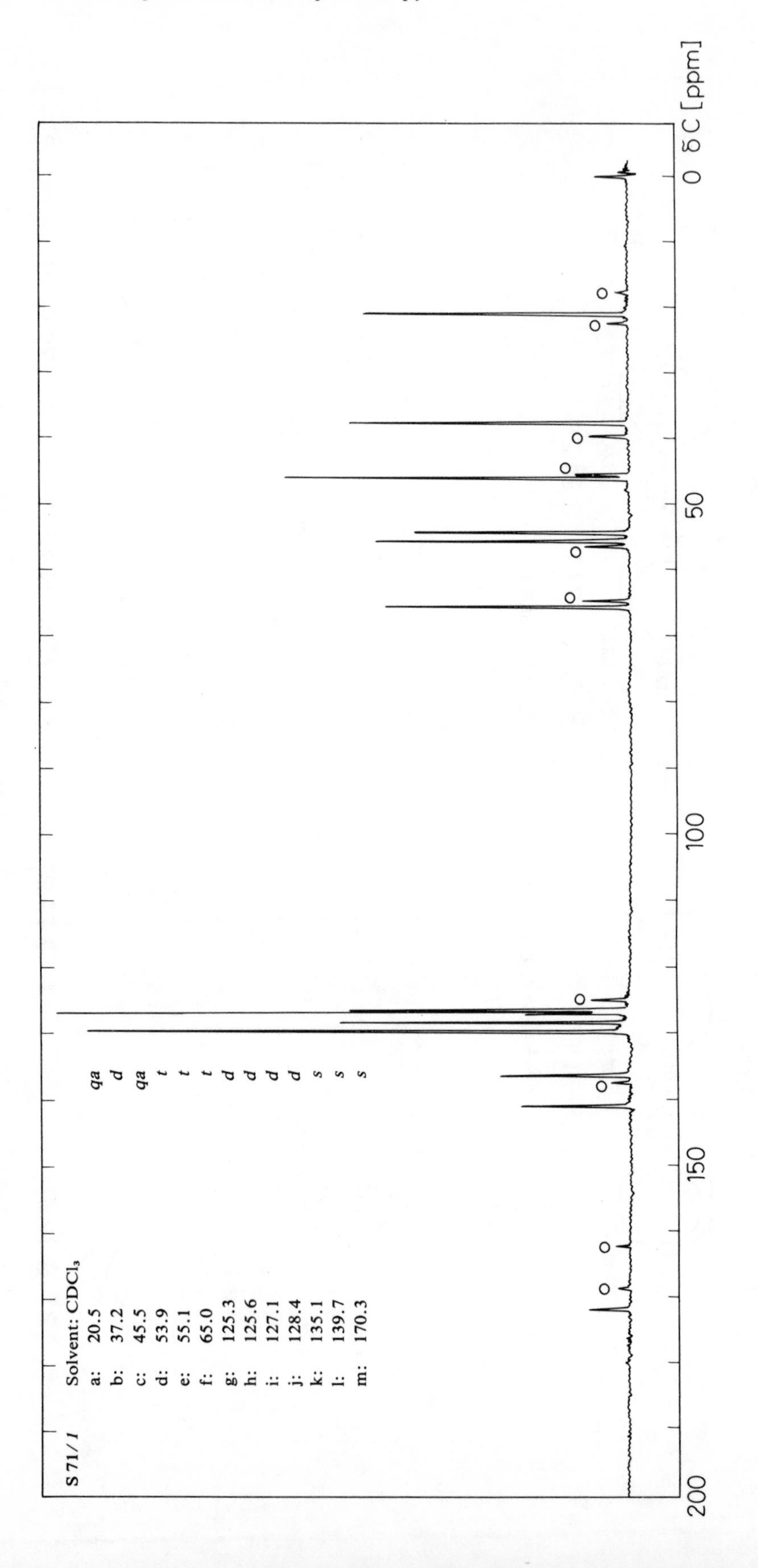
S 71/1
Solvent: $CDCl_3$
a: 20.5 qa
b: 37.2 d
c: 45.5 qa
d: 53.9 t
e: 55.1 t
f: 65.0 t
g: 125.3 d
h: 125.6 d
i: 127.1 d
j: 128.4 d
k: 135.1 s
l: 139.7 s
m: 170.3 s
0
50
100
150
200
δC [ppm]

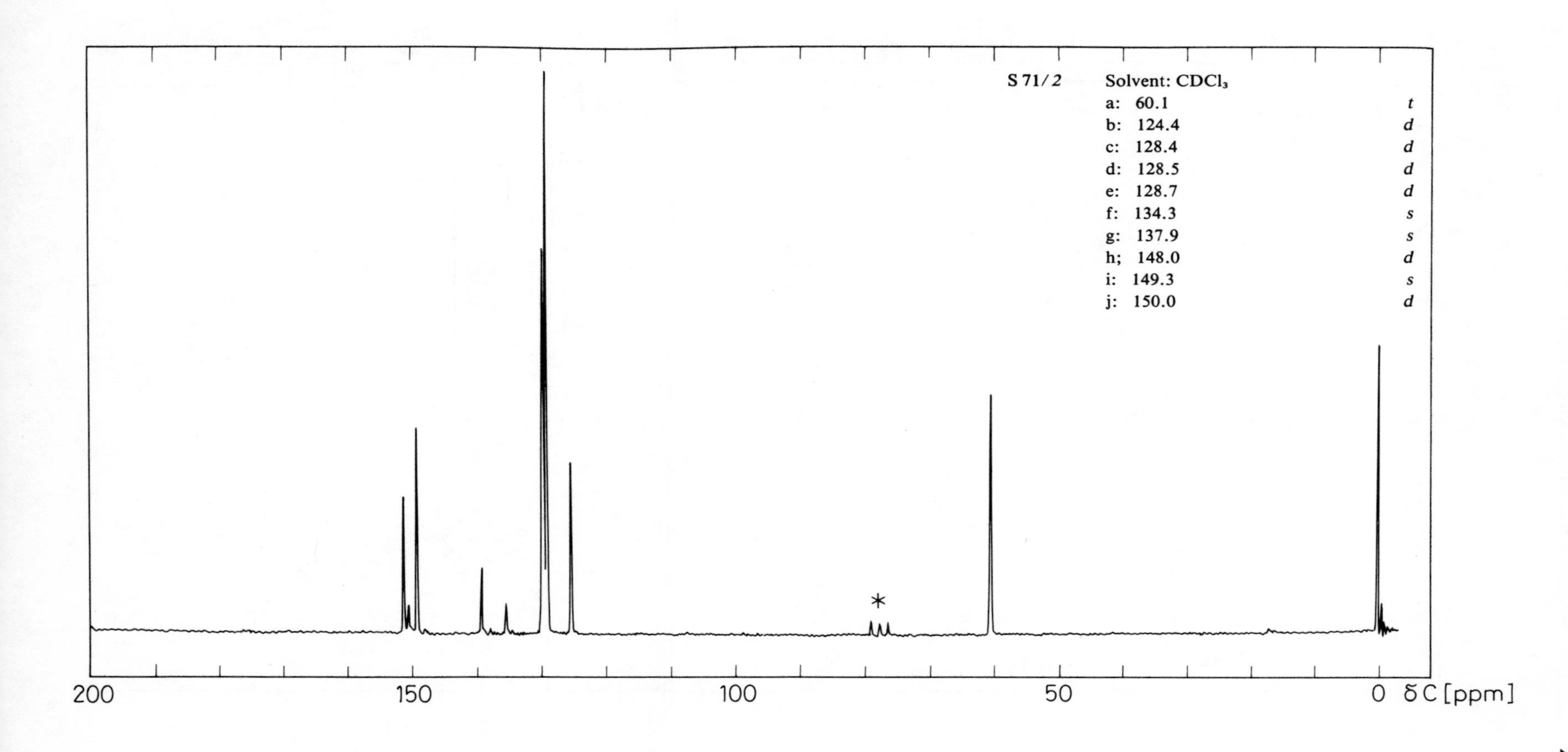

S 71/2
Solvent: CDCl3
a: 60.1 t
b: 124.4 d
c: 128.4 d
d: 128.5 d
e: 128.7 d
f: 134.3 s
g: 137.9 s
h; 148.0 d
i: 149.3 s
j: 150.0 d
*
200
150
100
50
0 δC [ppm]

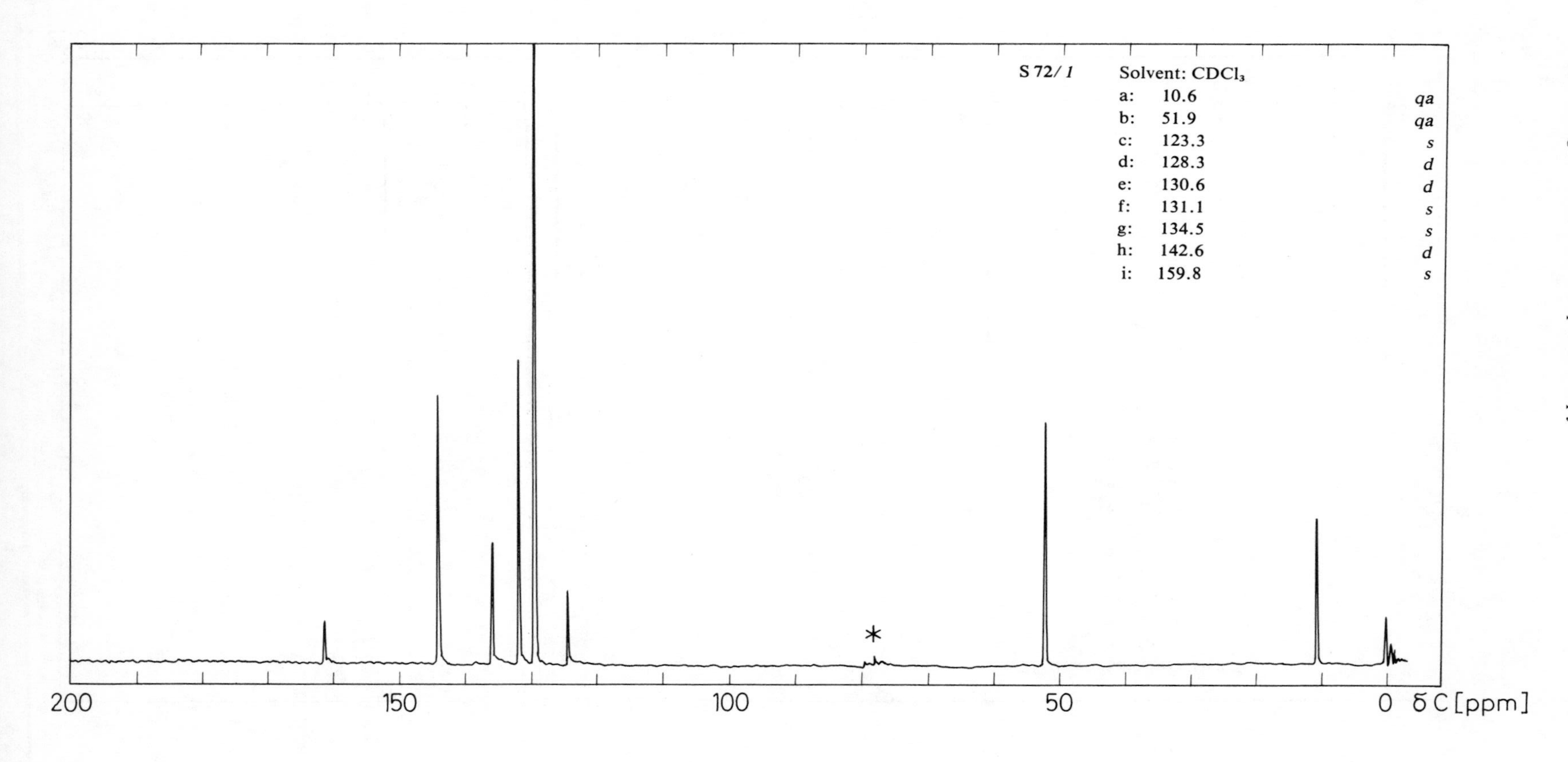
S 72/1
Solvent: $CDCl_3$
a: 10.6 qa
b: 51.9 qa
c: 123.3 s
d: 128.3 d
e: 130.6 d
f: 131.1 s
g: 134.5 s
h: 142.6 d
i: 159.8 s
*
200
150
100
50
0 δC [ppm]

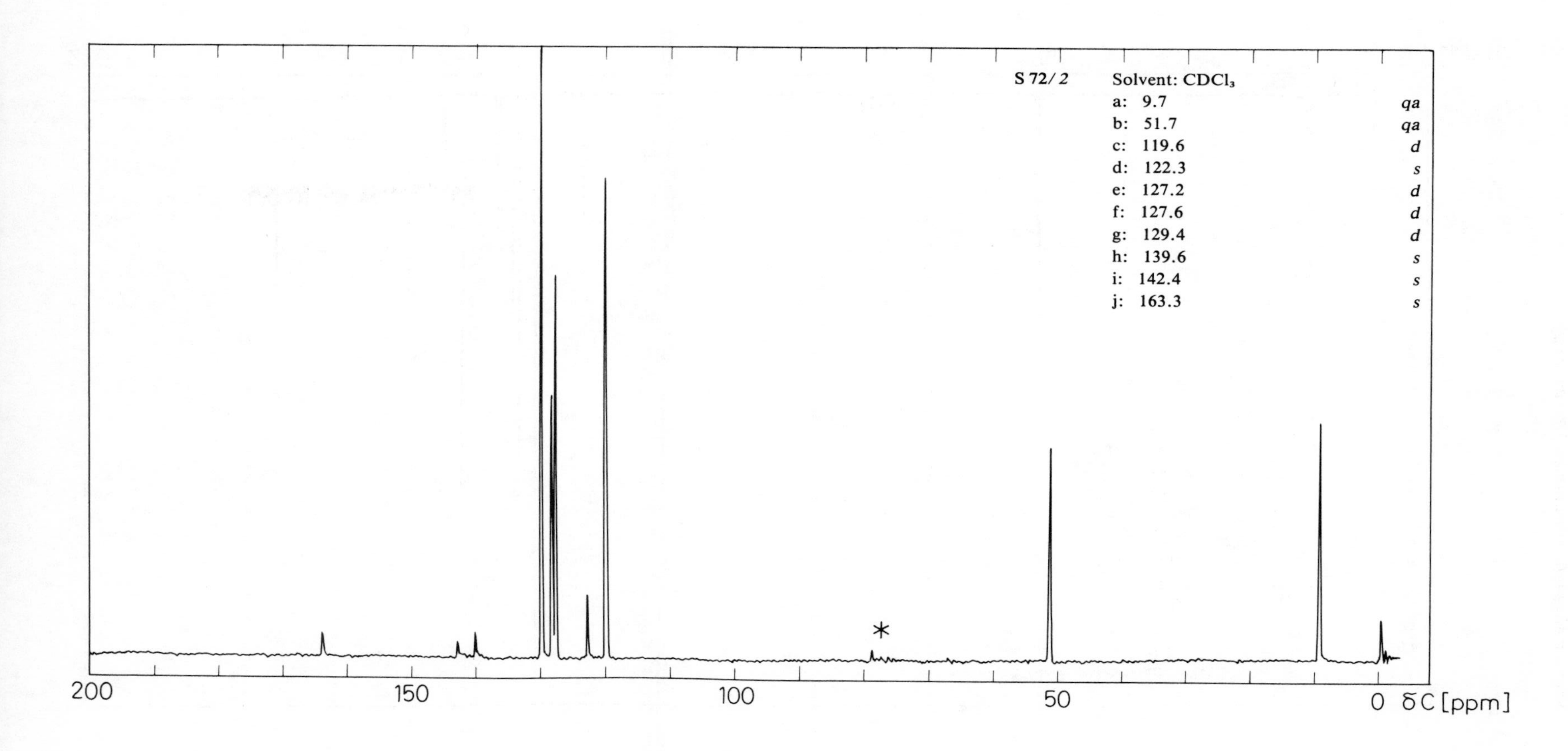
S 72/2
Solvent: $CDCl_3$
a: 9.7 qa
b: 51.7 qa
c: 119.6 d
d: 122.3 s
e: 127.2 d
f: 127.6 d
g: 129.4 d
h: 139.6 s
i: 142.4 s
j: 163.3 s
*
200
150
100
50
0 δC [ppm]

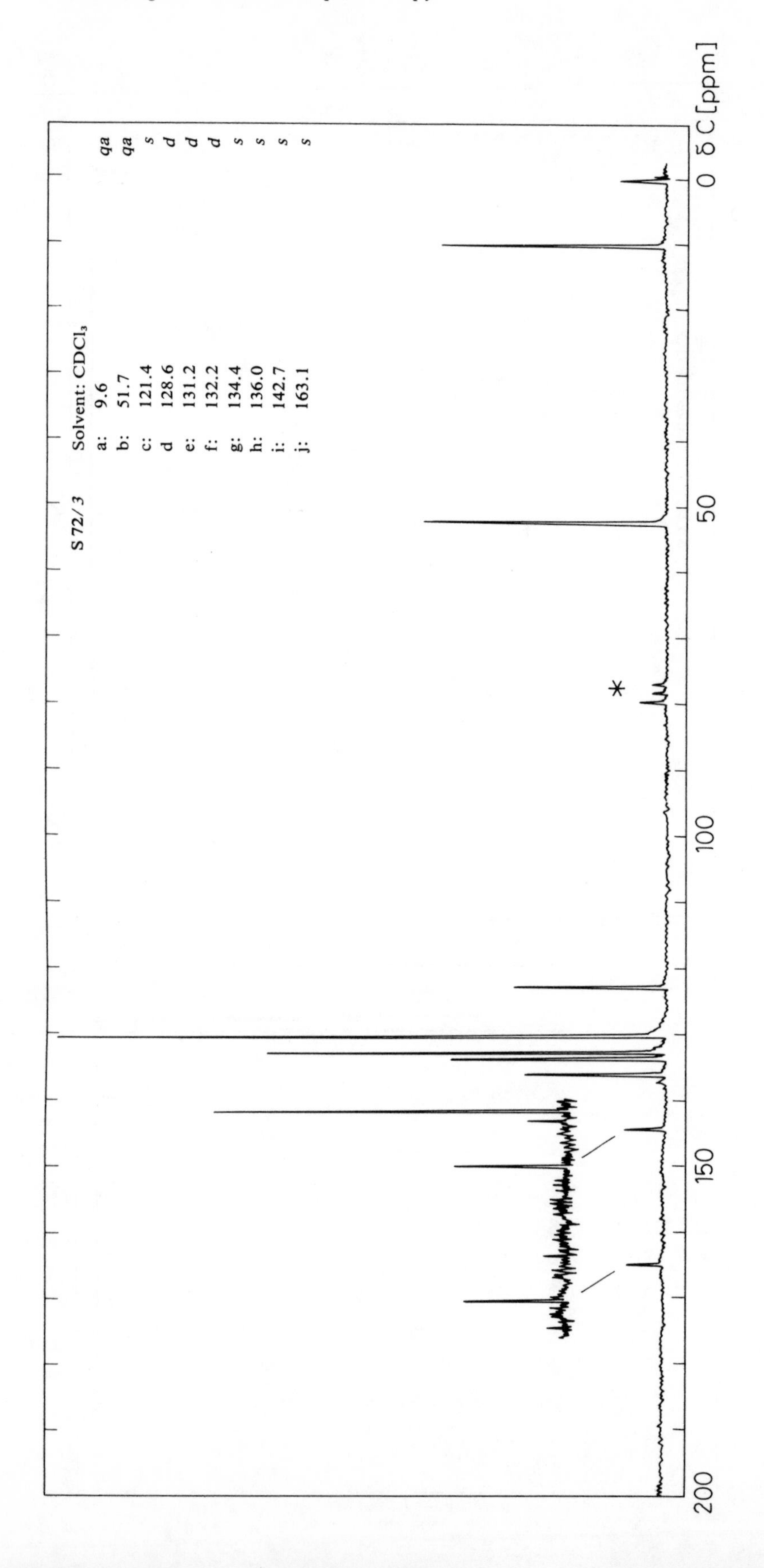
S 72/3
Solvent: $CDCl_3$
a: 9.6 qa
b: 51.7 qa
c: 121.4 s
d 128.6 d
e: 131.2 d
f: 132.2 d
g: 134.4 s
h: 136.0 s
i: 142.7 s
j: 163.1 s
*
200
150
100
50
0 δ C [ppm]

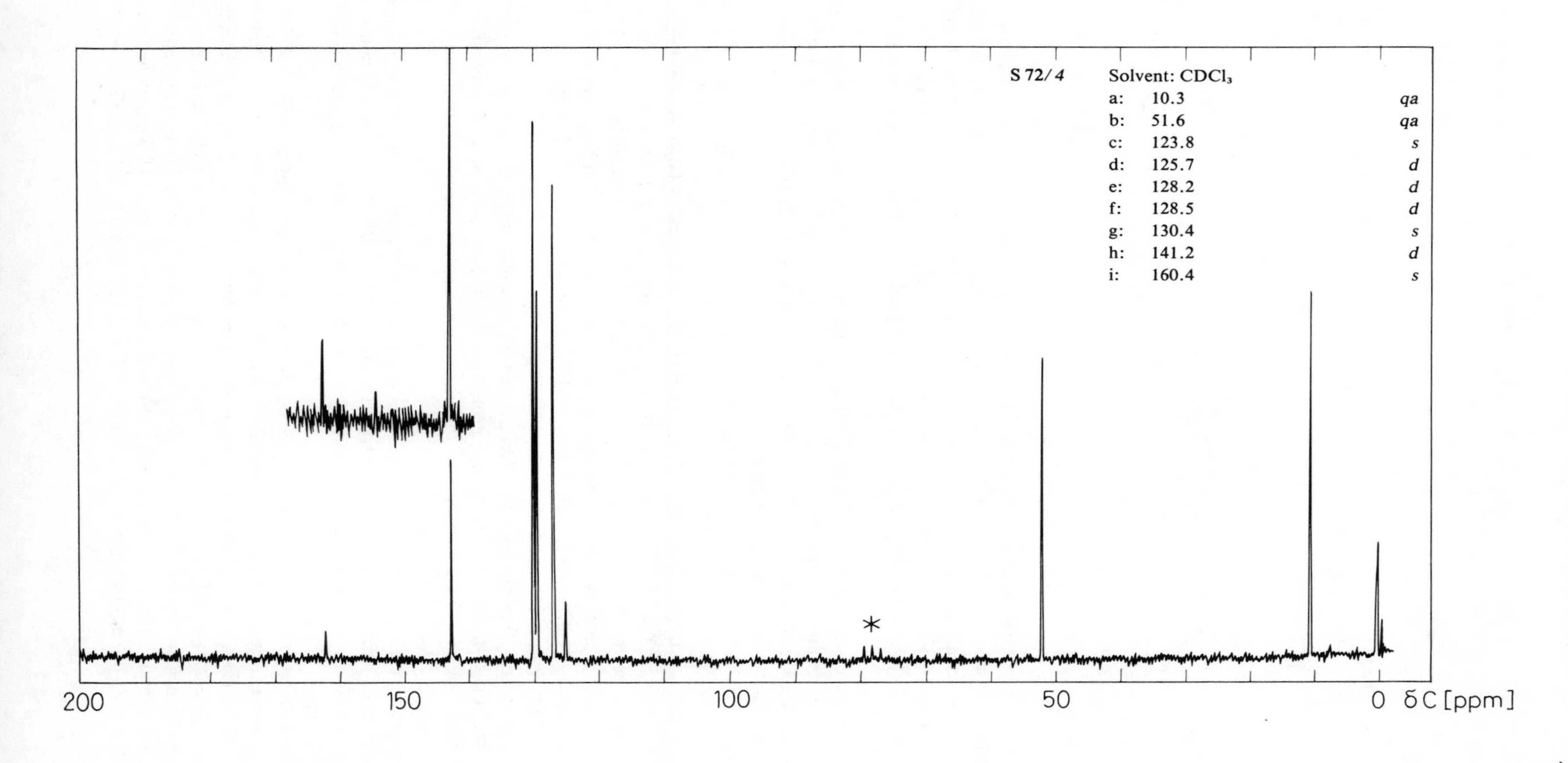

S 72/4
Solvent: CDCl3
a: 10.3 qa
b: 51.6 qa
c: 123.8 s
d: 125.7 d
e: 128.2 d
f: 128.5 d
g: 130.4 s
h: 141.2 d
i: 160.4 s
*
200
150
100
50
0 δC [ppm]

(**I,III**) > δC_{Me} (**II,IV**) and δC=O (**I,III**) < δC=O (**II,IV**). These relationships may be used to distinguish unknown isomers of analogous structure.

PROBLEM **73**[1326]

Assign Spectrum **73**/*1* of compound **I** and then identify the simple derivative from which Spectrum **73**/*2* was recorded.

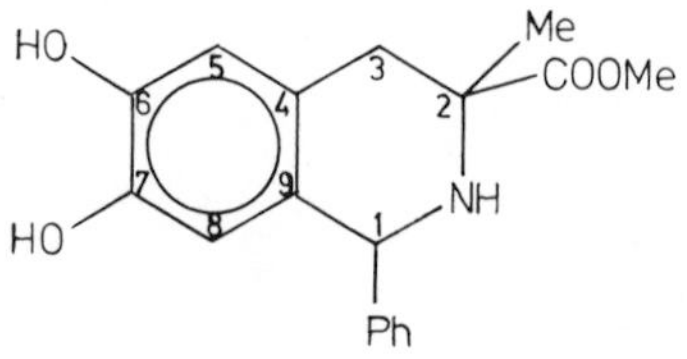

HO, HO, Me, COOMe, NH, Ph, 1, 2, 3, 4, 5, 6, 7, 8, 9

73/I

ANSWER

Of the five signals of saturated carbons, the upfield quartet belongs to the *C*-methyl group. The next signal, on the basis of its triplet splitting, is due to the methylene carbon. The next signal (quartet) arises from the methoxy group. Of the two signals at 60.2 and 60.4 ppm, on the basis of intensities as well as of multiplicities, the latter is the C-1 signal and the former (weaker and also singlet on the proton coupled spectrum) is the signal of quaternary C-2 atom.

Taking into account the anticipated effect of hydroxy groups (compare Table 74), the two, strong aromatic signals at the smallest shift belong to the C-5,8 atom pair and the next two, weaker signals belong to C-4,9, of which the downfield one at 128.5 ppm is the signal of C-9 adjacent to the tertiary carbon.

On the basis of intensities, the next three lines arise from the *meta, para* and *ortho* carbon atoms of the phenyl ring. Of the three lines in the downfield region of aromatic carbons, the two with the largest shifts probably belong to C-6,7 on the basis of their similar intensities, and the weaker line at 144.9 ppm arises from the substituted carbon atom of the phenyl ring. With the former, the relation δC-6 < δC-7, i.e., the assignment δC-6 = 147.7 and δC-7 = 145.2 ppm, is the more probable, since in *para* position with respect to C-6 there is a tertiary carbon, whereas to C-7 a methylene group can be found. The most deshielded line at 177.8 ppm is the carbonyl signal (Scheme 7).

On comparing Spectra **73**/*1* and **73**/*2*, it can be seen that atoms C-1,2,3 of the saturated ring and the *C*-methyl and *O*-methyl groups of the substituents on C-2 have practically invariant signals, and the four lines of the phenyl ring can also be found at almost the same frequencies.

However, the chemical shifts of the lines of the tetrasubstituted aromatic ring attached to the saturated heteroring are significantly different, and the difference is similar pairwise for the C-6,7, C-5,8, and C-4,9 atoms. Consequently, the structure of the compound is different in this ring.

Since among the saturated carbon signals two new upfield quartets occur in Spectrum **73**/*2* at almost identical chemical shifts (21.5 and 21.6 ppm) and there is a new line in the carbonyl region, too, at 169.4 ppm, it is clear that this spectrum corresponds to the 7,8-diacetoxy analogue (**II**) of **I**. The two quartets are the methyl signals, and the carbonyl signals coincide. The shifts of the tetrasubstituted aromatic ring are also in agreement with this assumption. The C-6,7 atoms are more shielded (141.5 and 141.2 ppm), whereas the C-5,8 and C-4,9 atoms more deshielded (123.1 and 124.2 and 133.4 and 137.4 ppm) than in the dihydroxy derivative, like with acetoxybenzene (compare Table 74). Signals of C-5 and C-8, similar to the spectrum of **I**, cannot be distinguished, owing to the hardly different shifts.

PROBLEM **74**[1327]

In the benzalation of 1,6-dibromomannitol (**I**), the formation of various diastereomeric 5,6-, or 7-membered cyclic mono- and dibenzal derivatives is possible. Of the four main products of this reaction, two proved to be identical to known compounds (**II** and **III**). Using the chemical shifts of the acetal carbon atoms of the diacetyl derivatives (**IV** and **V**) of these compounds (see on the formulas), determine the structures of the two other main products having Spectra **74**/*1* and **74**/*2*.

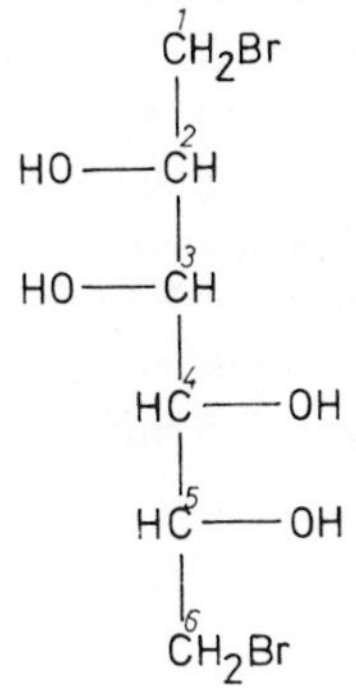

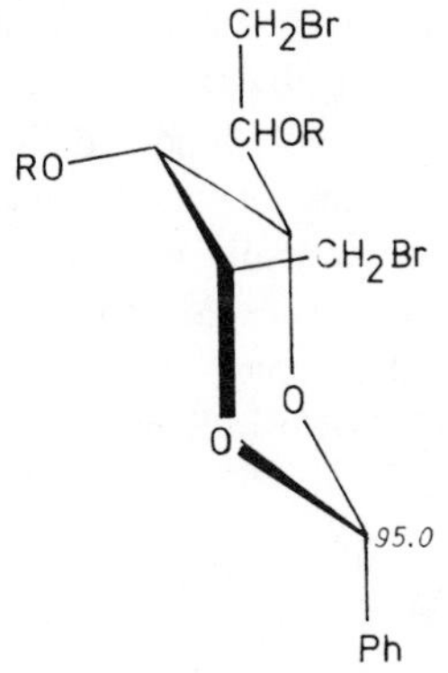

74/II (R : H)
74/IV (R : Ac)

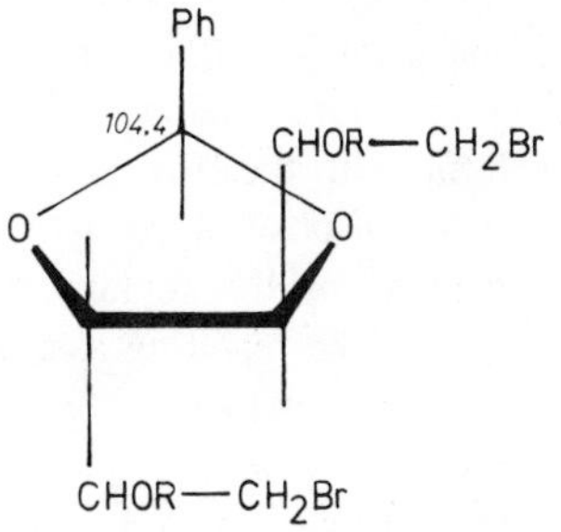

74/V (R : Ac)

ANSWER

Both spectra consist of eight lines, of which four correspond to the aromatic carbons of benzal group and the fifth corresponds to the acetal carbon. Thus, the three remaining lines can be assigned to the chemically equivalent atom pairs C-1,6, C-2,5, and C-3,4, i.e., both spectra arise from symmetrical molecules.

From the relative intensities of benzal and skeletal carbons, respectively, it is certain that both spectra correspond to dibenzal derivatives. Due to molecular symmetry, of the theoretically possible (5 + 5)-, (6 + 6)-, and (5 + 7)-membered cyclic dibenzal structures, the latter and the asymmetric diasteromers of the former two can be excluded. This leaves us with the two pairs of symmetrical 5 + 5 and 6 + 6 diastereomers, of which the ones corresponding to Spectra **74**/*1* and **74**/*2* should be selected. This requires the assignment of the spectra.

Of the four aromatic lines, the most downfield-shifted weak line belongs to the substituted carbon atom, and of the two strongest lines, the one closest to the shift of benzene* (i.e., the 128.6- and 128.0-ppm line in the two spectra) can be assigned to the *meta* and the other can be assigned to the *ortho* carbon atoms. The fourth aromatic line, which occupies a medium position both in shift and in intensity between the signals of substituted and unsubstituted (*ortho* and *meta*) carbon atoms, is the signal of *para* carbon (129.6 and 128.9 ppm).

* See Table 49 and Volume II, p. 187.

On the basis of chemical shift (96.1 and 94.6 ppm), the selection of acetal carbon is simple and the line of the C-1,6 secondary carbon atom pair is unambiguously assigned on the basis of their upfield position and the triplet splitting. The relatively strong shielding (32.4 and 29.1) may be attributed to the heavy atom effect.* The remaining two lines of the spectra, the upfield and downfield ones, belong to the atom pairs C-2,5 and C-3,6, respectively, since the latter is adjacent to two tertiary carbons, whereas the former has one secondary and one tertiary neighbor.

The closely similar line positions in the two spectra suggest that the two compounds are diastereomers with identical ring sizes. Regarding that the shift of acetal carbon (96.1 and 94.6 ppm) is very close to that measured for compound **IV** (95.0) and significantly different from that of compound **V** (104.4 ppm), it is also evident that the diasteromers containing two dioxane rings, i.e., compounds **VI** and **VII** with 6 + 6 rings, were formed. The remaining problem is the pairing of spectra with these structures. Comparing the shifts measured in the two spectra, C-1,6, C-2,5, and acetal carbons are more shielded and C-3,4 are less shielded in the case of Spectrum **74**/*2*. Since the field effect** increases the shielding of carbons bearing sterically hindered substituents, it is clear that Structure **VII** corresponds to Spectrum **74**/*2* (the *axial* and thus sterically hindered C-1,6 atoms and C-2,5 attached to them are more shielded). The opposite shift of C-3,4 signal may be interpreted so that the skeleton with the two dioxane rings is sterically more compressed in Structure **VI** (owing to the *axial* position of hydrogens, the ''H-inside'' form is less stable, and this causes the field effect on the signals of anellated carbons, since the latter bear the C-2,5 and acetal carbons). The resulting assignment is shown in Scheme 8.

* Compare Section 4.1.1.3, point 3, pp. 152—154.

** Compare Section 4.1.1.3, point 4, pp. 154—155.

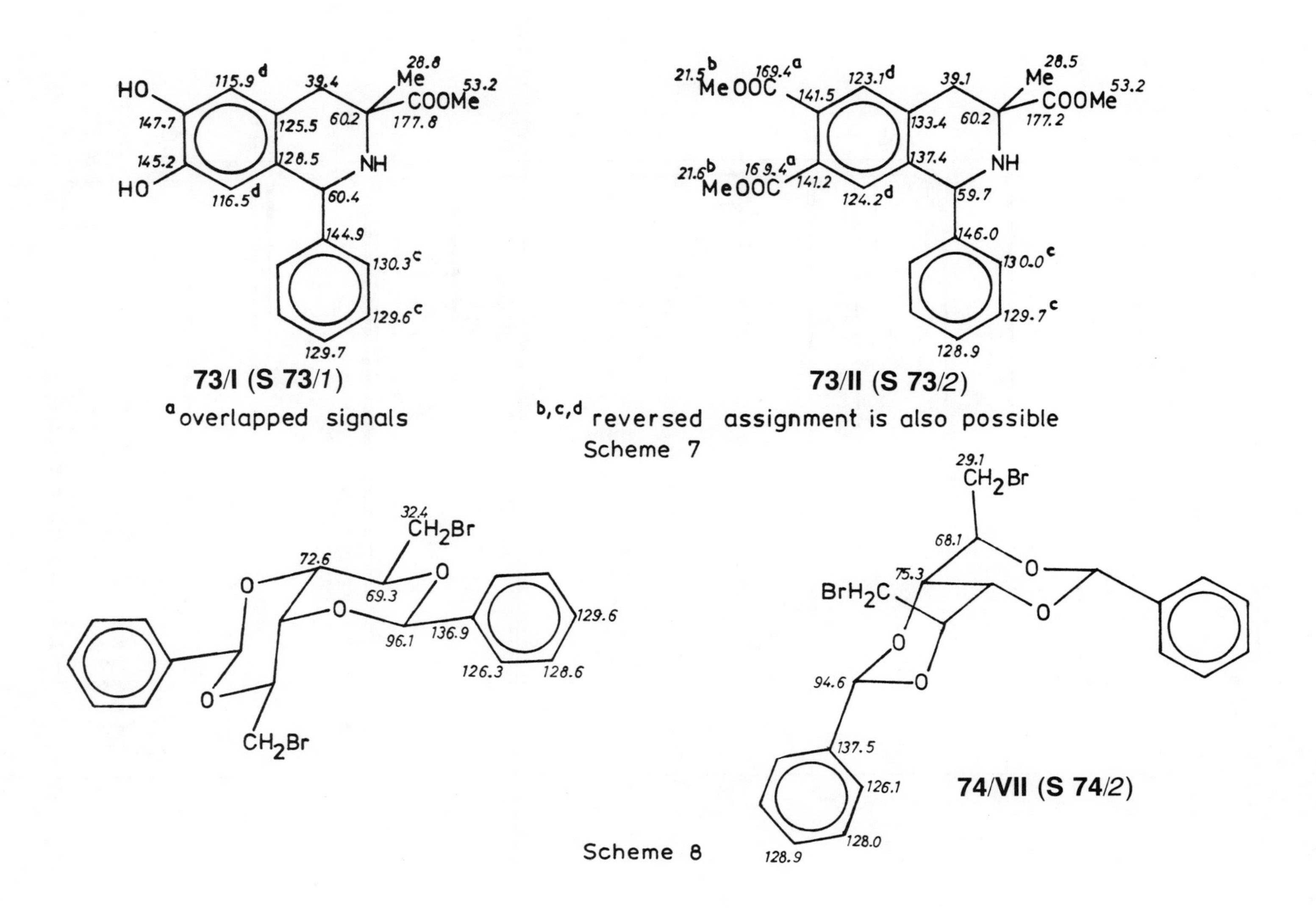

73/I (S 73/1)

[a] overlapped signals

73/II (S 73/2)

[b,c,d] reversed assignment is also possible

Scheme 7

74/VII (S 74/2)

Scheme 8

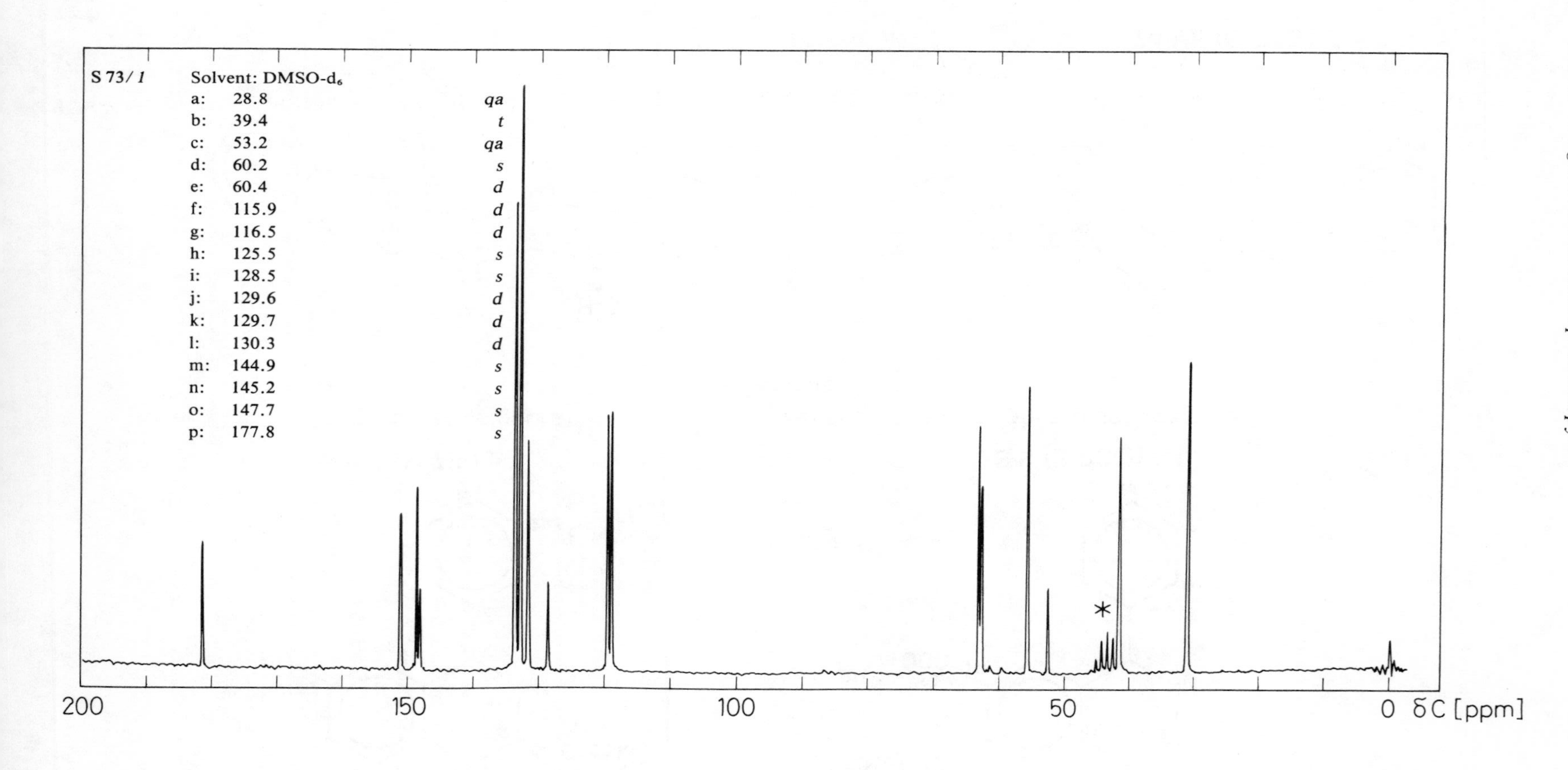
S 73/1
Solvent: DMSO-d6
a: 28.8 qa
b: 39.4 t
c: 53.2 qa
d: 60.2 s
e: 60.4 d
f: 115.9 d
g: 116.5 d
h: 125.5 s
i: 128.5 s
j: 129.6 d
k: 129.7 d
l: 130.3 d
m: 144.9 s
n: 145.2 s
o: 147.7 s
p: 177.8 s
*
200
150
100
50
0 δC [ppm]

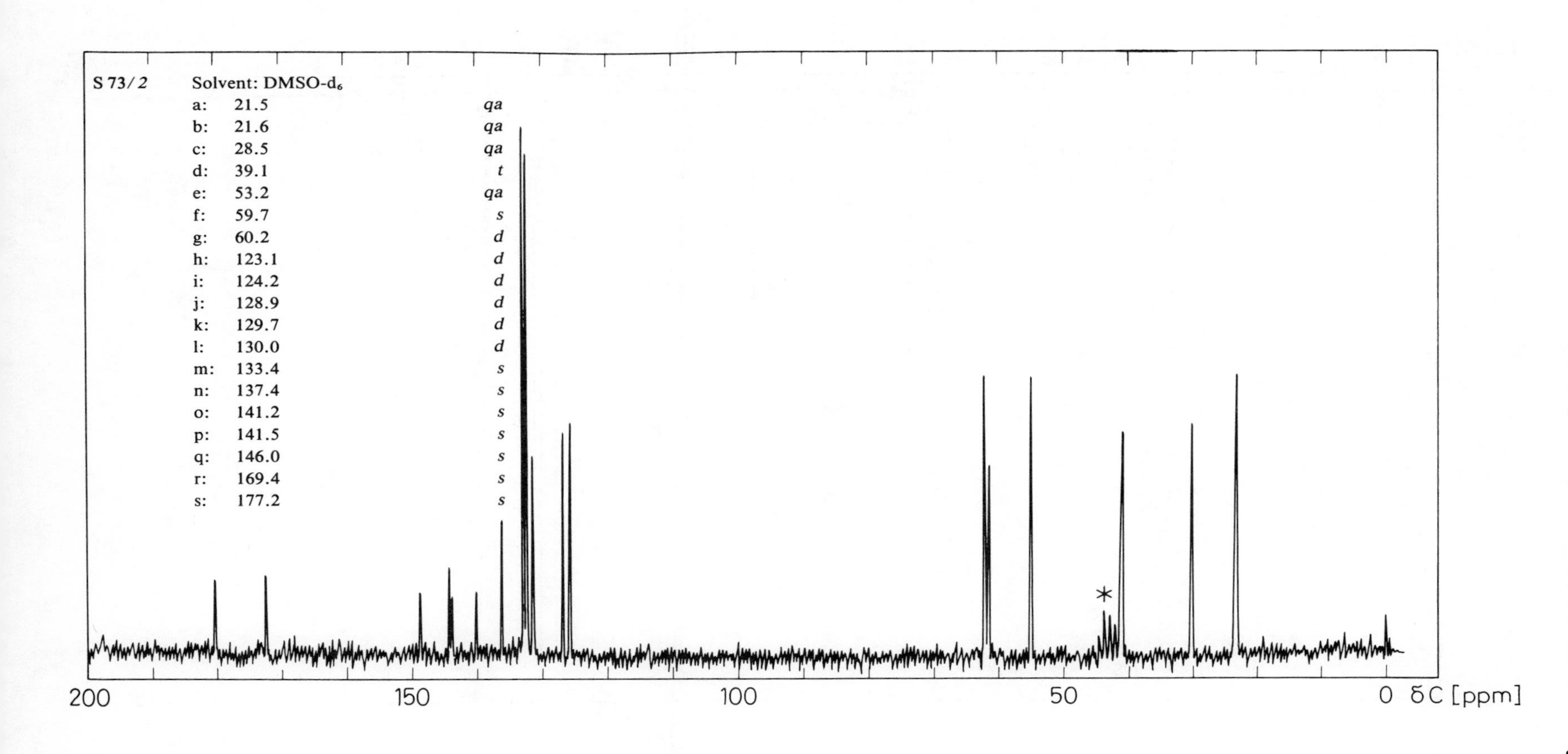
S 73/2
Solvent: DMSO-d6
a: 21.5 qa
b: 21.6 qa
c: 28.5 qa
d: 39.1 t
e: 53.2 qa
f: 59.7 s
g: 60.2 d
h: 123.1 d
i: 124.2 d
j: 128.9 d
k: 129.7 d
l: 130.0 d
m: 133.4 s
n: 137.4 s
o: 141.2 s
p: 141.5 s
q: 146.0 s
r: 169.4 s
s: 177.2 s
*
200
150
100
50
0 δC [ppm]

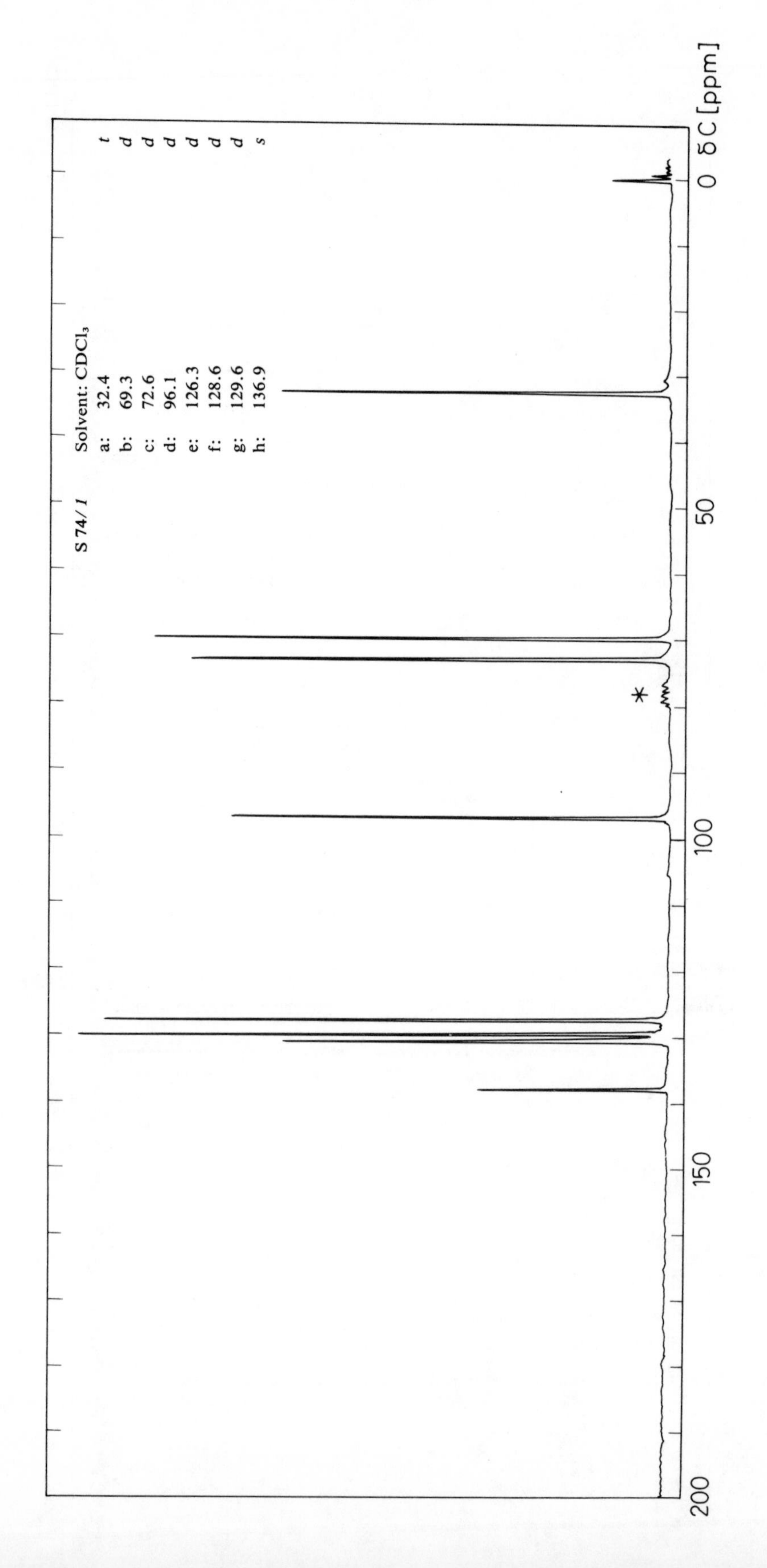

S 74/1
Solvent: CDCl3
a: 32.4 t
b: 69.3 d
c: 72.6 d
d: 96.1 d
e: 126.3 d
f: 128.6 d
g: 129.6 d
h: 136.9 s
*
0
50
100
150
200
δC [ppm]

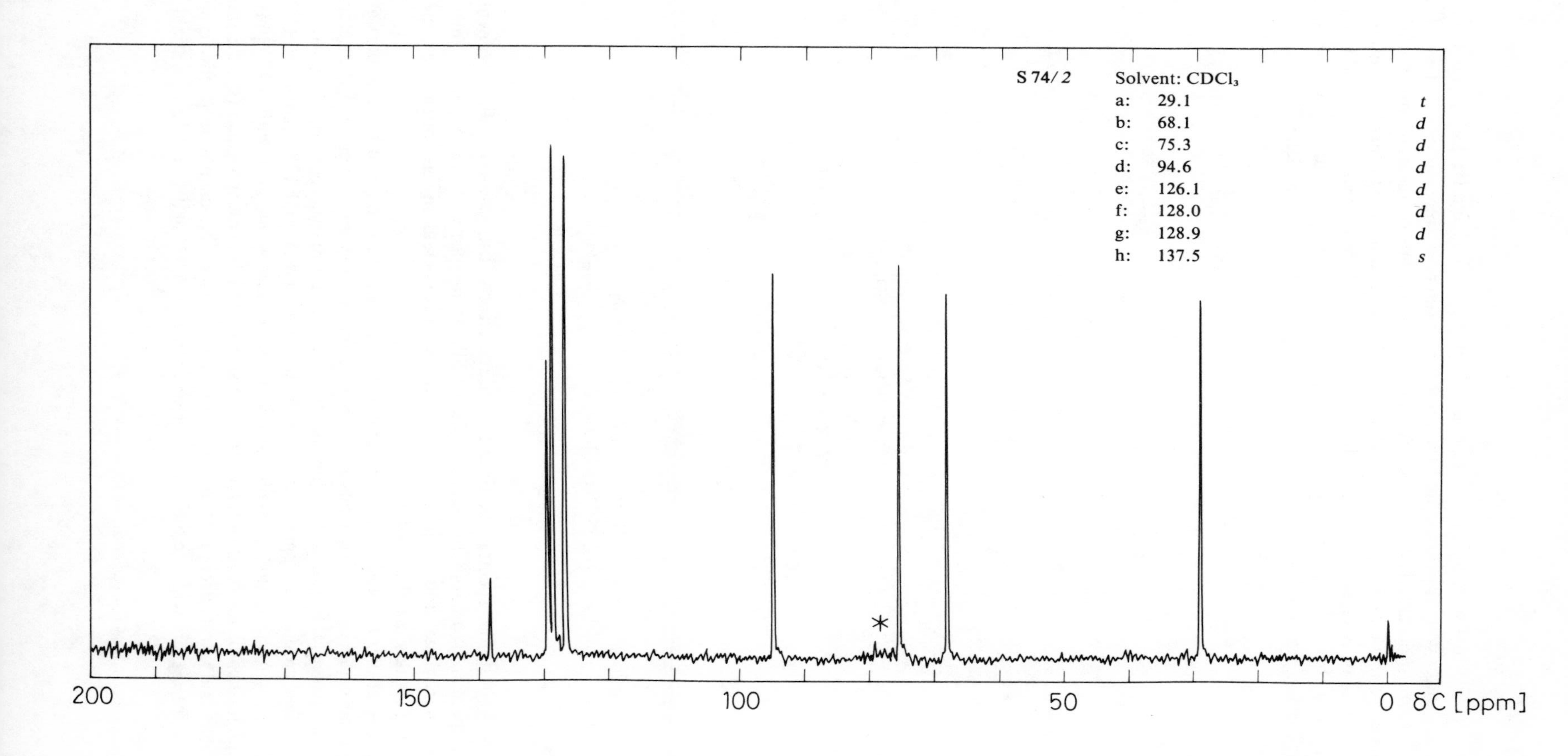
S 74/2
Solvent: $CDCl_3$
a: 29.1 t
b: 68.1 d
c: 75.3 d
d: 94.6 d
e: 126.1 d
f: 128.0 d
g: 128.9 d
h: 137.5 s
200
150
100
50
0
δC [ppm]

PROBLEM **75**[1325]

Spectra **75**/*1* to *4* were taken from *cis-trans* isomer pairs **I-II** and **III-IV**. Assign the lines, pair the spectra with structures, and interpret the differences, making use of the shift data of *cis* and *trans* decalines.* Is it possible to draw conclusions on the conformation of these molecules, if from the temperature invariance of the spectra it follows that the conformation of these systems is homogeneous? In the determination of steric structure it should be taken into account that with compound **III** in the more, crowded C* conformation (see below) the aryl group may not be *axial*, as also proved by the molecule model.

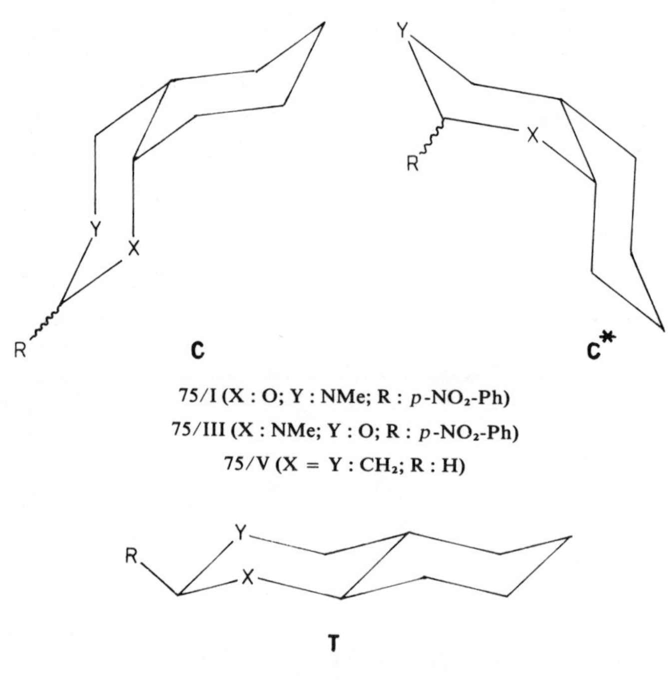

75/I (X : O; Y : NMe; R : *p*-NO_2-Ph)

75/III (X : NMe; Y : O; R : *p*-NO_2-Ph)

75/V (X = Y : CH_2; R : H)

75/II (X : O; Y : NMe; R : *p*-NO_2-Ph)

75/IV (X : NMe; Y : O; R : *p*-NO_2-Ph)

ANSWER

In the four similar spectra, the lines form three clearly distinct groups. In the downfield region, there are four aromatic signals in the middle of the spectra the two signals of methine and one methylene group adjacent to the heteroatom, whereas in the upfield part, the six remaining signals may be observed.

The two strong aromatic signals belong to C-3′,5′ and C-2′,6′, and since a nitro group shields the *ortho* carbons, but has only weak effect on the *meta* ones, and the saturated carbon influences the aromatic carbons neither in *ortho* nor in *meta* position (compare Table 74), the line around 123 ppm is the C-3′,5′ and the one around 128 ppm is the C-2′,6′ signal. The two substituted aromatic carbon signals of almost identical shift (147 ppm) can be distinguished on the basis of intensities; the signal of C-4′ is much weaker (see the spectra of toluene and nitrobenzene).[729] In Spectrum **75**/*3*, the C-4′ line cannot be identified at all.

The three signals with medium shifts arise from C-1, C-2, and C-4. The downfield line

* Compare the data in Equations 262 and 264.

arises from the C-1 atom situated between two heteroatoms. The two other lines can be identified on the basis of multiplicities. Oxygen causes a larger downfield shift in the adjacent carbon than nitrogen. Therefore, Spectra **75**/*2* and **75**/*4* belong to isomer pair **I-II**, since the triplet of C-2 is upfield from the C-4 doublet, i.e., the oxygen atom is adjacent to the C-4 atom. In Spectra **75**/*1* and **75**/*3*, arising from isomers **III** and **IV**, the triplet is downfield shifted.

Of the remaining six lines, the identification of the C-3 signal and the methyl signal is unambiguous on the basis of multiplicities, but the lines of C-5,6,7,8 may be assigned only on the basis of the steric structure of the molecule.

The *cis* and *trans* isomers may be selected on the basis of the C-3 and C-4 shifts. Taking into account the field effects observed for decaline,* it may be anticipated that the C-3,4 atoms are much more shielded in the *cis* isomers. Indeed, in Spectra **75**/*2* and **75**/*4* of isomers **I** and **II** the shifts of these two signals are different: in **75**/*4* they are downfield by 2.1 and 7.4 ppm, respectively. Consequently, this spectrum can be assigned to the *cis* isomer (**I**) and Spectrum **75**/*2* can be assigned to the *trans* (**II**). Likewise, on the basis of upfield shifts in **75**/*3* by 8.1 and 5.6 ppm, this spectrum corresponds to the *cis* isomer **III** and **75**/*1* to the *trans* isomer **IV** (Scheme 9).

Concerning the conformations it can be assumed that the *trans* isomers are evidently in the stable chair-chair conformation. Moreover, it should be taken into account that the two stable chair-chair conformations, of the *cis* isomers, which are equivalent in the case of decaline, are significantly different for compounds **I** and **III**.

In *cis* decaline, there are two pairs of α- and β-methylene carbons in sterically less or more hindered positions with both conformers, and via ring inversion the two groups of carbons exchange roles. Thus, the spectrum gives an average of the shifts corresponding to sterically hindered and not hindered positions, respectively. If in the case of compounds **I** and **III** either of conformations C or C* dominates, the field effect is much stronger for certain signals and absent for others (depending on the hindered or nonhindered position of the corresponding carbons in the preferred conformation).

In conformation C* large field effects may be anticipated on the C-1 and C-2 signals. In Spectrum **75**/*3* of compound **III** this is really observable, with respect to the Spectrum **75**/*1* of the isomer pair (**IV**). The shift of C-1 line decreases by 8.9 ppm and that of C-2 decreases by 5.7 ppm. (The field effect on both carbons is stronger than in decaline.**) Consequently, compound **III** is in conformation C* (*equatorial* NMe group and *axial* heterocyclic methylene group with respect to the cyclohexyl ring). In contrast, the *cis* isomer (**I**) has an C conformation (*axial* oxygen and *equatorial* heterocyclic methylene group with respect to the cyclohexyl ring). Here, the shifts of C-1 and C-2 are lower in Spectrum **75**/*2* of the *trans* isomer, and in the C conformer these carbons are in free position (the hindered positions are occupied by the heteroatoms).

Since in compound **III**, for steric reasons, the aryl group cannot be *axial*, and with respect to the shift of C-1 measured in Spectrum **75**/*3* (85.7 ppm) no field effect can be observed in the other three spectra (the C-1 atom is deshielded in all the other compounds), it is certain that the aryl group is *equatorial* in all four derivatives. By this the problems of configuration and conformation for the four compounds are settled; the remaining question is the assignment of signals C-5,6,7,8 and the interpretation of the chemical shifts on the basis of the steric structures assumed.

It may be anticipated that C-5 and C-8 are deshielded, since one of their neighbors is tertiary carbon. In the spectra of *trans* isomers this effect is not disturbed by other effects, and thus the signals of C-6 and C-7 atoms are at 25.3 and 25.8 ppm in Spectrum **75**/*1* and

* Compare the data in Equations 262 and 264.

** Compare Volume II, p. 264.

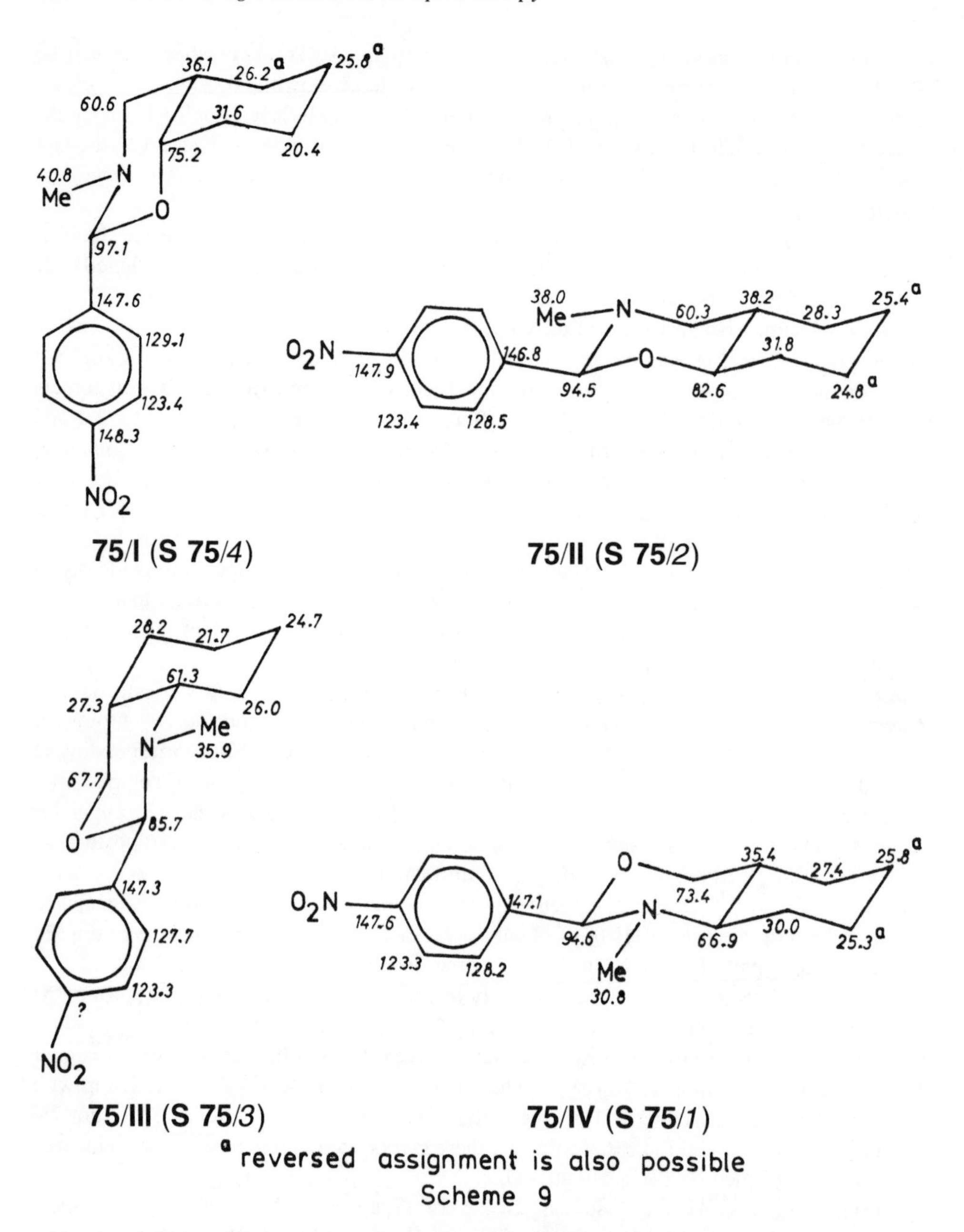

75/I (S 75/*4*) **75/II (S 75/*2*)**

75/III (S 75/*3*) **75/IV (S 75/*1*)**

[a] reversed assignment is also possible

Scheme 9

at 24.8 and 25.4 in Spectrum **75**/*2*. The two other lines can also be distinguished from one another, assuming that the neighborhood of tertiary carbon bearing a heteroatom causes a larger deshielding, and thus δC-5 > δC-8. The difference ought to be larger in Spectrum **75**/*2* (in the corresponding structure the heteroatom is oxygen, whereas in Structure **IV** giving Spectrum **75**/*1* it is a nitrogen). Accordingly, the C-5 and C-8 signals of compound **II** appear at 31.8 and 28.3 ppm, while those of compound **IV** at 30.0 and 27.4 ppm.

With the *cis* isomers, the field effect must also be taken into account. In compound **I** the shifts of C-6 and C-8 decrease; in the case of **III** the C-5 and C-7 atoms are expected to be more shielded. Thus, in Spectrum **75**/*4* the C-5 signal is at 31.6 ppm and the C-7 line is at 25.8 ppm (hardly different from that of **75**/*2*; the C-8 signal shifts to 26.2 and the C-6 signal shifts to 20.4 ppm. (An alternative, reverse assignment of C-7 and C-8 may not be excluded, but this does not affect our conclusions on structure.) It is also probable that in Spectrum **75**/*2*, δC-7 = 25.4 and δC-6 = 24.8 ppm (it is unlikely that the shift of the former increases by 1 ppm, and such a difference may not arise from experimental error). In Spectrum **75**/*3* the C-7 signal is at 21.7 ppm, and the C-6 and C-8 signals are hardly different in chemical shift from the corresponding lines of **75**/*1* (24.7 and 28.2 ppm), and thus the C-5 shift is 26.0 ppm. The difference (4.0 ppm) is about the same as for the C-6 signal of the isomer pair **I-II** (4.4 ppm).

PROBLEM **76**[1335]

In a reaction of **I** with vinyl-methyl ketone, two products (**II** and **III**) are obtained, which transform into the same compound (**IV**) on heating. Determine the Structures **II** and **III** on the basis of the Spectra (**76**/*1* and **76**/*2*) and explain the differences in the number and chemical shift of the signals. What is the Structure **IV** with Spectrum **76**/*3*? Acylation of **II** and **III** with acetic anhydride yields two pairs of product (**V** and **VI** and **VII** and **VIII**, respectively), which are all different, but have pairwise analogous structures (**V** and **VII** and **VI** and **VIII**). The chemical shifts of compounds **V** to **VIII** are as follows:

V: 19.9, 22.2, 28.9, 40.7, 52.4, 92.6, 115.1, 122.5, 138.4, 152.8, and 170.4.

VI: 19.5, 22.3, 27.2, 28.9, 40.5, 50.8, 92.9, 113.5, 120.4, 124.2, 124.6, 131.7, 142.3, 155.4, 169.2, and 170.1.

VII: 22.7, 23.2, 28.8, 40.6, 54.7, 90.6, 116.6, 123.1, 136.8, 153.1, and 171.2.

VIII: 21.4, 23.8, 25.2, 27.1, 39.0, 49.7, 88.8, 113.2, 120.5, 124.1, 124.4, 131.7, 142.5, 155.5, 169.3, and 169.7.

What are the structures **V** to **VIII**? Knowing these structures can their spectra help in the determination of the configuration of compound **II** and **III**?

CH_2SH
76/I
+
$CH_2{=}CH{-}CO{-}CH_3$

76/II → **76/V** + **76/VI**
76/III → **76/VII** + **76/VIII**
76/II, **76/III** —heating→ **76/IV**

ANSWER

The extremely similar spectra of **II** and **III** and their derivatives, moreover the fact that on heating they convert into the same substance, suggest that the structures of **II** and **III** are very similar, presumably isomeric. Since in the spectra of both compounds there are five lines, each with chemical shifts characteristic of saturated carbons, it seems obvious that the olefinic bond has been saturated, i.e., an addition of the mercaptyl group took place onto the C=C double bond, with the formation of a $-CH_2-S-CH_2-CH_2-CO-CH_3$ chain. Since one of the above five lines is a quartet (25.2 and 26.5 ppm), the presence of a methyl group is certain. The multiplicities of the four other signals (two triplets, one doublet and one singlet) indicate that one of the methylene groups of the adduct was deprotonated in the reaction and one of the sp^2 carbons of the reagent was saturated (otherwise there were only four signals in the region of saturated carbons).

Since there are not carbonyl signals in the spectra of **II** and **III** (the line of the most deshielded carbon is at 156.4 and 153.7 ppm, in the spectra), this group had to react with the methylene group adjacent to the ring, with the formation of imidazolo-tetrahydrothiophene derivatives (i.e., its *cis* and *trans* isomers, see Scheme 10).

The shifts and splittings are in agreement with these structures. In addition to the already-mentioned methyl line, the assignment of the other signals arising from saturated carbons is unambiguous on the basis of their multiplicity. Of the triplets, the upfield one belongs to C-5 of the *S*-methylene group, since the C-4 shift is increased more by the adjacent quaternary C-3 atom than by the shift of C-5 by the heteroatom. Therefore, the signals corresponding in turn to C-2,3,4,5 in the two spectra are 56.9 and 54.4 (doublets), 84.0 and 81.9 (singlets), 43.5 and 45.6 (triplets), and 30.7 and 30.8 ppm (triplets).

In the region of aromatic carbons two strong and two weak lines may be observed, indicating that the carbons of the benzene ring are pairwise equivalent. This shows, in turn, that there is a fast tautomery of benzimidazole ring* under the experimental conditions. A proof for this is that the 116.3-ppm line of Spectrum **76**/*1* and particularly the corresponding 113-ppm line of **III** in Spectrum **76**/*2* is broad, indicating that in the latter case the proton exchange is not fast enough for the averaging of chemical environments.

The chemical equivalence of C-2′,7′, C-3′,6′, and C-4′,5′ atom pairs also proves that the nitrogen atom of imidazole is not substituted, excluding the theoretically possible case that in the reaction leading to **II** and **III**, not the SH but the NH group reacts with the double bond. This would lead to the formation of *cis* and *trans* condensed benzimidazothiazepines not excluded by the number, shifts, and multiplicities of the saturated carbon signals.

The assignment of aromatic carbon lines is simple: the stronger upfield signals belong to C-3′,6′, the downfield ones belong to C-4′,5′ (owing to the substituent effect of nitrogen which decreases the chemical shifts of the *ortho* and increases that of the *para* carbons of the benzene ring), the weak upfield line belong to C-2′,7′, and the hardly significant downfield ones belong to C-1′ (Scheme 10). The relatively strong deshielding of C-3 and C-1′ atoms are worth of attention: the former is due to the adjacent oxygen and tertiery carbon atoms, and the latter is due to the adjacent two nitrogens.

The configuration of the two isomers, i.e., the pairing of the two spectra with the *cis* and *trans* structures given in Scheme 10, may be determined on the basis of the field effect.** Since the hydroxy group is "bulkier" than the methyl, the more crowded Structure **III** should be assigned to the spectrum in which C-2 and C-3 bearing the substituents are more shielded. In contrast, the methyl signal is subjected to this shielding field effect in Structure **II**, since here the methyl group is, instead of H-5, in *cis* position with the benzimidazole ring. The suggested steric structures also explain the stronger broadening of the C-3′,6′ signal in the spectrum of **III**. It may be assumed, namely, that the proton exchange is slowed down by an intramolecular hydrogen bond between the OH group and the nitrogen (see Figure 215), which is possible only when the hydroxy group and the benzimidazole ring are *cis*, i.e., with isomer **III**. Thus, the stronger broadening of the C-3′,6′ signal may be accepted as a proof for the postulated configuration.

* Compare Problem 56.

** Compare Section 4.1.1.3, point 4, pp. 154—155.

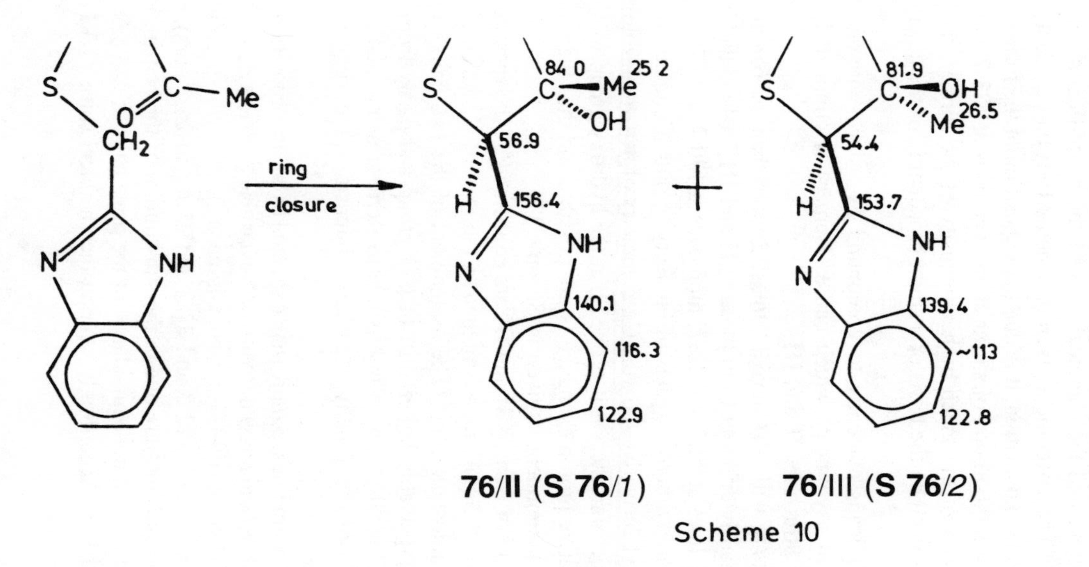

S
O
C
Me
CH2
N
NH
ring
closure
S
84 0
C
Me
25 2
OH
56.9
H
156.4
N
NH
140.1
116.3
122.9
76/II (S 76/1)
S
81.9
C
OH
Me
26.5
54.4
H
153.7
N
NH
139.4
~113
122.8
76/III (S 76/2)

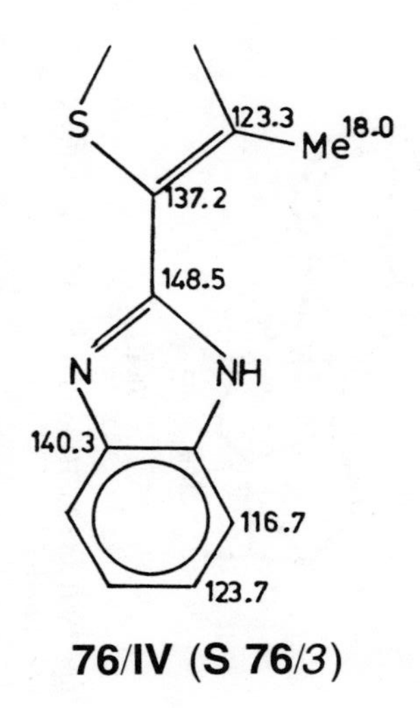

S
123.3
Me
18.0
137.2
148.5
N
NH
140.3
116.7
123.7
76/IV (S 76/3)

Scheme 10

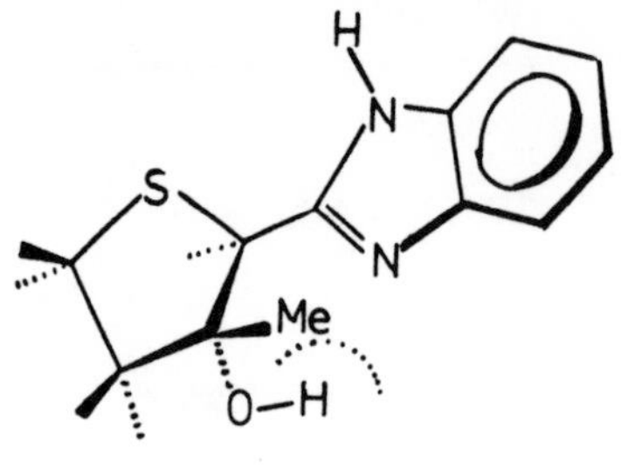

76/II 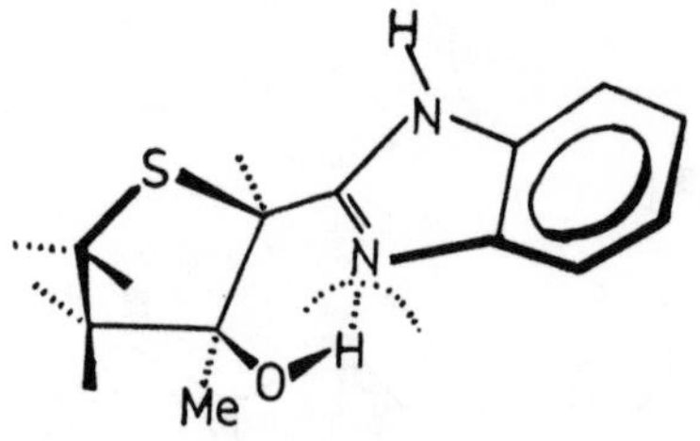 **76/III**

FIGURE 215. Steric structure of compounds **76/II** and **76/III** to illustrate the possibility in the latter to forming intramolecular hydrogen bond.

The derivative formed on heating from both isomers is the dehydrated dihydrothiophene analogue **IV**. Accordingly, the lines of imidazole carbons are practically unchanged, signals C-2 and C-3 shift into the olefinic region, and the chemical shift of the methyl carbon is characteristic of the $C(sp^2)CH_3$ groups.* The assignment given in Scheme 10 is thus evident, only the δC-2 > δC-3 relation needs some explanation: it is due to the position of the former inside the conjugation chain** and the neighborhood of sulfur atom. The lines at 123.7 and 123.3 ppm may be assigned on the basis of the much higher intensity of the former; the 137.2- and 140.3-ppm lines may be assigned on the basis of a slight broadening of the latter (proton exchange).

The dehydration of either **II** or **III** leads evidently to compound **IV**, since with the formation of double bond the center of asymmetry which causes isomery is eliminated. Thus, Structure **IV** is a final proof for Structures **II** and **III**.

The spectra of the acylation products **V** with **VII** have 11 lines, 2 more than the parent compounds. Nine of these signals are hardly different from those of **II** and **III**, making their assignment evident. The two remaining lines at 22.2 and 22.7, moreover at 170.4 and 171.2 ppm, may be assigned to the methyl and carbonyl signals of an acetyl group. Compounds **V** and **VII** are, therefore, the monoacetyl derivatives. By the C_{2v} symmetry of benzimidazole ring, following from the number of spectrum lines, it is proved that the hydroxy group was acylated; in the case of *N*-acetyl derivative proton exchange is not possible and thus the C-2′,7′ lines appear separately. The full assignment is given in Scheme 11.

The assumptions on the steric structure of isomers **II** and **III** are confirmed by the spectra of **V** and **VII**: the 3-methyl carbon is by 3.3 ppm more shielded in the case of **V**, which is due to the field effect of the *cis* benzimidazole ring. (The assignment of the two pairs of methyl signals arises partly from the expectably larger shift of C-3 group and partly from the fact that the shift of freely moving acetyl groups is insensitive to steric structure and is presumably hardly different for the two isomers. Finally, the relative intensities, too, suggest the given assignment.)

Compounds **VI** and **VIII** have 16 spectrum lines, 3 more than expected by the incorporation of 2 acetyl groups, i.e., the formation of *N,O*-diacetyl derivative. Nevertheless, diacetylation is evident from the two carbonyl and two acetyl-methyl signals each in the two spectra (169.2 and 170.1, 169.3 and 169.7, furthermore, 22.3 and 27.2 and 21.4 and 25.2 ppm).

On the basis of the above, the three extra lines are due to the substitution of the benzimidazole nitrogen and the consequently chemical nonequivalence of the three pairs of carbons equivalent in Structures **II**, **III**, **V**, and **VII**. Accordingly, among the chemical shifts of **VI**

* See Table 50.

** Compare Volume II, p. 176.

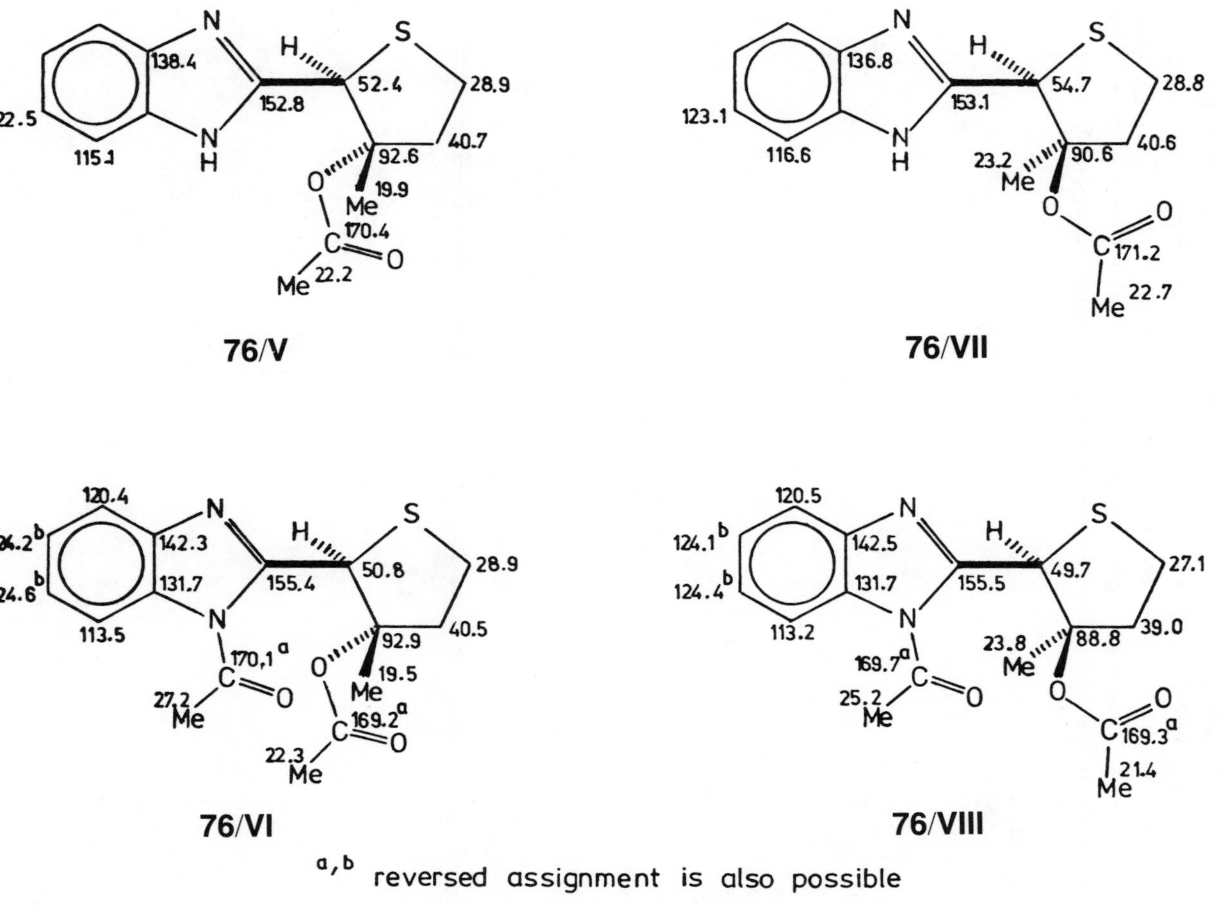

[a,b] reversed assignment is also possible

Scheme 11

and **VIII**, pairs of shifts occur close to the identical line pairs of the above compounds. Assuming the relations δC-2′ > δC-7′ and δC-3′ > δC-6′ arising from the strong $-I$ effect of sp^2 nitrogen, the assignment given an Scheme 11 can be made (the C-4′ and C-5′ lines cannot be distinguished).

It should be added that the assignment of the three methyl signals in each spectra arises in part from analogies with the monoacetyl derivatives **V** and **VII** and in part from the expected shift, approximately 27 ppm, for the *N*-acetyl group (the analogous signal of *N*-acetylbenzimidazole is at 27 ppm). There is a well-observable field effect on the C-3 methyl signal of isomer **VI** (the shift is 4.3 ppm lower than for **VIII**) and on the C-2,3,4, the acetyl methyl, and the acetyl carbonyl signals of **VIII**, which is more crowded in general. The steric compression shift arising from the two acetyl groups affects, therefore, 8 lines.

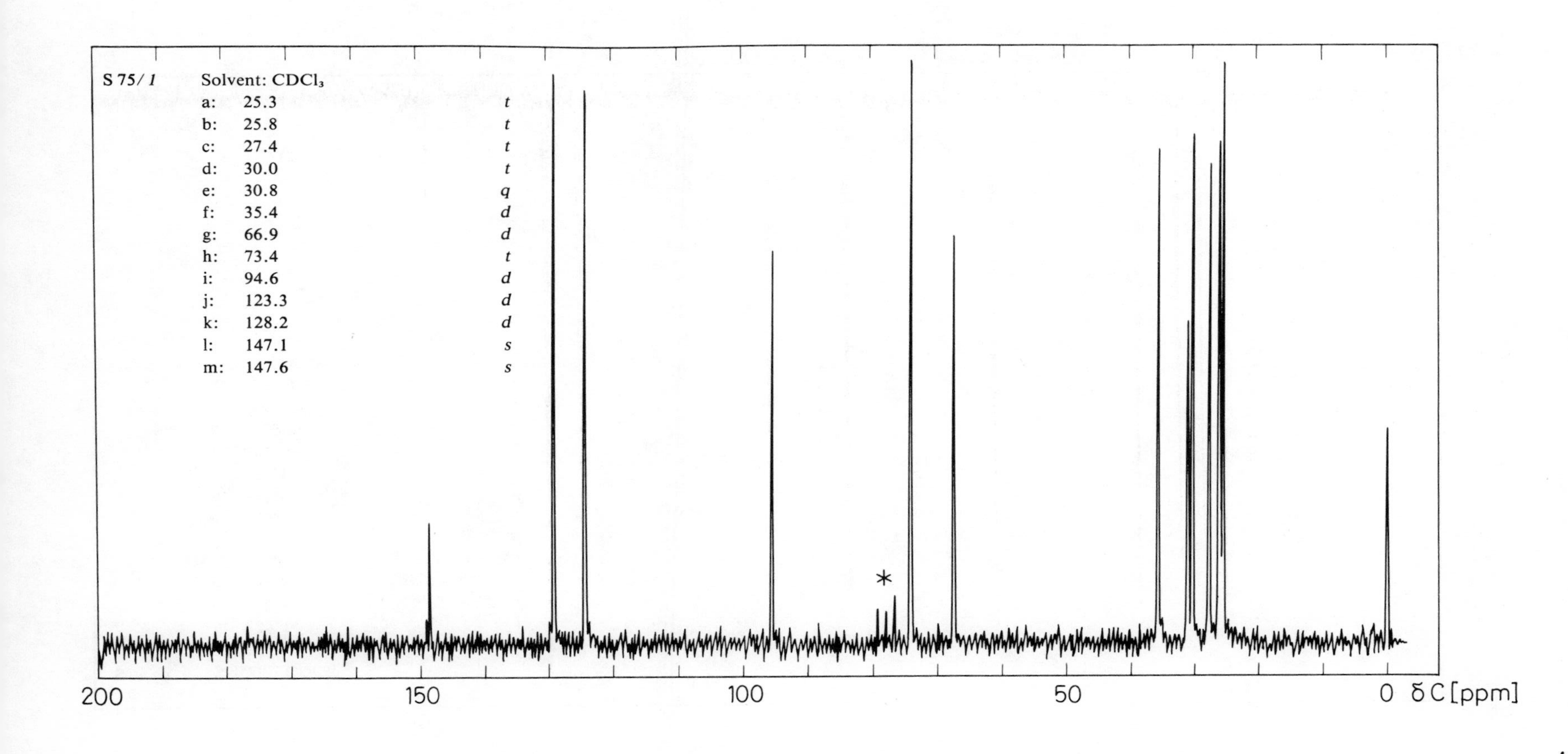
S 75/ 1
Solvent: $CDCl_3$
a: 25.3 t
b: 25.8 t
c: 27.4 t
d: 30.0 t
e: 30.8 q
f: 35.4 d
g: 66.9 d
h: 73.4 t
i: 94.6 d
j: 123.3 d
k: 128.2 d
l: 147.1 s
m: 147.6 s
*
200
150
100
50
0
δC [ppm]

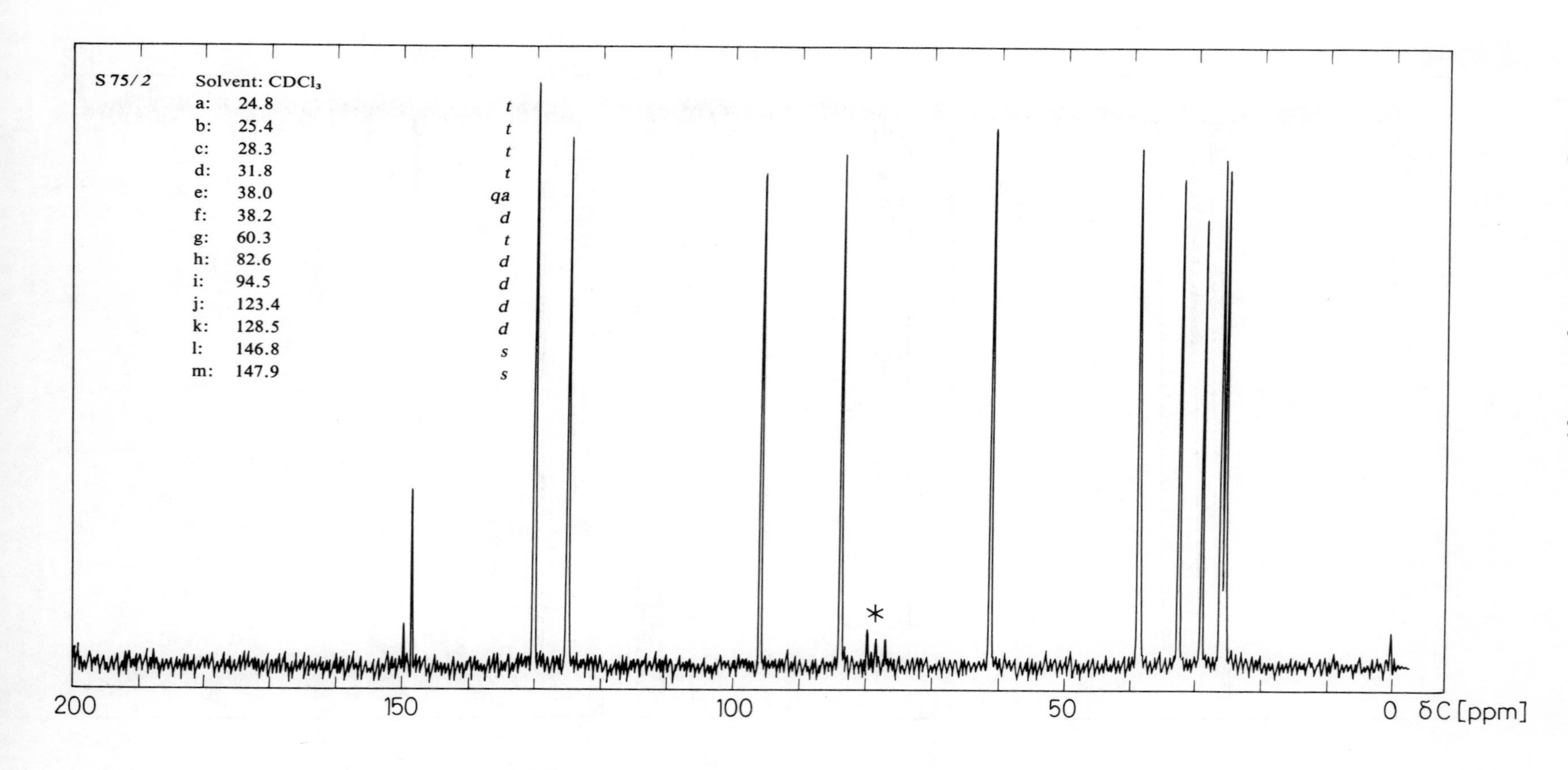
S 75/2
Solvent: CDCl₃
a: 24.8 t
b: 25.4 t
c: 28.3 t
d: 31.8 t
e: 38.0 qa
f: 38.2 d
g: 60.3 t
h: 82.6 d
i: 94.5 d
j: 123.4 d
k: 128.5 d
l: 146.8 s
m: 147.9 s
*
200
150
100
50
0
δC [ppm]

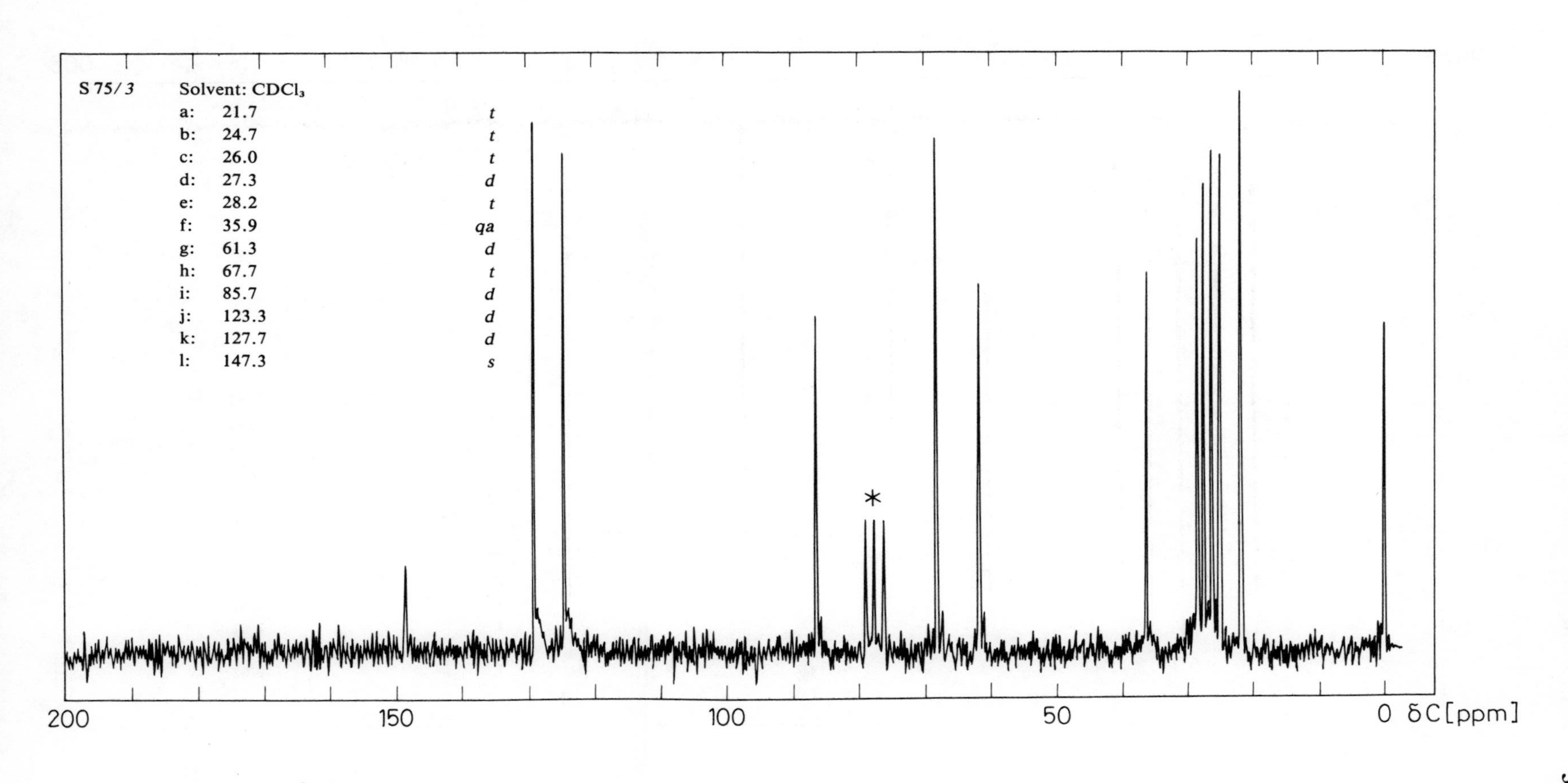

S 75/3
Solvent: CDCl3
a: 21.7 t
b: 24.7 t
c: 26.0 t
d: 27.3 d
e: 28.2 t
f: 35.9 qa
g: 61.3 d
h: 67.7 t
i: 85.7 d
j: 123.3 d
k: 127.7 d
l: 147.3 s
*
200
150
100
50
0 δC[ppm]

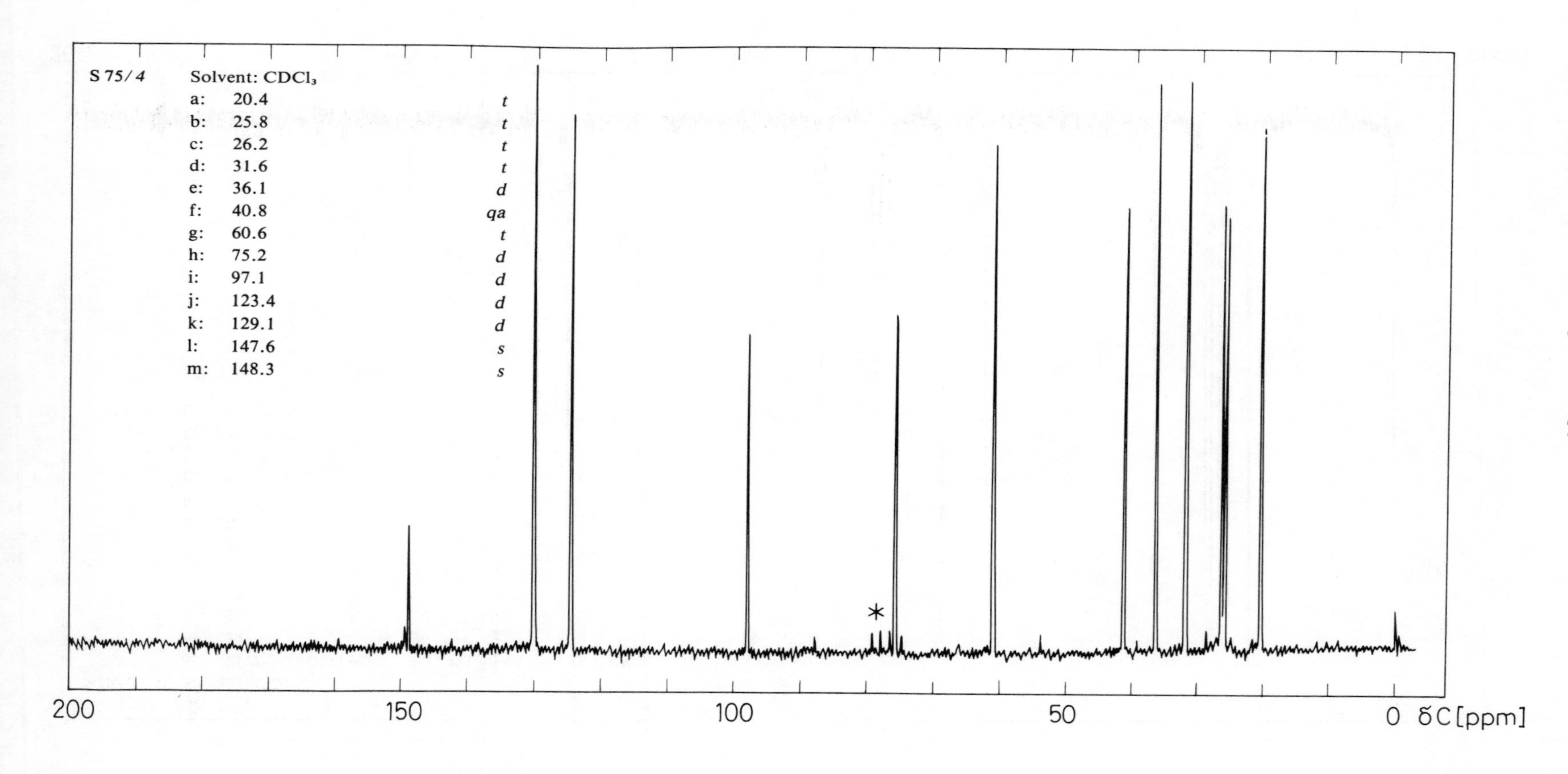
S 75/4
Solvent: $CDCl_3$
a: 20.4 t
b: 25.8 t
c: 26.2 t
d: 31.6 t
e: 36.1 d
f: 40.8 qa
g: 60.6 t
h: 75.2 d
i: 97.1 d
j: 123.4 d
k: 129.1 d
l: 147.6 s
m: 148.3 s
*
200
150
100
50
0
δC [ppm]

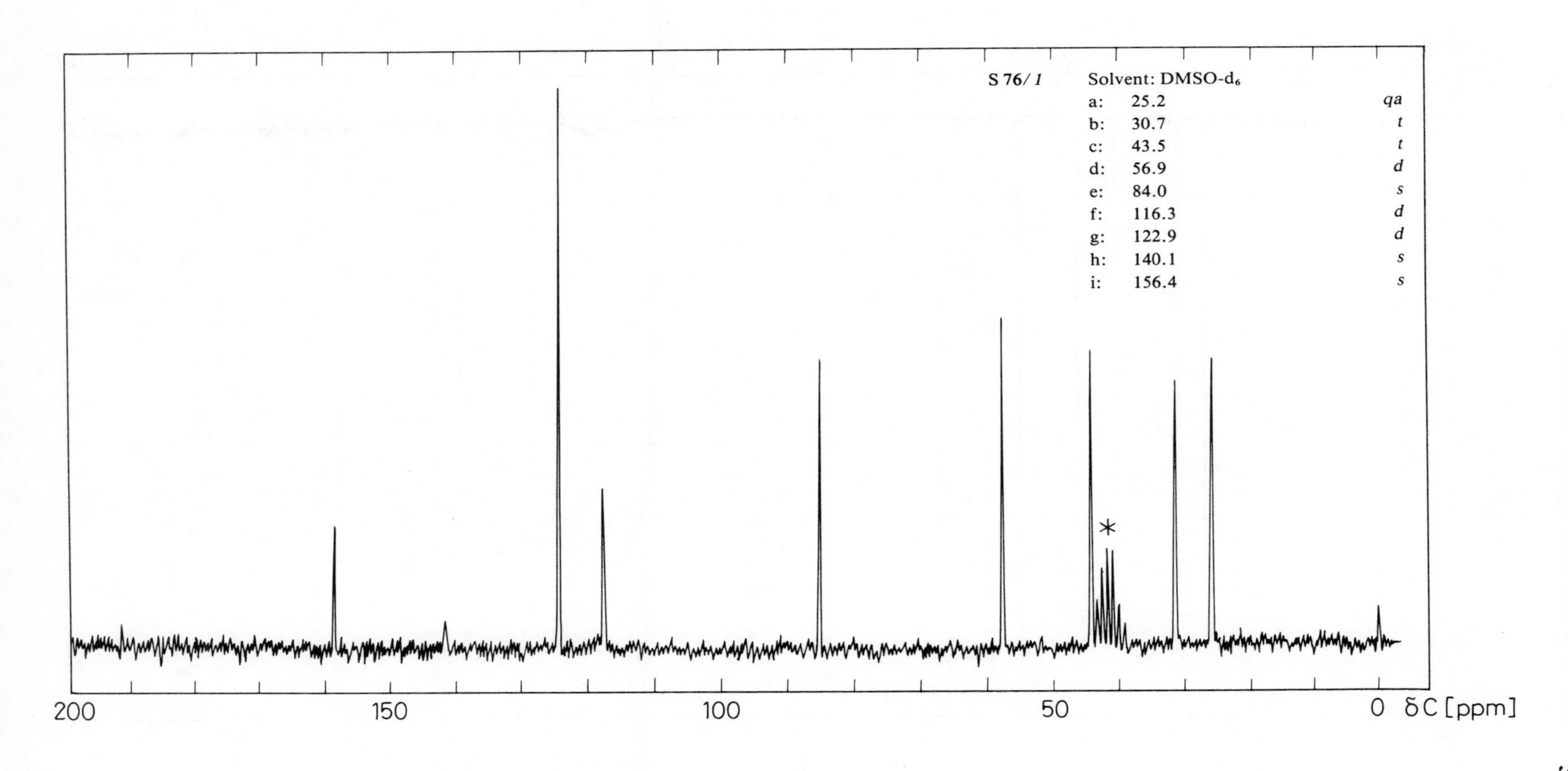
S 76/ 1
Solvent: DMSO-d6
a: 25.2 qa
b: 30.7 t
c: 43.5 t
d: 56.9 d
e: 84.0 s
f: 116.3 d
g: 122.9 d
h: 140.1 s
i: 156.4 s
*
200
150
100
50
0 δC [ppm]

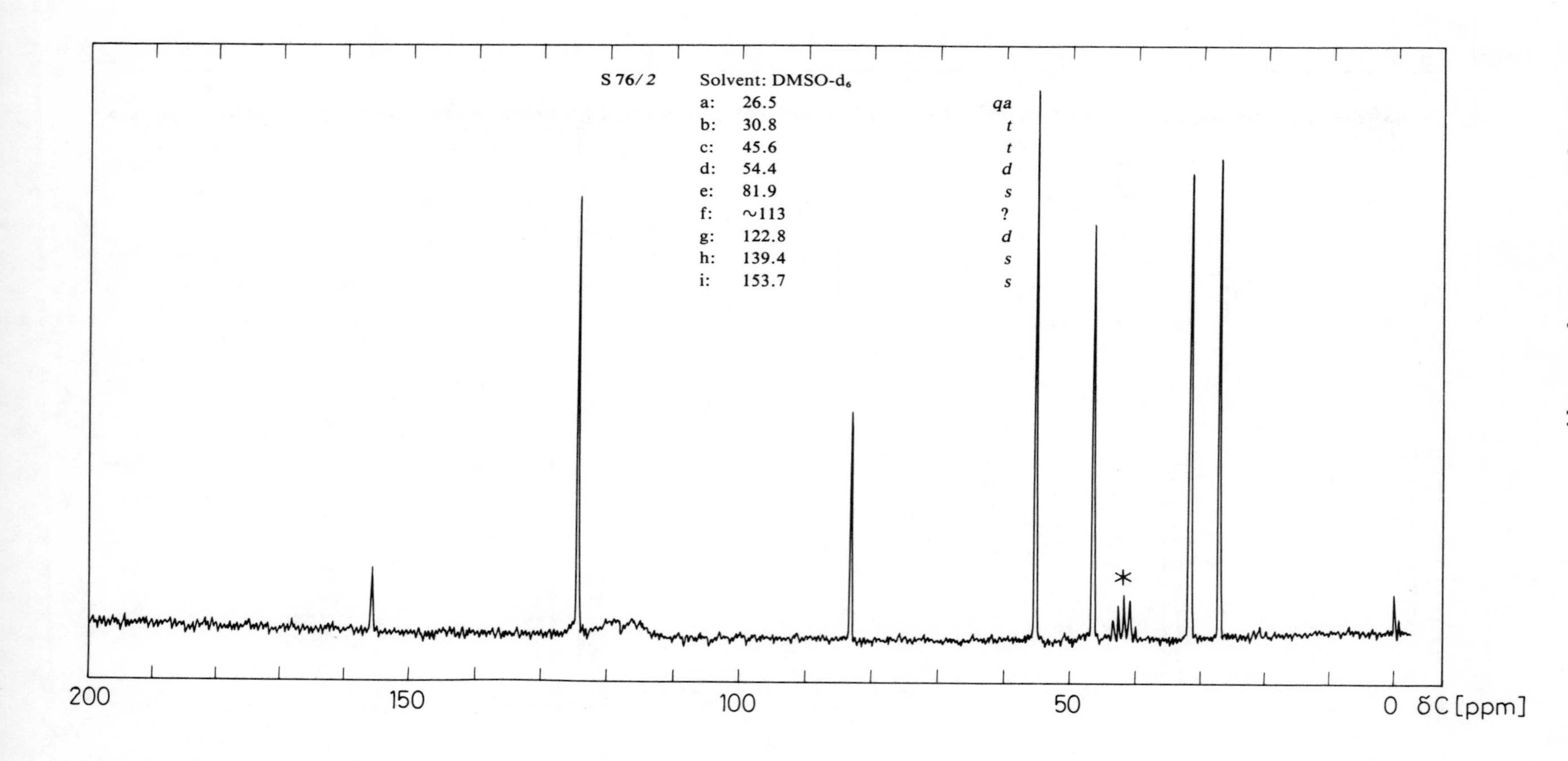
S 76/2
Solvent: DMSO-d_6
a: 26.5 *qa*
b: 30.8 *t*
c: 45.6 *t*
d: 54.4 *d*
e: 81.9 *s*
f: ~113 ?
g: 122.8 *d*
h: 139.4 *s*
i: 153.7 *s*
*
200
150
100
50
0 δC [ppm]

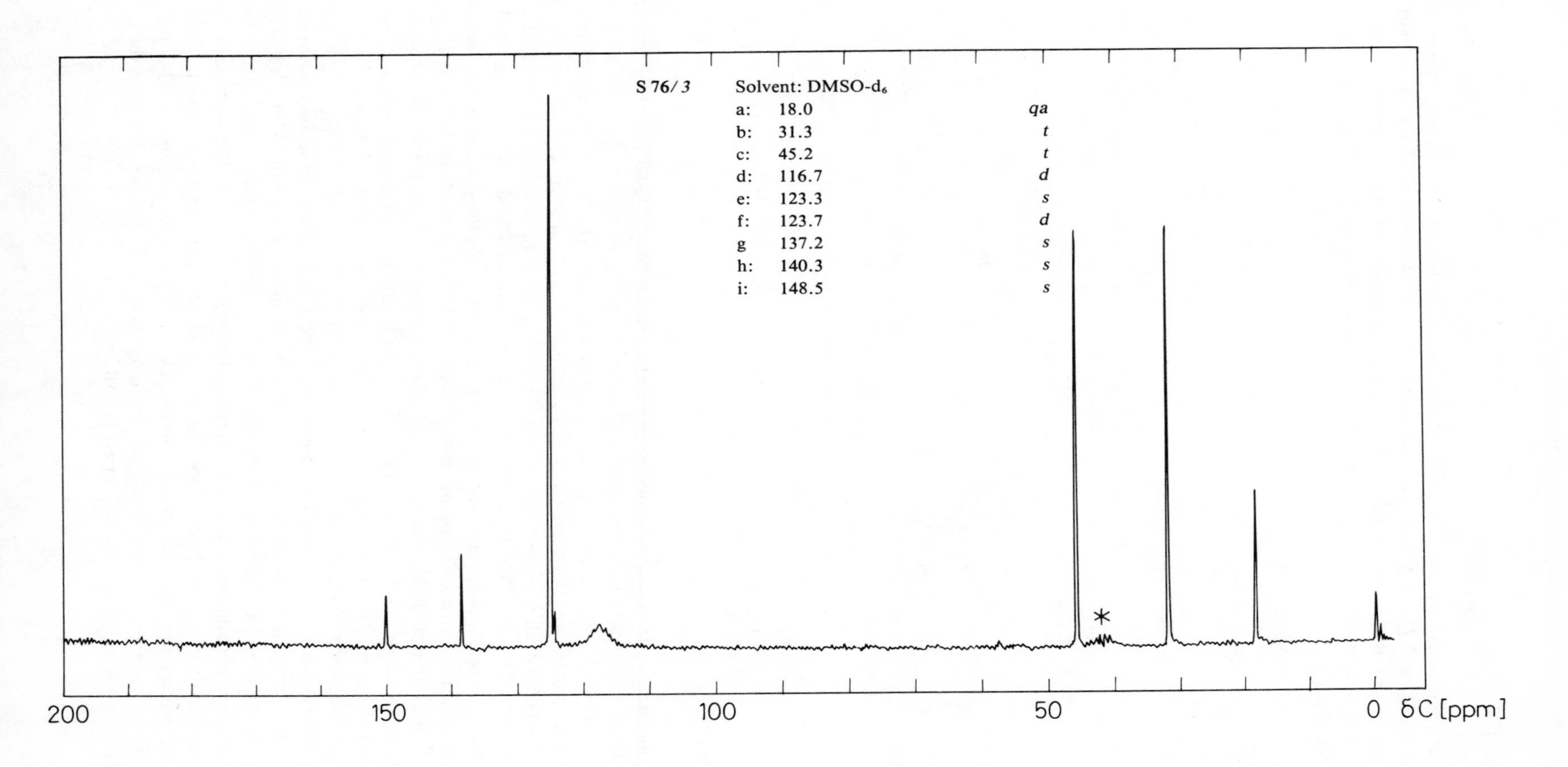
S 76/3
Solvent: DMSO-d₆
a: 18.0 qa
b: 31.3 t
c: 45.2 t
d: 116.7 d
e: 123.3 s
f: 123.7 d
g 137.2 s
h: 140.3 s
i: 148.5 s
*
200
150
100
50
0 δC [ppm]

PROBLEM 77[491]

By reacting benzothiazines **I** and **II** with chloro- and dichloroacetyl chloride, respectively, β-lactams **III** and **IV** or **V** and **VI** may be anticipated. Do Spectra **77**/*1* to *4* support this assumption, and if so, is it possible to determine the relative configuration of the substituents of the four-membered ring from the spectra? On the basis of the assignment, pair the structures with the spectra.

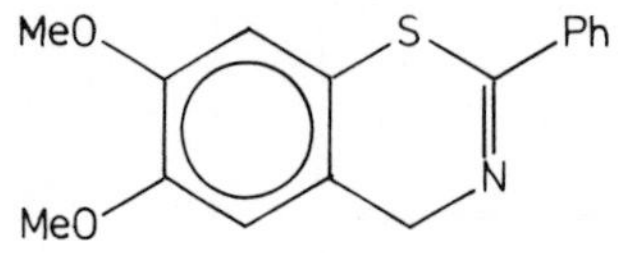

77/I

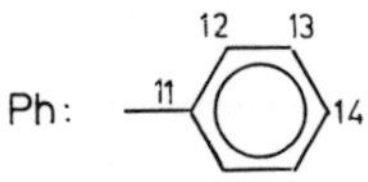

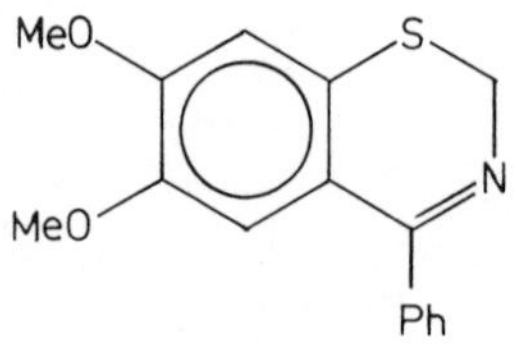

77/II

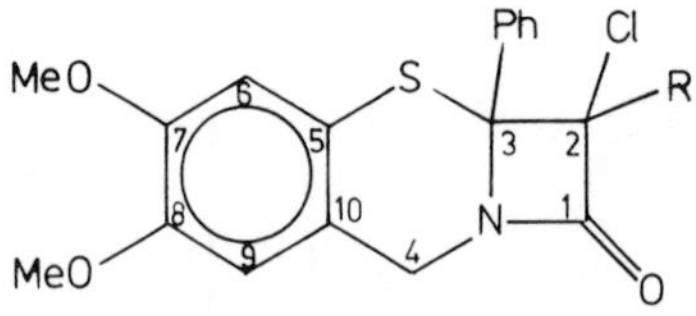

77/III (R : H)
77/IV (R : Cl)

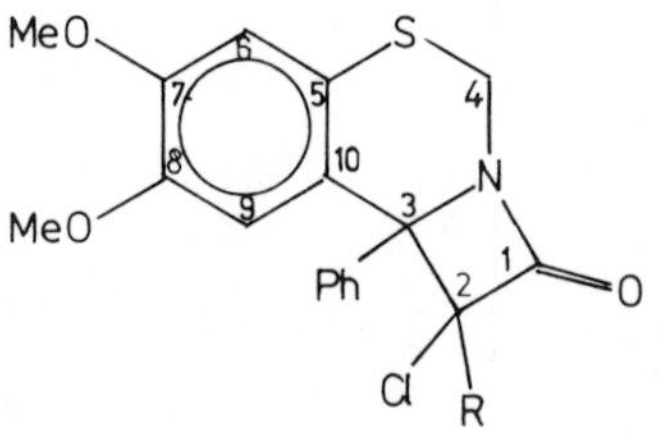

77/V (R : H)
77/VI (R : Cl)

ANSWER

In the spectra, 16 lines are expected to occur. The carbonyl signal is easy to assign as the outside downfield line in the spectra (160 to 165 ppm), as well as the lines of the five saturated carbons, which appear in the <100-ppm region. The methylene signal may be identified unambiguously on the basis of its upfield position and triplet splitting. The presumable chemical shift of methoxy carbons is around 56 ppm (compare Problem **72**), and since Spectra **77**/*2* and **77**/*3* do contain two quartets with similar shifts (55.9 and 56.4 and 57.4 and 57.6 ppm), it is reasonable to assume that the 56.0-ppm quartets in the two other spectra arise from two overlapping methyl signals. This also explains why these spectra have 15 lines instead of 16. The C-2 signal is doublet for Structures **III** and **V** and can thus be distinguished simply from the singlet of C-3 with similar shift. Since there is a doublet in Spectra **77**/*3* and **77**/*4* in the region of saturated carbons (67.3 and 68.3 ppm, respectively), it can be assigned to C-2, and thus the assignment of the 69.2- and 71.2-ppm signals to C-3 is straightforward, and, of course, it may be regarded as a proof that these spectra correspond to Structures **III** and **V**. Since the second chlorine atom in the case of **IV** and **VI** causes a strong deshielding for C-2, it is certain that δC-2 >> δC-3. The adjacence of quaternary C-2 atom replacing the tertiary carbon increases the shift of C-3 as well, but this effect may not significantly exceed 10 ppm. Thus, the 89.8- and 90.1-ppm singlets of Spectra **77**/*1* and **77**/*2* corresponding to the dichloro-derivatives **IV** and **VI** may be assigned to C-2 and also the singlet 82.1- and 73.6-ppm signals may be assigned to the C-3 atoms. The large difference between the two latter signals (8.5 ppm) is striking.

Of the ten aromatic signals, the two *ortho* and *meta* (C-12 and C-13) signals may be identified on the basis of their high intensity, the C-7,8 lines may be identified on the basis of large chemical shift (methoxy substitution), and the C-6,9 lines may be identified on the basis of small chemical shift (methoxy in *ortho* position). Since the signal of the *meta* phenyl carbons (C-13) is the least sensitive to substitution, the signals closest to the shift of benzene (128.5 ppm) are assigned to C-13.* They are in the sequence of spectra 128.6, 128.4★, 129.1★, and 128.4 ppm. Thus, the shifts of C-12 lines are 125.6, 128.1★, 129.5★, and 126.8 ppm. (In Spectra **77**/*2* and **77**/*3*, the opposite assignment is also possible.) The C-7,8 and C-6,9 signals cannot be distinguished among one another, too, but it is clear that the former appear in the four spectra at 148.2 and 149.1; 146.3 and 149.6; 149.4 and 150.5; and 148.1 and 148.7 ppm, respectively, whereas the latter appear at 111.4 and 112.6; 110.5 and 114.4; 113.1 and 114.4; and 111.2 and 112.4 ppm.

Of the remaining four lines two appear around 120 ppm and since the C-11 and C-14 atoms of the phenyl ring are more deshielded, they may be due to atoms C-5,10 only. The shifts in the four spectra are 121.3 and 125.0; 121.7 and 122.4; 122.9 and 132.6; and 120.1 and 121.9 ppm. The weak lines at 137.5, 136.0, 138.1, and 136.7 ppm, respectively, in the spectra belong to C-11, and the stronger lines at 129.2, 129.1, ?, and 128.9 ppm belong to C-14 (in Spectra **77**/*3* this line cannot be identified and probably overlaps with the 129.5-ppm signal).

Of the spectra of the two dichloro-derivatives, the signals of C-5,10 and particularly one of the C-6,9 signals are more downfield shifted in Spectrum **77**/*2* (110.5 and 122.4 ppm) than in Spectrum **77**/*1* (111.4 or 112.6 and 125.0 ppm). Consequently, this should be the spectrum of the most crowded, angularly anellated substance **VI**, in which the strongest field effect may be anticipated. Accordingly, Spectrum **77**/*1* corresponds to Structure **IV**. The field effect** is the strongest on C-3 and C-10, but with these atoms the recognition of it is difficult because of the different chemical environments of these carbons in **V** and **VI** or **IV** and **VI**, respectively. The coincidence of the two methoxy signals and the stronger shielding of C-12 suggests that Spectrum **77**/*4* is the pair of Spectrum **77**/*1*, and since the latter corresponds to the linearly fused derivative **IV**, Spectrum **77**/*4* may be assigned to Structure **III**. The full assignments are given in Scheme 12.

Taking into account the data given with the formulas, it may be stated that in compound **V** the phenyl and chlorine substituents are *cis*. Namely, in Spectrum **77**/*2* of dichloro derivative **VI**, a field effect may be observed in the signal of C-9,10 atoms. This is due to a steric compression between H-9 and the chlorine atom *trans* to the phenyl ring. Since this effect is absent in compound **V**, the chlorine and phenyl substituents must be in *cis* position. This conclusion can be drawn as well from the fact that the chemical shift of C-11 does not significantly differ in the spectra of **V** and **VI**. Since a chlorine in *cis* position should have a strong shielding effect, this configuration must be present in Structure **V**, too.

The same arguments may be applied in the case of **III**. Here, in the dichloro-derivative the shift of C-11 does not decrease, but even slightly increases.

In the possession of structures, some, previously ambiguous assignments may be cleared. With compounds **V** and **VI**, of the C-6,9 signals, evidently the line with the practically identical shift in the two spectra belongs to C-6 (114.4 ppm) and the other is the C-9 signal. Thus, the fact that the latter shift is 2.6 ppm lower in the spectrum of **VI** than in that of **V** is a further argument for the assumed configuration of the latter: there is a steric compression between the second chlorine *trans* to the phenyl ring and C-9, which decreases the shift of C-9 in the spectrum of **VI**. Similarly, the assignment of the 125.0-ppm line to C-5 in Spectrum **77**/*1*, and thus the assignment of the 121.3-ppm line to C-10 follows from the

* See Table 49 and Volume, p. 187.

** Compare Section 4.1.1.3, point 4, pp. 155—156.

77/III (S 77/4)

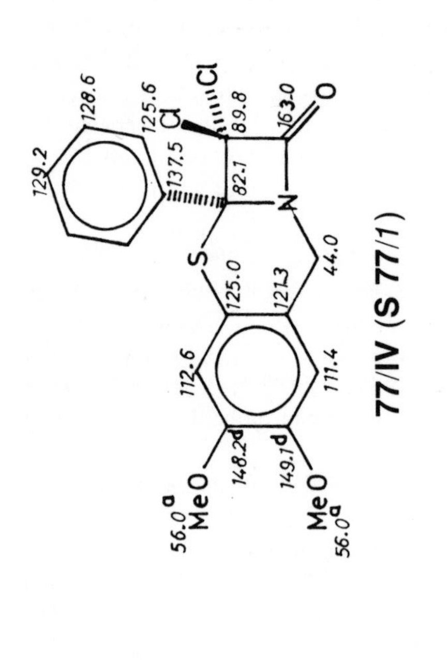

77/IV (S 77/1)

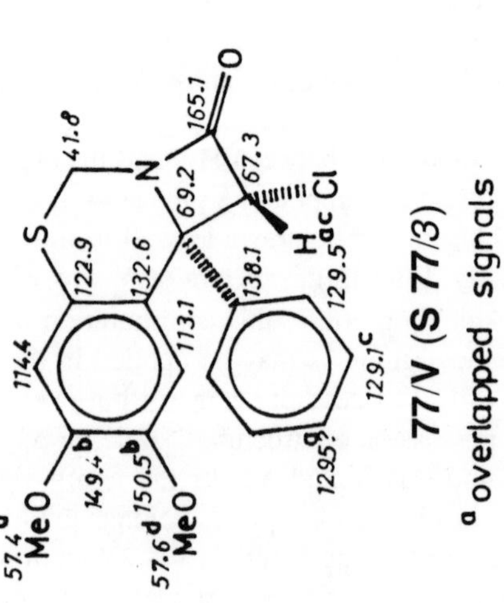

77/V (S 77/3)

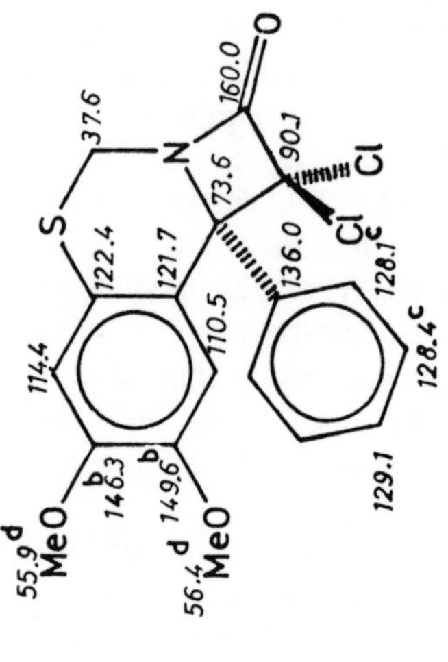

77/VI (S 77/2)

[a] overlapped signals

[b,c,d] reversed assignment is also possible

Scheme 12

effect of the second chlorine atom in compound **IV** on the adjacent C-5 atom, whereas the shift of C-10 may not differ significantly for **III** and **IV**. Therefore, for compound **III**, δC-5 = 120.1 and δC-10 = 121.9 ppm.

Note the very strong diamagnetic shift (10.9 ppm) of the C-10 signal in the spectrum of compound **VI** with respect to that of **V**, which justifies on its own the assignment of crowded Structure **VI** to Spectrum **77**/*2*. The field effect is also strong on the signal of C-3, if it is taken into account that the chlorination of the α-carbon (C-2) causes, instead of the expected approximately 11-ppm shift (compare Table 52), only a paramagnetic shift of 4.4 ppm for **VI** relatively to **V**.

PROBLEM **78**[803,1332]

The $LiAlH_4$ reduction of compound **78/I** yields a thiabicycloheptane derivative characterized by one of Structures **78/II** to **VII** according to IR, ^{1}H NMR, and mass spectra. On the basis of the Spectrum **78**/*1* decide among the structures, taking into account that according to the proton coupled spectrum the ^1J(C,H) couplings cause 167.5, 172, 132, 132, 177.5, and 141 Hz splitting, respectively, in the order of increasing chemical shifts for the six signals of the spectrum.

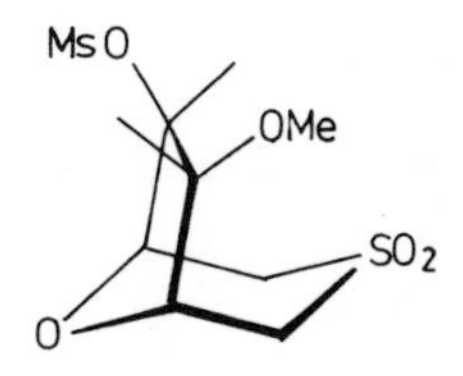

78/I (Ms : SO Me)

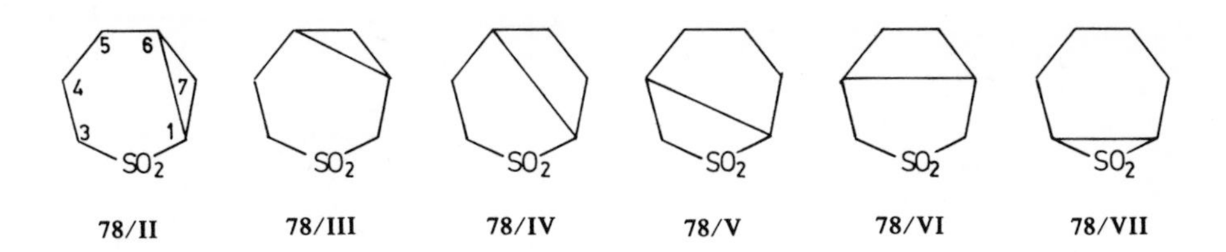

78/II **78/III** **78/IV** **78/V** **78/VI** **78/VII**

ANSWER

The four triplets and two doublets are in agreement with Structures **II** to **V** of the assumed ones, since compounds **VI** and **VII** may have only three signals (two triplets and one doublet) because of molecular symmetry. The 10.2-ppm signal excludes Structures **IV** and **V**, too, since so strongly shielded carbons may occur only in cyclopropane derivatives (compare Table 50). The chemical shift of cyclobutane carbons is 22.1 ppm (see Table 55), and this can only increase in the derivatives which contain heteroatom. Since the chemical shift of one doublet and triplet is much larger than the shifts of other signals, Structure **III** can also be discarded: in its spectrum two triplets would be separated according to the neighborhood of the SO_2 group. Therefore, Structure **II** remains, and this choice is also supported by the coupling constants. Cyclopropyl carbons have much higher ^1J(C,H) coupling constants (approximately 170 Hz, compare Table 82), and thus the assignment of the carbon which is most shielded beside the two doublets to cyclopropane cyclic carbons is supported by independent data. With Structure **III** the chemical shifts of the two doublets would not differ significantly, according to the same order of the neighbors. A relation $\delta C\text{-}4 < \delta C\text{-}5$ is probable on the basis of the tertiary C-6 neighbor of the latter carbon. The full assignment in given in Scheme 13.

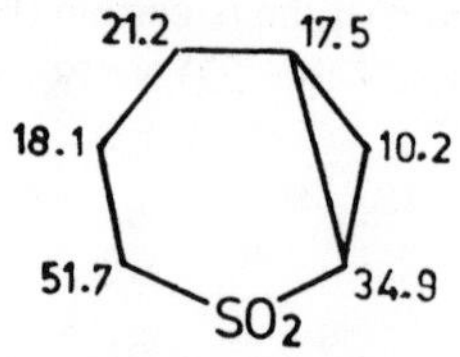

21.2 17.5 18.1 10.2 51.7 34.9 SO_2

78/II (S 78/1)

Scheme 13

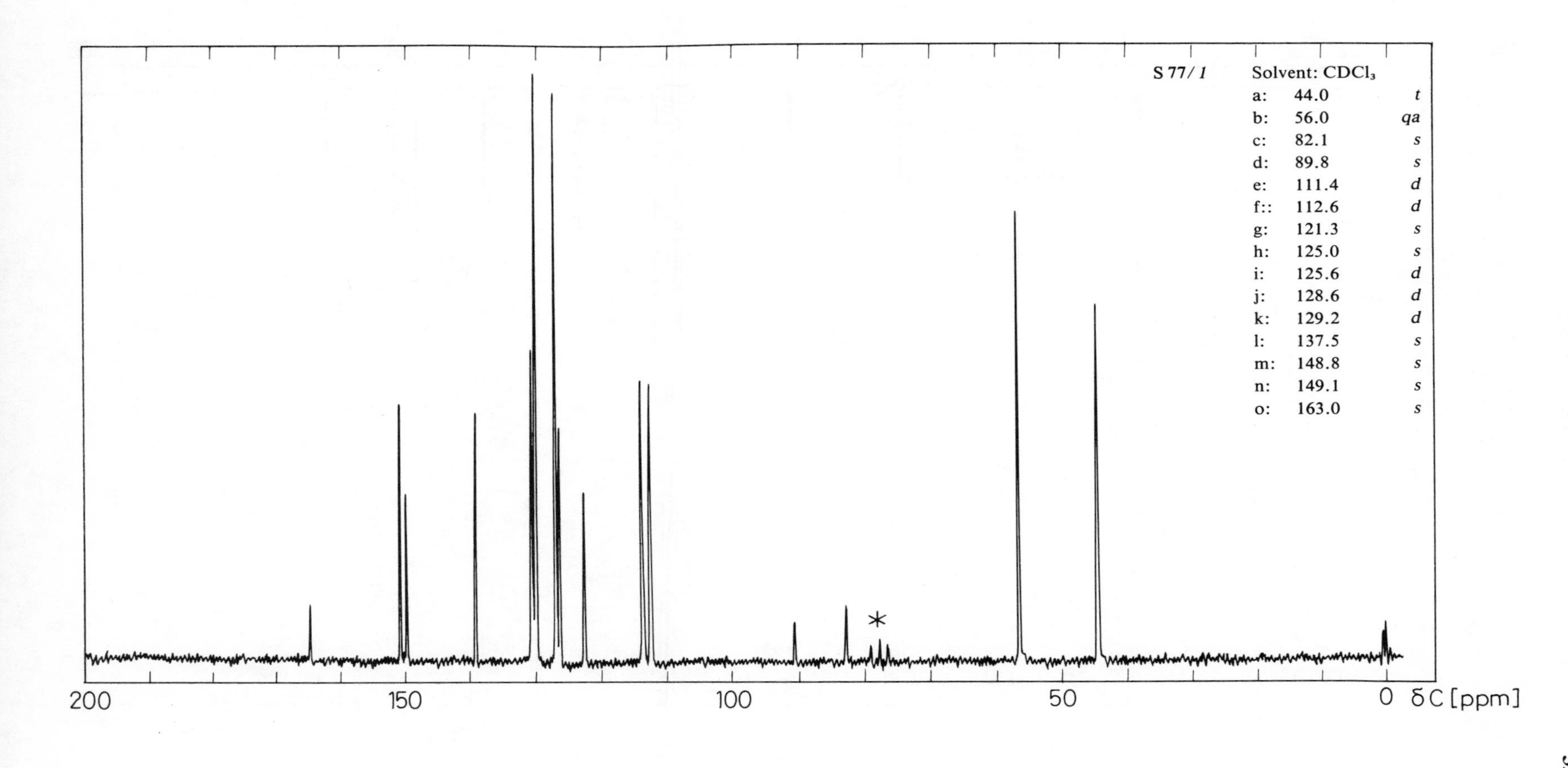

S 77/1
Solvent: $CDCl_3$
a: 44.0 t
b: 56.0 qa
c: 82.1 s
d: 89.8 s
e: 111.4 d
f:: 112.6 d
g: 121.3 s
h: 125.0 s
i: 125.6 d
j: 128.6 d
k: 129.2 d
l: 137.5 s
m: 148.8 s
n: 149.1 s
o: 163.0 s
*
200
150
100
50
0
δC [ppm]

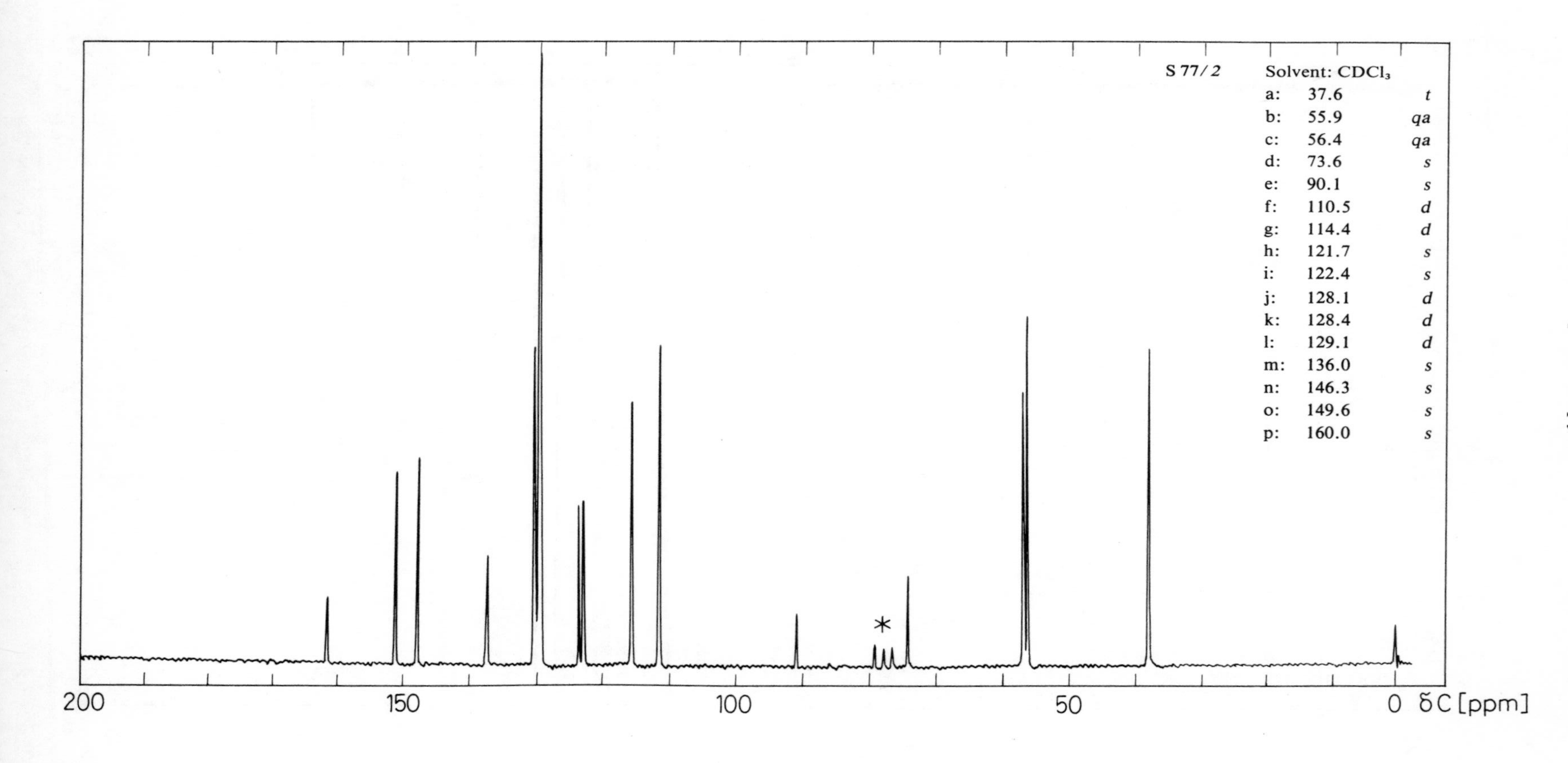
S 77/2
Solvent: CDCl3
a: 37.6 t
b: 55.9 qa
c: 56.4 qa
d: 73.6 s
e: 90.1 s
f: 110.5 d
g: 114.4 d
h: 121.7 s
i: 122.4 s
j: 128.1 d
k: 128.4 d
l: 129.1 d
m: 136.0 s
n: 146.3 s
o: 149.6 s
p: 160.0 s
*
200
150
100
50
0
δC [ppm]

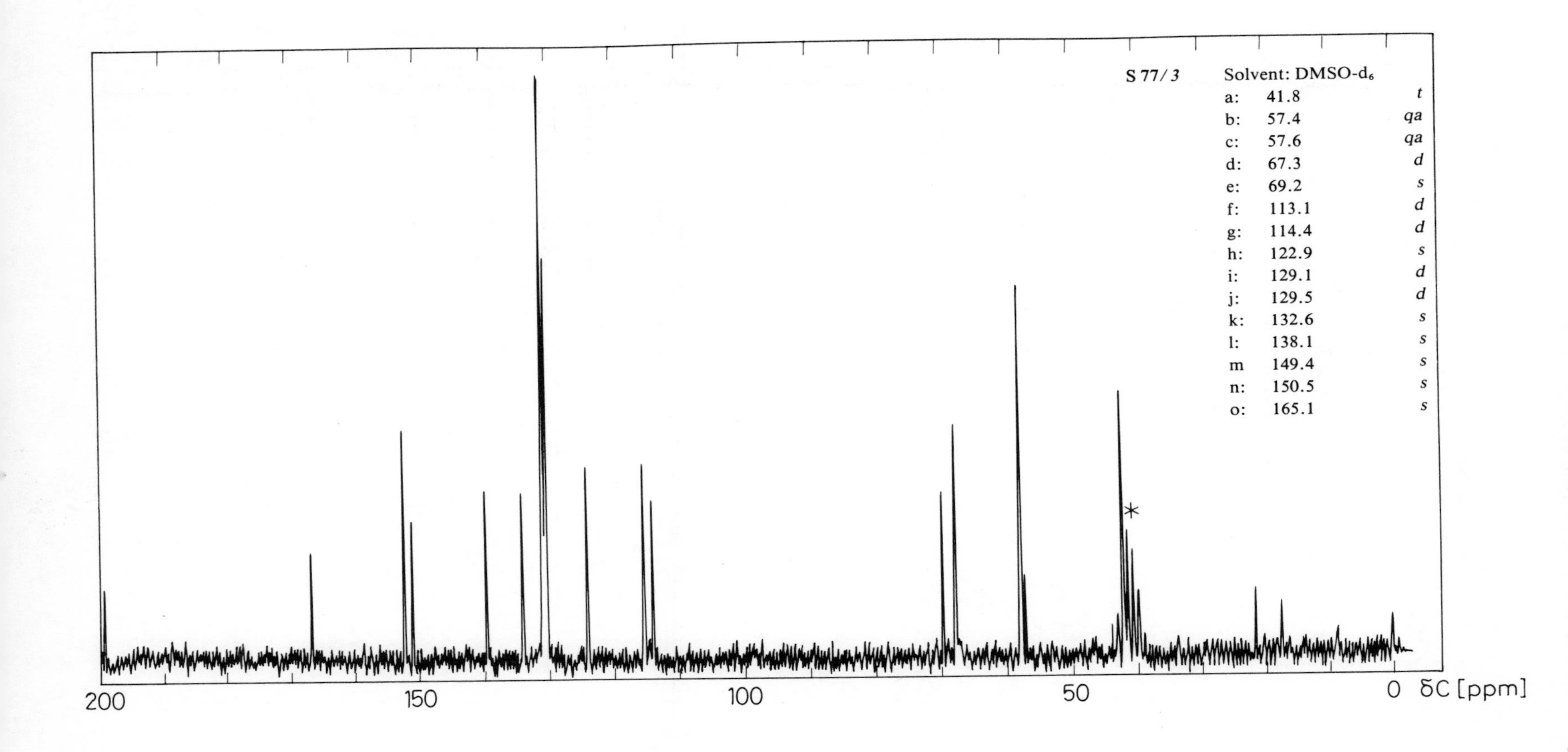
S 77/3
Solvent: DMSO-d₆
a: 41.8 t
b: 57.4 qa
c: 57.6 qa
d: 67.3 d
e: 69.2 s
f: 113.1 d
g: 114.4 d
h: 122.9 s
i: 129.1 d
j: 129.5 d
k: 132.6 s
l: 138.1 s
m 149.4 s
n: 150.5 s
o: 165.1 s
*
200
150
100
50
0
δC [ppm]

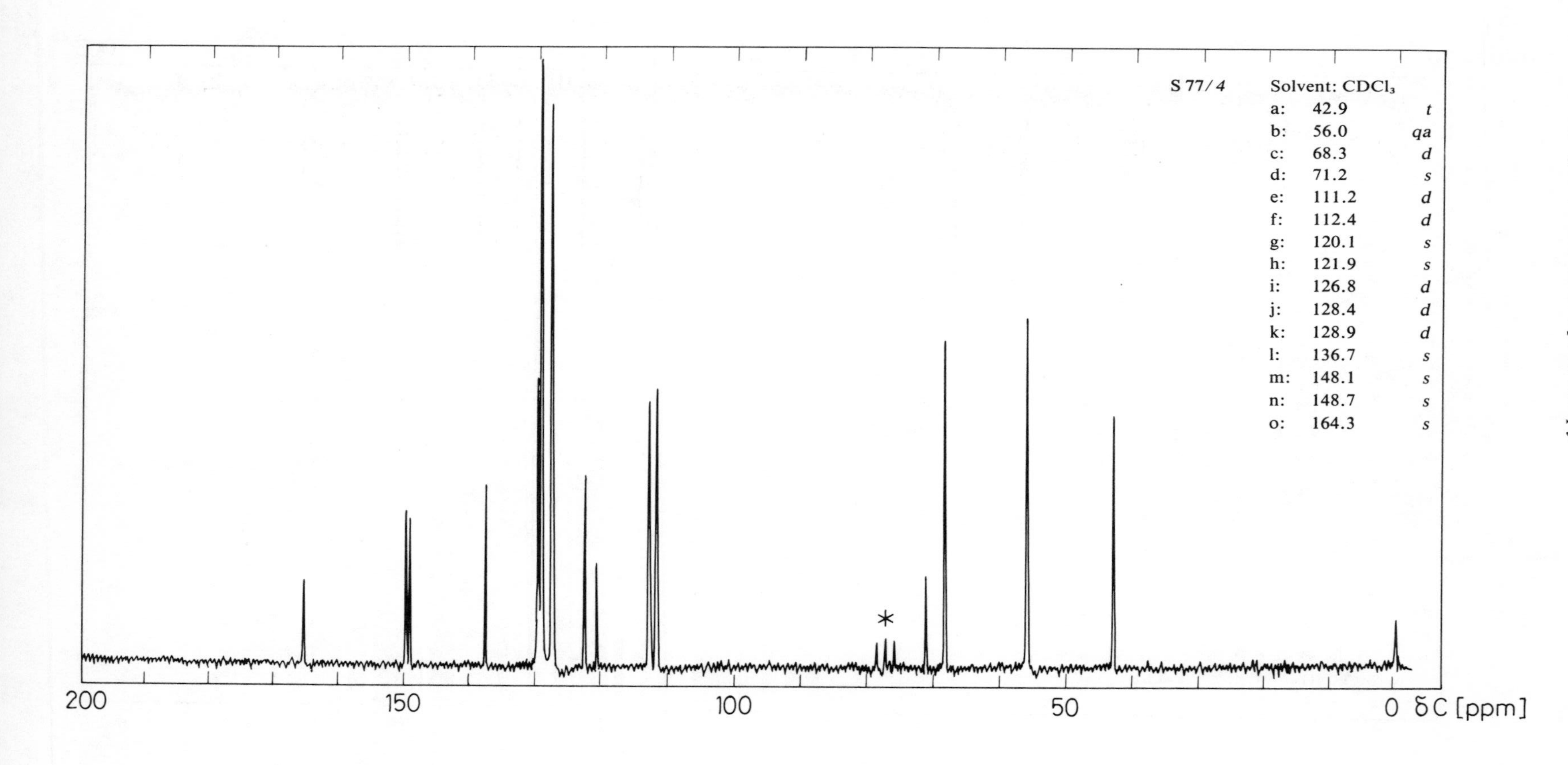
S 77/4
Solvent: $CDCl_3$
a: 42.9 t
b: 56.0 qa
c: 68.3 d
d: 71.2 s
e: 111.2 d
f: 112.4 d
g: 120.1 s
h: 121.9 s
i: 126.8 d
j: 128.4 d
k: 128.9 d
l: 136.7 s
m: 148.1 s
n: 148.7 s
o: 164.3 s
*
200
150
100
50
0 δC [ppm]

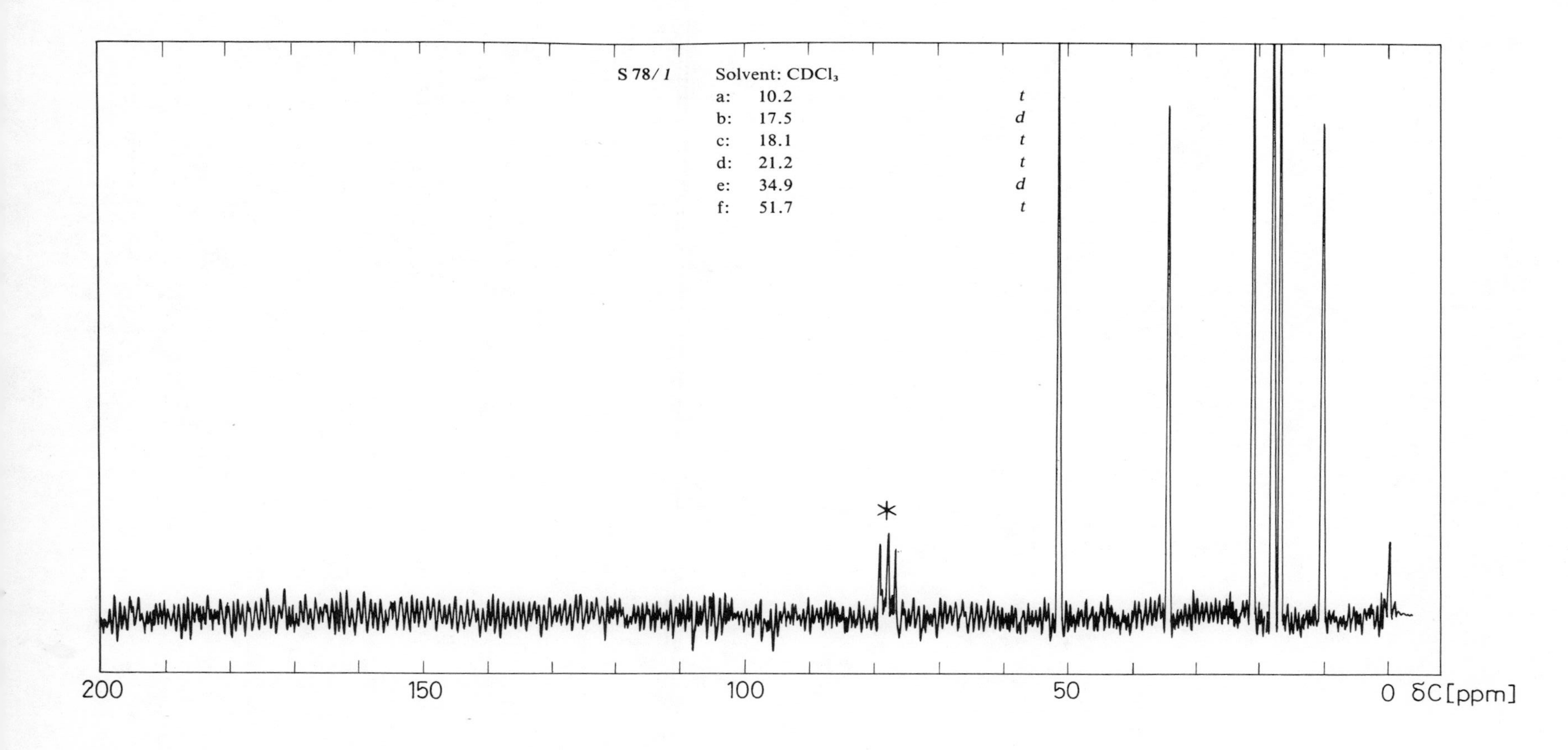

S 78/ 1
Solvent: $CDCl_3$
a: 10.2 t
b: 17.5 d
c: 18.1 t
d: 21.2 t
e: 34.9 d
f: 51.7 t
*
200
150
100
50
0 δC [ppm]

ABBREVIATIONS IN NMR THEORY AND METHODOLOGY

ADC	analog to digital converter
AF	audio frequency
ASIS	aromatic solvent induced shift
BBDR	broad band double resonance
CAT	computer of average transients
CIDEP	chemically induced dynamic electron polarization
CIDNP	chemically induced dynamic nuclear polarization
CW	continuous wave
DAC	digital to analog converter
DD	dipole-dipole, dipolar, relaxation mechanism or interaction
DEFT	driven-equilibrium FT
2DFTS	two-dimensional FT spectroscopy
DNMR	dynamic NMR
DNP	dynamic nuclear polarization
DRDS	double resonance difference spectroscopy
ESR	electron spin resonance
e.s.u.	electrostatic unit
FFT	fast Fourier transform
FID	free induction decay
FONAR	field focusing nuclear magnetic resonance
FT	Fourier transformation
HSP	homogeneity-spoiling pulse
INDOR	internuclear double resonance
IR	infrared
LCAO	linear combination of atomic orbitals (in quantum theory)
MO	— molecular orbital (in quantum theory) — master oscillator
NMR	nuclear magnetic resonance
NOE	nuclear Overhauser effect
NQR	nuclear quadrupole resonance
PFT	pulse Fourier transformation
ppm	part per million
RF	radio frequency
SA	chemical shift anisotropy (CSA), relaxation mechanism
SC	scalar, relaxation mechanism or interaction
SEFT	spin-echo FT
SPI	selective population inversion
SR	— spin rotation, relaxation mechanism or interaction — shift reagent
VCO	voltage controlled oscillator
WEFT	water-eliminated FT

CHEMICAL ABBREVIATIONS — COMPOUNDS AND FUNCTIONAL GROUPS

Ac	acetyl, CH_3CO–
ACAC	acetylacetone, $CH_3COCH_2COCH_3$
Ar	aryl (aromatic ring containing groups)
BHC	3-*t*-butylhydroxymethylene-*d*-camphor ≡ TBC

*n*Bu	*normal* butyl, $CH_3CH_2CH_2CH_2$–
*i*Bu	*iso*-butyl, $(CH_3)_2CHCH_2$–
*s*Bu	*sec.*-butyl, $CH_3CH_2CH(CH_3)$–
*t*Bu	*tert.*-butyl, $(CH_3)_3C$–
Bz	benzoyl, C_6H_5–CO–
Bzl	benzyl, C_6H_5–CH_2–
DMF, DMF-d_7	dimethyl-formamide, heptadeuterio-DMF
DMSO, DMSO-d_6	dimethyl-sulfoxide, hexadeuterio-DMSO
DPM	dipivaloyl methane, $(CH_3)_3C$–CO–CH_2–CO–$C(CH_3)_3$
DSS	2,2-dimethyl-2-silapentane-5-sulfonic acid sodium, $(CH_3)_3Si(CH_2)_3SO_3Na$
Et	ethyl, CH_3CH_2–
FACAM	3-trifluoroacetyl-*d*-camphor
FOD	1,1,1,2,2,3,3-heptafluoro-7,7-dimethyl-octane-4,6-dione
HFBC	3-heptafluorobutyryl-*d*-camphor ≡ HFC
HFC	see HFBC
HMPA (HMPT)	hexamethylphosphoramide
LIS	lanthanide (shift reagent) induced shift
LS	shift reagent — substrate complex
LSR	lanthanide shift reagent
Me	methyl, CH_3–
Ms	mesyl, CH_3SO_2–
Ph	phenyl, C_6H_5–
*n*Pr	*normal* propyl, $CH_3CH_2CH_2$–
*i*Pr	*iso*-propyl, $(CH_3)_2CH$–
Py, αPy, βPy, γPy	pyridyl, α-, β- and γ-pyridyl, C_5H_5N–
PVC	poly(vinyl chloride), $[CH_2Cl–CHCl_2]_n$
TBC	see BHC
TFA, TFA-d	trifluoro-acetic acid (CF_3COOH), deuterio-TFA (CF_3COOD)
TFC	see FACAM
TFM	trichloro fluoromethane, $CFCl_3$
THF	tetrahydrofurane, C_4H_8O
TMHD	2,2,6,6-tetramethylheptane-3,5-dione ≡ DPM
TMS	tetramethylsilane, $(CH_3)_4Si$
Ts	tosyl, *p*CH_3–C_6H_4–SO_2–

LIST OF NOTATIONS

Basic Physical Quantities

c	$(2.997925 \pm 0.000003) \cdot 10^8$ [m s^{-1}], velocity of light
e	$4.8024 \cdot 10^{-10}$ [e.s.u., CGS] = $1.60210 \cdot 10^{-19}$ [A s], [C], charge of electron
h	$6.6256 \cdot 10^{-34}$ [J s], Planck's constant
$\hbar$	$h/2\pi = 1.0545 \cdot 10^{-34}$ [J s]
k	$1.38054 \cdot 10^{-23}$ [J K^{-1}], Boltzmann's constant
m_e	$(9.1091 \pm 0.0004) \cdot 10^{-31}$ [kg], static electron mass
m_H	$(1.67252 \pm 0.00008) \cdot 10^{-27}$ [kg], static mass of proton
N	$6.02252 \cdot 10^{23}$ [M^{-1}], Avogadro's number
R	8.3143 [J K^{-1} M^{-1}], Rydberg's constant (universal molar gas constant)
β	$9.2732 \cdot 10^{-24}$ [J T^{-1}], Bohr magneton
β_N	$5.0505 \cdot 10^{-27}$ [J T^{-1}], nuclear magneton
μ_H	$2.67519 \cdot 10^8$ [s^{-1} T^{-1}], magnetogyric (gyromagnetic) ratio of proton
H	$1.41049 \cdot 10^{-26}$ [J T^{-1}], magnetic moment of proton

Symbols and Their Meanings

a	antisymmetric spin state, basic function transition, etc.
a, a'	*axial, quasi axial*
A	— signal intensity — hyperfine electron-nucleus interaction constant
$\mathbf{B}$	magnetic induction field (magnetic flux density), T [m kg s^{-2} A^{-2}], [m^{-2} s V]
$\mathbf{B}_0$	static external (polarizing) field of the spectrometer
$\mathbf{B}_l$	local magnetic field
$\mathbf{B}_\ell$	induced magnetic field
$\mathbf{B}_{1,2...}$	first, second, etc. exciting (RF) magnetic fields associated with ν_1, ν_2, etc.
C	constant, linear combination constant
$C_\parallel, C_\perp$	spin rotational coupling constants
D	diffusion coefficient
e, e'	*equatorial, quasi equatorial*
E	— energy, N [m^2 kg s^{-2}] — electronegativity
$\mathbf{E}$	electric field
E_z	z component of the electric field
$\mathcal{F}$	Fourier transformation operator
f	mole fraction
g	nuclear or electronic Lande- (g-) factor
$\mathcal{H}$	Hamilton operator (in energy or frequency unit)
$\mathcal{H}^0$, $\mathcal{H}'$	the spin-external field and spin-spin interaction terms in the Hamiltonian
$\mathbf{H}_{ij}$	element of matrix representation of $\mathcal{H}$
H	Hamilton function
$\mathcal{I}$	spin vector operator
$\mathcal{I}_w$	components of the spin vector operator $\mathcal{I}$

$\mathcal{I}^+$	raising ("absorption") spin operator
$\mathcal{I}^-$	lowering ("emission") spin operator
$\mathcal{I}^2$	square of the spin vector operator
I_w	eigenvalue of $\mathcal{I}_w$
I^2	eigenvalue of $\mathcal{I}^2$
$\mathbf{I}_{ij}$	elements of matrix representation of $\mathcal{I}^2$
i	$\sqrt{-1}$
I	— current intensity [A] — moment of inertia — nuclear spin quantum number
$\pm I$	inductive effect of atoms and functional groups
J	nuclear spin-spin coupling constant [Hz]
$^nJ(X,Y)$ or $^nJ_{XY}$	coupling constant between X and Y nuclei through n bonds [Hz]. Subscripts are given only as algebraic symbols for any interacting pairs of nuclei. Brackets are used for indicating the coupled nuclei, e.g., $J(^{19}F, ^1H)$ or $J(F, H)$ or the coupling path, e.g., $J(NCCH) \equiv {}^3J(N, H)$.
J^*	reduced coupling constant (in off-resonance)
J	— atomic quantum number representing the total angular moment — rotational quantum number
$^nK(X,Y)$	reduced nuclear spin-spin coupling constant [m^{-3} A^{-2} N]
K	— combination transition — constant, equilibrium constant of conformational motions and other exchange processes
$\mathcal{L}$	angular moment vector operator of electron orbitals
ℓ	— distance
ℓ	— line of force
$\mathbf{L}$	torque, N [m^2 kg s^{-2}]
m	— mass of particles (electrons, protons, etc.) [kg] — eigenvalue of I_z of a nucleus
m_T	total magnetic quantum number for a spin system: eigenvalues of I_{zT}
M	mole
$\pm M$	mesomeric effect of atoms and functional groups
$\mathbf{M}$	macroscopic magnetization of a spin system, T
$\mathbf{M}_0$	macroscopic magnetization in field $\mathbf{B}_0$
$\boldsymbol{M}_e$	equilibrium macroscopic magnetization
M_w	components of $\boldsymbol{M}$
n	population difference of magnetic quantum states
n_e	equilibrium population difference
N	— Newton — nucleus
N	number of magnetic nuclei
$\mathcal{O}$	operator, representing any physical parameters
$\mathcal{P}$	permutation operator
p	— p-orbital — mole fraction fractional population of rotamers, etc.
$\mathbf{P}$	angular impulse moment [kg m^2 s^{-1}]
P	nuclear polarization
q	electric field gradient

Q	— nuclear quadrupole moment
	— various quantities and line distances in spectra of different spin systems, consisting of chemical shifts and coupling constants of the system
	— contributions of σ- and π-electrons to shielding component σ^P in quantum theoretical calculations of the shielding
R_w	element of matrix representation of transition moment
r	— polar coordinate
	— distance of atoms and groups, atomic radius, bond length, radius of ring currents, etc. [m], [Å]
R	rate of relaxations
R_1, R_2, R^{DD}, R^{SC}, R^{SA}, R^{SR}, R^Q, see at T_1, etc.	
s	s-orbital
s	symmetric spin state, basic function, transition, etc.
S	electron density around a nucleus
S	singlet electron state
t	time [s]
t_p	pulse time
t_{aq}	acquisition time
t_d	delay time
t_{pi}	period of repetitive pulses
t_{si}	period of repetitive pulse sequences
T	— forbidden transition
	— triplet electron state
	— temperature [°C, °K]
T_c	coalescence temperature
T_s	spin temperature
T_1, T_{1N}	spin-lattice relaxation times (for nucleus N), [s]
T_2	spin-spin relaxation times [s]
$T_{1\rho}$, $T_{2\rho}$	spin-lattice and spin-spin relaxation times in the rotating frame
T_2^*	measurable spin-spin dephasing time shortened by field inhomogeneities
V	volume [m^3]
w	the common notation of Cartesian coordinates
W	— transition probability
	— signal width [Hz] for unresolved multiplet
Z	saturation factor
Z_0	saturation factor corresponding to maximal value of signal shape function
α, β	basic eigenfunction of spin corresponding $m = +1/2$ and $m = -1/2$
γ, γ_N	magnetogyric ratio of N nucleus [T^{-1} s^{-1}]
$\bar{\gamma}$	$\gamma/2\pi$
δN	chemical shift of nucleus N ppm
δ_{ij}	Krönecker-delta
Δ	difference
$\Delta\Delta$	difference of differences
ΔB_0, ΔB_1, etc.	inhomogeneity in fields B_0, B_1, etc.

Δ_c, Δ_d, Δ_p	Fermi-contact, diamagnetic and pseudocontact shifts in SR-technique
ΔE	— differences in electronegativity — average electronic excitation energy
$\Delta G\ddagger$	free enthalpy of activation [J M^{-1}]
Δ_i, Δ_{Ti}	SR's induced shift in the complex LS
$\Delta\nu$, $\Delta\nu XY$, $\Delta\delta$, $\Delta\delta XY$	chemical shift difference for X and Y nuclei, $\Delta\nu XY = \nu X - \nu Y$ [Hz], $\Delta\delta XY = \delta X - \delta Y$ [ppm]
$\Delta\nu_{1/2}$	half band width (band width at the half height of a signal)
$\Delta\nu_i$	spectral width
$\Delta\sigma$	anisotropy in σ, $\Delta\sigma = \sigma_{\parallel} - \sigma_{\perp}$
$\Delta\chi$	anisotropy in χ, $\Delta\chi = \chi_{\parallel} - \chi_{\perp}$
ϵ	sign-factor in CIDNP referring to reaction path
ζ	signal-to-noise ratio (''NMR sensitivity'', receptivity)
η	— viscosity, P [m^{-1} kg s^{-1}] — NOE-factor — assymmetry factor in $e^2qQ/\hbar$
θ, Θ	— dihedral or valence bond angle — polar coordinate — pulse angle
κ	rate constant of exchange processes
λ	— wavelength [m] — eigenvalues of various operators
$\boldsymbol{\mu}$	magnetic dipole moment, T
μ	— magnetic dipole moment in a given direction (e.g., μ_{B_0} or μ_w, etc.) — sign-factor in CIDNP referring to electronic state of radical pair precursor — rationalized permeability
μ_0	permeability of vacuum
μ'	nonrationalized permeability
ν	frequency, Hz [s^{-1}]
ν_0	measuring frequency (operating frequency of an NMR spectrometer)
ν_1, ν_2,	first, second, etc. excitation frequencies
ν_N	Larmor (precession) frequency of a nucleus N
ν_c	carrier frequency
π	π-electrons
ξ	steric correction substituent factors [ppm]
ρ	— electron density [C], [A s] — substituent constant [ppm] — L/S ratio in SR measurements
σ	— shielding constant — LS/L ratio in SR measurements — σ-electrons — sign-factor in CIDNP, referring to interacting nuclei in a common radical, or in different radicals
σ_N	shielding constant of a nucleus N
$\boldsymbol{\sigma}$	shielding constant tensor
σ_{vw}	shielding constant tensor components
$\sigma_{\parallel}$, $\sigma_{\perp}$	shielding constant components parallel or perpendicular to

	the molecular axis
σ^d, σ^p	diamagnetic and paramagnetic contributions to the shielding constant
τ	— time [s]
	— lifetime of individual species in exchange processes, and generally of a spin state
	— time between pulses in pulse sequences
	— chemical shift (thau scale) [ppm]
τ_c	correlation time of molecular tumblings
τ_J	correlation time of molecular rotation at a given ν_{rot} frequency
τ_{SC}	correlation time of a spin state in scalar relaxation
τ_e	correlation time in nuclear-electron relaxation
φ	— product eigen function
	— polar coordinate
	— dihedral or valence bond angle
χ	magnetic susceptibility
ψ	basic function of spin systems, eigenfunction of electronic quantum states
ω	angular velocity [s^{-1}]
	ω_N, ω_0, ω_1, ω_2, etc. see at ν_N,..., etc.

INDEX

A

B

C

D

E

F

G

H

I

J

K

L

M

N

O

P

Q

R

S

T

U

V